CISM - COURSES AND LECTURES No. 414

MATERIAL INSTABILITIES IN ELASTIC AND PLASTIC SOLIDS

ERRATA CORRIGE

Page 209

ELASTIC AND PSEUDO-ELASTIC INSTABILITY AND BIFURCATION

R.W. Ogden
University of Glasgow, Glasgow, UK

CISM COURSES AND LECTURES

The series presents lecture notes, monographs, edited works and proceedings in the field of Mechanics, Engineering, Computer Science and Applied Mathematics.
Purpose of the series is to make known in the international scientific and technical community results obtained in some of the activities organized by CISM, the International Centre for Mechanical Sciences.

INTERNATIONAL CENTRE FOR MECHANICAL SCIENCES

COURSES AND LECTURES - No. 414

MATERIAL INSTABILITIES IN ELASTIC AND PLASTIC SOLIDS

EDITED BY

HENRYK PETRYK
POLISH ACADEMY OF SCIENCES

Springer-Verlag Wien GmbH

This volume contains 85 illustrations

Originally published by Springer-Verlag Wien New York in 2000
SPIN 10768414

In order to make this volume available as economically and as rapidly as possible the authors' typescripts have been reproduced in their original forms. This method unfortunately has its typographical limitations but it is hoped that they in no way distract the reader.

ISBN 978-3-211-83328-5 ISBN 978-3-7091-2562-5 (eBook)
DOI 10.1007/978-3-7091-2562-5

PREFACE

Studies of material instabilities, representing basic tools for predicting fracture initiation and failure in various materials, have been the subject of a great amount of scientific work in the last years. The researchers in this field are often facing the difficulty of finding a firm and sufficiently general conceptual framework for examining the fundamental question of stability of solutions at the level of a material element. This volume is intended to fill the gap by collecting recent theoretical developments in the area of material instability in elastic and plastic solids along with related analytical and numerical methods and applications. The existing different approaches to instability phenomena in metal single crystals, polycrystals and in geomaterials are presented with the emphasis laid on mutual relations and on unifying concepts.

This monograph is made up of Lecture Notes for an Advanced School on "Material Instabilities in Elastic and Plastic Solids" held at the International Centre for Mechanical Sciences (CISM) in Udine, Italy, September 13-17, 1999. The course was presented by 6 lecturers from 6 different countries and attended by near forty participants from 8 European countries and US.

It was of course realized that the topic of material instabilities is so broad that could not be treated exhaustively by the six speakers during 4-5 days. The abstracts at the beginning of each Chapter of this volume specify the aspects upon which the attention has been focused. It can be seen that instabilities during isothermal deformation of rate-independent elastic and inelastic solids without an internal length scale constitute the main subject here. Regularization methods aimed at eliminating material instability and ill-posedness in analyses of strain localization are thus not discussed in detail. On the other hand, the diversity of material instabilities treated goes well beyond the typical problem of the localization of deformation.

The basic concepts of material instability in elastic solids are explained in the Chapter by R.W. Ogden and extended to elastic-plastic solids in other Chapters. The respective criteria such as the condition of ellipticity loss and the energy criterion are presented and related to the post-critical behaviour. In addition to quasi-static bifurcation and strain localization into single or multiple shear bands, the topics as wave propagation, moving shocks, internal snap-through and instability of flutter type are examined. Both associative and non-associative plasticity models are used, with due account for the effects of geometrical non-linearity and of vertex formation on the yield surface for a metal crystal or polycrystal.

Applications are presented to various problems, including initiation and evolution of shear bands, neck formation in sheets and bars, inflation of shells, ductile crack growth, behaviour of fluid-saturated, porous elastoplastic materials.

Finally, I wish to thank my colleagues D. Bigoni, W.J. Drugan, B. Loret, R.W. Ogden and K. Thermann for all their efforts in presenting the lectures and preparing the contributing chapters. I express my thanks to the participants of the course for showing a high level of interest and for stimulating discussions. I am also grateful to Professor S. Kaliszky, Rector of CISM, and his staff for the warm hospitality and assistance during the course, and to Professor C. Tasso, Executive Editor of CISM, for his encouragement to publish this volume.

Henryk Petryk

CONTENTS

BIFURCATION AND INSTABILITY OF NON-ASSOCIATIVE ELASTOPLASTIC SOLIDS

D. Bigoni
University of Trento, Trento, Italy

Abstract. Global and local uniqueness and stability criteria for elastoplastic solids with non-associative flow rules are presented. Hill's general theory is developed in the form generalized by Raniecki to non-associativity. Local stability criteria are presented and systematically discussed in a critical way. These are: positive definiteness and non-singularity of the constitutive operator, and positive definiteness (strong ellipticity) and non-singularity (ellipticity) of the acoustic tensor. The former criteria are particularly relevant for homogeneous deformation of solids subject to all-round controlled nominal surface tractions. Dually, the latter criteria are particularly relevant for homogeneous deformation of solids subject to displacements prescribed on the entire boundary. Flutter instability as related to complex conjugate eigenvalues of the acoustic tensor is also briefly discussed.

1 Introduction

Due to the activation of different micromechanisms at various structural levels -for instance, sliding on microfissures or on intergranular contact surfaces, and pore or joint interactions- a broad class of materials is characterized by an *internal friction*, roughly obeying a Coulomb law. As a consequence, pressure-sensitive yielding and plastic dilatancy become dominant phenomena (Nikolaevskii and Rice, 1979). This class of materials embraces porous and particulate reinforced metals, polymers, ceramics, powders, asphalts, granular materials, (traditional, fiber reinforced, and high-strength) concretes, rocks, and soils.

In the terminology of elastoplasticity, the Coulomb law of friction is intrinsically *non-associative*, in the sense that adopting the normality rule overestimates the plastic dilatancy (Drucker, 1954). Therefore, a non-associative flow-rule is generally considered the remedy for correctly modelling the inelastic response of the above-mentioned materials (Hill, 1967b; Mandel, 1966; Mróz, 1963; 1966)[1]. An important consequence of flow rule non-associativity is the lack of symmetry of the tangent constitutive operator, thus precluding the possibility of defining strain rate potentials.

Material instabilities start to grow at a point in a body and are strongly influenced by lack of symmetry of the tangent constitutive operator. For instance, flutter instability

[1] Lack of symmetry of the tangent constitutive operator may result also as a consequence of *elastoplastic coupling* (Hueckel, 1976). Also, in elastoplastic corner models symmetry is lost even for an associative flow-rule when the hardening moduli matrix is not symmetric.

is excluded in the case of symmetry. Bifurcations in diffuse modes, strain localization, and elastoplastic cavitation may occur at high strain hardening for non-associative flow laws.

There are different ways of modelling the occurrence of material instabilities. One is to consider the effects of inevitably existing small defects (imperfections) in a material which may be viewed as homogeneous at a certain scale. Another is to consider the macroscopic, inelastic, constitutive response of a homogeneous material element and seek specific instabilities and bifurcations. The latter description of material instabilities is due to Rice (1977) and will be followed here. Material instabilities may be detected with mechanical tests. However, mechanical tests on material elements often consist of imposing displacements (resultant forces) on certain external surfaces of a finite volume of material and measuring resultant forces (displacements) on the same surfaces. Under these conditions, stress and strain cannot be directly controlled. Therefore, instabilities occurring in a mechanical test should always be referred to a specific boundary value problem. As a consequence, material instabilities should be presented within the general context of bifurcation and instability theory for elastoplastic material. This theory was mainly developed by Hill (1958, 1959, 1978) for elastoplastic solids with an associative flow rule. In these notes, this theory will be presented with reference to the generalization to non-associative flow rules given by Raniecki (1979) and Raniecki and Bruhns (1981). In this general framework, several local criteria of stability and bifurcation will be introduced. These are: positive definiteness and non-singularity of the constitutive operator, strong ellipticity and ellipticity of the acoustic tensor and, finally, flutter instability (complex conjugate eigenvalues of the acoustic tensor). Emphasis is given to the connections between these criteria and to their mechanical interpretation.

2 Notation and preliminaries

We refer generally to Gurtin's (1972, 1981) notation. In particular, boldface minuscules $(\mathbf{a}, \mathbf{b}, ...)$ and majuscules $(\mathbf{A}, \mathbf{B}, ...)$ denote vectors (or vector fields) and second-order tensors (or tensor fields), respectively. The space of vectors is denoted by $\mathcal{V}$, the set of second-order tensors by Lin and its symmetric restriction by Sym. The inner products of two vectors $\mathbf{a}$ and $\mathbf{b}$ and two second-order tensors $\mathbf{A}$ and $\mathbf{B}$ are designated by

$$\mathbf{a} \cdot \mathbf{b} = \sum_{k=1}^{3} a_k b_k =: a_k b_k \quad \text{and} \quad \mathbf{A} \cdot \mathbf{B} = \sum_{h,k=1}^{3} A_{hk} B_{hk} =: A_{hk} B_{hk},$$

respectively. The product $\mathbf{AB}$ of two second-order tensors is defined by composition, namely, for every vector $\mathbf{a}$,

$$(\mathbf{AB})\mathbf{a} = \mathbf{A}(\mathbf{Ba}), \quad \text{or} \quad (\mathbf{AB})_{ij} = \sum_{k=1}^{3} A_{ik} B_{kj} =: A_{ik} B_{kj}.$$

The tensor product $\mathbf{a} \otimes \mathbf{b}$ of two vectors $\mathbf{a}$ and $\mathbf{b}$ is defined for every vector $\mathbf{v}$ by

$$(\mathbf{a} \otimes \mathbf{b})\mathbf{v} = \mathbf{a}(\mathbf{b} \cdot \mathbf{v}), \quad \text{or} \quad (\mathbf{a} \otimes \mathbf{b})_{ij} = a_i b_j.$$

Moreover, $|\cdot|$ denotes the euclidean norm, $\mathbf{I}$ the second-order identity tensor, a superscript T denotes transpose and tr the trace operator, i.e., for every $\mathbf{A}, \mathbf{B}$, $\mathrm{tr}(\mathbf{AB}) = \mathbf{A}^T \cdot \mathbf{B}$. We will use the symbol $\propto$ between second-order tensors, i.e. $\mathbf{A} \propto \mathbf{B}$, to mean that a scalar ρ exists such that $\mathbf{A} = \rho\mathbf{B}$.

Two symmetric, second-order tensors $\mathbf{A}$ and $\mathbf{B}$ are defined to be *coaxial* when their product commutes, $\mathbf{AB} = \mathbf{BA}$. Note that two coaxial tensors share at least one principal reference system (Appendix A).

Fourth-order tensors are denoted by sans-serif majuscules, as for instance, the elasticity tensor $\mathsf{E}[\cdot]$. These are linear mappings assigning to each second-order tensor $\mathbf{A}$ a second-order tensor $(\mathsf{E}[\mathbf{A}])_{ij} = \mathsf{E}_{ijhk}A_{hk}$.

The product EH of two fourth-order tensors is defined, analogously to second-order tensors, by composition, namely, for every $\mathbf{A} \in \mathrm{Lin}$,

$$(\mathsf{EH})[\mathbf{A}] = \mathsf{E}[\mathsf{H}[\mathbf{A}]], \quad \text{or} \quad (\mathsf{EH})_{ijhk} = \mathsf{E}_{ijst}\mathsf{H}_{sthk}.$$

Two tensorial products will be employed, denoted by symbols $\otimes$ and $\boxtimes$. These are defined, for every $\mathbf{A}, \mathbf{B}, \mathbf{C} \in \mathrm{Lin}$, as (Del Piero, 1979)

$$(\mathbf{A} \otimes \mathbf{B})[\mathbf{C}] = (\mathbf{B} \cdot \mathbf{C})\mathbf{A}, \quad \text{or} \quad (\mathbf{A} \otimes \mathbf{B})_{ijhk} = \mathbf{A}_{ij}\mathbf{B}_{hk},$$

and

$$(\mathbf{A} \boxtimes \mathbf{B})[\mathbf{C}] = \mathbf{A}\mathbf{C}\mathbf{B}^T, \quad \text{or} \quad (\mathbf{A} \boxtimes \mathbf{B})_{ijhk} = \mathbf{A}_{ih}\mathbf{B}_{jk}.$$

Note that, with the above definition, $\mathbf{I} \boxtimes \mathbf{I}$ is the fourth-order identity tensor. Defining the transpose of a fourth-order tensor (for every $\mathbf{A}, \mathbf{B} \in \mathrm{Lin}$) as $\mathbf{B} \cdot \mathsf{E}^T[\mathbf{A}] = \mathbf{A} \cdot \mathsf{E}[\mathbf{B}]$, we say that E has the major symmetry whenever $\mathsf{E} = \mathsf{E}^T$. Moreover, we note that

$$(\mathbf{A} \otimes \mathbf{B})^T = \mathbf{B} \otimes \mathbf{A} \quad \text{and} \quad (\mathbf{A} \boxtimes \mathbf{B})^T = \mathbf{A}^T \boxtimes \mathbf{B}^T.$$

The special symbol $\mathbf{I} \underline{\otimes} \mathbf{I}$ is reserved for the fourth-order tensor which associates to every second-order tensor $\mathbf{X}$ its symmetric part:

$$\mathbf{I} \underline{\otimes} \mathbf{I}[\mathbf{X}] = \frac{1}{2}\left(\mathbf{X} + \mathbf{X}^T\right).$$

Obviously $\mathbf{I} \underline{\otimes} \mathbf{I}$ is singular (because it associates the null tensor to every skew symmetric tensor), but its restriction to Sym is invertible and the inverse is the tensor itself.

For any given smooth vector field $\mathbf{a}$, the divergence of $\mathbf{a}$ is defined as

$$\mathrm{div}\,\mathbf{a} = \mathrm{tr}(\mathrm{grad}\,\mathbf{a}),$$

where $\mathrm{grad}\,\mathbf{a}$ is the gradient of $\mathbf{a}$ at $\mathbf{x}$ (in rectangular, cartesian components $(\mathrm{grad}\,\mathbf{a})_{ij} = a_{i,j}$ and $\mathrm{div}\,\mathbf{a} = a_{k,k}$). The divergence operator of a tensor field $\mathbf{A}$ is defined as the unique vector field $\mathrm{div}\mathbf{A}$ which for every constant vector $\mathbf{a}$ satisfies $(\mathrm{div}\mathbf{A}) \cdot \mathbf{a} = \mathrm{div}(\mathbf{A}^T\mathbf{a})$. In (rectangular, cartesian) components:

$$(\mathrm{div}\mathbf{A})_i = A_{ij,j}.$$

A dot over a symbol denotes rate of change (right hand derivative) of the quantity at a fixed material point with respect to a scalar time-like parameter (which will govern the deformation process).

Finally, we note that the symbol $\mathcal{C}(\cdot)$ is reserved for a nonlinear constitutive operator, assigning to each second-order tensor $\mathbf{A}$ the second-order tensor $\mathcal{C}(\mathbf{A})$.

3 Incremental boundary value problem

Quasi-static deformation of an inviscid solid body is assumed to be governed by a time-like parameter (abbreviated as 'time'). In a loading program, displacements and nominal surface tractions (mixed boundary conditions) are prescribed; these are assumed to be sufficiently regular functions of place and time over specific portions $\partial\Omega_\xi^0$ and $\partial\Omega_\sigma^0$ of the boundary in the reference configuration ($\partial\Omega^0 = \partial\Omega_\xi^0 \cup \partial\Omega_\sigma^0$). For simplicity, we limit the presentation to *controlled nominal surface tractions* on $\partial\Omega_\sigma^0$ (in other words, deformation-sensitive loadings are not considered), so that the boundary conditions are

$$\mathbf{x} = \boldsymbol{\xi}(\mathbf{x}_0, t), \quad \text{on} \quad \partial\Omega_\xi^0; \quad \mathbf{S}\mathbf{n}_0 = \boldsymbol{\sigma}(\mathbf{x}_0, t) \quad \text{on} \quad \partial\Omega_\sigma^0, \tag{1}$$

where $\mathbf{x}$ and $\mathbf{x}_0$ are the places occupied by the material points in the current and reference configuration, respectively, $\mathbf{n}_0$ is the outward unit vector to $\partial\Omega_\sigma^0$ and $\mathbf{S} \in$ Lin is the first Piola-Kirchhoff stress[2], defined with respect to the Cauchy stress $\mathbf{T} \in$ Sym by

$$\mathbf{S} = J\mathbf{T}\mathbf{F}^{-T}, \quad J = \det \mathbf{F}, \tag{2}$$

where $\mathbf{F} \in$ Lin is the deformation gradient.

At a generic stage of the loading program, i.e. at a generic time, we assume that the current geometry and the state of the body are known, and we analyze the response to a prescribed small perturbation of the boundary conditions. In other words, in a series expansion of all quantities specifying deformation of the body we analyze the first-order terms: velocity, stress and strain rates. This is the so-called velocity problem[3], in which velocities and traction rates are prescribed on complementary, regular subsurfaces of the boundary

$$\dot{\mathbf{x}} = \dot{\boldsymbol{\xi}}(\mathbf{x}_0, t), \quad \text{on} \quad \partial\Omega_\xi^0; \quad \dot{\mathbf{S}}\mathbf{n}_0 = \dot{\boldsymbol{\sigma}}(\mathbf{x}_0, t) \quad \text{on} \quad \partial\Omega_\sigma^0. \tag{3}$$

The superposed dot in eqn. (3) denotes the material time derivative (i.e. the time derivative at fixed $\mathbf{x}_0$), $\dot{\mathbf{x}}$ is the material description of the velocity. The velocity field is *assumed to be spatially continuous*. From the Lagrangean standpoint the first-order rate equations are, in the *absence of body forces*:

$$\operatorname{Div} \dot{\mathbf{S}} = \mathbf{0}, \quad \text{in} \quad \Omega^0 \setminus \Sigma^0, \tag{4}$$

where $\mathbf{S}$ is the first Piola-Kirchhoff stress. It may be important to note that the divergence operator Div in (4) is referred to material points $\mathbf{x}_0$. Moreover, Σ^0 represents any possible surface of discontinuity of $\dot{\mathbf{S}}$, in the reference description, and the notation in (4) indicates all of Ω^0 excluding Σ^0. Across this surface, the nominal traction rate must remain continuous (Hill, 1961; Chadwick and Powdrill, 1965):

$$[\![\dot{\mathbf{S}}]\!]\, \mathbf{m}_0 = \mathbf{0}, \quad \text{on} \quad \Sigma^0, \tag{5}$$

[2] The first Piola-Kirchhoff stress tensor is the transpose of the nominal stress tensor used, among others, by Hill (1978) and Ogden (1984). Note also that these authors use another definition of Div, namely $(\operatorname{Div}\mathbf{A})_j = A_{ij,i}$.

[3] Higher-order problems can also be analyzed; see Petryk and Thermann (1985), Nguyen and Triantafyllidis (1989), Cheng and Lu (1993), Bigoni (1996).

where the symbol $[\![\cdot]\!]$ denotes a jump of the relevant argument, i.e. $[\![\dot{\mathbf{S}}]\!] = \dot{\mathbf{S}}^+ - \dot{\mathbf{S}}^-$, and $\mathbf{m}_0$ is the unit normal vector to Σ^0 directed toward + and away from −. In weak form, eqs (3), (4), (5) are equivalent to

$$\int_{\Omega^0} \dot{\mathbf{S}} \cdot \nabla \mathbf{w} - \int_{\partial\Omega^0_\sigma} \dot{\boldsymbol{\sigma}} \cdot \mathbf{w} = 0\,, \tag{6}$$

for every (continuous and piecewise continuously twice differentiable) variation $\mathbf{w}$ of the velocity. In particular, $\nabla\mathbf{w} = \mathrm{Grad}\,\mathbf{w}$ is the gradient (with respect to material points $\mathbf{x}_0$) of a field $\mathbf{w}$ defined in the reference configuration and taking null values on the portions of the boundary where displacements (and velocities) are prescribed.

4 Constitutive Equations

Elastoplastic, isothermal and time independent material behaviour of a solid subject to large strains is described in this section. We present a broad constitutive framework in which many existing elastoplastic models may fit. In the interest of generality, certain details are intentionally left unspecified.

Adopting Ogden's (1984) notation, let us consider a pair of symmetric, Lagrangean, stress $\mathbf{T}^{(m)}$ and strain $\mathbf{E}^{(m)}$ measures, work-conjugate in the Hill sense (1968, 1978), so that

$$\mathbf{T}^{(m)} \cdot \dot{\mathbf{E}}^{(m)} \tag{7}$$

gives the stress power density (per unit volume of Ω^0) independently of the positive[4], integer exponent m. In particular, if we take

$$\mathbf{E}^{(m)} = \frac{1}{m}(\mathbf{U}^m - \mathbf{I}), \tag{8}$$

where $\mathbf{U}$ is the right stretch tensor, related to the deformation gradient $\mathbf{F}$ through $\mathbf{U} = (\mathbf{F}^T\mathbf{F})^{1/2}$, then the conjugate stress $\mathbf{T}^{(m)}$ is obtained by imposing the equality:

$$\mathbf{S} \cdot \dot{\mathbf{F}} = \mathbf{T}^{(m)} \cdot \dot{\mathbf{E}}^{(m)}. \tag{9}$$

For instance, when $m = 2$ we obtain the Green-Lagrange strain and the second Piola-Kirchhoff stress tensors, respectively:

$$\mathbf{E}^{(2)} = \frac{1}{2}(\mathbf{F}^T\mathbf{F} - \mathbf{I}), \qquad \mathbf{T}^{(2)} = J\mathbf{F}^{-1}\mathbf{T}\mathbf{F}^{-T}. \tag{10}$$

Following Hill and Rice (1973) and Hill (1978) but without introducing elastic potentials, inelastic materials are considered that may at any stage of deformation exhibit a purely elastic response for appropriate loading. For these materials, elastic response is

[4] Negative or null exponents can also be introduced (Ogden, 1984). The notation $\mathbf{T}^m = \underbrace{\mathbf{TT}...\mathbf{T}}_{m\ times}$ should not be confused with $\mathbf{T}^{(m)}$.

assumed to be a one-to-one relation between $\mathbf{T}^{(m)}$ and $\mathbf{E}^{(m)}$, though depending on the prior inelastic history, i.e.

$$\mathbf{T}^{(m)} = \hat{\mathbf{T}}^{(m)}(\mathbf{E}^{(m)}, \mathcal{K}), \quad \mathbf{E}^{(m)} = \hat{\mathbf{E}}^{(m)}(\mathbf{T}^{(m)}, \mathcal{K}), \tag{11}$$

where $\hat{\mathbf{T}}^{(m)}$ and $\hat{\mathbf{E}}^{(m)}$ are functionals of the prior history of inelastic deformation through the unspecified set $\mathcal{K}$ of variables of generic tensorial nature (thus embracing second-order tensors and scalars). For a purely elastic deformation rate (in other words, at fixed $\mathcal{K}$) we have

$$\dot{\mathbf{T}}^{(m)} = \mathsf{E}[\dot{\mathbf{E}}^{(m)}], \quad \dot{\mathbf{E}}^{(m)} = \mathsf{M}[\dot{\mathbf{T}}^{(m)}], \tag{12}$$

where

$$\mathsf{E}(\mathbf{E}^{(m)}, \mathcal{K}) = \frac{\partial \hat{\mathbf{T}}^{(m)}}{\partial \mathbf{E}^{(m)}}, \quad \mathsf{M}(\mathbf{T}^{(m)}, \mathcal{K}) = \frac{\partial \hat{\mathbf{E}}^{(m)}}{\partial \mathbf{T}^{(m)}}, \tag{13}$$

and obviously

$$\mathsf{E} = \mathsf{M}^{-1}. \tag{14}$$

For an increment involving elastic and inelastic strain rates, we may write

$$\dot{\mathbf{T}}^{(m)} = \mathsf{E}[\dot{\mathbf{E}}^{(m)}] - \dot{\Lambda}\mathsf{E}[\mathbf{P}], \quad \dot{\mathbf{E}}^{(m)} = \mathsf{M}[\dot{\mathbf{T}}^{(m)}] + \dot{\Lambda}\mathbf{P}, \tag{15}$$

where $\mathbf{P} \in \mathrm{Sym}$,

$$\dot{\Lambda}\mathbf{P} = \frac{\partial \hat{\mathbf{E}}^{(m)}}{\partial \mathcal{K}}[\dot{\mathcal{K}}] \tag{16}$$

and the scalar $\dot{\Lambda} \geq 0$, called the plastic multiplier, is null when $\dot{\mathcal{K}} = 0$. A yield surface is assumed at each $\mathcal{K}$. This may be alternatively expressed as $f_{\mathbf{T}^{(m)}}(\mathbf{T}^{(m)}, \mathcal{K}) \leq 0$ or as $f_{\mathbf{E}^{(m)}}(\mathbf{E}^{(m)}, \mathcal{K}) \leq 0$, thus defining regions of the $\mathbf{T}^{(m)}$ or $\mathbf{E}^{(m)}$ space, respectively, within which the response is elastic. Prager's consistency condition requires $\dot{f}_{\mathbf{T}^{(m)}} = \dot{f}_{\mathbf{E}^{(m)}} = 0$, when inelastic strain rate is different from zero. As a consequence, employing the stress space representation, the elastoplastic incremental constitutive equations can be written as

$$\boxed{\dot{\mathbf{T}}^{(m)} = \begin{cases} \mathsf{E}[\dot{\mathbf{E}}^{(m)}] - \dfrac{1}{g} < \mathbf{Q} \cdot \mathsf{E}[\dot{\mathbf{E}}^{(m)}] > \mathsf{E}[\mathbf{P}] & \text{if } f_{\mathbf{T}^{(m)}}(\mathbf{T}^{(m)}, \mathcal{K}) = 0, \\ \mathsf{E}[\dot{\mathbf{E}}^{(m)}] & \text{if } f_{\mathbf{T}^{(m)}}(\mathbf{T}^{(m)}, \mathcal{K}) < 0, \end{cases}} \tag{17}$$

where the operator $< \cdot >$ denotes the Macaulay brackets, i.e. $\forall \alpha \in \Re, < \alpha >= (\alpha + |\alpha|)/2$. Moreover, $\mathbf{Q} = \partial f_{\mathbf{T}^{(m)}} / \partial \mathbf{T}^{(m)} \in \mathrm{Sym}$ is the yield function gradient and the plastic modulus

$$g = h + \mathbf{Q} \cdot \mathsf{E}[\mathbf{P}], \tag{18}$$

is assumed to be strictly positive (a negative plastic modulus would correspond to a so-called locking material, not investigated here). In the Hill (1967b) notation, the hardening modulus h in (18) describes *hardening* when positive, *softening* when negative and *perfect plasticity* when null. It is defined as

$$\dot{\Lambda} h = -\frac{\partial f_{\mathbf{T}^{(m)}}}{\partial \mathcal{K}} \cdot \dot{\mathcal{K}}. \tag{19}$$

As Hill (1967b) remarks, hardening and softening are not measure-invariant concepts, in the sense that h depends on the choice of $\mathbf{T}^{(m)}$ and $\mathbf{E}^{(m)}$. Therefore, the nomenclature is, to some extent, arbitrary. Moreover, we remark that, in addition to h, also $\mathbf{Q}$, $\mathbf{P}$ and E are measure-dependent. On the contrary, the plastic modulus g can be shown to be measure-independent (Hill, 1967b; Petryk, 1999). Note also that all quantities appearing in the rate equations (17) fully depend on the entire path of deformation reckoned from some ground state.

The scalar product of the first equation in (17) with $\mathbf{Q}$ gives

$$\mathbf{Q}\cdot\dot{\mathbf{T}}^{(m)} = \mathbf{Q}\cdot\mathsf{E}[\dot{\mathbf{E}}^{(m)}] - \frac{\mathbf{Q}\cdot\mathsf{E}[\mathbf{P}]}{g} < \mathbf{Q}\cdot\mathsf{E}[\dot{\mathbf{E}}^{(m)}] > . \tag{20}$$

In the case when $h > 0$, we note that

$$\mathrm{sign}(\mathbf{Q}\cdot\mathsf{E}[\dot{\mathbf{E}}^{(m)}]) = \mathrm{sign}(\mathbf{Q}\cdot\dot{\mathbf{T}}^{(m)}).$$

Therefore, assuming $h > 0$ and using (20), we obtain the inverse constitutive equations

$$\dot{\mathbf{E}}^{(m)} = \begin{cases} \mathsf{M}[\dot{\mathbf{T}}^{(m)}] + \dfrac{1}{h} < \mathbf{Q}\cdot\dot{\mathbf{T}}^{(m)} > \mathbf{P} & \text{if } f(\mathbf{T}^{(m)},\mathcal{K}) = 0, \\ \mathsf{M}[\dot{\mathbf{T}}^{(m)}] & \text{if } f(\mathbf{T}^{(m)},\mathcal{K}) < 0, \end{cases} \tag{21}$$

It may be important to remark that *all possible choices of* $\mathbf{T}^{(m)}$ *and* $\mathbf{E}^{(m)}$ *in (17) or (21) are equivalent* and that *all resulting constitutive equations respect the requirement of material frame indifference* (Truesdell and Noll, 1965).

It is particularly convenient to write the constitutive equation (17) in terms of the material time derivative of the first Piola-Kirchhoff stress and of the deformation gradient. This can be done for any choice of $\mathbf{T}^{(m)}$ and $\mathbf{E}^{(m)}$. For instance, in the case of the second Piola-Kirchhoff stress tensor $\mathbf{T}^{(2)}$ and the Green-Lagrange strain tensor $\mathbf{E}^{(2)}$, the following relations hold true

$$\dot{\mathbf{S}} = \mathbf{F}\dot{\mathbf{T}}^{(2)} + J\mathbf{LTF}^{-T}, \quad \dot{\mathbf{E}}^{(2)} = \mathbf{F}^T\mathbf{DF}, \tag{22}$$

where $\mathbf{L} = \mathrm{grad}\,\mathbf{v}$ is the velocity gradient and $\mathbf{D}$ its symmetric part, i.e. the rate of deformation. The minor symmetries of E and the relation $\dot{\mathbf{F}} = \mathbf{LF}$ imply

$$\mathsf{E}[\mathbf{F}^T\mathbf{DF}] = \frac{1}{2}\mathsf{E}[\mathbf{F}^T\dot{\mathbf{F}} + (\mathbf{F}^T\dot{\mathbf{F}})^T] = \mathsf{E}[\mathbf{F}^T\dot{\mathbf{F}}]. \tag{23}$$

Moreover, from the equality $\mathbf{F}^{-T}\mathbf{B}\cdot\mathbf{FA} = \mathbf{B}\cdot\mathbf{A}$, holding for every $\mathbf{A},\mathbf{B} \in \mathrm{Lin}$, we conclude that (17) can be written as

$$\boxed{\dot{\mathbf{S}} = \begin{cases} \mathsf{G}[\dot{\mathbf{F}}] - \dfrac{1}{g} < (\mathbf{F}^{-T}\mathbf{Q})\cdot\mathsf{B}[\dot{\mathbf{F}}] > \mathsf{B}[\mathbf{F}^{-T}\mathbf{P}] & \text{if } f_{\mathbf{S}}(\mathbf{S},\mathcal{K}) = 0, \\ \mathsf{G}[\dot{\mathbf{F}}] & \text{if } f_{\mathbf{S}}(\mathbf{S},\mathcal{K}) < 0, \end{cases}} \tag{24}$$

where

$$\mathsf{B} = (\mathbf{F}\boxtimes\mathbf{I})\mathsf{E}(\mathbf{F}\boxtimes\mathbf{I})^T, \quad \mathsf{G} = \mathsf{B} + \mathbf{I}\boxtimes\mathbf{S}^T\mathbf{F}^{-T}. \tag{25}$$

It should be noted that neither B nor G have the minor symmetries, and that both B and G have the major symmetry only in the case of Green elasticity, i.e. when E has the major symmetry, too[5]. Moreover, the yield function $f_{\mathbf{S}}(\mathbf{S},\mathcal{K})$ in (24) has been expressed in terms of the first Piola-Kirchhoff stress tensor; its gradient with respect to $\mathbf{S}$ is $\mathbf{F}^{-T}\mathbf{Q}$. This follows from the chain rule of differentiation

$$\frac{\partial f_{\mathbf{S}}}{\partial \mathbf{S}} \cdot \mathbf{A} = \frac{\partial f_{\mathbf{T}^{(2)}}}{\partial \mathbf{T}^{(2)}} \cdot \frac{\partial \mathbf{T}^{(2)}}{\partial \mathbf{S}}[\mathbf{A}] = \mathbf{F}^{-T}\mathbf{Q} \cdot \mathbf{A}, \tag{26}$$

for every $\mathbf{A} \in \mathrm{Lin}$. When the yield criterion is satisfied, $f_{\mathbf{S}}(\mathbf{S},\mathcal{K}) = 0$, constitutive equations (24) define a piecewise-linear function of $\dot{\mathbf{F}}$

$$\dot{\mathbf{S}} = \mathcal{C}(\dot{\mathbf{F}}), \tag{27}$$

i.e. an incrementally nonlinear constitutive equation with two branches, corresponding to plastic loading and elastic unloading. This constitutive equation may obviously be written in the following form, useful for subsequent analysis:

$$\boxed{\dot{\mathbf{S}} = \mathsf{G}[\dot{\mathbf{F}}] - \frac{1}{g} < \mathbf{N} \cdot \dot{\mathbf{F}} > \mathbf{M},} \tag{28}$$

where $\mathbf{N} = \mathsf{B}^T[\mathbf{F}^{-T}\mathbf{Q}] \in \mathrm{Lin}$ and $\mathbf{M} = \mathsf{B}[\mathbf{F}^{-T}\mathbf{P}] \in \mathrm{Lin}$ are the yield surface and plastic potential normals, respectively, in strain space. In particular, expressing the yield function in terms of the deformation gradient, i.e. $f_{\mathbf{F}}(\mathbf{F},\mathcal{K})$, its gradient with respect to $\mathbf{F}$ is exactly $\mathbf{N}$; in fact

$$\frac{\partial f_{\mathbf{F}}}{\partial \mathbf{F}} \cdot \mathbf{A} = \frac{\partial f_{\mathbf{T}^{(2)}}}{\partial \mathbf{T}^{(2)}} \cdot \frac{\partial \mathbf{T}^{(2)}}{\partial \mathbf{E}^{(2)}} \left[\frac{\partial \mathbf{E}^{(2)}}{\partial \mathbf{F}}[\mathbf{A}]\right] = \mathsf{B}^T[\mathbf{F}^{-T}\mathbf{Q}] \cdot \mathbf{A}, \tag{29}$$

for every $\mathbf{A} \in \mathrm{Lin}$. Note also that $\mathbf{P} = \mathbf{Q}$ is equivalent to $\mathbf{M} = \mathbf{N}$ only when B has the major symmetry, i.e. for Green elasticity. In other words, simultaneous normality in stress and strain spaces is not assured for Cauchy elasticity (Hill, 1978). In the following, we will refer to associative flow rule when $\mathbf{P} = \mathbf{Q}$ and B has the major symmetry, so that normality is preserved in the strain space also, $\mathbf{M} = \mathbf{N}$.

As far as the choice of $\mathbf{P}$ and $\mathbf{Q}$ is concerned, this is to some extent arbitrary. However, experiments show that many materials exhibit a peculiar kind of non-associativity, involving only the volumetric part of plastic deformation. This case of special interest corresponds to so-called *deviatoric associativity*, where the deviatoric parts of $\mathbf{P}$ and $\mathbf{Q}$ are aligned. This may be generically defined as

$$\mathbf{P} = \chi_1 \hat{\mathbf{S}} + \frac{\chi_2}{3}\mathbf{I}, \quad \mathbf{Q} = \psi_1 \hat{\mathbf{S}} + \frac{\psi_2}{3}\mathbf{I}, \tag{30}$$

where $\hat{\mathbf{S}} \in \mathrm{Sym}$ is traceless, χ_1 and ψ_1 are assumed strictly positive and χ_2 and ψ_2 are assumed to be non-negative. The parameters ψ_2 and χ_2 describe the pressure-sensitivity

[5] Note that $\mathbf{S}^T\mathbf{F}^{-T} = J\mathbf{F}^{-1}\mathbf{T}\mathbf{F}^{-T} \in \mathrm{Sym}$. Thus, $\mathsf{G} - \mathsf{B} = \mathbf{I}\boxtimes\mathbf{S}^T\mathbf{F}^{-T}$ has always the major symmetry.

and the dilatancy of the material, respectively. For instance, in the case of the constitutive model proposed by Rudnicki and Rice (1975)

$$\chi_1 = \psi_1 = \frac{1}{2\sqrt{J_2}}, \quad \hat{\mathbf{S}} = \operatorname{dev}\mathbf{T}, \tag{31}$$

where $\operatorname{dev}\mathbf{T} = \mathbf{T} - (\mathrm{tr}\mathbf{T}/3)\mathbf{I}$ is the stress deviator and $J_2 = (\mathrm{dev}\mathbf{T}\cdot\mathrm{dev}\mathbf{T})/2$. Note that, obviously, deviatoric associativity implies coaxiality of $\mathbf{Q}$ and $\mathbf{P}$.

4.1 The infinitesimal theory

In the special case of the infinitesimal theory, equations (11) simplify to

$$\mathbf{T} = \mathsf{E}[\mathbf{E} - \mathbf{E}_p], \quad \mathbf{E} = \mathsf{M}[\mathbf{T}] + \mathbf{E}_p, \tag{32}$$

where $\mathbf{E} = \frac{1}{2}[\operatorname{grad}\mathbf{u} + (\operatorname{grad}\mathbf{u})^T]$ is the infinitesimal strain tensor with $\mathbf{E}_p$ being its plastic part. The fourth-order tensor $\mathsf{E} = \mathsf{M}^{-1}$ is now the elasticity tensor of the usual infinitesimal theory[6], which, in the well-known case of isotropy is

$$\mathsf{E} = \lambda\mathbf{I}\otimes\mathbf{I} + 2\mu\mathbf{I}\underline{\otimes}\mathbf{I}, \tag{33}$$

where $\mathbf{I}\underline{\otimes}\mathbf{I}$ is the fourth-order tensor which associates to every tensor $\mathbf{A} \in \mathrm{Lin}$ its symmetric part $\mathbf{I}\underline{\otimes}\mathbf{I}[\mathbf{A}] = (\mathbf{A} + \mathbf{A}^T)/2$ and λ and μ are the Lamé constants. In the stress space formulation, the yield function can be written as $f(\mathbf{T}, \mathcal{K}) \le 0$, where $\mathcal{K}$ is a generic set of internal variables, possibly depending on $\mathbf{E}_p$. Constitutive equations (27) therefore become (in the case $f(\mathbf{T}, \mathcal{K}) = 0$)

$$\dot{\mathbf{T}} = \mathsf{E}[\mathbf{D}] - \frac{1}{g} < \mathbf{Q}\cdot\mathsf{E}[\mathbf{D}] > \mathsf{E}[\mathbf{P}], \tag{34}$$

where $g = h + \mathbf{Q}\cdot\mathsf{E}[\mathbf{P}]$ is the plastic modulus (assumed strictly positive). Note that $\mathbf{Q}\cdot\mathsf{E}[\mathbf{D}] = \mathbf{D}\cdot\mathsf{E}^T[\mathbf{Q}]$; therefore the tangent operator is symmetric for the associative flow rule if and only if E has the major symmetry.

5 Global uniqueness and stability

5.1 Uniqueness

With reference to the Lagrangean description, the regular first-order rate problem can be stated with reference to the velocity field $\dot{\mathbf{x}}$, when the constitutive equation (27) is employed in $(3)_2$, $(4)_2$ and $(5)_2$. Therefore, we obtain:

$$\begin{aligned}
&\dot{\mathbf{x}} = \dot{\boldsymbol{\xi}}(\mathbf{x}_0, t), && \text{on } \partial\Omega^0_\xi, \\
&\mathcal{C}(\dot{\mathbf{F}})\mathbf{n}_0 = \dot{\boldsymbol{\sigma}}(\mathbf{x}_0, t), && \text{on } \partial\Omega^0_\sigma, \\
&\mathrm{Div}(\mathcal{C}(\dot{\mathbf{F}})) = \mathbf{0}, && \text{in } \Omega^0 \setminus \Sigma^0, \\
&[\![\mathcal{C}(\dot{\mathbf{F}})]\!]\,\mathbf{m}_0 = \mathbf{0}, && \text{on } \Sigma^0,
\end{aligned} \tag{35}$$

[6] Usually E is constant. In the case of elastoplastic coupling, E depends on $\mathbf{E}_p$ (Hueckel, 1976). In the following it is assumed constant.

where $\dot{\mathbf{F}} = \nabla\dot{\mathbf{x}}$ in $\Omega^0 \setminus \Sigma^0$ and $[\![\dot{\mathbf{x}}]\!] = \mathbf{0}$ across Σ^0.

The above equations refer to the part of the body at yielding, i.e. to the so-called plastic zone. In the elastic zone the governing equations are the same with $\boldsymbol{C}$ replaced by G.

We are now in a position to state the velocity problem, namely, given a certain state of a body, find a continuous and piecewise continuously twice differentiable (shortly, 'admissible') velocity field satisfying (35).

In order to obtain an exclusion condition for bifurcation in velocity, we follow the Hill (1950, 1958) argument[7], which generalizes to elastoplasticity the Kirchhoff theorem of linear elasticity. Suppose therefore that the velocity problem admits two solutions, say $\dot{\mathbf{x}}_1$ and $\dot{\mathbf{x}}_2$ both satisfying (35). Their difference $\Delta\dot{\mathbf{x}} = \dot{\mathbf{x}}_1 - \dot{\mathbf{x}}_2$ defines an admissible velocity field, with gradient $\Delta\dot{\mathbf{F}}$. Also the difference in stresses $\Delta\dot{\mathbf{S}}$ satisfies equilibrium equations (4), (5). Moreover, the difference fields satisfy homogeneous conditions on the boundary. On application of the divergence theorem it follows that

$$\int_{\Omega^0} \Delta\dot{\mathbf{S}} \cdot \Delta\dot{\mathbf{F}} = 0, \tag{36}$$

where $\Delta\dot{\mathbf{S}} = \boldsymbol{C}(\dot{\mathbf{F}}_1) - \boldsymbol{C}(\dot{\mathbf{F}}_2)$. It should be stressed that *due to the nonlinearity of* $\boldsymbol{C}$, $\Delta\dot{\mathbf{S}}$ does not, in general, coincide with $\boldsymbol{C}(\Delta\dot{\mathbf{F}})$.

Therefore, *a sufficient condition to exclude bifurcations of the velocity problem* is:

$$\boxed{\int_{\Omega^0} \Delta\dot{\mathbf{S}} \cdot \Delta\dot{\mathbf{F}} > 0,} \tag{37}$$

for all pairs of distinct, admissible velocity fields taking the given values $(35)_1$ on $\partial\Omega^0_\xi$.

Note that the exclusion condition (37) would be true even replacing '>'with '<'. We will see that this possibility can be excluded in terms of instability.

Raniecki comparison solids. The difficulty in proceeding with (37) is related to the nonlinearity of the constitutive operator $\boldsymbol{C}$. To overcome this problem, Hill (1958) proposed to introduce a *linear comparison solid*, which, when used to replace the original constitutive operator provides a lower bound to (37). Results of Hill were restricted to the associative flow rule, where the comparison solid turns out to coincide with a linear solid defined by the constitutive tensor corresponding to the loading branch of the elastoplastic operator (28). The Hill comparison theorem was later generalized by Raniecki (1979) and Raniecki and Bruhns (1981) to cover non-associative flow rules. In particular, Raniecki introduced a family of linear comparison solids (briefly, 'Raniecki comparison solids') defined by the following constitutive tensor having the major symmetry (for every $\psi \in \Re^+$)

$$\mathsf{R} = \mathsf{G} - \frac{1}{4\psi g}(\mathbf{M} + \psi\mathbf{N}) \otimes (\mathbf{M} + \psi\mathbf{N}), \tag{38}$$

[7] Hill's (1950) proof was restricted to the infinitesimal theory and was based on a theorem due to Melan (1938).

such that the following *comparison theorem* holds true

$$\boxed{\Delta\dot{\mathbf{S}} \cdot \Delta\dot{\mathbf{F}} \geq \Delta\dot{\mathbf{F}} \cdot \mathsf{R}[\Delta\dot{\mathbf{F}}],} \tag{39}$$

for every difference of tensors $\Delta\dot{\mathbf{F}} = \dot{\mathbf{F}}_1 - \dot{\mathbf{F}}_2$ and related difference $\Delta\dot{\mathbf{S}} = \mathcal{C}(\dot{\mathbf{F}}_1) - \mathcal{C}(\dot{\mathbf{F}}_2)$. Therefore, the exclusion condition (37) is necessarily satisfied when the stronger condition

$$\boxed{\int_{\Omega^0} \dot{\mathbf{F}} \cdot \mathsf{C}^c[\dot{\mathbf{F}}] > 0} \tag{40}$$

holds true for all (not identically zero) continuous and piecewise continuously twice differentiable velocity fields, satisfying homogeneous conditions on $\partial\Omega^0_\xi$. The comparison solid C^c in (40) is equal, by definition, to R in the current plastic zone and to G in the current elastic zone. Note that, due to the linearity of C^c, the difference fields denoted by Δ in (37) do not appear in (40).

In the case of hyperelasticity and associative flow-rule, $\mathbf{N} = \mathbf{M}$, the comparison solid (38) reduces to the Hill comparison solid by taking $\psi = 1$.

The Raniecki's comparison theorem can be proved as follows. Three cases must be analyzed:

$$\begin{array}{llll} 1)\ \mathbf{N}\cdot\dot{\mathbf{F}}_1 < 0 & \text{and} & \mathbf{N}\cdot\dot{\mathbf{F}}_2 < 0, & \text{(unloading/unloading)}, \\ 2)\ \mathbf{N}\cdot\dot{\mathbf{F}}_1 > 0 & \text{and} & \mathbf{N}\cdot\dot{\mathbf{F}}_2 > 0, & \text{(loading/loading)}, \\ 3)\ \mathbf{N}\cdot\dot{\mathbf{F}}_1 > 0 & \text{and} & \mathbf{N}\cdot\dot{\mathbf{F}}_2 < 0, & \text{(loading/unloading)}. \end{array}$$

In all cases, it suffices to prove that

$$\begin{gathered} \Delta\dot{\mathbf{S}} \cdot \Delta\dot{\mathbf{F}} - \Delta\dot{\mathbf{F}} \cdot \mathsf{R}[\Delta\dot{\mathbf{F}}] = \frac{1}{4\psi g} \\ \left(-4\psi < \mathbf{N}\cdot\dot{\mathbf{F}}_1 > \mathbf{M}\cdot\Delta\dot{\mathbf{F}} + 4\psi < \mathbf{N}\cdot\dot{\mathbf{F}}_2 > \mathbf{M}\cdot\Delta\dot{\mathbf{F}} + [(\mathbf{M}+\psi\mathbf{N})\cdot\Delta\dot{\mathbf{F}}]^2\right) \geq 0, \end{gathered} \tag{41}$$

which, taking into account that $\psi g > 0$ and analyzing Cases 1)-3), follows directly. It is important to remark that the comparison theorem (39) holds true for every G, in other words, regardless of the symmetries and the definiteness of the elastic tensor G.

Finally, it may be worth noting that the exclusion condition (40) for bifurcations of the velocity problem, may be shown to be sufficient to exclude second- and higher-order bifurcations (under specific regularity conditions, see Petryk and Thermann, 1985; Nguyen and Triantafyllidis, 1989; Cheng and Lu, 1993; Bigoni, 1996)[8].

Associative elastoplasticity. It may be interesting to keep contact with the case of the associative flow rule $\mathbf{Q} = \mathbf{P}$, and Green elasticity (E, B, and G have the major symmetry)[9]. In this case, constitutive equations admit a velocity-gradient potential. With

[8] Moreover, the techniques introduced in this section to exclude bifurcation can be exported to another, similar context. This is the problem of contact with friction of an elastic body with a constraint (Radi et al. 1999).

[9] A detailed presentation can be found in Petryk (1993a).

reference to (28), it follows[10] that

$$\dot{\mathbf{S}} = \frac{\partial U}{\partial \dot{\mathbf{F}}}, \quad U = \frac{\dot{\mathbf{F}} \cdot \mathsf{G}[\dot{\mathbf{F}}]}{2} - \frac{< \mathbf{N} \cdot \dot{\mathbf{F}} >^2}{2g}. \tag{42}$$

Let us consider now the following functional (Hill, 1958;1959), defined for every admissible velocity field $\mathbf{v}$ satisfying $(35)_1$ on $\partial\Omega^0_\xi$

$$J(\mathbf{v}) = \int_{\Omega^0} U(\mathbf{v}) - \int_{\partial\Omega^0_\sigma} \dot{\boldsymbol{\sigma}} \cdot \mathbf{v}. \tag{43}$$

The vanishing of the first weak (Gateaux[11]) variation of $J(\mathbf{v})$ with respect to every admissible variation $\mathbf{w}$ of $\mathbf{v}$ is equivalent to (6). Therefore, a velocity field is a solution of the rate problem if and only if it assigns to the functional (43) a stationary value. Moreover, when the uniqueness condition (37) holds true, the functional $J(\mathbf{v})$ can be proved to be strictly convex. It follows that in the range where (37) holds true, the unique solution assigns to the functional $J(\mathbf{v})$ a strict, absolute minimum (Hill, 1958; 1959).

For the linear comparison solid defined by $\psi = 1$, i.e. by the loading branch of the constitutive operator, we have:

$$\dot{\mathbf{S}} = \frac{\partial U^L}{\partial \dot{\mathbf{F}}}, \quad U^L = \frac{\dot{\mathbf{F}} \cdot \mathsf{G}[\dot{\mathbf{F}}]}{2} - \frac{(\mathbf{N} \cdot \dot{\mathbf{F}})^2}{2g}, \tag{44}$$

so that the exclusion condition (40) corresponds now to the positive definiteness of the quadratic functional (where U^L in the elastic zone has to be identified with the actual elastic potential)

$$I^L(\mathbf{w}) = \int_{\Omega^0} U^L(\mathbf{w}) > 0, \tag{45}$$

for every admissible field $\mathbf{w}$ vanishing on Ω^0_ξ.

The variational basis of the rate problem for an associative flow-rule has important consequences on bifurcation. In fact, for a given deformation path, let us assume that a series of configurations continuously evolves in parameter space satisfying $I^L > 0$, and that this is terminated by a configuration for which

$$\begin{cases} I^L(\mathbf{w}) \geq 0 \text{ for every } \mathbf{w} \\ I^L(\mathbf{w}^*) = 0 \text{ for some } \mathbf{w}^* \neq \mathbf{0}, \end{cases} \tag{46}$$

[10] Note that $\frac{\partial < \mathbf{N} \cdot \dot{\mathbf{F}} >^2}{\partial \dot{\mathbf{F}}} = < \mathbf{N} \cdot \dot{\mathbf{F}} > \frac{\partial (\mathbf{N} \cdot \dot{\mathbf{F}} + |\mathbf{N} \cdot \dot{\mathbf{F}}|)}{\partial \dot{\mathbf{F}}} = < \mathbf{N} \cdot \dot{\mathbf{F}} > (1 + \frac{\mathbf{N} \cdot \dot{\mathbf{F}}}{|\mathbf{N} \cdot \dot{\mathbf{F}}|})\mathbf{N} = 2 < \mathbf{N} \cdot \dot{\mathbf{F}} > \mathbf{N}$.

[11] The first weak variation of a functional J in the direction $\mathbf{w}$ at $\mathbf{v}$ is

$$\delta J(\mathbf{v}, \mathbf{w}) = \frac{d}{d\alpha} J(\mathbf{v} + \alpha\, \mathbf{w})|_{\alpha=0}.$$

where the admissible fields $\mathbf{w}$ vanish on Ω^0_ξ. As shown by Hill and Hutchinson (1975) and Young (1976), the first instant at which $I^L > 0$ fails to hold need not satisfy (46)[12]. However, for situations in which (46) holds true a *primary eigenstate* has been found and I^L is stationary at the minimum point $\mathbf{w}^*$. This is necessary and sufficient for $\mathbf{w}^*$ to be a solution of the homogeneous problem. Therefore, a *critical point* has been detected. This may represent *either* a true *bifurcation point* for the comparison solid *or* a so-called *limit point*. In the former case, i.e. when a bifurcation point in the comparison solid has been detected, the eigenmode $\mathbf{w}^*$ can be added to a solution $\mathbf{v}$ of the non-homogeneous problem to generate a bifurcated solution. In other words, if $\mathbf{v}$ is a solution, $\mathbf{v} + \gamma \mathbf{w}^*$, $\gamma \in \Re$ is a family of possible solutions. In this family, i.e. for certain values of γ, bifurcated solutions can usually be found that correspond to the plastic branch of the constitutive equations in the current plastic zone. Among these solutions, those which initiate a quasi-static post-bifurcation path represent genuine elastic-plastic bifurcations of the real elastoplastic solid, and may occur under broad hypotheses (Hutchinson, 1973).

'In loading comparison solid'. From the discussion relative to associative flow rules, it should be clear that failure of the Hill/Raniecki exclusion condition (40) is in general not sufficient for bifurcation even in the case of associative elastoplasticity. This becomes indeed more evident for non-associative elastoplasticity, where due to the lack of a variational structure of the governing incremental field equations, failure of (40) is far from implying bifurcation. We may note, however, that the comparison solid in the associative case plays a double role. On one hand it excludes bifurcation when used in (40) and, on the other hand, it provides a bifurcated field for the comparison solid when (40) fails, which can often be 'adjusted' for the real elastoplastic solid. In the non-associative case, the Raniecki comparison solids are effective for excluding bifurcation, but not to provide a bifurcated field useful for the real elastoplastic solid. Therefore, following Raniecki and Bruhns (1981), let us consider, even for non-associative flow rules a fictitious, incrementally-linear solid with tangent constitutive tensor

$$\mathsf{C} = \mathsf{G} - \frac{1}{g}\mathbf{M} \otimes \mathbf{N}, \tag{47}$$

corresponding to the loading branch of the constitutive operator (28) (briefly, 'in loading comparison solid'). Let us consider a deformation path in which an elastoplastic solid is deformed in the plastic range, so that the actual behaviour corresponds to the behaviour of the fictitious solid 'in loading'. The first bifurcation for this comparison solid may correspond (in the sense already explained for associative flow-rule) to a possible bifurcation of the real elastoplastic solid. Therefore, for non-associative elastoplasticity:

> *two bounds can be defined for bifurcation. The lower bound corresponds to failure of exclusion condition (40) for the optimal Raniecki comparison solid* R. *The upper bound corresponds to bifurcation in the 'in loading comparison solid' defined by the incremental tensor* C.

[12] In Hill and Hutchinson (1975) and Young (1976), this possibility arises in connection with the achivement of a situation where surface bifurcation modes exist of arbitrarily short wavelength.

Whether or not some bifurcation actually occurs in the real elastoplastic solid within the two bounds is still an open question. No examples are in fact yet known.

5.2 Stability

Let us consider a generic equilibrium configuration of a body at a fixed value of the loading parameter governing the deformation, so that the prescribed displacements on the boundary are fixed and the prescribed nominal surface tractions correspond, momentarily, to dead loading. Roughly speaking, the configuration is called stable when the effects of a small disturbance remain sufficiently small during the entire motion subsequent to the application of the disturbance itself. Therefore, stability analysis involves considerations about the dynamics of the system. The definition of stability becomes a precise concept when the measures of the distances and the class of perturbations are specified. However, path-dependence of inelastic material (together with certain mathematical difficulties connected to the analysis of a continuum problem) makes a rigorous analysis awkward. Therefore, we content ourselves with presenting the simple analysis which was proposed by Hill (1958) in the context of an associative flow rule. In that context, the Hill's analysis has a much more firm basis than in the case of a non-associative flow rule (Petryk, 1991; 1985b; 1993a;b).

Let us confine our attention to *direct* paths of departure from the equilibrium configuration (thus excluding arbitrary circuitous paths). In this way, a perturbed motion is a-priori assumed such that variations of the direction of the velocity field along the path are negligible. This is the so-called *directional stability*, which was analyzed by Hill in the framework of associative elastoplasticity. Along any admissible direct path starting from the equilibrium state under consideration, the work of deformation in the body can be written as

$$W = \int_{\Omega^0} \mathbf{S} \cdot \mathrm{Grad}\, \Delta\mathbf{u} + \frac{1}{2} \int_{\Omega^0} \Delta\mathbf{S} \cdot \mathrm{Grad}\, \Delta\mathbf{u} + o((\Delta t)^2), \tag{48}$$

where $\mathbf{S}$ is the first Piola-Kirchhoff stress at the equilibrium state, $\Delta\mathbf{u}$ and $\Delta\mathbf{S}$ are the increments in displacement and stress reached along the path. Finally, Δt is the increment in the time-like parameter measuring the length of the path. In view of the fact that $\Delta\mathbf{u}$ vanishes on $\partial\Omega^0_\xi$, from the principle of virtual power, the first integral in (48) is equal to the work done by external *dead* loads W^{load}. Therefore, the work difference can be written as:

$$W - W^{load} = \frac{(\Delta t)^2}{2} \int_{\Omega^0} \dot{\mathbf{S}} \cdot \nabla\dot{\mathbf{u}} + o((\Delta t)^2). \tag{49}$$

It follows from (27) that if

$$\boxed{\int_{\Omega^0} \nabla\mathbf{w} \cdot \boldsymbol{C}(\nabla\mathbf{w}) > 0, \quad \text{(stability in Hill's sense)}} \tag{50}$$

for every admissible (not identically zero) velocity field $\mathbf{w}$ taking zero values on $\partial\Omega^0_\xi$, then any movement from the equilibrium configuration requires some additional energy to be supplied to the system from external sources. In this sense (50) is a *sufficient condition for directional stability* of equilibrium. We will briefly refer to this condition as 'stability in Hill's sense', even if:

- it is only a sufficient condition (for the assumed class of paths),
- Hill never proposed to use this condition in a broader context than associative elastoplasticity.

With reference to elasticity, with '>' replaced with '≥', condition (50) was proposed by Hadamard (1903) and is usually called *infinitesimal stability*. As noticed by Truesdell and Noll (1965, Sect. 68) and Beatty (1987) the existence of a stored energy function is not essential in the above definition of stability. In the elastoplastic case, when the associative flow rule is assumed, condition (50), with '>' replaced with '≥', was proved also to be necessary for stability under broad hypotheses (Petryk, 1993a;b).

In the case of non-associative flow laws, two important points should be emphasized. First, instability as related to violation of (50) is not proven, so that (50) may not be a necessary condition for stability. Second, condition (50) should not be even considered sufficient. There are in fact certain instability phenomena, such as flutter, which may in principle occur even when (50) is satisfied.

It is worth noting that (40) implies (50), more explicitly[13]

$$\boxed{\text{exclusion condition for bifurcation} \Longrightarrow \text{stability in Hill's sense,}}$$

but the converse need not be true.

6 Local conditions for uniqueness and stability

From the global uniqueness and stability criteria considered in the previous section, local conditions may be derived, which are the subject of this section. The importance of local conditions lies in the connection to material instabilities, i.e. to instabilities which can develop from a point in a continuum. For instance, we will see that loss of ellipticity corresponds to strain localization into planar bands. Five local criteria will be analyzed, namely:

1) positive definiteness (PD) and
2) non-singularity (NS)
of the constitutive operator;
3) strong ellipticity (SE);
4) ellipticity (E);
5) flutter (F).

In the following we will assume for simplicity that *the constitutive equations evolve continuously with time*. Therefore, in a continuous deformation path initiating when (PD) [or (SE)] holds, failure of (PD) [or (SE)] will be shown to be simultaneous to failure of (NS) [or (E)] in the case of an *associative flow rule*. But this will be not the case of

[13] This statement is not equivalent to saying that uniqueness implies stability, because the Raniecki condition is a *sufficient* condition for uniqueness.

non-associative flow rules, where (PD) and (NS) as well as (SE) and (E) are different even in the case of a continuous loading path initiating when (PD) or (SE) hold. In particular, it will be shown that

$$
\boxed{(\mathrm{PD}) \Longrightarrow \begin{array}{l} \nearrow (\mathrm{SE}) \Longrightarrow (\mathrm{E}) \\ \searrow (\mathrm{NS}) \end{array}}
$$

from which it should be noted that, except for an associative flow law, (NS) has no relation with (SE) and (E).

6.1 A local sufficient condition for uniqueness: positive definiteness of the constitutive operator

It is easy to observe that, when R is positive definite at every point of the body, (40) and (50) are both satisfied. Positive definiteness of R implies positive definiteness of G, which is assumed in this section. Following Raniecki and Bruhns (1981), we will show that

$$
\boxed{\text{positive definiteness of } \mathsf{R}^{opt} \Longleftrightarrow \text{ positive definiteness of } \mathsf{C} \text{ (PD condition)},}
$$

where R^{opt} is the Raniecki solid corresponding to an optimal value of ψ to be defined later. In other words, positive definiteness of the best chosen Raniecki solid corresponds to positive definiteness of the 'in loading comparison solid'. Obviously, positive definiteness of $\boldsymbol{C}$ and C are equivalent (if $\mathbf{X} \cdot \boldsymbol{C}(\mathbf{X}) = \mathbf{X} \cdot \mathsf{G}[\mathbf{X}]$, it suffices to consider $-\mathbf{X}$ to obtain $\mathbf{X} \cdot \boldsymbol{C}(\mathbf{X}) = \mathbf{X} \cdot \mathsf{C}[\mathbf{X}]$)[14].

In order to show the equivalence between positive definiteness of R^{opt} and C, under the assumption that G is positive definite, let us consider, for every tensor $\mathbf{X} \in \mathrm{Lin}$, the following inequality:

$$
\mathbf{X} \cdot \mathsf{C}[\mathbf{X}] \geq \mathbf{X} \cdot \mathsf{R}[\mathbf{X}] = \mathbf{X} \cdot \tilde{\mathsf{G}}[\mathbf{X}] - \frac{(\mathbf{X} \cdot \mathbf{R})^2}{4\psi g}, \tag{51}
$$

where $\mathbf{R} = \mathbf{M} + \psi \mathbf{N}$ and

$$
\tilde{\mathsf{G}} = \frac{\mathsf{G} + \mathsf{G}^T}{2}
$$

is the symmetric (with respect to the major symmetry) part of G; only this part plays a role (in fact $\mathbf{X} \cdot \mathsf{G}[\mathbf{X}] = \mathbf{X} \cdot \tilde{\mathsf{G}}[\mathbf{X}]$). The Cauchy-Schwarz inequality in the metric $\tilde{\mathsf{G}}$ (Appendix B) can be used to yield

$$
(\mathbf{X} \cdot \mathbf{R})^2 = (\mathbf{X} \cdot \tilde{\mathsf{G}}[\tilde{\mathsf{G}}^{-1}[\mathbf{R}]])^2 \leq (\mathbf{X} \cdot \tilde{\mathsf{G}}[\mathbf{X}])(\mathbf{R} \cdot \tilde{\mathsf{G}}^{-1}[\mathbf{R}]), \tag{52}
$$

which, employed in (51), gives

$$
\mathbf{X} \cdot \mathsf{R}[\mathbf{X}] \geq \frac{\mathbf{X} \cdot \tilde{\mathsf{G}}[\mathbf{X}]}{4\psi g} \left(4\psi g - \mathbf{R} \cdot \tilde{\mathsf{G}}^{-1}[\mathbf{R}] \right). \tag{53}
$$

[14] Note that (only for a non-associative flow rule) for certain tensors $\mathbf{X}$ the elastoplastic response is stiffer than the elastic, namely, $\mathbf{X} \cdot \mathsf{G}[\mathbf{X}] < \mathbf{X} \cdot \mathsf{C}[\mathbf{X}]$ (Mróz, 1963, 1966; Runesson and Mróz, 1989).

From (53), we note that $\mathcal{C}$, C and R are positive definite when:

$$g > \frac{(\mathbf{M}+\psi\mathbf{N})\cdot\tilde{\mathsf{G}}^{-1}[\mathbf{M}+\psi\mathbf{N}]}{4\psi}. \tag{54}$$

However, when (54) is violated, R is not positive definite. In other words, (54) is a necessary and sufficient condition for positive definiteness of R. To show this it suffices to note that

$$g \leq \frac{\mathbf{R}\cdot\tilde{\mathsf{G}}^{-1}[\mathbf{R}]}{4\psi}, \qquad \mathbf{X} \propto \tilde{\mathsf{G}}^{-1}[\mathbf{R}] \quad \Longrightarrow \quad \mathbf{X}\cdot\mathsf{R}[\mathbf{X}] \leq 0.$$

We can consider now the dependence on ψ and state that

$$g > \min_{\psi>0}\left\{\frac{(\mathbf{M}+\psi\mathbf{N})\cdot\tilde{\mathsf{G}}^{-1}[\mathbf{M}+\psi\mathbf{N}]}{4\psi}\right\} \quad \Longrightarrow \quad \mathcal{C} \text{ and } \mathsf{C} \text{ positive definite.} \tag{55}$$

The minimum problem is solved by

$$\psi = \sqrt{\frac{\mathbf{M}\cdot\tilde{\mathsf{G}}^{-1}[\mathbf{M}]}{\mathbf{N}\cdot\tilde{\mathsf{G}}^{-1}[\mathbf{N}]}}.$$

which defines R^{opt} and yields the following proposition:

$$\boxed{g > g_{cr}^{PD} = \frac{1}{2}\left(\sqrt{(\mathbf{M}\cdot\tilde{\mathsf{G}}^{-1}[\mathbf{M}])(\mathbf{N}\cdot\tilde{\mathsf{G}}^{-1}[\mathbf{N}])} + \mathbf{M}\cdot\tilde{\mathsf{G}}^{-1}[\mathbf{N}]\right) \geq 0 \iff (\text{PD}).} \tag{56}$$

Note that $g_{cr}^{PD} = 0 \iff \mathbf{M} \propto -\mathbf{N}$, a condition that should never be satisfied for realistic constitutive models.

In order to complete the proof of (56), it remains to show that when $g \leq g_{cr}^{PD}$ tensor C is not positive definite. To this purpose, it suffices to note that

$$g \leq g_{cr}^{PD}, \quad \mathbf{X} \propto \sqrt{\mathbf{M}\cdot\tilde{\mathsf{G}}^{-1}[\mathbf{M}]}\,\tilde{\mathsf{G}}^{-1}[\mathbf{N}] + \sqrt{\mathbf{N}\cdot\tilde{\mathsf{G}}^{-1}[\mathbf{N}]}\,\tilde{\mathsf{G}}^{-1}[\mathbf{M}] \quad \Longrightarrow \quad \mathbf{X}\cdot\mathsf{C}[\mathbf{X}] \leq 0.$$

The condition (PD) has been expressed in terms of a critical value of the hardening modulus. In general, any condition expressed in this way is implicitly referred to a sufficiently regular deformation path in which the plastic modulus is a continuously varying function of the loading parameter.

A problem with the (PD) condition is that:

the hypothesis that G be positive definite is stronger than it may appear.

Let us analyze this point in detail. Let us express $\dot{\mathbf{S}}\cdot\dot{\mathbf{F}}$ going back to (22) and taking into account that $\dot{\mathbf{F}} = \mathbf{LF}$ as

$$\dot{\mathbf{S}}\cdot\dot{\mathbf{F}} = \mathbf{F}^T\mathbf{LF}\cdot\dot{\mathbf{T}}^{(2)} + \mathbf{LF}\cdot\mathbf{LKF}^{-T} = \mathbf{F}^T\mathbf{DF}\cdot\dot{\mathbf{T}}^{(2)} + \mathbf{LK}\cdot\mathbf{L}. \tag{57}$$

where $\mathbf{K} = J\mathbf{T} \in \text{Sym}$ is the Kirchhoff stress. Eqn. (57), using (17) but restricted to the loading branch of the constitutive operator, may be written as:

$$\mathsf{C}[\dot{\mathbf{F}}] \cdot \dot{\mathbf{F}} = \mathbf{D} \cdot \mathsf{H}[\mathbf{D}] + \frac{1}{2}\mathbf{D} \cdot \mathsf{W}[\mathbf{D}] \underbrace{- \mathbf{D} \cdot \mathsf{K}[\mathbf{W}] + \frac{1}{2}\mathbf{W} \cdot \mathsf{W}[\mathbf{W}]}_{\textit{terms depending on spin}} - \frac{\hat{\mathbf{Q}} \cdot \mathsf{H}[\mathbf{D}]}{g}\mathbf{D} \cdot \mathsf{H}[\hat{\mathbf{P}}], \quad (58)$$

where $\mathbf{W} = \mathbf{L} - \mathbf{D}$ is the spin, $\hat{\mathbf{Q}} = \mathbf{F}^{-T}\mathbf{Q}\mathbf{F}^{-1}$, $\hat{\mathbf{P}} = \mathbf{F}^{-T}\mathbf{P}\mathbf{F}^{-1}$ and

$$\mathsf{H} = (\mathbf{F}\boxtimes\mathbf{F})\mathsf{E}(\mathbf{F}\boxtimes\mathbf{F})^T, \quad \mathsf{W} = \mathbf{K}\boxtimes\mathbf{I} + \mathbf{I}\boxtimes\mathbf{K}, \quad \mathsf{K} = \mathbf{K}\boxtimes\mathbf{I} - \mathbf{I}\boxtimes\mathbf{K}, \quad (59)$$

where

- tensor H has always the minor symmetries (because E has) and has the major symmetry in the case of a Green elastic material.
- tensor W always has the major symmetry, but not the minor symmetries, and transforms symmetric tensors into symmetric tensors and skew tensors into skew tensors. Moreover, $\mathbf{W} \cdot \mathsf{W}[\mathbf{W}] = 2\mathbf{W} \cdot \mathbf{K}\mathbf{W}$.
- tensor K has always the major symmetry, but not the minor symmetries, and transforms symmetric tensors into skew tensors and skew tensors into symmetric tensors. Moreover, $\mathbf{W} \cdot \mathsf{K}[\mathbf{W}] = \mathbf{D} \cdot \mathsf{K}[\mathbf{D}] = \mathbf{0}$ and $-\mathbf{D} \cdot \mathsf{K}[\mathbf{W}] = 2\mathbf{D}\mathbf{K} \cdot \mathbf{W}$. Finally, when $\mathbf{A}$ and $\mathbf{K}$ are coaxial, i.e. $\mathbf{A}\mathbf{K} = \mathbf{K}\mathbf{A}$, $\mathsf{K}[\mathbf{A}] = \mathbf{0}$.

The spin $\mathbf{W}$ and the rate of deformation $\mathbf{D}$ are *independent* tensors. Therefore, a necessary condition for $\dot{\mathbf{S}} \cdot \dot{\mathbf{F}} > 0$, i.e. for C to be positive definite, is that $\mathbf{W} \cdot \mathsf{W}[\mathbf{W}] > 0$ for every $\mathbf{W}$, in other words:

$$\boxed{(\text{PD}) \implies (\text{tr}\mathbf{T})\mathbf{I} - \mathbf{T} \text{ pos. def.} \iff T_1 + T_2 > 0,\ T_1 + T_3 > 0,\ T_2 + T_3 > 0.} \quad (60)$$

Positive definiteness of $(\text{tr}\mathbf{T})\mathbf{I} - \mathbf{T}$ is equivalent to saying that the restriction of W to the space of skew tensors is positive definite. This condition was obtained by Hill (1967a) and it *holds true for every constitutive assumption*. In other words, (60) reflects a purely 'geometrical effect'. Now, let us assume positive definiteness of $(\text{tr}\mathbf{T})\mathbf{I} - \mathbf{T}$. Under this condition, the function (at fixed $\mathbf{D}$)

$$z(\mathbf{W}) = -\mathbf{D} \cdot \mathsf{K}[\mathbf{W}] + \frac{1}{2}\mathbf{W} \cdot \mathsf{W}[\mathbf{W}], \quad (61)$$

is strictly convex and therefore it admits a minimum point, which corresponds to

$$\frac{\partial z(\mathbf{W})}{\partial \mathbf{W}} = \mathsf{W}[\mathbf{W}] - \mathsf{K}[\mathbf{D}] = \mathbf{0}. \quad (62)$$

Due to positive definiteness of $(\text{tr}\mathbf{T})\mathbf{I} - \mathbf{T}$, the restriction of W to the space of all skew tensors is invertible[15] and therefore condition (62) defines the minimum point of $z(\mathbf{W})$ in terms of $\hat{\mathbf{W}}$

$$\hat{\mathbf{W}} = \mathsf{W}^{-1}\mathsf{K}[\mathbf{D}], \quad (63)$$

[15] This statement means that

$$\mathsf{W}\mathsf{W}^{-1} = \mathbf{I}\boxtimes\mathbf{I} - \mathbf{I}\,\overline{\underline{\otimes}}\,\mathbf{I},$$

so that $(\mathbf{I}\boxtimes\mathbf{I} - \mathbf{I}\,\overline{\underline{\otimes}}\,\mathbf{I})[\mathbf{A}] = (\mathbf{A} - \mathbf{A}^T)/2$ associates to every tensor its skew symmetric part (Del Piero, 1979).

as

$$z_{min} = -\frac{1}{2}\mathbf{D} \cdot \left(\mathsf{K}\mathsf{W}^{-1}\mathsf{K}\right)[\mathbf{D}]. \tag{64}$$

The components of tensor $\mathsf{K}\mathsf{W}^{-1}\mathsf{K}$ can be easily calculated in the principal reference system of $\mathbf{K}$. This calculation reveals that $\mathsf{K}\mathsf{W}^{-1}\mathsf{K}$ has both the major and minor symmetries and has only the following non-null components

$$(\mathsf{K}\mathsf{W}^{-1}\mathsf{K})_{ijij} = (\mathsf{K}\mathsf{W}^{-1}\mathsf{K})_{jiij} = \frac{(K_i - K_j)^2}{2(K_i + K_j)}, \quad i \neq j;\ i, j \in [1, 3]. \tag{65}$$

As a conclusion, the following inequality is obtained:

$$\mathsf{C}[\dot{\mathbf{F}}] \cdot \dot{\mathbf{F}} \geq \mathbf{D} \cdot \mathsf{L}[\mathbf{D}] - \frac{1}{g}(\mathbf{D} \cdot \mathsf{H}[\hat{\mathbf{P}}])(\mathbf{D} \cdot \mathsf{H}^T[\hat{\mathbf{Q}}]). \tag{66}$$

(with the equality holding true when $\mathbf{W} = \hat{\mathbf{W}}$) where

$$\mathsf{L} = \mathsf{H} + \frac{1}{2}\mathsf{W} - \frac{1}{2}\mathsf{K}\mathsf{W}^{-1}\mathsf{K}. \tag{67}$$

therefore, under condition that $(\mathrm{tr}\mathbf{T})\mathbf{I} - \mathbf{T}$ be positive definite,

> *positive definiteness of the restriction of* L *to the space of all symmetric tensors is equivalent to positive definiteness of* G *on* Lin.

Therefore, assuming L positive definite, it is possible to show that

$$\boxed{\begin{array}{l} T_1 + T_2 > 0,\ T_1 + T_3 > 0,\ T_2 + T_3 > 0,\ \&\ g > g_{cr}^{PD} \geq 0, \\ g_{cr}^{PD} = \frac{1}{2}\left(\sqrt{(\hat{\mathbf{P}} \cdot \mathsf{H}^T\tilde{\mathsf{L}}^{-1}\mathsf{H}[\hat{\mathbf{P}}])(\hat{\mathbf{Q}} \cdot \mathsf{H}\tilde{\mathsf{L}}^{-1}\mathsf{H}^T[\hat{\mathbf{Q}}])} + \hat{\mathbf{Q}} \cdot \mathsf{H}\tilde{\mathsf{L}}^{-1}\mathsf{H}[\hat{\mathbf{P}}]\right) \iff (\mathrm{PD}). \end{array}} \tag{68}$$

where $\tilde{\mathsf{L}} = (\mathsf{L} + \mathsf{L}^T)/2$ and $\tilde{\mathsf{L}}^{-1}\tilde{\mathsf{L}} = \mathbf{I} \underline{\boxtimes} \mathbf{I}$. Condition (68) was obtained by Raniecki and Bruhns (1981) in the case of L having the major symmetry. Condition (68) is equivalent to (56), but more explicit in the sense that it singles out the effect of $(\mathrm{tr}\mathbf{T})\mathbf{I} - \mathbf{T}$.

Uniaxial tension. In the case of uniaxial tension along axis 1, $T_1 = \sigma > 0$ and $T_2 = T_3 = 0$, thus $(\mathrm{tr}\mathbf{T})\mathbf{I} - \mathbf{T}$ is positive semi-definite and G is not invertible. Let us develop this point in detail. In this case, by direct calculation we obtain

$$z(\mathbf{W}) = \sigma\left(W_{12}^2 + W_{13}^2 - 2D_{12}W_{12} - 2D_{13}W_{13}\right), \tag{69}$$

which is zero when $\mathbf{D} = \mathbf{0}$ and the only non-null component of $\mathbf{W}$ is W_{23}, representing a spin about axis 1. The minimum of $z(\mathbf{W})$ is

$$z_{min} = -\sigma\left(D_{12}^2 + D_{13}^2\right). \tag{70}$$

It follows that

$$\dot{\mathbf{F}} \cdot \mathsf{C}[\dot{\mathbf{F}}] \geq \mathbf{D} \cdot \mathsf{H}[\mathbf{D}] - \frac{1}{g}(\mathbf{D} \cdot \mathsf{H}^T[\hat{\mathbf{Q}}])(\mathbf{D} \cdot \mathsf{H}[\hat{\mathbf{P}}]) + \sigma D_{11}^2, \tag{71}$$

and therefore, when the restriction of

$$\mathsf{H} + \sigma\, \mathbf{e}_1 \otimes \mathbf{e}_1 \otimes \mathbf{e}_1 \otimes \mathbf{e}_1 - \frac{1}{g}\mathsf{H}[\hat{\mathbf{P}}] \otimes \mathsf{H}^T[\hat{\mathbf{Q}}], \tag{72}$$

to Sym is positive definite, condition (PD) is verified, except for a velocity gradient having W_{23} as the unique non-null component. However, the integrals (40) and (50) can vanish only if $W_{23} \neq 0$ is the only non-null component of the velocity gradient everywhere in the body and this corresponds to a loss of uniqueness consisting in arbitrary rigid rotations about the axis of tension (which is also an axis of neutral stability). This circumstance becomes particularly clear if we consider a bar pulled in tension and therefore subject to homogeneous stress and all-round dead loading (Fig. 1 a). Analogously, $(\mathrm{tr}\mathbf{T})\mathbf{I} - \mathbf{T}$ is negative semi-definite (and thus G is indefinite) in the important case of a bar subject to dead loading of uniaxial compression, even for a vanishing small value of axial force. This correctly corresponds to a well-known instability due to rigid-body rotation (Fig. 1b).

Fig. 1. a-neutral rotational equilibrium of a rod under tensile dead load; b-rotational instability of a rod under compressive dead load.

Using (72), condition (PD) –except for rigid-body rotations about the axis of tension– can be written in explicit as:

$$\boxed{\begin{array}{l} T_1 > 0,\ T_2 = T_3 = 0,\ \&\ g > g_{cr}^{PD} \geq 0,\ \mathsf{L} = \mathsf{H} + \sigma\, \mathbf{e}_1 \otimes \mathbf{e}_1 \otimes \mathbf{e}_1 \otimes \mathbf{e}_1, \\ g_{cr}^{PD} = \frac{1}{2}\left(\sqrt{(\hat{\mathbf{P}} \cdot \mathsf{H}^T \tilde{\mathsf{L}}^{-1} \mathsf{H}[\hat{\mathbf{P}}])(\hat{\mathbf{Q}} \cdot \mathsf{H}\tilde{\mathsf{L}}^{-1} \mathsf{H}^T[\hat{\mathbf{Q}}])} + \hat{\mathbf{Q}} \cdot \mathsf{H}\tilde{\mathsf{L}}^{-1} \mathsf{H}[\hat{\mathbf{P}}]\right) \iff \text{(PD)} \end{array}} \tag{73}$$

where $\tilde{\mathsf{L}}$ is the symmetric part of L, assumed positive definite.

The infinitesimal theory. In the infinitesimal theory, where we refer to (34), it may be interesting to write g_{cr}^{PD} in terms of a critical value of the hardening modulus h_{cr}^{PD}. For a symmetric E, this becomes

$$h_{cr}^{PD} = \frac{1}{2}\left(\sqrt{(\mathbf{P}\cdot\mathsf{E}[\mathbf{P}])(\mathbf{Q}\cdot\mathsf{E}[\mathbf{Q}])} - \mathbf{P}\cdot\mathsf{E}[\mathbf{Q}]\right) \geq 0. \tag{74}$$

which is never negative and becomes null for associative elastoplasticity. Therefore, in the infinitesimal theory (and assuming E symmetric), (PD) is always lost before softening. Critical modulus (74) was obtained by Mróz (1963), Maier and Hueckel (1979) and Raniecki (1979). A derivation of the critical hardening modulus for plane strain and plane stress situations was given by Bigoni and Hueckel (1991a).

In closing this section, we note that in under the hypotheses of homogeneity and all-round controlled nominal surface tractions, failure of (PD) implies failure of the Hill sufficient conditions for stability (50) and uniqueness (40). This situation is critical in the case of the associative flow rule, where instability and bifurcation may occur at loss of (PD) under broad hypotheses (Hill, 1967; Miles, 1973). The situation is however still not clear both for stability and bifurcation in non-associative elastoplasticity. In that context, examples are not known in which any real bifurcation has been found in coincidence with failure of (PD).

6.2 Singularity of the constitutive operator

Loss of (PD) at a point of a body during a loading program is not in general a sufficient condition for bifurcation. There are however certain situations where loss of (PD) may become close to critical. These have been touched on in the closure of the previous section. Let us consider therefore a special class of problems where:
1) controlled nominal tractions are prescribed on the entire boundary and
2) material properties and deformation (and therefore stress) are homogeneous,
during a given loading path. These situations have been analyzed by Hill (1967a), Miles (1973), Raniecki and Bruhns (1981) and Ogden (1985). Under the above hypotheses, conditions (40) and (50) are equivalent to (PD). In other words, exclusion of bifurcation and stability in Hill's sense fail to hold when (PD) is lost, i.e. when at least one $\mathbf{X}^* \neq \mathbf{0}$ exists such that

$$\mathbf{X}^*\cdot\mathsf{C}[\mathbf{X}^*] = 0. \tag{75}$$

Except for the associative case, this condition does not mean that C is singular, in other words $\mathsf{C}[\mathbf{X}^*] \neq \mathbf{0}$ should be in general expected. Therefore, loss of (PD) is not directly connected to a bifurcation, even in the above special hypotheses. On the other hand, if we define the non-singularity condition

$$\boxed{\mathsf{C}[\mathbf{X}] \neq \mathbf{0}, \text{ for every (non-zero) } \mathbf{X} \in \mathrm{Lin} \quad \text{(NS condition).}} \tag{76}$$

we can understand that, in a sense, failure of this condition is critical for bifurcation of the homogeneous problem with controlled nominal surface tractions on the entire boundary.

We note, in passing, that obviously (PD) implies (NS) and that when (PD) fails the first time in a continuous loading path (and for continuous dependence of constitutive equations on time) (NS) also fails for associative, hyperelastic-plastic solids.

With respect to the constitutive operator (28) where G is assumed invertible[16], we note that, assuming $\mathbf{N} \cdot \mathsf{G}^{-1}[\mathbf{M}] > 0$

$$\dot{\mathbf{S}} = \mathbf{0} \iff \dot{\mathbf{F}} \propto \mathsf{G}^{-1}[\mathbf{M}], \; g = \mathbf{N} \cdot \mathsf{G}^{-1}[\mathbf{M}]. \tag{77}$$

Therefore

$$\boxed{g \neq g_{cr}^{NS} = \mathbf{N} \cdot \mathsf{G}^{-1}[\mathbf{M}], \quad \iff \quad \mathsf{C} \text{ is not singular.} \quad \text{(NS condition)}} \tag{78}$$

Note that $g_{cr}^{NS} < 0$ when $\mathbf{N} \cdot \mathsf{G}^{-1}[\mathbf{M}] < 0$. In this case, loss of (NS) does not occur (in the present constitutive framework).

Loss of (NS) is critical in the sense explained by Raniecki and Bruhns (1981), namely, assuming $\mathsf{C}[\mathbf{X}^*] = \mathbf{0}$, if $\dot{\mathbf{F}}$ is a solution of the problem in velocities (corresponding to plastic loading everywhere in the body) it satisfies:

$$\mathcal{C}(\dot{\mathbf{F}}) = \mathsf{C}[\dot{\mathbf{F}}] \text{ and } \mathsf{C}[\dot{\mathbf{F}}]\mathbf{n}_0 = \dot{\boldsymbol{\sigma}} \text{ on } \partial\Omega^0,$$

where $\mathbf{n}_0$ is the unit outward normal to $\partial\Omega^0$. It follows that if $\mathcal{C}(\dot{\mathbf{F}}+\gamma\mathbf{X}^*) = \mathsf{C}[\dot{\mathbf{F}}]+\gamma\mathsf{C}[\mathbf{X}^*]$ for at least some $\gamma \neq 0$,

$$\mathcal{C}(\dot{\mathbf{F}} + \gamma\mathbf{X}^*)\mathbf{n}_0 = \mathsf{C}[\dot{\mathbf{F}}]\mathbf{n}_0 = \dot{\boldsymbol{\sigma}} \text{ on } \partial\Omega^0,$$

and therefore either a *bifurcation* or a *load maximum* has been reached in the loading program (Hill, 1967a).

Condition (78) may be further elaborated, assuming $T_1 + T_2 > 0$, $T_1 + T_3 > 0$, $T_2 + T_3 > 0$, as follows. Let us observe that

$$\dot{\mathbf{S}}\mathbf{F}^T = \mathsf{H}[\mathbf{D}] + \mathbf{LK} - \frac{1}{g}(\hat{\mathbf{Q}} \cdot \mathsf{H}[\mathbf{D}])\mathsf{H}[\hat{\mathbf{P}}], \tag{79}$$

where we may note that

$$\mathbf{LK} = \frac{1}{2}(\mathsf{W} - \mathsf{K})[\mathbf{D}] + \frac{1}{2}(\mathsf{W} - \mathsf{K})[\mathbf{W}]. \tag{80}$$

For $\mathbf{W}$ in the form (63), i.e. $\mathbf{W} = \mathsf{W}^{-1}\mathsf{K}[\mathbf{D}]$, and L defined by (67), eqn. (79), using (80), becomes

$$\dot{\mathbf{S}}\mathbf{F}^T = \mathsf{L}[\mathbf{D}] - \frac{1}{g}(\hat{\mathbf{Q}} \cdot \mathsf{H}[\mathbf{D}])\mathsf{H}[\hat{\mathbf{P}}]. \tag{81}$$

Assuming L invertible, from (81) we conclude that, under hypothesis (63), $\dot{\mathbf{S}} = \mathbf{0} \iff g = \hat{\mathbf{Q}} \cdot (\mathsf{H}\mathsf{L}^{-1}\mathsf{H})[\hat{\mathbf{P}}]$ and $\mathbf{D} \propto \mathsf{L}^{-1}\mathsf{H}[\hat{\mathbf{P}}]$. Vice-versa, assuming $\dot{\mathbf{S}} = \mathbf{0}$ in (79), the minor symmetries of H imply $\mathbf{W} = \mathsf{W}^{-1}\mathsf{K}[\mathbf{D}]$. We obtain therefore the following proposition:

$$\boxed{\begin{array}{l} T_1 + T_2 > 0, \; T_1 + T_3 > 0, \; T_2 + T_3 > 0: \\ g \neq g_{cr}^{NS} = \hat{\mathbf{Q}} \cdot (\mathsf{H}\mathsf{L}^{-1}\mathsf{H})[\hat{\mathbf{P}}], \quad \iff \quad \mathsf{C} \text{ is not singular.} \quad \text{(NS condition)} \end{array}} \tag{82}$$

[16] Singularity of G implies singularity of $\mathcal{C}$.

Obviously, conditions (78) and (82) are equivalent, under the condition that $(\mathrm{tr}\mathbf{T})\mathbf{I} - \mathbf{T}$ is positive definite.

Uniaxial tension. For uniaxial tension along axis 1, $T_1 = \sigma > 0$ and $T_2 = T_3 = 0$. Direct calculations on (79) yield that the critical condition is again given by (82), but with

$$\mathsf{L} = \mathsf{H} + \sigma\, \mathbf{e}_1 \otimes \mathbf{e}_1 \otimes \mathbf{e}_1 \otimes \mathbf{e}_1 .$$

The infinitesimal theory. In the case of the infinitesimal theory (34) with E symmetric and positive definite, the condition (NS) becomes $h \neq 0$, and:

$$\text{for } h = 0 \quad \text{and} \quad \mathbf{X}^* \propto \mathbf{P}, \quad \mathsf{C}[\mathbf{X}^*] = \mathbf{0}.$$

As a conclusion and with the exception of associative plasticity, (PD) and (NS) clearly do not coincide in the infinitesimal theory, where (PD) is lost before softening and (NS) is always lost in the perfectly plastic case ($h = 0$).

6.3 Strong ellipticity

In the previous section we have analyzed the special condition of a homogeneous body subject to all-round controlled nominal surface tractions. In that case, the (PD) and (NS) conditions play a special role. Now we analyze a dual case in which a homogeneous body is deformed in an homogeneous way under prescribed displacements on the entire boundary. As a result, we will show that strong ellipticity (SE) and ellipticity (E) play a role similar to those shown for (PD) and (NS) in the previous section.

Let us begin by considering the 'in loading comparison solid', i.e. the fictitious solid defined by the constitutive tensor C. For this material, we will show the validity of the following *uniqueness theorem for the velocity problem, due to van Hove (1947)*:

For a homogeneous and homogeneously deformed body, characterized by an incrementally linear constitutive operator (here C) and subject to prescribed velocity over the entire boundary, the strong ellipticity condition

$$\boxed{\mathbf{g} \cdot \mathsf{C}[\mathbf{g} \otimes \mathbf{n}]\mathbf{n} > 0, \qquad \text{(SE condition)}} \tag{83}$$

for every non-zero unit vector[17] $\mathbf{n}$ and vector $\mathbf{g}$, implies that the velocity problem has at most one solution.

The condition of strong ellipticity may be expressed, in a different notation, as the positive definiteness of the acoustic tensor[18] $\mathbf{A}(\mathbf{n})$ defined, for every unit vector $\mathbf{n}$ and $\mathbf{g} \in \mathcal{V}$, as

$$\boxed{\mathbf{A}(\mathbf{n})\mathbf{g} = \mathsf{C}[\mathbf{g} \otimes \mathbf{n}]\mathbf{n}.} \tag{84}$$

[17] We prefer to introduce the acoustic tensor directly with respect to unit vectors $\mathbf{n}$.
[18] In our definition of acoustic tensor we refer to a unit mass density.

With reference to the 'in loading' comparison solid (47), the acoustic tensor becomes:

$$\boxed{\mathbf{A}(\mathbf{n}) = \mathbf{A}_E(\mathbf{n}) - \frac{1}{g}\mathbf{M}\mathbf{n} \otimes \mathbf{N}\mathbf{n},} \tag{85}$$

where $\mathbf{A}_E(\mathbf{n})$ is the elastic acoustic tensor

$$\mathbf{A}_E(\mathbf{n})\mathbf{g} = \mathsf{G}[\mathbf{g} \otimes \mathbf{n}]\mathbf{n},$$

assumed *positive definite* in the following.

For the proof of the van Hove theorem, we follow Hayes (1966). Let us consider the functional in (40) with C^c replaced by C. When this functional is positive, for every admissible velocity field $\dot{\mathbf{x}}$ vanishing on $\partial\Omega^0$, bifurcation is excluded (in the incrementally linear solid defined by C). We can extend the definition of $\dot{\mathbf{x}}$ from $\bar{\Omega}^0$ to all Euclidean space $\mathcal{E}$ by simply defining $\dot{\mathbf{x}} = \mathbf{0}$ on $\mathcal{E} \setminus \bar{\Omega}^0$. The resulting field is admissible and its gradient is discontinuous only on $\partial\Omega^0$. Therefore, both $\dot{\mathbf{x}}$ and $\nabla\dot{\mathbf{x}}$ possess the three-dimensional Fourier transforms

$$\dot{\mathbf{x}}^*(\mathbf{x}_0) = \left(\frac{1}{2\pi}\right)^{\frac{3}{2}} \int_{\mathcal{E}} e^{i(\mathbf{x}_0 \cdot \mathbf{y})}\dot{\mathbf{x}}(\mathbf{y})dv_{\mathbf{y}}, \tag{86}$$

$$\mathbf{X}(\mathbf{x}_0) = \left(\frac{1}{2\pi}\right)^{\frac{3}{2}} \int_{\mathcal{E}} e^{i(\mathbf{x}_0 \cdot \mathbf{y})}\nabla\dot{\mathbf{x}}(\mathbf{y})dv_{\mathbf{y}}. \tag{87}$$

Before continuing, it should be noted that (86), (87) and the divergence theorem yield

$$\mathbf{X} = -i\,\dot{\mathbf{x}}^* \otimes \mathbf{x}_0, \tag{88}$$

where $i = \sqrt{-1}$ is the imaginary unit. Since homogeneity is satisfied in Ω^0, the generalized Parseval theorem gives

$$\int_{\mathcal{E}} \nabla\dot{\mathbf{x}} \cdot \mathsf{C}[\nabla\dot{\mathbf{x}}] = C_{ijhk} \int_{\mathcal{E}} \dot{x}_{i,j}\dot{x}_{h,k} = C_{ijhk} \int_{\mathcal{E}} X_{ij}\overline{X}_{hk} = \int_{\mathcal{E}} \mathbf{X} \cdot \mathsf{C}[\overline{\mathbf{X}}], \tag{89}$$

where $\overline{\mathbf{X}}$ is the complex conjugate of $\mathbf{X}$. Now, we note that (88) implies

$$\mathbf{X}_R = \dot{\mathbf{x}}_I^* \otimes \mathbf{x}_0, \quad \mathbf{X}_I = -\dot{\mathbf{x}}_R^* \otimes \mathbf{x}_0, \tag{90}$$

where the indices R and I stand for the real and imaginary parts, respectively. Using (90) in (89), it may be found that

$$\int_{\Omega^0} \nabla\dot{\mathbf{x}} \cdot \mathsf{C}[\nabla\dot{\mathbf{x}}] = \int_{\mathcal{E}} (\dot{\mathbf{x}}_R^* \otimes \mathbf{x}_0) \cdot \mathsf{C}[\dot{\mathbf{x}}_R^* \otimes \mathbf{x}_0] + \int_{\mathcal{E}} (\dot{\mathbf{x}}_I^* \otimes \mathbf{x}_0) \cdot \mathsf{C}[\dot{\mathbf{x}}_I^* \otimes \mathbf{x}_0]. \tag{91}$$

As a conclusion, when (SE) holds, the above integral is greater than zero, and uniqueness follows.

Obviously, the van Hove theorem holds true for the Raniecki family of incrementally linear solids. Defining therefore the acoustic tensor relative to the generic Raniecki solid

$$\mathbf{A}_R(\mathbf{n}, \psi)\mathbf{g} = \mathsf{R}[\mathbf{g} \otimes \mathbf{n}]\mathbf{n}, \tag{92}$$

we may conclude that for a homogeneous elastoplastic body subject to prescribed velocities on the entire boundary, strong ellipticity of at least one of the Raniecki solids ($\mathrm{SE_R}$) is sufficient for uniqueness. As a conclusion, we can state that

$(\mathrm{SE_R}) \Longrightarrow$ uniqueness for an *elastoplastic solid* for van Hove's b.v.p.

$(\mathrm{SE}) \Longrightarrow$ uniqueness for the 'in loading comparison solid' for van Hove's b.v.p.

where by 'van Hove b.v.p.' we mean boundary value problems satisfying van Hove's conditions. An immediate consequence of the Raniecki comparison theorem is the following implication

$$(\mathrm{PD}) \quad \Longrightarrow \quad (\mathrm{SE_R}) \quad \Longrightarrow \quad (\mathrm{SE}). \tag{93}$$

Now we will determine the critical plastic moduli for loss of (SE) and loss of ($\mathrm{SE_R}$).

In order to write (SE) of C in terms of a critical value of the plastic modulus, let us consider, for every vector $\mathbf{x}$ and unit vector $\mathbf{n}$, the following inequality:

$$\mathbf{x}\cdot\mathbf{A}(\mathbf{n})\mathbf{x} \geq \mathbf{x}\cdot\mathbf{A}_R(\mathbf{n},\psi)\mathbf{x} = \mathbf{x}\cdot\tilde{\mathbf{A}}_E(\mathbf{n})\mathbf{x} - \frac{(\mathbf{x}\cdot\mathbf{Rn})^2}{4\psi g}, \tag{94}$$

where $\tilde{\mathbf{A}}_E(\mathbf{n}) = [\mathbf{A}_E(\mathbf{n}) + \mathbf{A}_E^T(\mathbf{n})]/2$, is the symmetric part of the elastic acoustic tensor.

We proceed analogously to the proof of Raniecki and Bruhns for (PD), using the Cauchy-Schwarz inequality in the metric $\tilde{\mathbf{A}}_E(\mathbf{n})$ (Appendix B) (the dependence of $\tilde{\mathbf{A}}_E$ on $\mathbf{n}$ is omitted for simplicity in the following)

$$(\mathbf{x}\cdot\mathbf{Rn})^2 = (\mathbf{x}\cdot\tilde{\mathbf{A}}_E\tilde{\mathbf{A}}_E^{-1}\mathbf{Rn})^2 \leq (\mathbf{x}\cdot\tilde{\mathbf{A}}_E\mathbf{x})(\mathbf{Rn}\cdot\tilde{\mathbf{A}}_E^{-1}\mathbf{Rn}). \tag{95}$$

After calculations that parallel those reported in Sect. 6.1, we find that

$$g > \min_{\psi>0}\left\{\frac{(\mathbf{Mn}+\psi\mathbf{Nn})\cdot\tilde{\mathbf{A}}_E^{-1}(\mathbf{Mn}+\psi\mathbf{Nn})}{4\psi}\right\} \quad \Longrightarrow \quad \mathbf{A}(\mathbf{n}) \text{ pos. def.} \tag{96}$$

The minimum problem is solved by

$$\psi(\mathbf{n}) = \sqrt{\frac{\mathbf{Mn}\cdot\tilde{\mathbf{A}}_E^{-1}\mathbf{Mn}}{\mathbf{Nn}\cdot\tilde{\mathbf{A}}_E^{-1}\mathbf{Nn}}}, \tag{97}$$

which yields the following proposition:

$$\boxed{g > g_{cr}^{SE}(\mathbf{n}) \quad \Longleftrightarrow \quad \mathbf{A}(\mathbf{n}) \text{ pos. def.} \qquad (\text{at fixed } \mathbf{n}\ ),} \tag{98}$$

where

$$\boxed{g_{cr}^{SE}(\mathbf{n}) = \frac{1}{2}\left(\sqrt{(\mathbf{Mn}\cdot\tilde{\mathbf{A}}_E^{-1}\mathbf{Mn})(\mathbf{Nn}\cdot\tilde{\mathbf{A}}_E^{-1}\mathbf{Nn})} + \mathbf{Mn}\cdot\tilde{\mathbf{A}}_E^{-1}\mathbf{Nn}\right) \geq 0.} \tag{99}$$

Note that $g_{cr}^{SE} = 0 \Longleftrightarrow \mathbf{N} \propto -\mathbf{M}$, a condition which should never be satisfied for realistic constitutive models.

In order to complete the proof of (98), it may be checked that at $g = g_{cr}^{SE}(\mathbf{n})$, tensor $\mathbf{A}(\mathbf{n})$ loses positive definiteness for vectors $\mathbf{x}^*$ defined as

$$\mathbf{x}^* \propto (\sqrt{\mathbf{Mn} \cdot \tilde{\mathbf{A}}_E^{-1}\mathbf{Mn}})\, \tilde{\mathbf{A}}_E^{-1}\mathbf{Nn} + (\sqrt{\mathbf{Nn} \cdot \tilde{\mathbf{A}}_E^{-1}\mathbf{Nn}})\, \tilde{\mathbf{A}}_E^{-1}\mathbf{Mn},$$

and that for $g \le g_{cr}^{SE}(\mathbf{n})$, $\mathbf{x}^* \cdot \mathbf{A}(\mathbf{n})\mathbf{x}^* \le 0$.

It may be important to note from (99) that for every $\mathbf{n}$, failure of (SE) necessarily occurs for $g_{cr}^{SE}(\mathbf{n})$ positive.

All the above holds at $\mathbf{n}$ fixed, and proves that for a given $\mathbf{n}$, loss of (PD) in the 'in loading' comparison solid C and in the optimal Raniecki solid defined by (97) are equivalent. However, loss of (SE) in C will actually occur when g equals the maximum of $g_{cr}^{SE}(\mathbf{n})$ as a function of $\mathbf{n}$

$$\boxed{g > g_{cr}^{SE} = \max_{\mathbf{n},|\mathbf{n}|=1} g_{cr}^{SE}(\mathbf{n}) \Longleftrightarrow \text{(SE condition).}} \tag{100}$$

Therefore, differently from g_{cr}^{PD}, the critical value of the plastic modulus for loss of strong ellipticity, g_{cr}^{SE}, is not given in explicit terms, but as the solution of a constrained maximization problem.

Let us consider now strong ellipticity for Raniecki solids. From (94) and (95) we obtain that

$$\boxed{\mathbf{A}_R(\mathbf{n},\psi) \;\; \text{pos. def.} \Longleftrightarrow \;\; g > g_{cr}^{SE_R}(\mathbf{n},\psi) = \frac{(\mathbf{Mn} + \psi\mathbf{Nn}) \cdot \tilde{\mathbf{A}}_E^{-1}(\mathbf{Mn} + \psi\mathbf{Nn})}{4\psi}}, \tag{101}$$

Taking into consideration the optimal ψ, we conclude that (SE$_\text{R}$) can be written as

$$\boxed{g > g_{cr}^{SE_R} = \inf_{\psi>0} \max_{\mathbf{n},|\mathbf{n}|=1} g_{cr}^{SE_R}(\mathbf{n},\psi), \quad \Longleftrightarrow \quad \text{(SE}_\text{R} \text{ condition)}} \tag{102}$$

The above critical plastic moduli were obtained by Bigoni and Zaccaria (1992a;b). It may be worth noting that (93) is equivalent to

$$g_{cr}^{PD} \ge g_{cr}^{SE_R}(\mathbf{n},\psi) \ge g_{cr}^{SE}(\mathbf{n}), \tag{103}$$

holding for every ψ and $\mathbf{n}$. Moreover, we have seen that

$$g_{cr}^{SE}(\mathbf{n}) = \min_{\psi>0} g_{cr}^{SE_R}(\mathbf{n},\psi). \tag{104}$$

Therefore, (SE) and (SE$_\text{R}$) *are equivalent criteria whenever*

$$\inf_{\psi>0} \max_{\mathbf{n},|\mathbf{n}|=1} g_{cr}^{SE_R}(\mathbf{n},\psi) = \max_{\mathbf{n},|\mathbf{n}|=1} \min_{\psi>0} g_{cr}^{SE_R}(\mathbf{n},\psi), \tag{105}$$

a condition which was proved only under very restrictive assumptions (infinitesimal theory, isotropic elasticity and coaxiality of $\mathbf{P}$ and $\mathbf{Q}$).

Untill this point we have given an interpretation of the (SE_R) and (SE) conditions as sufficient conditions for uniqueness of a special class of problems (van Hove hypotheses)[19]. In particular, the former condition is sufficient for uniqueness of the velocity problem for the elastoplastic solid, and the latter for the fictitious solid corresponding to the loading branch of the constitutive tangent operator. However, as stressed by Bigoni and Zaccaria (1992 b),

the strong ellipticity condition has a meaning even when a generic situation of inhomogeneous deformation and mixed boundary conditions is considered.

In order to explain this point, let us introduce the semi-strong ellipticity condition (SSE), which is defined as in (83) except that '>' is replaced with '≥'. In other words, (SSE) is the condition of semi-positive definiteness of the acoustic tensor. In nonlinear elasticity, the theorem of Cattaneo (1946) (see also Truesdell and Noll, 1965, Sect. 68bis) proves that (SSE) is a necessary condition for infinitesimal stability. In non-associative elastoplasticity the theorem of Cattaneo was generalized by Ryzhak (1987)[20]. Assuming positive definiteness of $\mathbf{A}_E(\mathbf{n})$, the Ryzhak theorem states that (SSE) is a necessary condition for semi-stability in Hill's sense, namely

$$\boxed{\text{semi-stability in Hill sense} \Longrightarrow \mathsf{C} \text{ is (SSE),}}$$

where with the term 'semi-stability', we intend the sufficient stability condition (50) where '>' is replaced with '≥'. The (SSE) is a local criterion and when it fails during a loading program of a generic (inhomogeneous) boundary value problem (with mixed boundary conditions) Hill's sufficient stability condition does not hold.

The infinitesimal theory. In the special case of infinitesimal theory (34) and Green elasticity, the critical plastic modulus (99) simplifies to

$$g_{cr}^{SE}(\mathbf{n}) = \frac{1}{2}\left(\sqrt{(\mathsf{E}[\mathbf{P}]\mathbf{n}\cdot\mathbf{A}_E^{-1}\mathsf{E}[\mathbf{P}\mathbf{n}])(\mathsf{E}[\mathbf{Q}]\mathbf{n}\cdot\mathbf{A}_E^{-1}\mathsf{E}[\mathbf{Q}]\mathbf{n})} + \mathsf{E}[\mathbf{P}]\mathbf{n}\cdot\mathbf{A}_E^{-1}\mathsf{E}[\mathbf{Q}]\mathbf{n}\right) \geq 0, \tag{106}$$

where once again we note that $g_{cr}^{SE} = 0$ is equivalent to $\mathbf{P} \propto -\mathbf{Q}$, a condition which is excluded for realistic constitutive models. Assuming isotropic elasticity (33), the acoustic tensor can be written in the well-known form

$$\mathbf{A}_E(\mathbf{n}) = (\lambda+\mu)\mathbf{n}\otimes\mathbf{n} + \mu\mathbf{I}, \tag{107}$$

with the inverse

$$\mathbf{A}_E^{-1}(\mathbf{n}) = -\frac{\lambda+\mu}{\mu(\lambda+2\mu)}\mathbf{n}\otimes\mathbf{n} + \frac{1}{\mu}\mathbf{I}. \tag{108}$$

Under the above hypotheses, and assuming in addition coaxiality of $\mathbf{P}$ and $\mathbf{Q}$, Bigoni and Zaccaria (1992, a,b) have proved (105) so that:

---

[19] The van Hove theorem has been generalized in special cases by Ryzhak (1993; 1994)

[20] A simpler, alternative proof can be inferred from (Petryk, 1985a; 1992), as an application of Graves' theorem. The discussion there was confined to symmetry of the constitutive operator, but in the (SSE) condition only the symmetric part of the constitutive operator plays a role.

in the case of the infinitesimal theory, with isotropic elasticity and $\mathbf{P}$ *and* $\mathbf{Q}$ *coaxial, (SE) and (SE*$_{\mathrm{R}}$*) are equivalent criteria.*

Moreover, they obtained an explicit solution for (100).

6.4 Strain localization

For associative hyperelastic-plastic behaviour, $\mathbf{N} = \mathbf{M}$, in a continuous deformation evolution initiating when (SE) holds, the acoustic tensor becomes singular as soon as (SE) fails. This is no longer true in the non-associative case, where (SE) is usually lost while the acoustic tensor is still non-singular. Therefore, we can introduce a condition analogous to (NS), but on the acoustic tensor (defined here relatively to the 'in loading' comparison solid). This is the condition of ellipticity

$$\boxed{\det \mathbf{A}(\mathbf{n}) \neq 0, \quad \text{for all } \mathbf{n} \in \mathcal{V}, |\mathbf{n}| = 1 \quad \text{(E condition).}} \tag{109}$$

In a (sufficiently regular) deformation path, starting from a situation in which $\det \mathbf{A}(\mathbf{n}) > 0$ is satisfied, the acoustic tensor $\mathbf{A}(\mathbf{n})$ becomes singular when the plastic modulus reaches a critical value $g_{cr}^{E}(\mathbf{n})$. This critical hardening modulus was derived by Rice (1977) as follows. The condition

$$\det\left(\mathbf{A}_E(\mathbf{n}) - \frac{1}{g}\mathbf{M}\mathbf{n} \otimes \mathbf{N}\mathbf{n}\right) > 0,$$

assuming $\mathbf{A}_E$ *positive definite*, can be written as

$$\det \mathbf{A}_E(\mathbf{n}) \det\left(\mathbf{I} - \frac{1}{g}\mathbf{A}_E^{-1}(\mathbf{n})\mathbf{M}\mathbf{n} \otimes \mathbf{N}\mathbf{n}\right) > 0.$$

Making use of the identity $\det(\mathbf{I} + \mathbf{a} \otimes \mathbf{b}) = 1 + \mathbf{a} \cdot \mathbf{b}$, holding for every $\mathbf{a}$ and $\mathbf{b} \in \mathcal{V}$, we obtain the critical hardening modulus for loss of (E) in the direction $\mathbf{n}$

$$\boxed{g_{cr}^{E}(\mathbf{n}) = \mathbf{N}\mathbf{n} \cdot \mathbf{A}_E^{-1}(\mathbf{n})\mathbf{M}\mathbf{n},} \tag{110}$$

which is greater than zero for associative elastoplasticity, but may exceptionally result negative for a non-associative flow rule[21]. When the acoustic tensor is singular, i.e. when $g = g_{cr}^{E}$, the eigenvector corresponding to the null eigenvalue is

$$\mathbf{g} \propto \mathbf{A}_E^{-1}(\mathbf{n})\mathbf{M}\mathbf{n}. \tag{111}$$

As for the case of the (SE) condition, the critical plastic modulus for loss of (E) is the solution of the constrained maximization problem

$$\boxed{g > g_{cr}^{E} = \max_{\mathbf{n},|\mathbf{n}|=1} g_{cr}^{E}(\mathbf{n}) \iff \det \mathbf{A}(\mathbf{n}) > 0, \text{ for all } \mathbf{n} \in \mathcal{V}, |\mathbf{n}| = 1 \implies \text{(E condition).}} \tag{112}$$

[21] In the case $g_{cr}^{E} < 0$, localization is excluded for strictly positive values of g, i.e. in the present context.

The condition of loss of ellipticity admits a particularly nice mechanical interpretation, namely, *localization of deformation into a planar band* becomes possible at failure of (E). This has been known since Hadamard (1903), but was investigated in the case of elastoplasticity by Nadai (1931, 1950), Hill (1952, 1962), Prager (1954), Thomas (1953, 1961), Mandel (1966), Rudnicki and Rice (1975), Rice (1977), Rice and Rudnicki (1980). Strain localization may be directly linked to the initiation and growth of slip mechanisms and fractures in solids. It is experimentally observed in a wide range of materials (including metals, polymers, concretes, geomaterials) and is one of the most explored research fields, since Rice's (1977) paper.

It is important to understand how failure of ellipticity is connected to the emergence of localized deformations. To this purpose, let us consider an infinite body, subject to remote boundary conditions sufficient to impose continued quasi-static, homogeneous deformation. At any instant of the deformation process, the uniform stress field trivially satisfies the equilibrium equations. At a certain point of the deformation process, let us assume that a non-trivial incremental solution becomes possible, consisting of a velocity gradient that is uniform except across a planar band, where it is discontinuous. Inside and outside the band the incremental stress and strain fields remain uniform, so that equilibrium and compatibility are satisfied. If the band has normal $\mathbf{n}_0$ in the material description, the nominal traction must remain continuous, (5), across the band:

$$\left[\left[\dot{\mathbf{S}}\right]\right] \mathbf{n}_0 = \mathbf{0}; \tag{113}$$

moreover, the jump in velocity gradient across the band must satisfy the *Maxwell compatibility conditions*[22]

$$\left[\left[\dot{\mathbf{F}}\right]\right] = \mathbf{g} \otimes \mathbf{n}_0. \tag{114}$$

If we express via constitutive equations (28) the Piola-Kirchhoff stress rate in (113) in terms of jump of the gradient of velocity written using (114), we arrive at

$$\left(\mathcal{C}(\dot{\mathbf{F}} + \mathbf{g} \otimes \mathbf{n}_0) - \mathcal{C}(\dot{\mathbf{F}})\right) \mathbf{n}_0 = \mathbf{0}. \tag{115}$$

This is a necessary condition for strain localization into a planar band. Four cases need to be examined, corresponding to conditions of plastic loading (or elastic unloading) inside and outside the band, and plastic loading inside (or outside) and elastic unloading outside

[22] The compatibility conditions follow from continuity of velocity across the band, $[\![\dot{\mathbf{x}}]\!] = \mathbf{0}$. This can be easily shown, defining as $[\![\nabla\dot{\mathbf{x}}]\!]\,\mathbf{m}$ the directional derivative of the velocity jump in the generic direction singled out by the unit vector $\mathbf{m}$. Due to continuity of $\dot{\mathbf{x}}$, the derivative along $\mathbf{n}_0$, i.e. $[\![\nabla\dot{\mathbf{x}}]\!]\,\mathbf{n}_0$, remains unrestricted, but the derivative orthogonal to $\mathbf{n}_0$ must vanish: $[\![\nabla\dot{\mathbf{x}}]\!]\,\mathbf{t} = \mathbf{0}$, for every unit vector $\mathbf{t}$ orthogonal to $\mathbf{n}_0$. This condition implies that, when projected onto a basis $\mathbf{n}_0$, $\mathbf{t}$, $\mathbf{s} = \mathbf{n}_0 \times \mathbf{t}$, tensor $[\![\nabla\dot{\mathbf{x}}]\!]$ has only the three non-null components $[\![\nabla\dot{x}]\!]_{n_0 n_0}$, $[\![\nabla\dot{x}]\!]_{t n_0}$ and $[\![\nabla\dot{x}]\!]_{s n_0}$. Therefore, defining a vector $\mathbf{g}$ as

$$\mathbf{g} = [\![\nabla\dot{x}]\!]_{n_0 n_0}\,\mathbf{n}_0 + [\![\nabla\dot{x}]\!]_{t n_0}\,\mathbf{t} + [\![\nabla\dot{x}]\!]_{s n_0}\,\mathbf{s},$$

it is readily obtained that $[\![\nabla\dot{\mathbf{x}}]\!] = \mathbf{g} \otimes \mathbf{n}_0$, eqn. (114).

(or inside) the band:

$$1)\ \Big(\mathcal{C}(\dot{\mathbf{F}}+\mathbf{g}\otimes\mathbf{n}_0)-\mathcal{C}(\dot{\mathbf{F}})\Big)\,\mathbf{n}_0=\mathsf{C}[\mathbf{g}\otimes\mathbf{n}_0]\mathbf{n}_0, \qquad \text{(plastic/plastic)},$$

$$2)\ \Big(\mathcal{C}(\dot{\mathbf{F}}+\mathbf{g}\otimes\mathbf{n}_0)-\mathcal{C}(\dot{\mathbf{F}})\Big)\,\mathbf{n}_0=\mathsf{G}[\mathbf{g}\otimes\mathbf{n}_0]\mathbf{n}_0, \qquad \text{(elastic/elastic)},$$

$$3)\ \Big(\mathcal{C}(\dot{\mathbf{F}}+\mathbf{g}\otimes\mathbf{n}_0)-\mathcal{C}(\dot{\mathbf{F}})\Big)\,\mathbf{n}_0=\mathsf{C}[\mathbf{g}\otimes\mathbf{n}_0]\mathbf{n}_0+(\mathsf{C}-\mathsf{G})[\dot{\mathbf{F}}]\mathbf{n}_0, \quad \text{(plastic/elastic)},$$

$$4)\ \Big(\mathcal{C}(\dot{\mathbf{F}}+\mathbf{g}\otimes\mathbf{n}_0)-\mathcal{C}(\dot{\mathbf{F}})\Big)\,\mathbf{n}_0=\mathsf{G}[\mathbf{g}\otimes\mathbf{n}_0]\mathbf{n}_0-(\mathsf{C}-\mathsf{G})[\dot{\mathbf{F}}]\mathbf{n}_0, \quad \text{(elastic/plastic)}.$$

It is important to note that only Case 1) corresponds to violation of (E), eqn. (109), i.e. $\det \mathbf{A}(\mathbf{n}) = 0$. Case 2) corresponds to a purely elastic loss of (E), $\det \mathbf{A}_E(\mathbf{n}) = 0$. However, we assume for simplicity $\mathbf{A}_E(\mathbf{n})$ positive definite and therefore Case 2) is not considered. We report now the Rice and Rudnicki (1980) proof that

> if $\det \mathbf{A}(\mathbf{n}) > 0$, so that Case 1) is excluded, Cases 3) and 4) are also excluded.

Therefore, we assume that (the index 0 of $\mathbf{n}_0$ is omitted for conciseness)

$$\det \mathbf{A}(\mathbf{n}) > 0 \iff 1-\frac{\mathbf{Nn}\cdot\mathbf{A}_E^{-1}(\mathbf{n})\mathbf{Mn}}{g} > 0. \tag{116}$$

Let us analyze Case 3). If (115) is satisfied under Conditions 3), we have:

$$\mathbf{N}\cdot\dot{\mathbf{F}} \le 0,\quad \mathbf{N}\cdot\dot{\mathbf{F}}+\mathbf{g}\cdot\mathbf{Nn}\ge 0,\quad \mathbf{g}=\frac{\mathbf{N}\cdot\dot{\mathbf{F}}+\mathbf{g}\cdot\mathbf{Nn}}{g}\mathbf{A}_E^{-1}(\mathbf{n})\mathbf{Mn}, \tag{117}$$

from which it is immediate that $\mathbf{g}\cdot\mathbf{A}_E^{-1}(\mathbf{n})\mathbf{Mn}\ge 0$. But (excluding the trivial case $\mathbf{N}\cdot\dot{\mathbf{F}}+\mathbf{g}\cdot\mathbf{Nn}=0$) the determinant

$$\det\left\{\mathbf{I}-\frac{\mathbf{A}_E^{-1}(\mathbf{n})\mathbf{Mn}}{g}\otimes\left(\mathbf{Nn}+\frac{\mathbf{N}\cdot\dot{\mathbf{F}}}{\mathbf{g}^2}\mathbf{g}\right)\right\}$$

$$=1-\frac{\mathbf{Nn}\cdot\mathbf{A}_E^{-1}(\mathbf{n})\mathbf{Mn}}{g}-\frac{\mathbf{N}\cdot\dot{\mathbf{F}}}{g\mathbf{g}^2}\mathbf{g}\cdot\mathbf{A}_E^{-1}(\mathbf{n})\mathbf{Mn}, \tag{118}$$

is positive, so that (115) is not satisfied. Let us analyze Case 4). If (115) is satisfied under Conditions 4), we have:

$$\mathbf{N}\cdot\dot{\mathbf{F}} \ge 0,\quad \mathbf{N}\cdot\dot{\mathbf{F}}+\mathbf{g}\cdot\mathbf{Nn}\le 0,\quad \mathbf{g}=-\frac{\mathbf{N}\cdot\dot{\mathbf{F}}}{g}\mathbf{A}_E^{-1}(\mathbf{n})\mathbf{Mn}. \tag{119}$$

Using $(119)_3$ and $(119)_1$ in $(119)_2$, (excluding the trivial case $\mathbf{N}\cdot\dot{\mathbf{F}}=0$) we get

$$1-\frac{\mathbf{Nn}\cdot\mathbf{A}_E^{-1}(\mathbf{n})\mathbf{Mn}}{g}\le 0, \tag{120}$$

which is a contradiction to (116). It follows from the above that, assuming $\mathbf{A}_E(\mathbf{n})$ positive definite and a continuous dependence of constitutive equations on time:

In a loading program controlled by a (regularly) varying parameter and starting from a situation of ellipticity with $\det \mathbf{A}(\mathbf{n}) > 0$, *the first possibility of localization always occurs at failure of ellipticty,* $\det \mathbf{A}(\mathbf{n}) = 0$, *in the comparison solid corresponding to the loading branch of the constitutive operator* C.

The infinitesimal theory. In the case of infinitesimal theory (34) and isotropic elasticity (33), eqn. (108) allows a simplification of (110), which becomes

$$g_{cr}^{E} = \max_{\mathbf{n},\ |\mathbf{n}|=1} \left\{ -\frac{\lambda+\mu}{\mu(\lambda+2\mu)} (\mathbf{n}\cdot\mathsf{E}[\mathbf{Q}]\mathbf{n})(\mathbf{n}\cdot\mathsf{E}[\mathbf{P}]\mathbf{n}) + \frac{1}{\mu}\mathsf{E}[\mathbf{Q}]\mathbf{n}\cdot\mathsf{E}[\mathbf{P}]\mathbf{n} \right\}, \tag{121}$$

where E has the isotropic form (33). The constrained maximization problem, (121), was solved under various hypotheses on the form of $\mathbf{P}$ and $\mathbf{Q}$ by Rudnicki and Rice (1975), Needleman and Ortiz (1991), Ottosen and Runesson (1991). It was solved under the sole hypothesis of coaxiality of $\mathbf{P}$ and $\mathbf{Q}$ by Boehler and Willis (1991) and Bigoni and Hueckel (1990, 1991a,b). Moreover, generalization of the above solution to incorporate the effects of geometrical terms were given by Rudnicki and Rice (1975) and Szabó (1994), and to incorporate anisotropic elasticity by Bigoni and Loret (1999) and Bigoni et al. (2000).

6.5 Flutter instability

In the case of non-associative elastoplasticity (or when the elastic tensor does not possess the major symmetry), the acoustic tensor is non-symmetric and there is, in principle, the possibility of a particular type of instability. This is the so-called *flutter instability* (Rice, 1977) and corresponds to the occurrence of two complex conjugate eigenvalues of the acoustic tensor or, in other words, when (for at least one unit vector $\mathbf{n}$):

$$\boxed{\mathbf{A}(\mathbf{n}) \text{ has complex eigenvalues} \Longleftrightarrow \text{flutter instability} \quad \text{(F condition).}}$$

This instability is for many reasons not still completely understood. In particular, the following points will be addressed:
1) when flutter instability may occur,
2) what is known about its mechanical interpretation.

Onset of flutter. There are two possible approaches to the problem. One is simply to consider flutter instability to be excluded when the eigenvalues of the acoustic tensor are real. Another is to consider the coalescence of two eigenvalues as a critical condition. In fact, when two eigenvalues are coincident, an appropriate, infinitesimally small disturbance may induce complex eigenvalues. The latter point of view was suggested by An and Schaeffer (1990). Proceeding with it, we should restrict the class of possible disturbances, otherwise we will end up considering unstable a linear, isotropic elastic material (in which the two eigenvalues μ of the acoustic tensor always coincide).

Different classes of disturbances have been considered: Loret (1992) analyzed perturbations in the plastic flow direction, Bigoni and Zaccaria (1994a) and Bigoni (1995) considered perturbations due to effects of large strains. Recently, Bigoni and Loret (1999) have analyzed perturbations in terms of a small (hyper)elastic anisotropy superimposed on the usual isotropic-elastic, plastic models. The latter perturbation is perhaps the more convincing, because it is *symmetric* and has a clear physical interpretation. The result is that flutter may be triggered by such a vanishing small perturbation (for non-associative flow rules), when two eigenvalues of the acoustic tensor coincide. The same perturbation has no effects when superimposed on a linear elastic law or on an associative elastoplastic model. These results are not still conclusive, but *suggest* that:

> *for non-associative elastoplasticity the condition of coalescence of eigenvalues of the acoustic tensor could be considered critical for flutter even in cases in which complex eigenvalues are excluded in the absence of any perturbation.*

As a conclusion, the onset of flutter can be defined simply by finding the conditions of coalescence of eigenvalues. However, this may be not an easy task for elastoplasticity at finite strain. We will therefore limit the investigation to elastic-plastic solids with isotropic elasticity and subject to small strains. The results are mainly due to Loret et al. (1990); however, we follow Bigoni and Zaccaria (1994a) and consider the acoustic tensor corresponding to the plastic branch of the constitutive equation (34)

$$\mathbf{A}(\mathbf{n}) = (\lambda + \mu)\mathbf{n} \otimes \mathbf{n} + \mu \mathbf{I} - \frac{1}{g}\mathbf{p} \otimes \mathbf{q}, \tag{122}$$

where

$$\mathbf{q} = \lambda(\mathrm{tr}\mathbf{Q})\mathbf{n} + 2\mu\mathbf{Q}\mathbf{n}, \quad \mathbf{p} = \lambda(\mathrm{tr}\mathbf{P})\mathbf{n} + 2\mu\mathbf{P}\mathbf{n}, \tag{123}$$

are linear functions of $\mathbf{n}$. Assuming that $\mathbf{n} \times \mathbf{q} \neq \mathbf{0}$[23] let us consider the following non-orthogonal dual bases of $\mathcal{V}$, so that $\mathbf{e}^i \cdot \mathbf{e}_j = \delta^i_j$

$$\begin{aligned} &\mathbf{e}_1 = \mathbf{n}, && \mathbf{e}_2 = \mathbf{q}, && \mathbf{e}_3 = \frac{\mathbf{n}\times\mathbf{q}}{|\mathbf{n}\times\mathbf{q}|}, \\ &\mathbf{e}^1 = \frac{(\mathbf{q}^2)\mathbf{n} - (\mathbf{q}\cdot\mathbf{n})\mathbf{q}}{\mathbf{q}^2 - (\mathbf{q}\cdot\mathbf{n})^2}, && \mathbf{e}^2 = \frac{(\mathbf{q}\cdot\mathbf{n})\mathbf{n} - \mathbf{q}}{\sqrt{\mathbf{q}^2 - (\mathbf{q}\cdot\mathbf{n})^2}}, && \mathbf{e}^3 = \mathbf{e}_3. \end{aligned} \tag{124}$$

Projected onto the bases (124), the eigenvalue problem for (122) gives the characteristic equation

$$\det \begin{pmatrix} \lambda + 2\mu - \eta & -\frac{1}{g}\mathbf{p}\cdot\mathbf{n} & 0 \\ (\lambda+\mu)\mathbf{q}\cdot\mathbf{n} & \mu - \frac{1}{g}\mathbf{p}\cdot\mathbf{q} - \eta & 0 \\ 0 & -\frac{1}{g}\mathbf{p}\cdot\mathbf{e}_3 & \mu - \eta \end{pmatrix} = 0, \tag{125}$$

where η is the generic eigenvalue of the acoustic tensor (122). The solutions of the characteristic equation are the eigenvalue μ and the two roots of the polynomial equation:

$$\eta^2 - \left(\lambda + 3\mu - \frac{1}{g}\mathbf{p}\cdot\mathbf{q}\right)\eta + (\lambda + 2\mu)\left(\mu - \frac{1}{g}\mathbf{p}\cdot\mathbf{q}\right) + \frac{1}{g}(\lambda+\mu)(\mathbf{p}\cdot\mathbf{n})(\mathbf{q}\cdot\mathbf{n}) = 0. \tag{126}$$

[23] In the special case $\mathbf{n} \times \mathbf{q} = \mathbf{0}$ the final results do not change. This particular case is straightforward and analyzed in Bigoni and Zaccaria (1994a) and Bigoni (1995).

Flutter corresponds to negative values of the discriminant of the second-order polynomial in (126). This can be written as

$$\Delta = \left(\lambda + 3\mu - \frac{1}{g}\mathbf{p}\cdot\mathbf{q}\right)^2 - 4\mu(\lambda + 2\mu)\left(1 - \frac{g^E_{cr}(\mathbf{n})}{g}\right), \tag{127}$$

where $g^E_{cr}(\mathbf{n})$ is the critical plastic modulus for strain localization (121) at fixed $\mathbf{n}$, i.e.

$$g^E_{cr}(\mathbf{n}) = -\frac{\lambda + \mu}{\mu(\lambda + 2\mu)}(\mathbf{p}\cdot\mathbf{n})(\mathbf{q}\cdot\mathbf{n}) + \frac{\mathbf{p}\cdot\mathbf{q}}{\mu}. \tag{128}$$

Therefore, we conclude that (Bigoni and Zaccaria, 1994a):

> *for a given direction* $\mathbf{n}$, *flutter is always excluded for values of the plastic modulus less than or equal to the critical plastic modulus for localization in a band orthogonal to that direction* $\mathbf{n}$[24].

Straightforward manipulation of the discriminant (127) allows one to obtain the necessary and sufficient conditions for flutter

$$\boxed{(\mathbf{n}\cdot\mathbf{p})(\mathbf{n}\cdot\mathbf{q}) > 0, \quad \& \quad (\mathbf{n}\cdot\mathbf{p})(\mathbf{n}\cdot\mathbf{q}) - \mathbf{p}\cdot\mathbf{q} > 0, \quad \& \quad g \in (g_1, g_2).} \tag{129}$$

where

$$\boxed{\left.\begin{matrix} g_1 \\ g_2 \end{matrix}\right\} = \frac{1}{\lambda + \mu}\left(\sqrt{(\mathbf{n}\cdot\mathbf{p})(\mathbf{n}\cdot\mathbf{q})} \pm \sqrt{(\mathbf{n}\cdot\mathbf{p})(\mathbf{n}\cdot\mathbf{p}) - \mathbf{p}\cdot\mathbf{q}}\right)^2.} \tag{130}$$

If we assume deviatoric associativity (30), a simple calculation shows that $(129)_2$ is never satisfied, therefore we reach the conclusion (Loret et al., 1990):

> *For elastic-plastic solids in the presence of isotropic elasticity and deviatoric associativity (30) with* χ_1 *and* ψ_1 *strictly positive, complex eigenvalues of the acoustic tensor are excluded.*

However, coincident eigenvalues are possible. These may be determined by requiring that the discriminant (127) be null[25]. This occurs when one of the following two conditions is satisfied

$$(\mathbf{n}\cdot\mathbf{p})(\mathbf{n}\cdot\mathbf{q}) = 0, \quad \& \quad g = g^C_{cr}(\mathbf{n}) = -\frac{\mathbf{p}\cdot\mathbf{q}}{\lambda + \mu}, \tag{131}$$

or

$$(\mathbf{n}\cdot\mathbf{p})(\mathbf{n}\cdot\mathbf{q}) = \mathbf{p}\cdot\mathbf{q}, \quad \& \quad g = g^C_{cr}(\mathbf{n}) = \frac{\mathbf{p}\cdot\mathbf{q}}{\lambda + \mu}. \tag{132}$$

Assuming isotropic elasticity (33) and deviatoric associativity (30), it is easy to obtain that

$$\mathbf{p}\cdot\mathbf{q} - (\mathbf{n}\cdot\mathbf{p})(\mathbf{n}\cdot\mathbf{q}) = 4\mu^2\chi_1\psi_1\left(\hat{\mathbf{S}}\mathbf{n}\cdot\hat{\mathbf{S}}\mathbf{n} - (\mathbf{n}\cdot\hat{\mathbf{S}}\mathbf{n})^2\right) \geq 0. \tag{133}$$

---

[24] This does not mean that flutter always occurs before strain localization. Rather, it means that flutter at a certain $\mathbf{n}$ always occurs before localization corresponding to that $\mathbf{n}$.

[25] We are thus interested in determining coincident roots of (126) and do not consider the situation when (126) has different roots but one coincides with μ.

Therefore

$$(\mathbf{n}\cdot\mathbf{p})(\mathbf{n}\cdot\mathbf{q}) = 0 \implies \mathbf{p}\cdot\mathbf{q} \geq 0 \implies g_{cr}^{C}(\mathbf{n}) \leq 0, \tag{134}$$

so that Case (131) is not interesting. In Case (132), (126) gives two coincident solutions equal to μ. Therefore:

The acoustic tensor (for certain $\mathbf{n}$*) has an eigenvalue equal to* μ*, with multiplicity 3, when condition (132) is satisfied.*

Now we determine the critical plastic modulus for such a coalescence. We begin noting that the following condition holds true at coalescence

$$\hat{\mathbf{S}}\mathbf{n}\cdot\hat{\mathbf{S}}\mathbf{n} = (\mathbf{n}\cdot\hat{\mathbf{S}}\mathbf{n})^2, \tag{135}$$

and is verified if and only if $\mathbf{n}$ is an eigenvector of $\hat{\mathbf{S}}$, but in this case $\mathbf{n}$ is also an eigenvector of $\mathsf{E}[\mathbf{P}]$ and $\mathsf{E}[\mathbf{Q}]$[26]. Therefore, the critical plastic modulus for coalescence of eigenvalues is

$$\boxed{g_{cr}^{C} = \max_{i=1,2,3} \frac{(\mathsf{E}[\mathbf{P}])_i(\mathsf{E}[\mathbf{Q}])_i}{\lambda+\mu}}, \tag{136}$$

where E has the isotropic form (33) and the index i, not summed, denotes principal components of $\mathsf{E}[\mathbf{P}]$ and $\mathsf{E}[\mathbf{Q}]$ (in the same reference system).

To summarize with the above specific example at hand, we may note that for deviatoric associativity, complex eigenvalues of the acoustic tensor are excluded, but coalescence of three eigenvalues may occur. When coalescence occurs and with reference to the above example, Bigoni and Loret (1999) have shown that a perturbation in terms of a small (appropriate) elastic anisotropy superimposed on the isotropic elastic law is sufficient to trigger flutter. Therefore, even if for this model complex eigenvalues are excluded, flutter as induced by physically motivated perturbations is possible and the critical condition corresponds to coalescence of the three eigenvalues of the acoustic tensor.

When coalescence is considered, Bigoni and Loret (1999) have shown that this may occur without relation to the other criteria, namely (PD), (NS), (SE) and (E). In conclusion:

coalescence of eigenvalues of the acoustic tensor and therefore flutter may occur even when (PD) is verified. Therefore, flutter instability is not excluded by the Hill sufficient condition of stability.

Physical meaning and consequences of flutter. As far as the mechanical interpretation of flutter instability is concerned, results are scarce. In particular, Bigoni and Willis (1994) have analyzed a particular wave propagation problem, showing that flutter may correspond to an oscillating motion of material particles, which blows up with time. However, the analysis is valid only for an incrementally *linear* material. As pointed out by Rice (1977), in the case of an elastoplastic solid, an oscillation may yield a crossing

[26] Vice-versa, all eigenvectors of $\mathsf{E}[\mathbf{P}]$ and $\mathsf{E}[\mathbf{Q}]$ are also eigenvectors of $\hat{\mathbf{S}}$ because χ_1 and ψ_1 have been assumed strictly positive.

of the loading-branch constitutive cone and may thus invalidate an analysis based on a linear material. Only a partial answer has been given to this point. This is the numerical analysis of Simões (1997), in which a growth and decay of an oscillation is found in a plane strain intentation of a Drucker-Prager, ideally-plastic material with non-associative flow law. In this problem, flutter is excluded in the sense intended in the present section, but it may occur as connected to the presence of a free boundary. However, we may conclude that the real mechanical meaning of flutter is still not completely understood and more investigation is needed.

6.6 Other types of instabilities

The reader should be aware at this point that there may occur many types of instabilities in elastoplastic solids, with different mechanical consequences. However, a number of these instabilities were not analyzed for simplicity here. Some of these are briefly mentioned below.

Occurrence of a particular condition was noted by Ottosen and Runesson (1991), Brannon and Drugan (1993) and Bigoni (1995). It corresponds to the situation in which the acoustic tensor has two coincident eigenvalues with geometric multiplicity smaller than algebraic multiplicity. This is possible when the acoustic tensor is not symmetric. As an example, let us consider the Jordan block

$$\begin{bmatrix} 1 & 1 \\ 0 & 1 \end{bmatrix},$$

which has an eigenvalue equal to 1 with geometric multiplicity 1 and algebraic multiplicity 2 (in other words the matrix is defective). This possibility is again excluded in the case of isotropic-elastic, plastic solid with deviatoric associativity. But it may occur in other, more general, circumstances. Whether or not this occurrence may correspond to a material instability is presently not known.

A number of material instabilities may occur at a point of a boundary. These were investigated by Benallal et al. (1990), as related to the possibility of a surface instability (Biot, 1965; Hill and Hutchinson, 1975). The possibility of flutter instability in terms of complex conjugate velocities of propagation of Rayleigh waves at a free boundary of an elastoplastic solid was discovered by Loret et al. (1995).

As far as the author is aware, internal instability in the sense of Biot (1965) was not systematically investigated for elastoplastic solids.

Finally, cavitation has been thoroughly investigated for nonlinear elastic solids [see Horgan and Polignone (1995) and references cited therein], but only scarcely analyzed in the case of elastoplastic solids (Huang et al., 1991). For elastoplasticity, the effects of flow rule non-associativity should be dominant [see the related analysis by Bigoni and Laudiero (1989)].

7 Examples

Examples are presented relative to the application of the criteria described in the previous sections. We begin with the simple case of the infinitesimal theory and then we analyze a situation of uniaxial tension and compression of a solid bar, subject to finite strains.

7.1 The infinitesimal theory

We begin with some words of caution as to the use of small strain assumption in any bifurcation and stability analysis. Usually, 'geometrical effects' are crucially important in these problems. To convince oneself of this fact it suffices to recall Euler buckling of rods. Roughly speaking, with respect to the small strain approximation, the various local stability thresholds presented in Sect. 6 contain 'geometrical terms' on the order of stress over elastic shear modulus. These terms become important when the critical hardening moduli become comparable to a representative stress level (Hill, 1958; 1978; Rudnicki and Rice, 1975).

Proceeding now with the case in which all 'geometrical terms' are neglected in the equations, we assume, for simplicity, isotropic elasticity (33) and deviatoric associativity in the form (30)-(31). In other words, $\mathbf{Q}$ and $\mathbf{P}$ are the normals to yield and plastic potential surfaces, respectively, of the Drucker-Prager type. The model reduces therefore to von Mises plasticity when $\text{tr}\mathbf{P} = \text{tr}\mathbf{Q} = 0$. The threshold for loss of (PD) is given by (74) and the threshold for loss of (NS) is simply[27] $h_{cr}^{NS} = 0$. The critical hardening modulus for coalescence of the three eigenvalues of the acoustic tensor, condition (C), can be easily calculated using (136). Evaluation of thresholds for loss of (SE) and (E) requires solution of a constrained maximization problem for (106) and (121). As already mentioned, these maximization problems admit an analytical solution, which can be found in (Rudnicki and Rice, 1975; Bigoni and Hueckel, 1990; 1991a; Ortiz and Needleman, 1991; Ottosen and Runesson, 1991; Bigoni and Zaccaria, 1992a;b). A few values of the critical hardening moduli (divided by the elastic shear modulus μ) are reported in Tab. 1, relative to uniaxial tension. A material with Poisson's ratio $\nu = 0$ and 0.3 is considered, for several values of pressure-sensitivity $\text{tr}\mathbf{Q}$ and dilatancy $\text{tr}\mathbf{P}$.

Note that in the column (E), corresponding to loss of ellipticity, the angle (in degrees) has been reported in parenthesis between the normal to the band and the direction orthogonal to the axis of tension. Obviously, in the uniaxial stress problem analyzed, infinite bands become possible at strain localization, with normals describing a cone about the tension axis. Additional numerical results can be found in (Rudnicki and Rice, 1975[28]; Bigoni and Zaccaria, 1992a,b; Bigoni and Loret, 1999) and need not be reported here.

The results presented in the table are sufficient to draw the following conclusions.

---

[27] A complete discussion on the eigenvalues of the elastoplastic tangent operator for isotropic elasticity may be found in Bigoni and Zaccaria (1994b).

[28] Rudnicki and Rice (1975) neglected a term in the analytical solution for h_{cr}^{E}, thus obtaining imprecise numerical results. The complete soultion can be found in (Bigoni and Hueckel, 1990).

- As expected, in the case of an associative flow rule it is $h_{cr}^{PD} = h_{cr}^{NS} = 0$. Moreover, loss of (E), coincident with loss of (SE), is always excluded at positive hardening.
- For non-associative flow laws, the critical hardening modulus for (PD) is always positive. The critical hardening moduli for loss of (SE) and (E) are different and may take any sign, even positive. Strain localization may therefore occur at positive hardening, before loss of (NS), occurring at $h = 0$.
- Loss of (PD), (NS) and (SE) always occur for a positive plastic modulus, i.e. before the snap-back limit $g = 0$ (not considered in the present notes). On the other hand, loss of (E) *could* be excluded for strictly positive values of the plastic modulus g (even if this does not occur in the case treated by the table).

Table 1. Uniaxial tension. Values of h_{cr}/μ defining loss of positive definiteness (PD), strong ellipticity (SE) and ellipticity (E) of the elastic-plastic constitutive operator and coalescence of three eigenvalues of the acoustic tensor (C), for Poisson's ratio $\nu = 0$ and 0.3, and several values of pressure-sensitivity $\mathrm{tr}\,\mathbf{Q}$, and plastic dilatancy $\mathrm{tr}\,\mathbf{P}$. θ in degrees is given in parenthesis.

$\mathrm{tr}\,\mathbf{Q}$	$\mathrm{tr}\,\mathbf{P}$	ν	(PD)	(SE)	(E) (θ)	(C)
0	0	0	0	-0.167	-0.167 (54.7°)	0.333
		0.3	0	-0.217	-0.217 (48.8°)	-0.467
0.3	0	0	0.015	-0.101	-0.104 (58.3°)	0.564
		0.3	0.047	-0.106	-0.130 (53.2°)	-0.166
0.3	0.3	0	0	-0.071	-0.071 (62.2°)	0.775
		0.3	0	-0.093	-0.093 (57.8°)	0.108
0.6	0	0	0.057	-0.021	-0.031 (62.2°)	0.795
		0.3	0.167	0.054	-0.018 (57.8°)	0.133
0.6	0.3	0	0.013	-0.027	-0.028 (66.4°)	0.986
		0.3	0.034	-0.020	-0.031 (62.9°)	0.382
0.6	0.6	0	0	-0.016	-0.016 (71.3°)	1.177
		0.3	0	-0.020	-0.020 (68.6°)	0.630
0.9	0	0	0.120	0.070	0.052 (66.4°)	1.026
		0.3	0.330	0.239	0.117 (62.9°)	0.434
0.9	0.3	0	0.049	0.029	0.024 (71.3°)	1.197
		0.3	0.115	0.082	0.054 (68.6°)	0.656
0.9	0.6	0	0.011	0.007	0.007 (77.8°)	1.368
		0.3	0.022	0.017	0.015 (76.1°)	0.878
0.9	0.9	0	0	-0.0005	-0.0005 (90.0°)	1.539
		0.3	0	-0.001	-0.001 (90.0°)	1.101

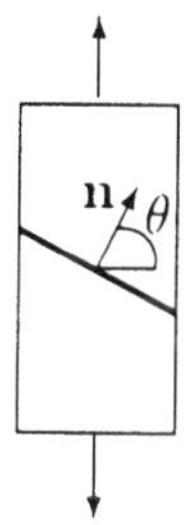

- Coalescence of eigenvalues of the acoustic tensor seems to have no relation to the other stability criteria. It may occur at positive or negative hardening, depending strongly on the value of the Poisson's ratio. It may occur before (PD), or after (E), or between these thresholds.

- All stability thresholds are strongly influenced by
  - state of stress,
  - constitutive parameters (dilatancy, pressure-sensitivity, Poisson's ratio),
  - constitutive features not investigated in the example [yield surface curvature or vertex, non-coaxiality of **P** and **Q**, elastic or plastic anisotropy (Bigoni and Loret, 1999; Bigoni et al. 2000)].

7.2 A simple elastoplastic model including 'geometrical effects'

Bifurcations in simple boundary value problems such as compression or tension of cylinders or blocks deformed in plane strain have been thoroughly analyzed from an analytical point of view for hyperelastic materials (see for instance, Biot, 1965; Hill and Hutchinson, 1975). These analyses are often relevant to associative elastoplasticity, where they represent bifurcations of the comparison solid corresponding to the loading branch of the constitutive operator. As explained in Sect. 5, bifurcations detected in such a comparison solid may also represent genuine elastoplastic bifurcations under broad hypotheses (Hutchinson, 1973). As an example, we consider the analysis of Hutchinson and Miles (1974) in which axisymmetric bifurcations are investigated during uniaxial tension of a solid bar. The bar is incrementally linear and hyperelastic, but the results are also pertinent to associative elastoplasticity.

The situation of non-associative elastoplasticity is much more complicated. In particular, we have seen that the search for bifurcations in an elastoplastic solid with a non-associative flow rule is replaced by the search for bifurcations in two linear comparison solids. One can be any member of the family of Raniecki comparison solids and the other is simply the linear solid defined by the loading branch of the constitutive operator. Bifurcation in the former solid determines a lower bound to bifurcation stresses and bifurcation in the latter provides an upper bound. Obviously, it will be convenient to determine the optimal Raniecki solid as a function of parameter ψ with respect to the bifurcation problem under consideration.

Needleman (1979), Vardoulakis (1981; 1983), Chau and Rudnicki (1990), and Chau (1992; 1995) have investigated bifurcations in the 'in loading comparison solid' of an elastoplastic material with non-associative flow rule, similar, in essence, to the Rudnicki and Rice (1975) model. The sole analytical analysis of the Raniecki bounds (of which the author is aware) is that given by Bruhns and Raniecki (1982), whereas numerical results were presented by Tvergaard (1982), Kleiber (1984; 1986) and Tomita et al. (1988).

We develop below a simple example involving uniaxial tension and compression of a cylindrical bar. For this example, we give the (PD), (NS), (SE) and (E) thresholds and investigate the Raniecki bounds for assigned, axisymmetric mode of bifurcation. It is important to realize that *priority is given to mathematical simplicity*, therefore the example should be considered as a prototype, not intended to properly model any real material. Nevertheless, the main constitutive features such as pressure-sensitivity and plastic dilatancy are taken into account. We base the analysis on the simple constitutive model suggested by Hill (1962) for an associative flow rule (Hutchinson, 1973; Neale, 1981). This model has the structure (24) in the relative Lagrangean description (i.e.

using the current configuration as reference, $\mathbf{F} = \mathbf{I}$)

$$\dot{\mathbf{S}} = \mathsf{E}[\mathbf{D}] + \mathbf{LK} - \frac{1}{g} < \mathbf{D} \cdot \mathsf{E}[\mathbf{Q}] > \mathsf{E}[\mathbf{P}], \tag{137}$$

where E has the isotropic form (33). When λ and μ are not considered *constants* and have a special dependence on deformation, Christoffersen (1991) has proved hyperelastic behaviour for the elastic part of (137). However, we will consider λ and μ constant in the following. Moreover, tensors $\mathbf{P}$ and $\mathbf{Q}$ are selected in the form (30)-(31) proposed by Rudnicki and Rice (1975), with *isochoric plastic deformation*

$$\mathbf{Q} = \frac{\operatorname{dev}\mathbf{T}}{2\sqrt{J_2}} + \frac{\alpha}{3}\mathbf{I}, \quad \mathbf{P} = \frac{\operatorname{dev}\mathbf{T}}{2\sqrt{J_2}}. \tag{138}$$

Model (137) differs from that analyzed in the case of the infinitesimal theory only by the presence of the 'geometrical term' $\mathbf{LK}$. The elastic acoustic tensor $\mathbf{A}_E(\mathbf{n})$ corresponding to (137) is

$$\mathbf{A}_E(\mathbf{n}) = (\lambda + \mu)\mathbf{n} \otimes \mathbf{n} + \bar{\mu}\mathbf{I}, \tag{139}$$

where $\bar{\mu} = \mu + \mathbf{n} \cdot \mathbf{Kn}$ and has the inverse

$$\mathbf{A}_E^{-1}(\mathbf{n}) = -\frac{\lambda + \mu}{\bar{\mu}(\lambda + \mu + \bar{\mu})}\mathbf{n} \otimes \mathbf{n} + \frac{1}{\bar{\mu}}\mathbf{I}. \tag{140}$$

Flutter instability for the model (137) was analyzed by Bigoni and Zaccaria (1994a). They found that complex eigenvalues are excluded for deviatoric associativity (30)-(31), but coalescence of eigenvalues may still occur. This will not be examined below.

Axisymmetric bifurcations of a solid bar under uniaxial stress. Let us consider a circular cylindrical specimen of radius R and height L subject to axisymmetric deformation with uniaxial stress $\mathbf{T} = \sigma \mathbf{e}_z \otimes \mathbf{e}_z$ aligned along the cylinder axis $\mathbf{e}_z$ and traction-free lateral surface. The relative Lagrangean description is assumed, in which the current configuration is taken as reference ($\mathbf{F} = \mathbf{I}$). A cylindrical coordinate system (r, θ, z) is adopted with the z-axis coincident with the axis of the cylinder. The ends $(z = 0, L)$ are subject to flat, frictionless, rigid constraints keeping null the nominal tangential traction, i.e. $S_{rz} = 0$. The material constitutive model is assumed to correspond to the loading branch of constitutive equation (137) and is therefore given by

$$\dot{\mathbf{S}} = \mathsf{D}[\mathbf{D}] + \mathbf{LK}, \quad \mathsf{D} = \mathsf{E} - \frac{1}{g}\mathsf{E}[\mathbf{Q}] \otimes \mathsf{E}[\mathbf{P}]. \tag{141}$$

We will also analyze bifurcations in the Raniecki solid, which can be obtained simply by replacing D in (141) with

$$\mathsf{D}^R = \mathsf{E} - \frac{1}{4\psi g}(\mathsf{E}[\mathbf{P}] + \psi\mathsf{E}[\mathbf{Q}]) \otimes (\mathsf{E}[\mathbf{P}] + \psi\mathsf{E}[\mathbf{Q}]). \tag{142}$$

Incremental equilibrium equations *for axisymmetric deformations* (in cylindrical co-ordinates) are:

$$\dot{S}_{rr,r} + \dot{S}_{rz,z} + \frac{1}{r}(\dot{S}_{rr} - \dot{S}_{\theta\theta}) = 0,$$
$$\dot{S}_{zr,r} + \dot{S}_{zz,z} + \frac{1}{r}\dot{S}_{zr} = 0. \tag{143}$$

Bifurcations are sought in terms of incremental fields satisfying

$$v_z = 0, \quad \dot{S}_{rz} = 0 \quad \text{at} \quad z = 0, L,$$
$$\dot{S}_{rr} = 0, \quad \dot{S}_{zr} = 0 \quad \text{at} \quad r = R. \tag{144}$$

The method of solution parallels that proposed by Chau (1995) and is based on the introduction of the velocity potential $\Phi(r, z)$ so that

$$v_r = \Phi_{,rz}, \quad v_z = -\frac{1}{A}[B\mathcal{N}(\Phi) + C\Phi_{,zz}], \tag{145}$$

where $\mathcal{N}(\Phi) = (r\Phi_{,r})_{,r}/r$ and

$$A = \mathsf{D}_{rrzz} + \mathsf{D}_{rzrz}, \quad B = \mathsf{D}_{rrrr}, \quad C = \mathsf{D}_{rzrz} + \sigma. \tag{146}$$

For subsequent reference, we introduce three additional quantities:

$$D = \mathsf{D}_{zzzz} + \sigma, \quad E = \mathsf{D}_{rzrz}, \quad F = \mathsf{D}_{zzrr} + \mathsf{D}_{rzrz}\,. \tag{147}$$

We note, in passing, that different but equivalent potentials can be introduced in the bifurcation analysis, as discussed by Miles and Nuwahyid (1985). In the axisymmetric problem under consideration, the non-vanishing components of velocity gradient $\mathbf{L}$ are:

$$L_{rr} = v_{r,r}, \quad L_{\theta\theta} = \frac{v_r}{r}, \quad L_{zz} = v_{z,z}, \quad L_{rz} = v_{r,z}, \quad L_{zr} = v_{z,r}. \tag{148}$$

Substitution of the constitutive law (141) into the equilibrium equations (143) reveals that $(143)_1$ is identically satisfied, while $(143)_2$ gives:

$$\left[\mathcal{N}(\cdot) - \rho_1^2(\cdot)_{,zz}\right]\left[\mathcal{N}(\cdot) - \rho_2^2(\cdot)_{,zz}\right]\Phi = 0, \tag{149}$$

where ρ_1 and ρ_2 satisfy the condition:

$$EB\rho_i^4 + (EC + DB - AF)\rho_i^2 + DC = 0, \quad (i = 1, 2). \tag{150}$$

The nature of the roots of (150) defines the regime classification as follows

- two ρ_i complex conjugate pairs in the elliptic complex regime,
- four ρ_i pure imaginary in the elliptic imaginary regime,
- four ρ_i real in the hyperbolic regime,
- two ρ_i real and two pure imaginary in the parabolic regime.

The interest in the above classification lies in the fact that, when the boundary of the elliptic regime is touched, strain localization may occur. In other words, loss of ellipticity, which is completely equivalent to failure of condition (109), occurs when at least two roots ρ_i become real. Diffuse bifurcated solutions are sought now *in the elliptic regime* of the form:

$$\Phi(r, z) = \phi(r) \sin \eta z, \tag{151}$$

where $\eta = k\pi/L$ and $k = 1, 2, \dots n$. The field (151) satisfies boundary conditions (144) at $z = 0, L$. The equilibrium equation (149) becomes:

$$\left[\overline{\mathcal{N}}(\cdot) + \rho_1^2\eta^2\right]\left[\overline{\mathcal{N}}(\cdot) + \rho_2^2\eta^2\right]\phi = 0, \tag{152}$$

where $\overline{\mathcal{N}}(\phi) = (d/dr)(r(d\phi/dr))/r$. The general solution of (152) for the cylinder can be expressed in terms of the Bessel function of order 0, $J_0(x)$:

$$\phi(r) = c_1 J_0(\eta r \rho_1) + c_2 J_0(\eta r \rho_2), \tag{153}$$

where constants c_1 and c_2 are, in general, complex.

Imposing boundary conditions at $r = R$ yields a 2×2 linear, homogeneous system. Bifurcation occurs when non-trivial solutions of this system are possible. This condition simply means that bifurcation is possible when $\det M_{ij} = 0$, where M_{ij} is defined as

$$M_{1j} = (\mathsf{D}_{rrrr} - \mathsf{D}_{rr\theta\theta})\rho_j J_1(\eta R\rho_j) + (\mathsf{D}_{rrzz}(\frac{B}{A}\rho_j^2 + \frac{C}{A}) - \mathsf{D}_{rrrr}\rho_j^2)\eta R J_0(\eta R\rho_j)$$
$$M_{2j} = \rho_j(1 - \frac{C}{A} - \frac{B}{A}\rho_j^2)J_1(\eta R\rho_j), \qquad (j = 1, 2). \tag{154}$$

Results. In the interest of simplicity, we assume $\lambda = 0$ (corresponding to a null Poisson's ratio). Therefore, the model (137) reduces to

$$\frac{\dot{\mathbf{S}}}{2\mu} = \mathbf{D} + \frac{\mathbf{LK}}{2\mu} - \frac{2\mu}{g} < \mathbf{D} \cdot \mathbf{Q} > \mathbf{P}, \tag{155}$$

where $\mathbf{P}$ and $\mathbf{Q}$ take the form (138), in which the stress is uniaxial

$$\mathbf{P} = \frac{\operatorname{sign}\sigma}{2\sqrt{3}}\left(2\mathbf{e}_z \otimes \mathbf{e}_z - \mathbf{e}_r \otimes \mathbf{e}_r - \mathbf{e}_\theta \otimes \mathbf{e}_\theta\right), \quad \mathbf{Q} = \mathbf{P} + \frac{\alpha}{3}\mathbf{I}. \tag{156}$$

The Raniecki comparison solids are defined by

$$\frac{\dot{\mathbf{S}}}{2\mu} = \mathbf{D} + \frac{\mathbf{LK}}{2\mu} - \frac{\mu}{2\psi g}(\mathbf{D} \cdot \mathbf{P} + \psi \mathbf{D} \cdot \mathbf{Q})(\mathbf{P} + \psi \mathbf{Q}), \tag{157}$$

moreover, we select for simplicity the value of ψ suggested by Bruhns and Raniecki (1982)

$$\psi = \sqrt{\frac{\mathbf{P} \cdot \mathbf{P}}{\mathbf{Q} \cdot \mathbf{Q}}} = \sqrt{\frac{1}{1 + \frac{2}{3}\alpha^2}}, \tag{158}$$

which is independent of the current stress level, and it is therefore not optimal both for (PD) and (SE_R).

The critical plastic modulus for failure of (PD) can be obtained from (73) by noting that

$$\mathsf{L} = \mathbf{I}\underline{\overline{\otimes}}\mathbf{I} + \frac{\sigma}{2\mu}\mathbf{e}_z \otimes \mathbf{e}_z \otimes \mathbf{e}_z \otimes \mathbf{e}_z, \tag{159}$$

has the inverse

$$\mathsf{L}^{-1} = \mathbf{I}\underline{\overline{\otimes}}\mathbf{I} - \frac{\sigma}{2\mu + \sigma}\mathbf{e}_z \otimes \mathbf{e}_z \otimes \mathbf{e}_z \otimes \mathbf{e}_z. \tag{160}$$

For uniaxial tension, σ takes positive values and L is always positive definite. Note also that, due to coaxiality of $\mathbf{P}$, $\mathbf{Q}$ and $\mathbf{T}$, failure of (NS) occurs for a mode with null spin and always corresponds to *a maximum load*.

Failure of (SE) and (E) can be obtained from (98) and (110), on the basis of the elastic acoustic tensor (139) and its inverse (140). Note that, for $\lambda = 0$, the acoustic tensor has eigenvalues equal to $\bar{\mu}$ and $\mu + \bar{\mu}$, which are always positive for tension and vanish in compression for the first time when $\mathbf{n} = \mathbf{e}_z$ and $\sigma = -\mu$. When these values are reached, a purely elastic loss of ellipticity occurs. In order to find the (SE) and (E) thresholds, the constrained maximization problems (100) and (112) have to be solved. Due to axial symmetry, this is an easy task in the present situation, where the unit vector $\mathbf{n}$ can be chosen without loss of generality in the form $\mathbf{n} = n_z\mathbf{e}_z + n_r\mathbf{e}_r$. Failure of ($SE_R$) can be obtained from (102) by solving the inf-max problem for $\mathbf{n}$ and ψ. However, for the present case of uniaxial tension, we have numerically proven the coincidence of (SE) and (SE_R), which holds for a certain optimal parameter ψ, which is a function of the current stress. Therefore, only (SE) will be reported in the figures. It is important to remark that both (PD) and (SE)≡(SE_R) are not relevant (except at $\sigma = 0$) for the Raniecki solid defined by the fixed value of ψ given by (158). This solid has its own peculiar curves for loss of (PD) and (SE) –not reported in the figures–, which, by the way, are close to the curves (PD) and (SE) relative to the solid 'in loading' (coincident with the optimal Raniecki solid).

Curves corresponding to failure of (PD), (NS), (SE) and (E) are reported in Figs. 2-4, together with the curves corresponding to axisymmetric bifurcation for a given bifurcation mode ηR. These curves are referred to a plane defined by axes $\sigma/2\mu$ (the axial stress divided by 2μ) and μ/g (the inverse of the plastic modulus, multiplied by μ). Note that the curves corresponding to (PD) and (NS) are reported only for tension, because tensor G is not positive definite in compression. Note also that the reported range of variation of $\sigma/(2\mu)$ is very large, and probably not sensible with the assumed simplified model. However, we prefer to show a picture as complete as possible. The range of variation of μ/g goes from 0, representative of purely elastic behaviour, i.e. $g \to \infty$, to softening, i.e. beyond the perfectly plastic limit $h = 0$, corresponding to $\mu/g = 1$. Note that the relation between h and g in this simple model is just

$$\frac{g}{\mu} = 1 + \frac{h}{\mu}.$$

Let us begin the discussion with Fig. 2, relative to the associative case, $\alpha = 0$. In this case, the plasticity is referred to a von Mises yield surface. Here the (PD) and (NS) thresholds and also the (SE) and (E) thresholds coincide. The (PD)≡(NS) curve approaches the horizontal axis at $\mu/g = 1$, corresponding to $h = 0$, which is the value of loss of (PD)≡(NS) for the infinitesimal theory. This is obtained when the 'geometrical term' $\sigma/(2\mu)$ is set equal to zero. The curve relative to loss of (E)≡(SE) crosses the horizontal axis at $\mu/g = 1.2$, corresponding to $h/\mu = -0.167$, again the value relative to the infinitesimal theory (see Tab. 1). The curve continues indefinitely in tension, but terminates at $\sigma/(2\mu) = -0.5$ in compression, where failure of ellipticity of the elastic tensor occurs. Two other curves are repoted in the graph, relative to axisymmetric bifurcations of the cylindrical bar with prescribed modes $\eta R = 1$ and $\eta R = 2$. These cross (continuously) the horizontal axis a few decimals on the right side of the value corresponding to loss of (PD)≡(NS). As sketched in the figure, the bifurcation modes are necking modes for tension and a barrelling modes for compression. Points situated at positive values of $\sigma/(2\mu)$ and on the left side of curve (PD)≡(NS) are representative of states for which (PD) is satisfied and *bifurcation is a-priori excluded.* Points situated at $\sigma/(2\mu) > -0.5$ and on the left side of curve (SE)≡(E) are representative of *elliptic* states, for which *strain localization is excluded.*

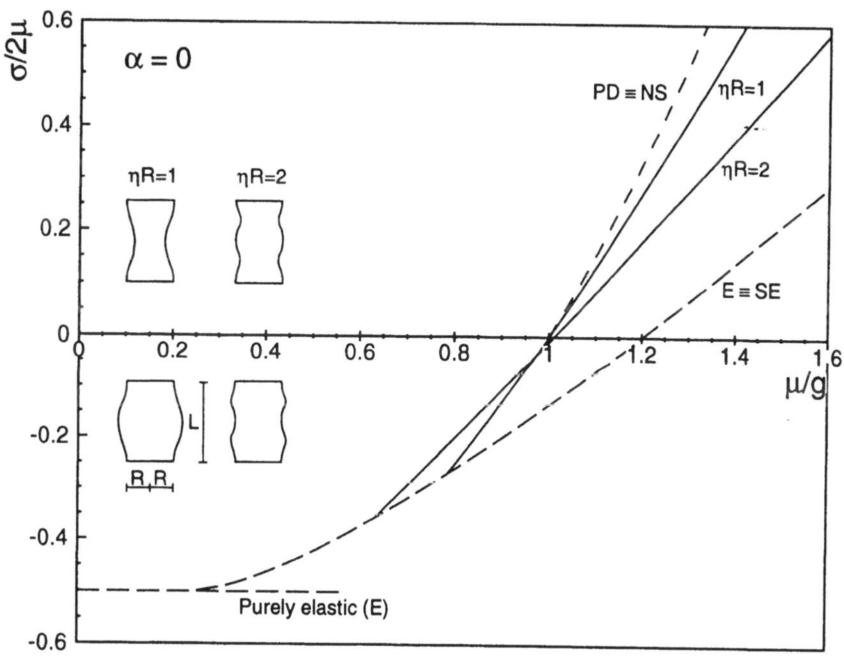

Fig. 2. Axisymmetric bifurcations and local stability criteria for a cylindrical bar subject to uniaxial stress. Associative flow-rule.

Let us assume that g is some continuous function of the stress state, which is infinity at the beginning of plastic flow and continuously decreases when the stress increases (or decreases, in compression). Let us consider now a monotonic loading program of uniaxial tension of a cylindrical specimen starting from some unloaded state, represented by the origin in the diagram of Fig. 2.

At the beginning of the loading program, the material is in the elastic range and the point representative of the state simply moves up along the axis $\sigma/(2\mu)$, at $\mu/g = 0$. When the yield surface is reached, plastic deformation starts and the point moves toward right in the diagram. Before the point touches the (PD)≡(NS) curve, the response is unique. As soon as the point reaches the (PD)≡(NS) curve, a maximum load occurs, and the loading program can be continued only if axial displacement is prescribed (for a situation in which axial dead loading would be prescribed, instability would occur). After the (PD)≡(NS) curve is passed, the point representative of the state will touch the curve corresponding to necking bifurcation (with mode $\eta R = 1$). This bifurcation terminates the homogeneous deformation of the specimen. This might not be the first bifurcation encountered, because an earlier bifurcation might occur in another -not investigated- mode (represented by a curve lying between the two curves (PD)≡(NS) and that corresponding to the mode $\eta R = 1$). The other bifurcation mode, $\eta R = 2$ occurs after that corresponding to $\eta R = 1$ in tension (but before in compression). It is clear from the graph that there is no way of touching the (SE)≡(E) curve in tension without encountering a bifurcation into a diffuse mode. Now, let us analyze compression. Here (PD) is not defined, and in

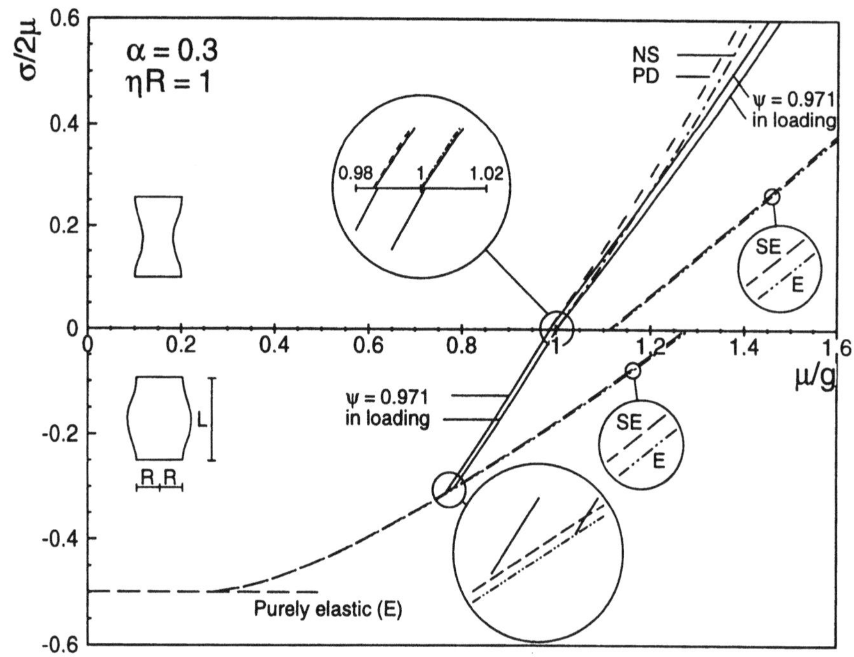

Fig. 3. Bounds to axisymmetric bifurcations and local stability criteria for a cylindrical bar subject to uniaxial stress. Non-associative flow-rule.

fact bifurcation can occur immediately, even at vanishing small stress. This is the well-known case of Euler buckling of a very slender beam, which would be recovered here by analyzing *antisymmetric bifurcation modes*. However, if we concentrate on axisymmetric bifurcation at fixed modes $\eta R = 1, 2$ and consider a loading program at prescribed axial displacements, we see that there is only a range in which bifurcation with $\eta R = 2$ can be attained. Outside this range, loss of (SE)≡(E) will terminate the homogeneous response of the specimen. Once again, we stress that the picture of bifurcation is not complete, because we expect that other axisymmetric or antisymmetric modes can be encountered before that corresponding to $\eta R = 2$.

We are now in a position to analyze the case of non-associativity. Here, the (PD), (NS), (SE), and (E) curves are separated and (SE) and (E) do not cross the horizontal axis with continuity[29]. This reflects the fact that the assumed model has a Drucker-Prager yield surface. The normal to this surface in tension has a different inclination than in compression, and different values of thresholds (SE) and (E) result. For the same reason, the curves representative of diffuse bifurcation modes also do not cross the horizontal axis with continuity (even if the jump is so small that it cannot really be appreciated at the scale of the figure, see detail in Fig. 3). Fig. 3 is relative to $\alpha = 0.3$ and Fig. 4

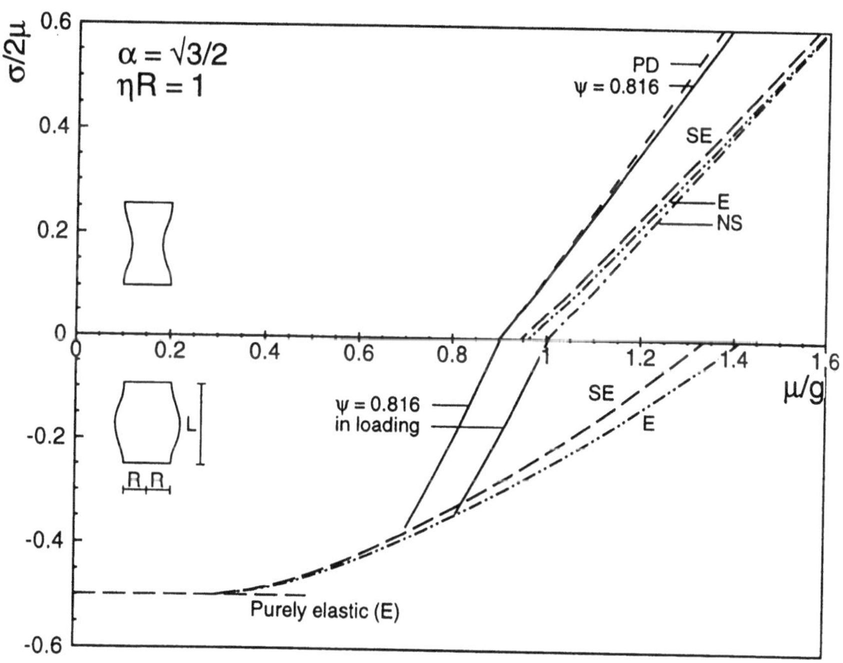

Fig. 4. Bounds to axisymmetric bifurcations and local stability criteria for a cylindrical bar subject to uniaxial stress. Non-associative flow-rule.

[29] This would also occur for the associative case with pressure-sensitivity.

to $\alpha = \sqrt{3}/2$. As for the case of an associative flow rule, the points where the (PD), (NS), (SE) and (E) curves approach the horizontal axis correspond to the values known from the infinitesimal theory (and reported in Tab. 1, for $\alpha = 0.3$ and tension stress state). Points situated at positive values of $\sigma/(2\mu)$ and on the left side of the curve (PD) correspond to situations in which bifurcation is a-priori excluded. The curve (NS) signals that a maximum load has been reached. Curve (E) is relative to strain localization and is terminated in the negative part of the graph by the horizontal line corresponding to $\sigma/(2\mu) = -0.5$. The same line also terminates the curve (SE). When the curve (SE) is crossed, the sufficient stability condition of Hill is certainly lost, independently of the specific boundary conditions.

In Fig. 3, loss of (PD) occurs before (NS), which precedes (SE) and (E). This is completely different in Fig. 4, where failure of (PD) occurs before (SE) and (E), but (NS) is a line and follows (E) [the curve (NS) crosses the curve (E) where $\sigma/(2\mu) = 1$]. It should also be noted that the curves representative of the Raniecki lower bounds terminate in compression when loss of ellipticity occurs in the specific comparison solid (defined by $\psi = 0.971$ for Fig. 3 and $\psi = 0.816$ for Fig. 4), which does not coincide with the curve (SE) (relative to an optimal choice of ψ, see detail in Fig. 3).

Both the Raniecki bounds are reported in Figs. 3 and 4, for fixed mode $\eta R = 1$. They are separated and are observed both in tension and compression.

Regarding Fig. 3, we note that in tension, the lower bound is represented by a curve starting between (PD) and (NS) and later crossing (NS). The upper bound is close to the right side of (NS). These bifurcations in tension occur well before the loss of (SE) and (E). In compression, just considering the fixed mode of bifurcation $\eta R = 1$, diffuse modes may go beyond the elliptic range, so that there is a region in which (SE) and (E) are lost before barrelling bifurcation.

The curve corresponding to the lower bound $\psi = 0.816$ in Fig. 4 initiates in compression close (but does not touch) to the (SE) curve and continues (with a discontinuity crossing the horizontal axis) in tension, between (PD) and (NS). The bifurcation curve relative to the upper bound, with $\eta R = 1$, is not found in tension in the elliptic regime.

8 A concluding remark

Deformations of solids are limited by the occurrence of different failure modes taking place at different scales. As an example, let us recall the behaviour of a metallic, ductile bar pulled in tension. Necking bifurcation is the first threshold for which homogeneity is lost. Subsequently, elastoplastic cavitation occurs in the necked zone. This is a prelude to strain localization and, eventually, fracture propagates along the localized bands yielding complete failure. All these phenomena may be described as instabilities which interact and may be in cooperation/competition to reach global failure.

Many of the instabilities occurring in laboratory tests, such as necking, barrelling, shear banding, and cavitation, can be at least qualitatively described in terms of the classical theory of elastoplasticity, referring to smooth yield surfaces and the normality rule. Behaviour of non-metallic materials, however, is more complicated than the classical theory of plasticity may predict. Therefore, flow rule non-associativity was advocated as a more accurate constitutive description. The result is that non-associativity not only promotes instabilities which may be already qualitatively described in the framework of normality, but opens possibility of new types of instabilities, such as flutter. These instabilities are not yet fully understood. If, on one hand, these may merely reflect a general deficiency of the non-associative models, on the other hand, these may model new phenomena, not yet fully experimentally detected.

References

An, L. and Schaeffer, D. (1990). The flutter instability in granular flow. *J. Mech. Phys. Solids* 40:683-698.

Beatty, M.F. (1987). Topics in finite elasticity: hyperelasticity of rubber, elastomers, and biological tissues-with examples. *Appl. Mech. Rev.* 40:1699-1734.

Benallal, A. Billardon, R. and Geymonat, G. (1990). Phènoménes de localisation á la frontière d'un solide. *C. R. Acad. Sci., Paris* 310:670-684.

Bigoni, D. (1995). On flutter instability in elastoplastic constitutive models. *Int. J. Solids Structures* 32:3167-3189.

Bigoni, D. (1996). On smooth bifurcations in non-associative elastoplasticity. *J. Mech. Phys. Solids* 44:1337-1351.

Bigoni, D. and Hueckel, T. (1990). A note on strain localization for a class of non-associative plasticity rules. *Ingenieur-Archiv* 60:491-499.

Bigoni, D. and Hueckel, T. (1991a). Uniqueness and localization-I. Associative and non-associative elastoplasticity. *Int. J. Solids Structures* 28:197-213.

Bigoni, D. and Hueckel, T. (1991b). Uniqueness and localization-II. Coupled elastoplasticity. *Int. J. Solids Structures* 28:215-224.

Bigoni, D. and Laudiero, F. (1989). The quasi-static finite cavity expansion in a non-standard elastoplastic medium. *Int. J. Mech. Sci.* 31:825-837.

Bigoni, D. and Loret, B. (1999). Effects of elastic anisotropy on strain localization and flutter instability in plastic solids. *J. Mech. Phys. Solids* 47:1409-1436.

Bigoni, D., Loret, B. and Radi, E. (2000). Localization of deformation in plane elastic-plastic solids with anisotropic elasticity. *J. Mech. Phys. Solids , Special issue dedicated to Prof. J.R. Willis*, in press.

Bigoni, D. and Willis, J.R. (1994). A dynamical interpretation of flutter instability. In: Chambon, R., Desrues, J. and Vardoulakis, I., eds., *Localisation and Bifurcation of Rocks and Soils* Rotterdam: A.A. Balkema Scientific Publishers. 51-58.

Bigoni, D. and Zaccaria, D. (1992a). Strong ellipticity of comparison solids in elastoplasticity with volumetric non-associativity. *Int. J. Solids Structures* 29:2123-2136.

Bigoni, D. and Zaccaria, D. (1992b). Loss of strong ellipticity in non-associative elastoplasticity. *J. Mech. Phys. Solids* 40:1313-1331.

Bigoni, D. and Zaccaria, D. (1994a). On eigenvalues of the acoustic tensor in elastoplasticity. *Eur. J. Mechanics-A/Solids* 13:621-638.

Bigoni, D. and Zaccaria, D. (1994b). Eigenvalues of the elastoplastic constitutive operator. *ZAMM* 74:355-357.

Biot, M.A. (1965) *Mechanics of incremental deformations.* New York:Wiley.

Boehler, J.P. and Willis J.R. (1991). An analysis of localization in highly pre-deformed sheet steel. *Unpublished.*

Brannon, R.M. and Drugan, W.J. (1993). Influence of non-classical elastic-plastic constitutive features on shock wave existence and spectral solutions. *J. Mech. Phys. Solids* 41:297-330.

Bruhns, O. and Raniecki, B. (1982). Ein Schrankenverfahren bei Verzweigungsproblemen inelastischer Formänderungen. *ZAMM* 62:T111-T113.

Cattaneo, C. (1946). Su un teorema fondamentale nella teoria delle onde di discontinuità. *Atti Acad. Naz. Lincei* (parts I and II) I:67-72 and 728-734.

Chadwick, P. and Powdrill B. (1965). Singular surfaces in linear thermoelasticity. *Int. J. Eng. Science* 3:561-595.

Chau, K.T. (1992). Non-normality and bifurcation in a compressible pressure-sensitive circular cylinder under axisymmetric tension and compression. *Int. J. Solids Structures* 29:801-824.

Chau, K.T. (1995). Buckling, barrelling, and surface instabilities of a finite, transversely isotropic circular cylinder. *Quart. Appl. Math.* 53:225-244.

Chau, K.T. and Rudnicki, J.W. (1990). Bifurcations of compressible pressure- sensitive materials in plane strain tension and compression. *J. Mech. Phys. Solids* 38:875-898.

Cheng, Y.S. and Lu, W.D. (1993). Uniqueness and bifurcation in elastic-plastic solids. *Int. J. Solids Structures* 30:3073-3084.

Christoffersen, J. (1991). Hyperelastic relations with isotropic rate forms appropriate for elastoplasticity. *Eur. J. Mechanics-A/Solids* 10:91-99.

Del Piero, G. (1979). Some properties of the set of fourth-order tensors, with applications to elasticty. *J. Elasticity* 9:245-261.

Drucker, D.C. (1954). Coulomb friction plasticity and limit loads. *J. Appl. Mech.* 76:71-74.

Gurtin, M.E. (1972). The linear theory of Elasticity. In Flügge, S., ed., *Encyclopedia of Physics* VIa/2. Berlin:Springer. 1-295.

Gurtin, M.E. (1981). *An introduction to continuum mechanics.* New York:Academic Press.

Hayes, M. (1966). On the displacement boundary-value problem in linear elastostatics. *Quart. J. Mech. Appl. Math.* XIX:151-155.

Hadamard, J. (1903). *Leçons sur la Propagation des Ondes et les Équations de l' Hydrodynamique.* Paris:Hermann.

Hill, R. (1950). *The mathematical theory of plasticity.* Oxford:Clarendon Press.

Hill, R. (1952). On discontinuous plastic states, with special reference to localized necking in thin sheets. *J. Mech. Phys. Solids* 1:19-30.

Hill, R. (1958). A general theory of uniqueness and stability in elastic-plastic solids. *J. Mech. Phys. Solids* 6:236-249.

Hill, R. (1959). Some basic principles in the mechanics of solids without a natural time. *J. Mech. Phys. Solids* 7:209-225.

Hill, R. (1961). Discontinuity relations in mechanics of solids. In Sneddon, I.N. and Hill, R., eds., *Progress in Solid Mechanics* II. Amsterdam:North-Holland. 247-276.

Hill, R. (1962). Acceleration waves in solids. *J. Mech. Phys. Solids* 10:1-16.

Hill, R. (1967a). Eigenmodal deformations in elastic/plastic continua. *J. Mech. Phys. Solids* 15:371-386.

Hill, R. (1967b). On the classical constitutive laws for elastic/plastic solids. In Broberg, B., ed., *Recent Progress in Applied Mechanics, The Folke Odkvist Volume* Stockholm:Almqvist & Wiksell. 241-249.

Hill, R. (1968). On constitutive inequalities for simple materials. *J. Mech. Phys. Solids* 16:229-242.

Hill, R. (1978) Aspects of invariance in solid mechanics. In Yih, C.-S., ed., *Advances in Applied Mechanics* 18. New York:Academic Press. 1-75.

Hill, R. and Hutchinson, J. W. (1975). Bifurcation phenomena in the plane tension test. *J. Mech. Phys. Solids* 23:239-264.

Hill, R. and Rice, J. R. (1973). Elastic potentials and the structure of inelastic constitutive laws. *SIAM J. Appl. Math.* 25:448-461.

Horgan, C.O. and Polignone, D.A. (1995). Cavitation in nonlinearly elastic solids: A review. *Appl. Mech. Rev.* 48:471-485.

Huang, K., Hutchinson, J.W. and Tvergaard, V. (1991). Cavitation instabilities in elastic-plastic solids. *J. Mech. Phys. Solids* 39:223-241.

Hueckel, T. (1976). Coupling of elastic and plastic deformation of bulk solids. *Meccanica* 11:227-235.

Hutchinson, J. W. (1973). Post-bifurcation behavior in the plastic range. *J. Mech. Phys. Solids* 21:163-190.

Hutchinson, J. W. and Miles, J.P. (1974). Bifurcation analysis of the onset of necking in an elastic/plastic cylinder under uniaxial tension. *J. Mech. Phys. Solids* 22:61-71.

Kleiber, M. (1984) Numerical study on necking-type bifurcations in void-containing elastic-plastic material. *Int. J. Solids Structures* 20:191-210.

Kleiber, M. (1986) On plastic localization and failure in plane strain and round void containing tensile bars. *Int. J. Plasticity* 2:205-221.

Loret, B. (1992). Does deviation from deviatoric associativity lead to the onset of flutter instability? *J. Mech. Phys. Solids* 40:1363-1375.

Loret, B., Martins, J.A.C. and Simões, F.M.F. (1995). Surface boundary conditions trigger flutter instability in non-associative elastic-plastic solids. *Int. J. Solids Structures* 32:2155-2190.

Loret, B., Prevost, J.H. and Harireche, O. (1990). Loss of hyperbolicity in elastic-plastic solids with deviatoric associativity. *Eur. J. Mechanics-A/Solids* 9:225-231.

Maier, G. and Hueckel, T. (1979). Non associated and coupled flow-rules of elastoplasticity for rock-like materials. *Int. J. Rock Mech. Min. Sci.* 16:77-92.

Mandel, J. (1966). Conditions de stabilité et postulat de Drucker. In Kravtchenko, J. and Sirieys, P.M., eds., *Rheology and Soil Mechanics*. Berlin:Springer. 58-68.

Melan, E. (1938). Zur Plastizität des räumliche Kontinuums. *Ingenieur-Archiv* 9:116-126.

Miles, J.P. (1973). Fluid-pressure eigenstates and bifurcation in tension specimens under lateral pressure. *J. Mech. Phys. Solids* 21:145-162.

Miles, J.P. and Nuwayhid, U.A. (1985). Bifurcation in compressible elastic/plastic cylinders under uniaxial tension. *Appl. Sci. Res.* 42:33-54.

Mróz, Z. (1963). Non-associated flow laws in plasticity *J. de Mechanique* 2:21-42.

Mróz, Z. (1966). On forms of constitutive laws for elastic-plastic solids. *Arch. Mech. Stosowanej* 18:1-34.

Nadai, A. (1931) *Plasticity*. New York:McGraw-Hill.

Nadai, A. (1950) *Theory of flow and fracture of solids*. New York:McGraw-Hill.

Neale, K.W. (1981). Phenomenological constitutive laws in finite plasticity *SM Archives* 6:79-128.

Needleman, A. (1979). Non-normality and bifurcation in plane strain tension or compression. *J. Mech. Phys. Solids* 27:231-254.

Needleman, A. and Ortiz, M. (1991). Effects of boundaries and interfaces on shear-band localization. *Int. J. Solids Structures* 28:859-877.

Nguyen, S.Q. and Triantafyllidis, N. (1989). Plastic bifurcation and postbifurcation analysis for generalized standard continua. *J. Mech. Phys. Solids* 37:545-566.

Nikolaevskii, V.N. and Rice, J.R. (1979). Current topics in non-elastic deformation of geological materials. In Timmerhaus, K.D. and Barber, M.S., eds., *Proceedings of the Sixth AIRAPT Conference: High Pressure Science and Technology*. New York:Plenum. 2:455-464.

Ogden, R.W. (1984). *Non-linear elastic deformations*. Chichester:Ellis Horwood.

Ogden, R. W. (1985). Local and global bifurcation phenomena in plane-strain finite elasticity. *Int. J. Solids Structures* 21:121-132.

Ottosen, N.S. and Runesson, K. (1991). Acceleration waves in elastoplasticity. *Int. J. Solids Structures* 28:135-159.

Petryk, H. (1985a). On energy criteria of plastic instability. In *Plastic Instability, Proc. Considère Memorial*. Paris:Ecole Nat. Ponts Chauss. Press. 215-226.
Petryk, H. (1985b). On stability and symmetry conditions in time-independent plasticity. *Arch. Mech.* 37:503-520.
Petryk, H. (1991). The energy criteria of instability in time-independent inelastic solids. *Arch. Mech.* 43:519-545.
Petryk, H. (1992). Material instability and strain-rate discontinuities in incrementally nonlinear continua. *J. Mech. Phys. Solids* 40:1227-1250.
Petryk, H. (1993a). Theory of bifurcation and instability in time-independent plasticity. In Nguyen, Q.S., ed., *CISM Lecture Notes No. 327*, Udine 1991. Wien:Springer. 95-152.
Petryk, H. (1993b). Stability and constitutive inequalities in plasticity. In Muschik, W., ed., *CISM Lecture Notes No. 336*, Udine 1992. Wien:Springer. 255-329.
Petryk, H. (1999). General conditions for uniqueness in materials with multiple mechanisms of inelastic deformation. *J. Mech. Phys. Solids in press.*
Petryk, H. and Thermann, K. (1985). Second-order bifurcation in elastic-plastic solids. *J. Mech. Phys. Solids* 33:577-593.
Prager, W. (1954). Discontinuous fields of plastic stress and flow. In *2nd Nat. Congr. Appl. Mech.*, Ann Arbor, Michigan, 21-32.
Radi, E., Bigoni, D. and Tralli, A. (1999). On uniqueness for frictional contact rate problems. *J. Mech. Phys. Solids* 47:275-296.
Raniecki, B. (1979). Uniqueness criteria in solids with non-associated plastic flow laws at finite deformations, *Bull. Acad. Polon. Sci. ser. sci. tech.* XXVII:391-399.
Raniecki, B. and Bruhns, O.T. (1981). Bounds to bifurcation stresses in solids with non-associated plastic flow law at finite strain. *J. Mech. Phys. Solids* 29:153-171.
Rice, J. R. (1977). The localization of plastic deformation. In Koiter, W.T., ed., *Theoretical and Applied Mechanics*. Amsterdam:North-Holland. 207-220.
Rice, J.R. and Rudnicki, J.W. (1980). A note on some features of the theory of localization of deformation. *Int. J. Solids Structures* 16:597-605.
Rudnicki, J.W. and Rice, J.R. (1975). Conditions for the localization of deformations in pressure-sensitive dilatant materials. *J. Mech. Phys. Solids* 23:371-394.
Runesson, K. and Mróz, Z. (1989). A note on non-associated plastic flow rules. *Int. J. Plasticity* 5:639-658.
Ryzhak, E. I. (1987). Necessity of Hadamard conditions for stability of elastic-plastic solids. *Izv. AN SSSR MTT* (Mechanics of Solids) 99-102.
Ryzhak, E. I. (1993). On stable deformation of "unstable" materials in a rigid triaxial testing machine. *J. Mech. Phys. Solids* 41:1345-1356.
Ryzhak, E. I. (1994). On stability of homogeneous elastic bodies under boundary conditions weaker than displacement conditions. *Q. Jl. Mech. appl. Math.* 47:663-672.
Simões, F.M.F. (1997). Instabilities in non-associated problems of solid mechanics. Ph.D. Thesis, Technical University of Lisbon, *in Portuguese.*
Szabó, L. (1994). Shear band formulation in finite elastoplasticity. *Int. J. Solids Structures* 31:1291-1308.
Thomas, T.Y. (1953). The effect of compressibility on the inclination of plastic slip bands in flat bars. *Proc. Nat. Acad. Sci.* 39:266-273.
Thomas, T.Y. (1961) *Plastic flows and fracture of solids.* New York:Academic Press.
Tomita, Y., Shindo, A. and Fatnassi, A. (1988). Bounding approach to bifurcation point of annular plates with nonassociated flow law subjected to uniform tension at their outer edges. *Int. J. Plasticity* 4:251-263.
Truesdell, C. and Noll, W. (1965). The non-linear field theories of mechanics. In Flügge, S., ed., *Encyclopedia of Physics*:III/3. Berlin:Springer-Verlag.

Tvergaard, V. (1982). Influence of void nucleation on ductile shear fracture at a free surface. *J. Mech. Phys. Solids* 30:399-425
van Hove, L. (1947). Sur l'extension de la condition de Legendre du calcul des variations aux intégrales multiples à plusieurs fonctions inconnues. *Proc. Sect. Sci. K. Akad. van Wetenschappen*, Amsterdam, 50:18-23.
Vardoulakis, I. (1981). Bifurcation analysis of the plane rectilinear deformation on dry sand samples. *Int. J. Solids Structures* 11:1085-1101.
Vardoulakis, I. (1983). Rigid granular plasticity model and bifurcation in the triaxial test. *Acta Mechanica* 49:57-79.
Young, N.J.B. (1976). Bifurcation phenomena in the plane compression test. *J. Mech. Phys. Solids* 24:77-91.

APPENDIX A. A note on coaxiality.

We prove the following *proposition*:

Two symmetric second-order tensors **A**, **B** commute, i.e. **AB** = **BA** if and only if they possess at least three common, linearly independent, eigenvectors (thus defining a principal reference system).

Proof. The sufficiency is trivial. In fact, let us assume that the two tensors share a principal reference system. Represented in this system, the two tensors evidently commute. Assume now that the two tensors commute. Represented in the principal reference system of **A** these two tensors write

$$\mathbf{A} = \sum_{i=1}^{3} \alpha_i \mathbf{a}_i \otimes \mathbf{a}_i\,, \quad \mathbf{B} = \sum_{i,j=1}^{3} \beta_{ij} \mathbf{a}_i \otimes \mathbf{a}_j\,. \tag{A.1}$$

where $\mathbf{a}_i$, $i \in [1,3]$ are the unit eigenvectors of **A** and α_i the corresponding eigenvalues, β_{ij} are the components of **B** onto $\mathbf{a}_i$, which, due to symmetry, satisfy $\beta_{ij} = \beta_{ji}$.

Imposing the condition **AB** = **BA** gives:

$$(\alpha_i - \alpha_j)\beta_{ij} = 0\,, \quad (i \neq j \in [1,3]). \tag{A.2}$$

Condition (A.2) implies that either $\mathbf{a}_i$ and $\mathbf{a}_j$ are eigenvectors of **B**, so that $\beta_{ij} = 0$, or the characteristic space corresponding to $\mathbf{a}_i$ and $\mathbf{a}_j$ is a plane, so that $\alpha_i = \alpha_j$. In the latter case, it is always possible to choose in this plane a reference system which is principal for **B**.

□

Remark. 1 The above proof is restricted to dimension 3. Its generalization to dimension n is straightforward.

Remark. 2 It is important to realize that the above proposition *does not imply* that two coaxial tensors share *all* eigenvectors. Take for instance the identity tensor. This is coaxial to every tensor, but shares only three eigenvectors with any symmetric second-order tensor with distinct eigenvalues.

APPENDIX B. On the metric induced by semi-positive definite tensors.

Let us consider the positive scalar valued function f defined over the space of nth-order tensors as

$$f(\mathbf{A}) = \sqrt{\mathbf{A} \cdot \mathsf{H}[\mathbf{A}]} \geq 0, \tag{B.1}$$

where $\mathbf{A}$ is a generic nth-order tensor and H is any semi-positive definite $(2 \times n)$th-order tensor (symmetries are not required). In particular, the cases in which H is a fourth-order ($n = 2$) or second-order tensor ($n = 1$) are particularly relevant here. We begin with a proof of the Cauchy-Schwarz inequality. Take any $\lambda \in \Re$ and consider, for every nth-order tensors $\mathbf{A}$ and $\mathbf{B}$

$$0 \leq [f(\lambda\mathbf{A} + \mathbf{B})]^2 = \lambda^2 \mathbf{A} \cdot \mathsf{H}[\mathbf{A}] + \lambda(\mathbf{A} \cdot \mathsf{H}[\mathbf{B}] + \mathbf{B} \cdot \mathsf{H}[\mathbf{A}]) + \mathbf{B} \cdot \mathsf{H}[\mathbf{B}]. \tag{B.2}$$

It may be observed from (B.2) that the discriminant of the second-order polynomial in λ must be negative or null. This condition yields the Cauchy-Schwarz inequality

$$\frac{|\mathbf{A} \cdot \mathsf{H}[\mathbf{B}] + \mathbf{B} \cdot \mathsf{H}[\mathbf{A}]|}{2} \leq \sqrt{\mathbf{A} \cdot \mathsf{H}[\mathbf{A}]}\sqrt{\mathbf{B} \cdot \mathsf{H}[\mathbf{B}]}. \tag{B.3}$$

Taking now $\lambda = 1$ in (B.2) and using (B.3), the triangle inequality is readily obtained:

$$(\mathbf{A} + \mathbf{B}) \cdot \mathsf{H}[\mathbf{A} + \mathbf{B}] \leq \mathbf{A} \cdot \mathsf{H}[\mathbf{A}] + |\mathbf{A} \cdot \mathsf{H}[\mathbf{B}] + \mathbf{B} \cdot \mathsf{H}[\mathbf{A}]| + \mathbf{B} \cdot \mathsf{H}[\mathbf{B}] \tag{B.4}$$

$$\leq \left(\sqrt{\mathbf{A} \cdot \mathsf{H}[\mathbf{A}]} + \sqrt{\mathbf{B} \cdot \mathsf{H}[\mathbf{B}]}\right)^2. \tag{B.5}$$

As a consequence of (B.4), we note that

$$f(\mathbf{A}) = f(\mathbf{A} \pm \mathbf{B} \mp \mathbf{B}) \leq f(\mathbf{A} \pm \mathbf{B}) + f(\mathbf{B}), \quad f(\mathbf{B}) = f(\mathbf{B} \pm \mathbf{A} \mp \mathbf{A}) \leq f(\mathbf{A} \pm \mathbf{B}) + f(\mathbf{A}), \tag{B.6}$$

which imply

$$\sqrt{(\mathbf{A} \pm \mathbf{B}) \cdot \mathsf{H}[\mathbf{A} \pm \mathbf{B}]} \geq |\sqrt{\mathbf{A} \cdot \mathsf{H}[\mathbf{A}]} - \sqrt{\mathbf{B} \cdot \mathsf{H}[\mathbf{B}]}|. \tag{B.7}$$

Finally, we note that for every $\lambda \in \Re$

$$f(\lambda\mathbf{A}) = |\lambda| f(\mathbf{A}), \tag{B.8}$$

therefore, function f defines a seminorm on the space of nth order-tensors (a norm when H is positive definite).

THERMODYNAMIC ANALYSIS OF SHOCK WAVES IN GENERAL MEDIA

APPLICATIONS TO ELASTIC-PLASTIC SHOCK WAVES AND DUCTILE CRACK GROWTH

W.J. Drugan
University of Wisconsin-Madison, Madison, WI, USA

Abstract: A thermodynamic comparison is made between propagating shock waves and smooth waves, in arbitrary media and for arbitrary deformations, for two classes of shock wave: weak shocks; and shocks of arbitrary strength that propagate under steady-state conditions. It is shown that the thermodynamic differences between shocks and smooth waves are slight for weak shocks, and nil for steady-state shocks. This legitimizes both the derivation of restrictions on steadily propagating shock waves by the analysis of suitably-chosen smooth waves, and the use of a purely mechanical constitutive model in the derivation of these restrictions. An important specific application of this approach to deriving shock wave restrictions is provided by analyzing a very broad class of rate-independent elastic-plastic materials. It is shown that positive-definiteness of the elastic modulus tensor and normality of the plastic deformation-rate tensor to the yield surface, together with standard weak conservation laws and compatibility requirements, greatly restrict the permissibility of shock waves. This is especially so in quasi-static deformations, but strong restrictions on the permissible types and speeds of dynamically propagating shock waves are also shown to result. These analyses are first performed within a small-displacement-gradient formulation, for which the results are highly restrictive and easily interpreted. The finite deformation versions of the calculations are then performed. Some examples of the implications of the results to problems in quasi-static and dynamic elastic-plastic crack growth are also provided, for isotropic materials and for ductile single crystals.

1 Introduction and Motivation

There are very important reasons for needing to understand the conditions under which shock waves can exist and the restrictions governing them when they do. Most continuum mechanics boundary-value problems are formulated in terms of purely mechanical constitutive equations, because the determination of physically-based, fully thermomechanical constitutive equations is extremely difficult. Furthermore, within the context of elastoplasticity, the additional assumption of rate-independence of the constitutive model is often made. While such constitutive simplifications are often essential for tractability of the problem under consideration, one does not wish to exclude from the problem's solution physical phenomena that can actually occur in nature, nor does one want to permit the solution to exhibit phenomena that are physically impossible. These statements apply especially to the case of shock waves.

In actual materials, the large gradients in material velocity and temperature associated with a shock wave produce marked effects of friction and heat conduction, causing irreversible thermodynamic processes and hence (additional) entropy production. Thus, in general in such large gradient regions, purely mechanical constitutive equations are not sufficient to capture the physical processes occurring in the shock wave. However, these regions are usually observed to be very narrow; thus, an alternative to the use of a full thermomechanical constitutive formulation for the boundary-value problem at hand is to employ purely mechanical constitutive equations, but permit the existence of propagating sharp (jump) discontinuities in the appropriate field variables to capture the physical phenomenon of a shock wave. When this approach is adopted, which is usually the case as noted above, it is obviously of crucial importance that a careful determination be made of the conditions under which such propagating jump discontinuities should be admitted, and of the types of discontinuities allowed, when solutions are sought to boundary value problems involving purely mechanical constitutive equations. This is crucially important since if only continuous (i.e., smooth) solutions are sought to purely mechanical boundary-value problems, important physical solutions may be missed, whereas if propagating jump discontinuities are inappropriately permitted in the analysis of such problems, "solutions" may be found that lack physical counterparts.

It is the purpose of these notes to show how to analyze shock waves, in order to determine the conditions under which they can exist, and when they do, what restrictions they must obey. The analysis of shock waves under absolutely arbitrary conditions is an exceedingly difficult task, especially in solids, in part because this necessitates a realistic (thermomechanical) constitutive model of the material of interest that is valid under the extreme conditions experienced during shock wave passage. As these lecture notes will show, there are certain conditions under which accurate analysis of shock waves is possible without the need for such sophisticated constitutive modeling: namely, when the shock wave is weak, or when it propagates under steady-state conditions. When such conditions hold, a very general thermodynamic analysis – to be explained here – shows that the change of state experienced by a material particle during shock wave passage is thermodynamically very similar (weak shocks) or identical (steady-state shocks) to that experienced by a particle during smooth wave passage. This means that a shock wave under these conditions can be characterized by the analysis of a suitably chosen smooth wave, for which use of a purely mechanical constitutive model is sufficient. This procedure will be illustrated here for a large class of rate-independent elastic-plastic materials, and several specific applications of the results will be described.

The notes are organized as follows: Section 2 will introduce notation and review fundamental principles applicable to general media. It will also show how to derive the forms of these principles applicable to the steady-state propagation of a smooth wave, and to the propagation of a general shock wave. Section 3 will make a thermodynamic comparison between a shock wave and a smooth wave in an arbitrary medium for two cases: general unsteady weak shock waves, and steady-state shock waves of arbitrary strength. The resulting thermodynamic equivalence of these waves will be exploited in Section 4 to derive restrictions on equilibrium (quasi-static) and dynamic shock waves in elastic-plastic materials, within a small-displacement-gradient formulation for maximum clarity. Section 5 will reconsider the derivation of restrictions on steady-state shock waves within a full finite deformation formulation. Finally, Section 6 will present several applications of the general results derived to problems in elastic-plastic fracture mechanics, for isotropic materials and for ductile single crystals.

2 Notation; Governing Equations and Their Forms Appropriate for Smooth Wave Analysis and for Shock Waves

2.1 Deformation

We will analyze a material body experiencing arbitrary three-dimensional deformations. It is useful to consider two configurations of the body: the actual or *current* configuration at time t, and a fixed *reference* configuration that coincides with the current configuration at time $t = t_0$. Let $\mathbf{x}$ be the current position vector of a material particle P that had position vector $\mathbf{X}$ in the reference configuration, both with respect to the same Cartesian coordinate axes, as illustrated in Figure 1. We assume there exists a mapping, often called the motion, such that

$$\mathbf{x} = \mathbf{x}(\mathbf{X}, t) \qquad [\text{so by definition } \mathbf{x}(\mathbf{X}, t_0) = \mathbf{X}]. \tag{2.1}$$

[To avoid excessive notation, the same symbol $\mathbf{x}$ is employed for the function.] The displacement vector field $\mathbf{u}$ is thus the difference between each particle's current and reference positions:

$$\mathbf{u} = \mathbf{x}(\mathbf{X}, t) - \mathbf{X}. \tag{2.2}$$

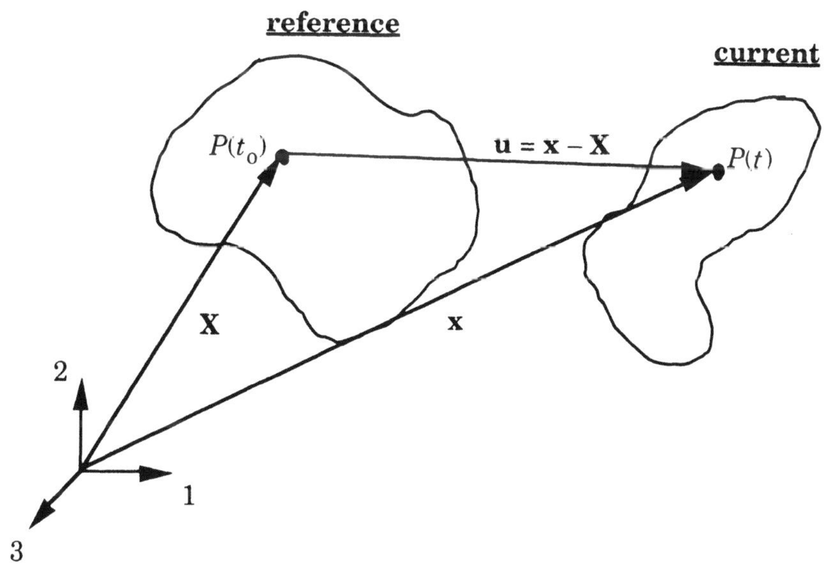

Figure 1. Position and displacement of a material particle in a general motion: reference and current configurations of a material body.

Here and throughout the notes, boldface will denote vectors and tensors, and Cartesian components of these, with respect to the **X** or **x** system, as appropriate, will be indicated by Latin subscripts *i, j, k, l, m, n*, which have range 1, 2, 3. The Einstein summation convention will be adopted: a repeated subscript in a term implies summation over all permitted values of that subscript.

The velocity vector field, **v**, is obtained by taking the material time derivative (that is, the time derivative holding **X** fixed) of the displacement field; equivalent notations for this time derivative are indicated:

$$\mathbf{v} = \left.\frac{\partial \mathbf{u}}{\partial t}\right|_{\mathbf{X}} \equiv \frac{\mathrm{d}\mathbf{u}}{\mathrm{d}t} \equiv \dot{\mathbf{u}}. \tag{2.3}$$

An infinitesimal material line element in the reference configuration, d**X**, is mapped into its current configuration image, d**x**, via the *deformation gradient* tensor **F**(**X**, *t*):

$$\mathrm{d}\mathbf{x} = \mathbf{F} \bullet \mathrm{d}\mathbf{X} \quad \Leftrightarrow \quad \mathrm{d}x_i = F_{ij}\mathrm{d}X_j, \tag{2.4}$$

where both the direct and index notation forms are shown. [Throughout the notes, a single dot will indicate contraction over adjacent indices, as in (2.4), while a double dot will indicate contraction over two sets of indices, e.g. $\mathbf{A}:\mathbf{B} = A_{ij}B_{ji}$.] Thus, the deformation gradient tensor is given by, in regions where the motion is differentiable:

$$\mathbf{F} = \frac{\partial \mathbf{x}}{\partial \mathbf{X}} \equiv \mathbf{x}\overleftarrow{\nabla}_0 \quad \Leftrightarrow \quad F_{ij} = \frac{\partial x_i}{\partial X_j}. \tag{2.5}$$

2.2 Traction Vectors and Stress Tensors

Consider an infinitesimal smooth patch of a material surface within or on a body. As illustrated in Figure 2, we denote the unit normal vector to this surface as **N** and its infinitesimal area as $\mathrm{d}A_0$ in the reference configuration, while these quantities are denoted **n** and d*A* in the current configuration. The total vector of force acting across this area patch in the current configuration is d**f**, which has an arbitrary orientation in general. By choosing infinitesimal material line elements aligned with the sides of the area element and employing (2.4), it is straightforward to show that the change of the material area element in a general deformation is given by

$$\mathbf{n}\mathrm{d}A = J\left(\mathbf{F}^{-1}\right)^T \bullet \mathbf{N}\,\mathrm{d}A_0, \tag{2.6}$$

where superscripts -1 and T denote tensor inverse and transpose, respectively, and $J = \det(\mathbf{F})$.

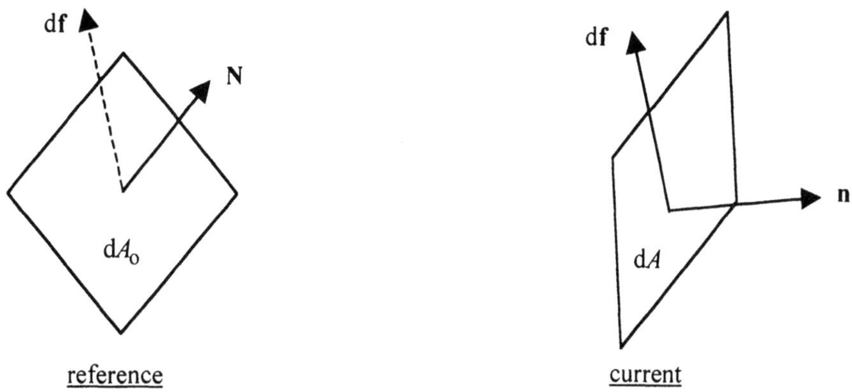

Figure 2. Infinitesimal smooth patch of a material surface inside or on a body in the reference and current configurations, with respective unit normal vectors, and the total vector of force transmitted across it in the current configuration.

The Cauchy traction vector, $\mathbf{t}$, is defined to be the total current configuration force acting across the infinitesimal surface area per unit *current* area, while the nominal or First Piola-Kirchhoff traction vector, $\mathbf{t}_0$, is the same force but per unit *reference* area:

$$\mathrm{d}\mathbf{f} = \mathbf{t}\mathrm{d}A = \mathbf{t}_0\mathrm{d}A_0, \tag{2.7}$$

so that both traction vectors have the same direction but different magnitudes. The Cauchy stress tensor $\boldsymbol{\sigma}$ linearly relates the current configuration surface element unit normal vector to the Cauchy traction vector, while the nominal or First Piola-Kirchhoff stress tensor $\mathbf{T}$ is defined similarly with respect to the reference configuration surface element unit normal:

$$\mathbf{t} = \mathbf{n} \bullet \boldsymbol{\sigma}, \qquad \mathbf{t}_0 = \mathbf{N} \bullet \mathbf{T}. \tag{2.8}$$

Combining (2.7) and (2.8) and using (2.6), one deduces that the Cauchy and nominal stress tensors are related as:

$$\boldsymbol{\sigma} = \frac{1}{J}\mathbf{F} \bullet \mathbf{T} \quad \Leftrightarrow \quad \mathbf{T} = J\,\mathbf{F}^{-1} \bullet \boldsymbol{\sigma}. \tag{2.9}$$

2.3 Integral (Weak) Forms of the Conservation Laws

Here we shall summarize some of the fundamental postulates of continuum mechanics that are valid for arbitrary materials and which will be needed in the upcoming shock wave analysis. These will be expressed in integral (weak) form, because this form is valid regardless of whether the field quantities are smooth or contain jumps. We shall then review how the specific smooth and jump forms of these postulates are obtained from the integral forms.

In terms of a fixed reference configuration, the postulates of conservation of mass, linear momentum, angular momentum and energy (First Law of thermodynamics) are:

$$\frac{\mathrm{d}}{\mathrm{d}t}\int_{V_0} \rho_0 \mathrm{d}V_0 = 0 \tag{2.10}$$

$$\frac{\mathrm{d}}{\mathrm{d}t}\int_{V_0} \rho_0 \mathbf{v}\,\mathrm{d}V_0 = \int_{S_0} \mathbf{t}_0 \mathrm{d}S_0 + \int_{V_0} \rho_0 \mathbf{b}\,\mathrm{d}V_0 \tag{2.11}$$

$$\frac{\mathrm{d}}{\mathrm{d}t}\int_{V_0} \mathbf{x}\times\rho_0 \mathbf{v}\,\mathrm{d}V_0 = \int_{S_0} \mathbf{x}\times\mathbf{t}_0 \mathrm{d}S_0 + \int_{V_0} \mathbf{x}\times\rho_0 \mathbf{b}\,\mathrm{d}V_0 \tag{2.12}$$

$$\begin{aligned}\frac{\mathrm{d}}{\mathrm{d}t}\int_{V_0} \left(\tfrac{1}{2}\rho_0 \mathbf{v}\bullet\mathbf{v} + \rho_0 u\right)\mathrm{d}V_0 &= \int_{S_0} \mathbf{t}_0\bullet \mathbf{v}\,\mathrm{d}S_0 + \int_{V_0} \rho_0 \mathbf{b}\bullet\mathbf{v}\,\mathrm{d}V_0 \\ &\quad + \int_{V_0} \rho_0 r\,\mathrm{d}V_0 - \int_{S_0} \mathbf{Q}\bullet\mathbf{N}\,\mathrm{d}S_0\,. \end{aligned} \tag{2.13}$$

Here, $\mathrm{d}/\mathrm{d}t$ denotes material time derivative as before; ρ_0 is reference mass density; V_0 is an arbitrary reference configuration subvolume having surface S_0 and unit outward normal $\mathbf{N}$; $\mathbf{b}$ is the body force vector per unit mass; $\times$ denotes vector cross product; u is internal energy and r is heat source, both per unit reference mass; and $\mathbf{Q}$ is the nominal heat flux vector.

2.4 Local (Smooth) Forms of the Conservation Laws

In reference configuration regions where all field variables have continuous first derivatives with respect to both position and time, the integral forms of the conservation laws reviewed in Section 2.3 can be simplified. This is accomplished by proceeding through (2.10) – (2.13) in order, performing the following steps: take the material time derivatives inside the integrals (the integration regions are fixed with respect to material time differentiation since they are reference configuration regions), apply the divergence theorem where necessary to convert surface to volume integrals, employ the local forms of the preceding equations [for example, (2.11) simplifies by use of the local form of (2.10); (2.12) simplifies by use of the local forms of (2.10) and (2.11), etc.], and require that the resulting volume integrals indeed apply to arbitrary subvolumes. The resulting local forms of the conservation laws are, in order:

$$\dot{\rho}_0 = 0 \tag{2.14}$$

$$\nabla_0 \bullet \mathbf{T} + \rho_0 \mathbf{b} = \rho_0 \dot{\mathbf{v}} \tag{2.15}$$

$$\mathbf{F} \bullet \mathbf{T} = (\mathbf{F} \bullet \mathbf{T})^T \tag{2.16}$$

$$\rho_0 \dot{u} = \mathbf{T} : \dot{\mathbf{F}} + \rho_0 r - \nabla_0 \bullet \mathbf{Q}. \tag{2.17}$$

To these we append the material time rate of (2.5):

$$\dot{\mathbf{F}} = \dot{\mathbf{x}} \overleftarrow{\nabla}_0 = \mathbf{v} \overleftarrow{\nabla}_0. \tag{2.18}$$

2.5 Steady-State Local Forms of the Conservation Laws

Shortly we shall wish to analyze the fields associated with a smooth wave that propagates under steady-state conditions with respect to an observer moving with the wave. Specifically, we shall consider a smooth wave whose reference configuration image propagates without rotating under steady-state conditions with constant speed c_0 in the direction of its unit normal $\mathbf{N}$. Under these conditions, the material time rate of any field quantity $\psi(\mathbf{X}, t)$ within the wave can be calculated by introducing a set of coordinates $\boldsymbol{\xi}$ that propagates with the reference configuration image of the wave. Then, employing the chain rule to write the material time derivative in terms of this coordinate system gives

$$\dot{\psi} = \frac{\mathrm{d}\psi(\boldsymbol{\xi}, t)}{\mathrm{d}t} = \left.\frac{\partial \psi(\boldsymbol{\xi}, t)}{\partial t}\right|_{\boldsymbol{\xi}} + \frac{\partial \psi(\boldsymbol{\xi}, t)}{\partial \boldsymbol{\xi}} \bullet \frac{\partial \boldsymbol{\xi}}{\partial t} = \frac{\partial \psi(\boldsymbol{\xi}, t)}{\partial \boldsymbol{\xi}} \bullet (-c_0 \mathbf{N}) = -c_0 \psi \overleftarrow{\nabla}_0 \bullet \mathbf{N}, \tag{2.19}$$

where the third equality is due to the fact that the time derivative holding $\boldsymbol{\xi}$ fixed vanishes for steady-state propagation and the fourth equality is due to the spatial normal-direction derivative in terms of $\boldsymbol{\xi}$ being the same as that in terms of $\mathbf{X}$. As noted, we also assume

$$\dot{\mathbf{N}} = \mathbf{0}, \quad \dot{c}_0 = 0. \tag{2.20}$$

Because our goal will be to analyze a smooth wave that is chosen to emulate a shock wave, we shall further restrict consideration to smooth waves having the feature that field quantities may vary *through* the wave (i.e., in the $\mathbf{N}$-direction), but *not* parallel to the wave. This assumption means that the reference configuration del operator simplifies to

$$\nabla_0 = \mathbf{N} \frac{\mathrm{d}}{\mathrm{d}\nu}, \tag{2.21}$$

where ν measures distance in the $\mathbf{N}$ direction.

Let us now employ (2.19) – (2.21) to simplify the local forms of the conservation laws given in Section 2.4 to forms that will apply to the special class of shock-emulating steady-state smooth waves just described. Combining (2.19) with (2.21) gives

$$\dot{\psi} = -c_0 \frac{d\psi}{d\nu}. \tag{2.22}$$

The steady-state shock-emulating smooth wave form of compatibility (2.18) is determined by application of (2.21) to find

$$\dot{\mathbf{F}} = \mathbf{v}\overleftarrow{\nabla}_0 = \frac{d\mathbf{v}}{d\nu}\mathbf{N}, \tag{2.23}$$

from which it is immediately evident that

$$\boxed{\dot{\mathbf{F}} = (\dot{\mathbf{F}} \bullet \mathbf{N})\mathbf{N}.} \tag{2.24}$$

A useful alternative form is obtained by choosing $\psi = \mathbf{v}$ in (2.22) and applying this to the dot product of (2.23) with $\mathbf{N}$:

$$\dot{\mathbf{F}} \bullet \mathbf{N} = \frac{d\mathbf{v}}{d\nu} = -\frac{1}{c_0}\dot{\mathbf{v}}. \tag{2.25}$$

To derive the steady-state shock-emulating smooth wave form of linear momentum conservation (2.15), first apply (2.21) and then (2.22) with the choice $\psi = \mathbf{T}$ to show:

$$\nabla_0 \bullet \mathbf{T} = \mathbf{N} \bullet \frac{d\mathbf{T}}{d\nu} = -\frac{1}{c_0}\mathbf{N} \bullet \dot{\mathbf{T}}. \tag{2.26}$$

Now, assuming that no body forces act, application of (2.25) and (2.26) to (2.15) gives its steady-state smooth wave form:

$$\boxed{\mathbf{N} \bullet \dot{\mathbf{T}} = \rho_0 c_0^2 \, \dot{\mathbf{F}} \bullet \mathbf{N}.} \tag{2.27}$$

It will also prove useful in the upcoming analysis to have the material time rates of (2.24) and (2.27) which respectively are, using (2.14) and (2.20)

$$\ddot{\mathbf{F}} = (\ddot{\mathbf{F}} \bullet \mathbf{N})\mathbf{N} \tag{2.28}$$

$$\mathbf{N} \bullet \ddot{\mathbf{T}} = \rho_0 c_0^2 \, \ddot{\mathbf{F}} \bullet \mathbf{N}. \tag{2.29}$$

2.6 Shock Wave Forms of the Conservation Laws

As noted in the Introduction, shock waves in actual materials are almost always observed to be extremely narrow. Thus, the standard way to derive the conditions imposed by the conservation laws governing the jumps in fundamental field quantities (such as stress, velocity, etc.) across a shock is to idealize the shock as being a mathematical surface. In this section, we review this procedure, showing how to apply the integral forms of the conservation laws presented in Section 2.3 to derive the jump forms of these laws that must be satisfied across propagating surfaces of discontinuity under arbitrary conditions.

Figure 3(a) illustrates the reference configuration image of a surface of strong discontinuity, Σ_0, which propagates with speed c_0 in the direction of its unit normal vector $\mathbf{N}$. Consider a fixed subvolume of material V_0 enclosed by surface S_0 through which Σ_0 is currently moving. Since this is a fixed reference configuration, S_0 does not vary with time. Let V_0^+, which is enclosed by $S_0^+ \cup \Sigma_0$, denote the portion of V_0 that lies ahead of Σ_0, and similarly for V_0^- and $S_0^- \cup \Sigma_0$. We assume that all field quantities contained in the conservation laws (2.10) – (2.13) are differentiable in space and time everywhere except possibly across Σ_0.

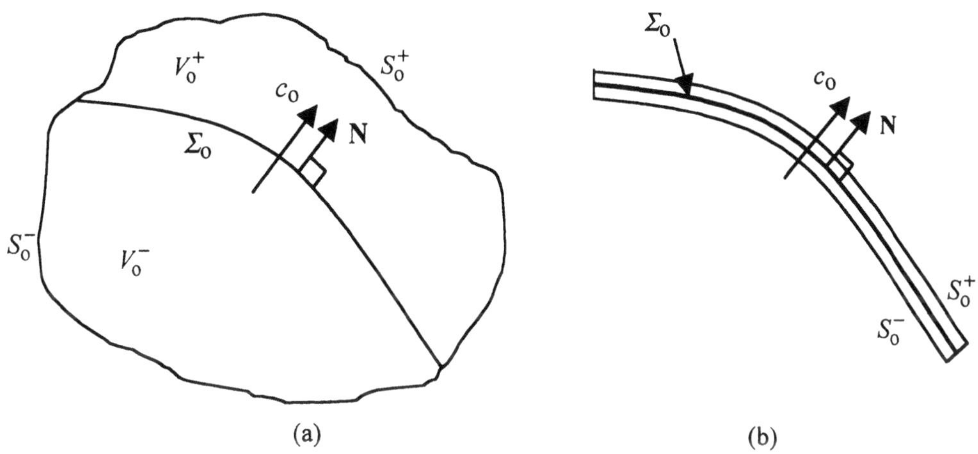

Figure 3. (a) Reference configuration material subvolume V_0 enclosed by surface S_0 through which Σ_0 is currently moving; (b) limit as surfaces collapse onto Σ_0.

Notice that all the integral forms of the conservation laws, (2.10) – (2.13), have a left-hand side that may be written generally as

$$\frac{\mathrm{d}}{\mathrm{d}t}\int_{V_0} \rho_0 \psi \mathrm{d}V_0 , \tag{2.30}$$

where here ψ denotes a tensor field of any order (scalar, vector, etc.). If the integral in (2.30) is applied to a fixed reference configuration subvolume of material V_0 inside which all field quantities are differentiable in space and time [that is, ***not*** containing a surface of strong discontinuity], then, since the integration region is fixed, we have

$$\frac{\mathrm{d}}{\mathrm{d}t}\int_{V_0} \rho_0 \psi \mathrm{d}V_0 = \int_{V_0} \frac{\mathrm{d}}{\mathrm{d}t}(\rho_0 \psi)\mathrm{d}V_0 . \tag{2.31}$$

However, if (2.30) is written for the subvolumes V_0^+ and V_0^- of Figure 3(a), (2.31) changes because of the motion of the surface portion Σ_0 of these subvolumes; the results are, respectively:

$$\frac{\mathrm{d}}{\mathrm{d}t}\int_{V_0^+}\rho_0\psi\mathrm{d}V_0 = \int_{V_0^+}\frac{\mathrm{d}}{\mathrm{d}t}\left(\rho_0\psi\right)\mathrm{d}V_0 - \int_{\Sigma_0}\left(\rho_0\psi\right)^+ c_0\mathrm{d}S \tag{2.32a}$$

$$\frac{\mathrm{d}}{\mathrm{d}t}\int_{V_0^-}\rho_0\psi\mathrm{d}V_0 = \int_{V_0^-}\frac{\mathrm{d}}{\mathrm{d}t}\left(\rho_0\psi\right)\mathrm{d}V_0 + \int_{\Sigma_0}\left(\rho_0\psi\right)^- c_0\mathrm{d}S. \tag{2.32b}$$

Adding these shows how (2.31) must be modified if V_0 contains a moving surface of strong discontinuity:

$$\frac{\mathrm{d}}{\mathrm{d}t}\int_{V_0}\rho_0\psi\mathrm{d}V_0 = \int_{V_0}\frac{\mathrm{d}}{\mathrm{d}t}\left(\rho_0\psi\right)\mathrm{d}V_0 - \int_{\Sigma_0}[\![\rho_0\psi]\!] c_0\mathrm{d}S, \tag{2.33}$$

where the volume in the right-side integral *excludes* Σ_0. Here and throughout the notes, ψ^+ and ψ^- denote the limiting values of ψ just ahead of and just behind Σ_0, respectively. The jump in these limiting values across Σ_0 is denoted as

$$[\![\psi]\!] \equiv \psi^+ - \psi^-, \tag{2.34}$$

and the average of these limiting values as

$$\overline{\psi} \equiv \frac{\psi^+ + \psi^-}{2}. \tag{2.35}$$

Since we consider the possibility that field quantities may jump across Σ_0, to obtain the jump forms of the integral conservation laws, we apply them to the subvolume V_0 of Figure 3(a) and employ the identity (2.33) to rewrite their left-hand sides. Applying this procedure to (2.10), (2.11) and (2.13) gives, respectively, where the volume integrals below *exclude* Σ_0:

$$\int_{V_0}\frac{\mathrm{d}}{\mathrm{d}t}\left(\rho_0\right)\mathrm{d}V_0 = \int_{\Sigma_0}[\![\rho_0]\!] c_0\mathrm{d}S \tag{2.36}$$

$$\int_{V_0}\left[\frac{\mathrm{d}}{\mathrm{d}t}\left(\rho_0\mathbf{v}\right) - \rho_0\mathbf{b}\right]\mathrm{d}V_0 = \int_{S_0}\mathbf{t}_0\mathrm{d}S_0 + \int_{\Sigma_0}[\![\rho_0\mathbf{v}]\!] c_0\mathrm{d}S \tag{2.37}$$

$$\int_{V_0} \left\{ \frac{\mathrm{d}}{\mathrm{d}t}\left[\rho_0\left(\frac{1}{2}\mathbf{v}\bullet\mathbf{v}+u\right)\right]-\rho_0\left(\mathbf{b}\bullet\mathbf{v}+r\right)\right\}\mathrm{d}V_0 = \int_{S_0}\left(\mathbf{t}_0\bullet\mathbf{v}-\mathbf{Q}\bullet\mathbf{N}\right)\mathrm{d}S_0 + \int_{\Sigma_0}\left[\!\left[\rho_0\left(\frac{1}{2}\mathbf{v}\bullet\mathbf{v}+u\right)\right]\!\right]c_0\mathrm{d}S. \tag{2.38}$$

Finally, these equations are evaluated in the limit as the surfaces S_0^+, S_0^- collapse onto Σ_0 in the manner illustrated in Figure 3(b); the results are:

$$\int_{\Sigma_0} [\![\,\rho_0]\!]\, c_0 \mathrm{d}S = 0 \tag{2.39}$$

$$\int_{\Sigma_0} [\![\,\mathbf{t}_0 + \rho_0 c_0 \mathbf{v}]\!]\,\mathrm{d}S = \mathbf{0} \tag{2.40}$$

$$\int_{\Sigma_0} \left[\!\left[\rho_0 c_0\left(\frac{1}{2}\mathbf{v}\bullet\mathbf{v}+u\right)+\mathbf{t}_0\bullet\mathbf{v}-\mathbf{Q}\bullet\mathbf{N}\right]\!\right]\mathrm{d}S = 0. \tag{2.41}$$

Since the above arguments must hold for any material subvolume V_0 containing a portion of Σ_0, the integrands of (2.39) – (2.41) must vanish pointwise along Σ_0. Thus, the jump requirements of (2.10) – (2.13) are:

$$\boxed{c_0[\![\,\rho_0]\!] = 0} \tag{2.42}$$

$$\boxed{[\![\,\mathbf{t}_0]\!] = -\rho_0 c_0 [\![\,\mathbf{v}]\!]} \tag{2.43}$$

$$\left[\!\left[\rho_0 c_0\left(\frac{1}{2}\mathbf{v}\bullet\mathbf{v}+u\right)+\mathbf{t}_0\bullet\mathbf{v}-\mathbf{Q}\bullet\mathbf{N}\right]\!\right] = 0. \tag{2.44}$$

Exercise: Show that (2.12) does not lead to any additional jump restriction.

It will be convenient to use a simplified form of (2.44). To derive this, we will need two compatibility results based on the requirement that the displacement field be continuous across Σ_0. The reasons for this requirement are physical: normal displacement continuity is required so that material cracking or interpenetration does not occur; tangential displacement continuity so that only finite work is produced by finite motion of the discontinuity surface.

The first compatibility result, apparently due to J. C. Maxwell in 1873, is derived as follows: since we require $[\![\mathbf{u}]\!] = \mathbf{0}$ all along Σ_0, it is clear that

$$[\![\mathbf{u}]\!]\overleftarrow{\nabla}_0 \bullet \mathrm{d}\mathbf{X} = [\![\mathbf{u}\overleftarrow{\nabla}_0 \bullet \mathrm{d}\mathbf{X}]\!] = \mathbf{0} \quad \text{for all } \mathrm{d}\mathbf{X} \text{ such that } \mathrm{d}\mathbf{X} \bullet \mathbf{N} = 0. \tag{2.45}$$

This obviously implies

$$[\![\mathbf{u}\overleftarrow{\nabla}_0]\!] = \boldsymbol{\lambda}\mathbf{N}, \tag{2.46}$$

where the right side is a dyad with $\boldsymbol{\lambda}$ an undetermined vector. Using (2.2) and (2.5), this may be rewritten as:

$$\underline{[\![\mathbf{F}]\!] = \left([\![\mathbf{F}]\!] \bullet \mathbf{N}\right)\mathbf{N}}. \tag{2.47}$$

The second result needed is Hadamard's (1903) kinematical compatibility condition:

$$\underline{[\![\mathbf{v}]\!] = -c_0[\![\mathbf{F}]\!] \bullet \mathbf{N}}. \tag{2.48}$$

This is another implication of the requirement that the displacement field be continuous across Σ_0. To derive (2.48), we introduce a set of coordinates $\boldsymbol{\xi}$ that propagates with Σ_0. Making the assumption that the space and time derivatives of $\mathbf{u}$ exist and approach finite limits as Σ_0 is approached, the chain rule then allows one to write the material time rate of the displacement vector field $\mathbf{u}(\boldsymbol{\xi}, t)$ on either side of Σ_0 as:

$$\begin{aligned}\mathbf{v} &= \frac{\mathrm{d}\mathbf{u}(\boldsymbol{\xi},t)}{\mathrm{d}t} = \left.\frac{\partial\mathbf{u}(\boldsymbol{\xi},t)}{\partial t}\right|_{\boldsymbol{\xi}} + \frac{\partial\mathbf{u}(\boldsymbol{\xi},t)}{\partial\boldsymbol{\xi}} \bullet \frac{\partial\boldsymbol{\xi}}{\partial t} = \left.\frac{\partial\mathbf{u}(\boldsymbol{\xi},t)}{\partial t}\right|_{\boldsymbol{\xi}} + \frac{\partial\mathbf{u}(\boldsymbol{\xi},t)}{\partial\boldsymbol{\xi}} \bullet \left(-c_0\mathbf{N}\right) \\ &= \left.\frac{\partial\mathbf{u}(\boldsymbol{\xi},t)}{\partial t}\right|_{\boldsymbol{\xi}} - c_0\mathbf{u}\overleftarrow{\nabla}_0 \bullet \mathbf{N} = \left.\frac{\partial\mathbf{u}(\boldsymbol{\xi},t)}{\partial t}\right|_{\boldsymbol{\xi}} - c_0\left(\mathbf{F}-\mathbf{I}\right) \bullet \mathbf{N},\end{aligned} \tag{2.49}$$

having employed (2.2) and (2.5); $\mathbf{I}$ is the second-rank identity tensor. Now, since $\mathbf{u}$ remains continuous across Σ_0, so must its time derivative following Σ_0, i.e., its time derivative holding $\boldsymbol{\xi}$ fixed; thus writing (2.49) just ahead of and just behind Σ_0 and subtracting gives (2.48).

Now, to simplify (2.44), observe that

$$[\![\tfrac{1}{2}\rho_0 c_0 \mathbf{v} \bullet \mathbf{v} + \mathbf{t}_0 \bullet \mathbf{v}]\!] = [\![\mathbf{t}_0 \bullet \mathbf{v}]\!] + \rho_0 c_0 [\![\mathbf{v}]\!] \bullet \bar{\mathbf{v}} = [\![\mathbf{t}_0 \bullet \mathbf{v}]\!] - [\![\mathbf{t}_0]\!] \bullet \bar{\mathbf{v}} = \bar{\mathbf{t}}_0 \bullet [\![\mathbf{v}]\!], \tag{2.50}$$

where (2.43) was employed to go from the second to third expression.

Thus, employing (2.8), (2.42), (2.48) and (2.50), (2.44) simplifies to:

$$\rho_0 c_0 [\![u]\!] = c_0 \mathbf{N} \bullet \overline{\mathbf{T}} \bullet [\![\mathbf{F}]\!] \bullet \mathbf{N} + \mathbf{N} \bullet [\![\mathbf{Q}]\!], \tag{2.51}$$

which, employing (2.47) can be written finally as

$$\llbracket u \rrbracket = \frac{1}{\rho_0}\overline{\mathbf{T}} : \llbracket \mathbf{F} \rrbracket + \frac{\mathbf{N} \bullet \llbracket \mathbf{Q} \rrbracket}{\rho_0 c_0}. \tag{2.52}$$

2.7 The Hugoniot

The Hugoniot function H is defined to be

$$H(\mathbf{T}, \mathbf{F}, u, \mathbf{Q}) \equiv (u - u_1) - \frac{1}{2\rho_0}(\mathbf{T} + \mathbf{T}_1) : (\mathbf{F} - \mathbf{F}_1) - \frac{\mathbf{N} \bullet (\mathbf{Q} - \mathbf{Q}_1)}{\rho_0 c_0}. \tag{2.53}$$

Regarding $\{\mathbf{T}, \mathbf{F}, u, \mathbf{Q}\}$ as the set of all states on one side of a shock front associated with a given STATE 1 $\{\mathbf{T}_1, \mathbf{F}_1, u_1, \mathbf{Q}_1\}$ on the other side, the shock wave form of the conservation of energy, (2.52), is then expressed by the condition

$$H = 0. \tag{2.54}$$

The set of solutions to the equation (2.54) is generally regarded as the set of all possible constrained equilibrium states achievable by a material particle once it passes through a shock.

3 Comparison of a Shock Wave with a Smooth Wave in Arbitrary Media

3.1 General Unsteady Weak Shock Waves

Here we generalize to three-dimensional finite deformations of arbitrary media an idea first proposed in the context of one-dimensional deformations of inviscid fluids by Courant and Friedrichs (1948). Their fascinating, pioneering idea was that one could explore how a shock wave and a smooth wave differed thermodynamically by taking increments of the Hugoniot function (valid for shock waves) and comparing them with increments of the smooth wave form of the first law.

Our related but generalized approach, due to Brannon *et al.* (1995), will be to calculate the order at which the First Law of thermodynamics differs between a shock wave and a smooth wave. This will be accomplished as follows. Recall that the Hugoniot function being equal to zero is the shock wave form of the First Law. To explore how a smooth wave differs thermodynamically from a shock wave, we shall apply the Hugoniot function to a smooth wave, and evaluate the order at which, and the manner in which, this differs from the shock wave value (zero). First, all the field variables appearing in the Hugoniot function are parameterized as functions of time. Then, we perform a Taylor expansion of the Hugoniot function about the state on one side of the smooth wave (STATE 1, attained at time $t = t_1$), to obtain its value at the other side of the wave, attained at time t:

$$H(t) = H_1 + (t - t_1)\dot{H}_1 + \tfrac{1}{2}(t - t_1)^2 \ddot{H}_1 + \tfrac{1}{6}(t - t_1)^3 \dddot{H}_1 + \dots, \tag{3.1}$$

where a subscript 1 denotes evaluation of the subscripted quantity at STATE 1, and superposed dots denote material time derivatives. This is illustrated schematically in Figure 4. This Taylor expansion approach is restricted to a comparison between *weak* shocks and smooth waves, since one would expect that a smooth wave chosen to emulate a strong shock would need to be far wider (and hence take much longer to pass a material point for a given shock speed) than one emulating a weak shock.

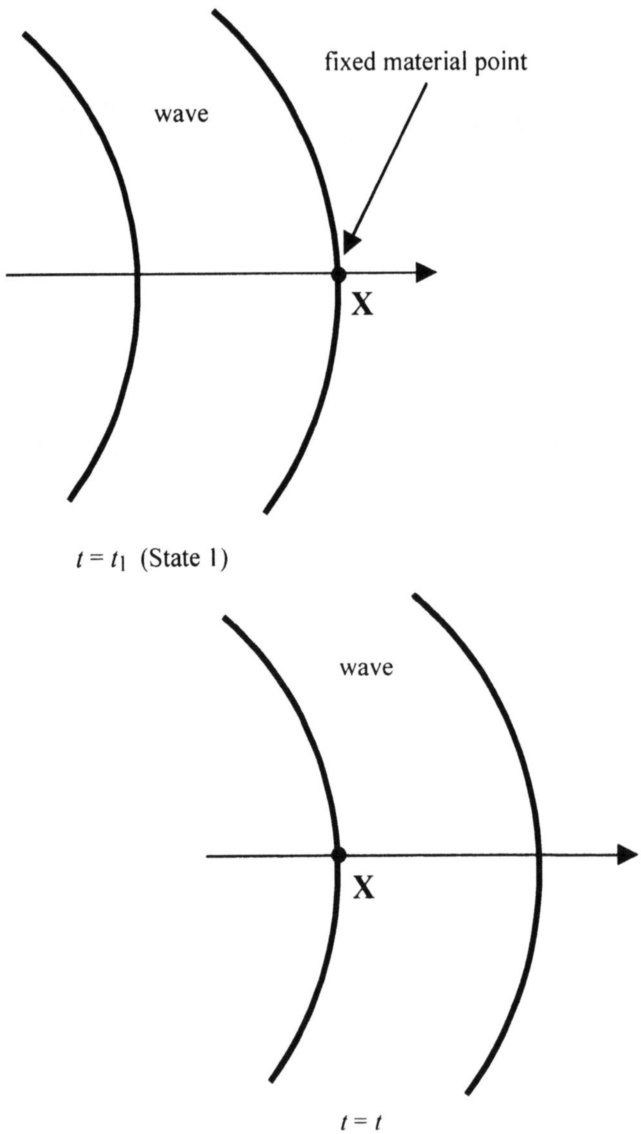

Figure 4. Reference configuration images at times t_1 and t of a smooth wave portion passing a fixed material point at which the Hugoniot function is evaluated.

The material time rate of the Hugoniot function is calculated from (2.53). Since we are evaluating the application of this to a smooth wave, the local forms (2.14) – (2.17) of the conservation laws apply; using (2.14) and (2.17) in particular (the latter to substitute for $\dot{u}$), we find

$$\dot{H}(t) = \frac{1}{2\rho_0}\left[\left(\mathbf{T}-\mathbf{T}_1\right):\dot{\mathbf{F}} - \dot{\mathbf{T}}:\left(\mathbf{F}-\mathbf{F}_1\right)\right] + r - \frac{1}{\rho_0 c_0}\left[c_0\,\nabla_0 \bullet \mathbf{Q} + \mathbf{N}\bullet\dot{\mathbf{Q}} + \left(\dot{\mathbf{N}} - \frac{\dot{c}_0}{c_0}\mathbf{N}\right)\bullet\left(\mathbf{Q}-\mathbf{Q}_1\right)\right]. \tag{3.2}$$

Higher-order material time rates of H are calculated from (3.2), applying (2.14).

Courant and Friedrichs (1948) assumed that the heat source and the heat flux were both zero; we shall assume zero heat source and at most constant heat flux. Under these assumptions, the Hugoniot function and its material time rates through third order are found from (2.53), (3.2) and its derivatives to be:

$$H(t) \equiv \left(u - u_1\right) - \frac{1}{2\rho_0}\left(\mathbf{T}+\mathbf{T}_1\right):\left(\mathbf{F}-\mathbf{F}_1\right) \tag{3.3a}$$

$$\dot{H}(t) = \frac{1}{2\rho_0}\left[\left(\mathbf{T}-\mathbf{T}_1\right):\dot{\mathbf{F}} - \dot{\mathbf{T}}:\left(\mathbf{F}-\mathbf{F}_1\right)\right] \tag{3.3b}$$

$$\ddot{H}(t) = \frac{1}{2\rho_0}\left[\left(\mathbf{T}-\mathbf{T}_1\right):\ddot{\mathbf{F}} - \ddot{\mathbf{T}}:\left(\mathbf{F}-\mathbf{F}_1\right)\right] \tag{3.3c}$$

$$\dddot{H}(t) = \frac{1}{2\rho_0}\left[\dot{\mathbf{T}}:\ddot{\mathbf{F}} - \ddot{\mathbf{T}}:\dot{\mathbf{F}} + \left(\mathbf{T}-\mathbf{T}_1\right):\dddot{\mathbf{F}} - \dddot{\mathbf{T}}:\left(\mathbf{F}-\mathbf{F}_1\right)\right]. \tag{3.3d}$$

At STATE 1, (3.3) show directly that H and its rates through third order are:

$$H_1 \equiv H\left(t_1\right) \equiv 0 \tag{3.4a}$$

$$\dot{H}_1 \equiv \dot{H}\left(t_1\right) = 0 \tag{3.4b}$$

$$\ddot{H}_1 \equiv \ddot{H}\left(t_1\right) = 0 \tag{3.4c}$$

$$\dddot{H}_1 \equiv \dddot{H}\left(t_1\right) = \frac{1}{2\rho_0}\left(\dot{\mathbf{T}}_1:\ddot{\mathbf{F}}_1 - \ddot{\mathbf{T}}_1:\dot{\mathbf{F}}_1\right). \tag{3.4d}$$

Since STATE 1 is arbitrary, the above results show that a general weak shock wave is thermodynamically identical to a smooth wave up to third order in material time rates of fundamental field quantities, at which point contributions due to the shock itself first appear. We emphasize that the above analysis has not employed any constitutive assumptions, and thus is valid for *arbitrary materials*.

3.2 Weak Shock Waves Propagating Under Steady-State Conditions

The conclusions for general unsteady weak shock waves derived in the previous section undergo a surprising and dramatic simplification when the shock wave is propagating under steady-state conditions. In this case, one compares the steadily propagating shock wave with a steadily propagating smooth wave. For such a smooth wave, the general local forms of the conservation laws and the compatibility condition simplify as shown in Section 2.5, when we make the additional assumptions that no body forces act, that the wave propagates without rotating with constant reference configuration speed c_0 in the direction of its normal, and since the objective is to explore how closely a smooth wave can resemble a shock wave, that field quantities vary through the wave but not parallel to it.

Recall now from the previous section that for general unsteady weak shock waves, the Taylor expansion of the Hugoniot function, (3.1) with (3.4), is zero through second order, but that the third-order term contains the following generally-nonzero coefficient:

$$\dddot{H}_1 = \frac{1}{2\rho_0}\left(\dot{\mathbf{T}}_1 : \ddot{\mathbf{F}}_1 - \ddot{\mathbf{T}}_1 : \dot{\mathbf{F}}_1\right). \tag{3.5}$$

Now, however, application of the steady-state compatibility condition (2.24) and its material time derivative (2.28) permit the two right-side terms in (3.5) to be written as:

$$\ddot{\mathbf{T}}_1 : \dot{\mathbf{F}}_1 = \ddot{\mathbf{T}}_1 : \left[\left(\dot{\mathbf{F}}_1 \bullet \mathbf{N}\right)\mathbf{N}\right] = \left(\mathbf{N} \bullet \ddot{\mathbf{T}}_1\right) \bullet \left(\dot{\mathbf{F}}_1 \bullet \mathbf{N}\right) \tag{3.6a}$$

$$\dot{\mathbf{T}}_1 : \ddot{\mathbf{F}}_1 = \dot{\mathbf{T}}_1 : \left[\left(\ddot{\mathbf{F}}_1 \bullet \mathbf{N}\right)\mathbf{N}\right] = \left(\mathbf{N} \bullet \dot{\mathbf{T}}_1\right) \bullet \left(\ddot{\mathbf{F}}_1 \bullet \mathbf{N}\right). \tag{3.6b}$$

Then, substitution of these into (3.5), followed by application of the steady-state form of linear momentum conservation (2.27) and its material time rate (2.29) gives:

$$\begin{aligned} \dddot{H}_1 &= \frac{1}{2\rho_0}\left[\left(\mathbf{N} \bullet \dot{\mathbf{T}}_1\right) \bullet \left(\ddot{\mathbf{F}}_1 \bullet \mathbf{N}\right) - \left(\mathbf{N} \bullet \ddot{\mathbf{T}}_1\right) \bullet \left(\dot{\mathbf{F}}_1 \bullet \mathbf{N}\right)\right] \\ &= \frac{1}{2\rho_0}\left[\left(\rho_0 c_0^2\, \dot{\mathbf{F}}_1 \bullet \mathbf{N}\right) \bullet \left(\ddot{\mathbf{F}}_1 \bullet \mathbf{N}\right) - \left(\rho_0 c_0^2\, \ddot{\mathbf{F}}_1 \bullet \mathbf{N}\right) \bullet \left(\dot{\mathbf{F}}_1 \bullet \mathbf{N}\right)\right] = 0. \end{aligned} \tag{3.7}$$

That is, under the steady-state propagation conditions described, the coefficient of the third-order term in the Taylor expansion of the Hugoniot function vanishes identically. Furthermore, it is straightforward to show, by extending the procedures illustrated above and in Section 2.5, that the fourth-order coefficient, and, by induction, the coefficients of all higher-order terms in the Taylor expansion, vanish identically for the class of steadily propagating smooth waves described. The conclusion is that for *steady-state* propagation of weak shock waves, a suitably chosen steadily-propagating smooth wave has a Taylor expansion of the Hugoniot function that vanishes identically. At the very least, this implies that a steady-state weak shock wave is much more closely approximated thermodynamically by a smooth wave than is a general nonsteady weak shock wave. However, we have not yet proven that a suitably-chosen smooth wave is thermodynamically *identical* to a steady-state shock wave, since smooth functions exist which are zero at a point and whose Taylor expansions are identically zero at that point, but which nevertheless become nonzero away from that point; a simple example is the function $\exp\left[-1/(t-t_1)^2\right]$.

3.3 Steady-State Shock Waves of Arbitrary Strength

The analysis of the previous section strongly suggests that under conditions of steady-state propagation, a suitably-chosen smooth wave may be thermodynamically *identical* to a shock wave. Here we shall examine the smooth wave – shock wave correspondence for steady-state conditions *without* use of a Taylor expansion; thus, the analysis will be valid for shocks of arbitrary strength. Our approach (Drugan, 1998) will be to integrate the smooth (local) form of the First Law of thermodynamics across a smooth wave, and compare the resulting jump condition with the shock wave form of the First Law.

From (2.17), the local form of the First Law, valid throughout a smooth wave under our assumptions of zero heat source and at most constant heat flux, is

$$\rho_0 \dot{u} = \mathbf{T} : \dot{\mathbf{F}}. \tag{3.8}$$

Let us integrate this at a fixed material point from the time of arrival ($t = t^+$) to the time of departure ($t = t^-$) of a smooth wave [see Figure 4, where we now call the departure time t^-], applying (2.14):

$$\int_{t^+}^{t^-} \dot{u}\,\mathrm{d}t = \frac{1}{\rho_0} \int_{t^+}^{t^-} \mathbf{T} : \dot{\mathbf{F}}\,\mathrm{d}t. \tag{3.9}$$

Note that for a general nonsteady wave, (3.9) is not directly integrable in general: more specific information is needed, and the result will differ for different waves. However, under the special conditions described in Section 2.5, namely steady-state propagation of a smooth wave with a non-rotating reference configuration image and field quantities that vary through the wave but not along it (and absent body forces), (3.9) *is* directly integrable for finite deformations and arbitrary materials, as will now be shown.

First, employing the steady-state form (2.24) of compatibility, the right-side integrand of (3.9) becomes

$$\mathbf{T}:\dot{\mathbf{F}}=\mathbf{T}:\left[\left(\dot{\mathbf{F}}\bullet\mathbf{N}\right)\mathbf{N}\right]=\left(\mathbf{N}\bullet\mathbf{T}\right)\bullet\left(\dot{\mathbf{F}}\bullet\mathbf{N}\right). \tag{3.10}$$

Next, we employ the steady-state form (2.27) of linear momentum conservation to substitute for $\dot{\mathbf{F}}\bullet\mathbf{N}$ in (3.10), which results in

$$\mathbf{T}:\dot{\mathbf{F}}=\frac{1}{\rho_0 c_0^2}\left(\mathbf{N}\bullet\mathbf{T}\right)\bullet\left(\mathbf{N}\bullet\dot{\mathbf{T}}\right). \tag{3.11}$$

Recall from (2.20) that $\dot{\mathbf{N}}=\mathbf{0}$ for the class of steady-state smooth waves considered, and define for convenience the vector $\mathbf{P}\equiv\mathbf{N}\bullet\mathbf{T}$. Then employing (2.14), (2.20) and (3.11), (3.9) becomes:

$$\int_{t^+}^{t^-}\dot{u}\,\mathrm{d}t=\frac{1}{\rho_0^2 c_0^2}\int_{t^+}^{t^-}\mathbf{P}\bullet\dot{\mathbf{P}}\,\mathrm{d}t. \tag{3.12}$$

This can now be integrated directly, giving:

$$u^- - u^+ = \frac{1}{2\rho_0^2 c_0^2}\left[\mathbf{P}^-\bullet\mathbf{P}^- - \mathbf{P}^+\bullet\mathbf{P}^+\right]=-\frac{1}{2\rho_0^2 c_0^2}\left(\mathbf{P}^+ + \mathbf{P}^-\right)\bullet\left(\mathbf{P}^+ - \mathbf{P}^-\right). \tag{3.13}$$

Next, we integrate linear momentum conservation (2.27) and compatibility (2.24) at a fixed material point just during passage of a smooth wave, using (2.14) and (2.20), to obtain:

$$\left(\mathbf{P}^+ - \mathbf{P}^-\right)=\rho_0 c_0^2\left(\mathbf{F}^+ - \mathbf{F}^-\right)\bullet\mathbf{N} \tag{3.14}$$

$$\left(\mathbf{F}^+ - \mathbf{F}^-\right)=\left[\left(\mathbf{F}^+ - \mathbf{F}^-\right)\bullet\mathbf{N}\right]\mathbf{N}. \tag{3.15}$$

Application of first (3.14) and then (3.15) to (3.13) leads to the final result

$$u^+ - u^- = \frac{1}{2\rho_0}\left[\mathbf{N}\bullet\left(\mathbf{T}^+ + \mathbf{T}^-\right)\right]\bullet\left[\left(\mathbf{F}^+ - \mathbf{F}^-\right)\bullet\mathbf{N}\right]=\frac{1}{2\rho_0}\left(\mathbf{T}^+ + \mathbf{T}^-\right):\left(\mathbf{F}^+ - \mathbf{F}^-\right). \tag{3.16}$$

Observe that this is identical to the shock wave form of the First Law, (2.52) when, as we have assumed, the heat flux is at most constant:

$$[\![u]\!] = \frac{1}{\rho_0} \overline{\mathbf{T}} : [\![\mathbf{F}]\!]. \tag{3.17}$$

Thus in this section we have shown directly that when the smooth form of the First Law of thermodynamics is integrated at a fixed material point just during passage of a steady-state non-rotating smooth wave of the type that emulates a shock wave (that is, one that has field quantities that vary through the wave but not along it, since for a shock any change in field quantities along the shock is negligible compared to their change across it), the result is *exactly* the *shock wave* form of the First Law. That is, we have proved here that the thermodynamic difference that exists in general between a shock wave and a smooth wave, as shown for example by the generally non-vanishing coefficient of the third-order term (3.4d) in the Taylor expansion of Section 3.1, *vanishes completely* for steady-state shock wave propagation *for general finite deformations and in arbitrary materials*, when the smooth wave is chosen as described in the preceding.

3.4 Implications for the Analysis of Restrictions on Shock Waves in Specific Material Types

We noted earlier that the analysis of shock waves under absolutely arbitrary conditions is an exceedingly difficult task, especially in solids, in part because this necessitates a realistic (fully thermomechanical) constitutive model of the material of interest that is valid under the extreme conditions experienced during shock wave passage. Such detailed and accurate thermomechanical constitutive models, especially those based on first principles rather than empiricisms, are scarce. We have shown in this section that there are at least two sets of circumstances under which the conditions governing the existence of, and restrictions on, permissible shock waves may be derived without the need for such sophisticated constitutive modeling. One such set of circumstances is when a general nonsteady shock wave is weak (i.e., involves only small jumps in the fundamental field variables such as stress, deformation gradient, velocity, internal energy); we showed in Section 3.1 that such a weak shock wave can be very accurately *approximated* by a suitably-chosen smooth wave, the thermodynamic difference between the two arising only at *third* order in the Taylor expansion of the Hugoniot function. The second set of circumstances is when a shock wave of arbitrary strength propagates under steady-state conditions; in this case, we showed in Section 3.3 that such a shock wave is thermodynamically *identical* to a suitably-chosen smooth wave. The importance of these findings is that, approximately in the weak-shock case and exactly in the steady-state shock case, it is sufficient to analyze a smooth wave to derive restrictions on shock waves. Since a material particle does not experience the extreme conditions it does during shock wave passage when it passes through a smooth wave, a fully thermomechanical constitutive model is not necessitated for the smooth wave case. Rather, a purely mechanical constitutive model suffices for smooth wave analysis. Thus, in the two sets of circumstances just reviewed, further shock wave restrictions [beyond those derived in Section 2.6 from the general conservation laws that apply to all media] due to specific types of constitutive response, can be derived by analyzing a smooth wave in a purely mechanical constitutive description of the material class under consideration. Restrictions of great specificity are possible especially in the steady-state shock wave case, because the analysis of the stress and deformation history of a material particle as it passes through the appropriately-chosen smooth wave model of the shock in this case is greatly aided by the application of the steady-state forms of the conservation laws and geometrical compatibility, as derived in Section 2.5.

In the following sections, we shall illustrate these concepts, and derive a multitude of explicit results, for steady-state shock wave propagation in a very broad class of rate-independent, purely mechanical elastic-plastic constitutive models.

4 Application to Steady-State Shock Waves in Elastic-Plastic Media: Small-Displacement-Gradient Formulation

We shall now illustrate the application of the just-proved principle that restrictions on a steady-state shock wave, beyond those required by the general conservation laws, can be derived for specific material classes by the analysis of a suitably-chosen smooth wave with a purely mechanical constitutive model describing the material class. The focus will be a very general class of rate-independent elastic-plastic solids that is inclusive of many structural materials of practical importance. In order to render the analysis as transparent as possible, to derive results whose meanings are lucid and easily interpreted, to employ well-accepted and exhaustively-studied constitutive restrictions that permit a vast range of material response including nearly arbitrary anisotropy, and to facilitate application of the results to several specific practical examples, we shall initially carry out the analysis within a small-displacement-gradient formulation.

4.1 Small-Displacement-Gradient Forms of the Local Conservation Laws and Their Steady-State Representations

Throughout this section, we shall employ a small-displacement-gradient formulation; namely, we shall simplify the governing equations by the assumption

$$\left|\mathbf{u}\overleftarrow{\nabla}_0\right| << 1 . \tag{4.1}$$

When this holds, the distinction between the reference and current configurations, described e.g. in Sections 2.1 and 2.2, is negligible, and it will be useful to recall that the deformation gradient admits the general representation [using (2.2) in (2.5)]

$$\mathbf{F} = \mathbf{I} + \mathbf{u}\overleftarrow{\nabla}_0 , \tag{4.2}$$

where $\mathbf{I}$ is the second-rank identity tensor. Using these equations, the conservation of mass, linear momentum and angular momentum, (2.14) (2.16), simplify to, assuming no body forces act:

$$\dot{\rho} = 0 \tag{4.3}$$

$$\nabla \bullet \boldsymbol{\sigma} = \rho \dot{\mathbf{v}} \tag{4.4}$$

$$\boldsymbol{\sigma} = \boldsymbol{\sigma}^T , \tag{4.5}$$

and the definition of the infinitesimal strain tensor, which may be regarded as the condition of geometrical compatibility, is:

$$\boldsymbol{\varepsilon} = \frac{1}{2}\left[\mathbf{u}\overleftarrow{\nabla} + \left(\mathbf{u}\overleftarrow{\nabla}\right)^T\right]. \tag{4.6}$$

In the above, $\boldsymbol{\sigma}$ is the stress tensor, $\boldsymbol{\varepsilon}$ the infinitesimal strain tensor, $\mathbf{u}$ the displacement vector, $\mathbf{v}$ the velocity vector, ρ the mass density and a superposed dot again denotes material time derivative.

Conditions (2.19) – (2.21) describing the steady-state, constant-speed propagation of a smooth, non-rotating wave with the shock-emulating property that quantities vary through but not along it become:

$$\dot{\psi} = -c\psi\overleftarrow{\nabla} \bullet \mathbf{n} \tag{4.7}$$

$$\dot{\mathbf{n}} = \mathbf{0}, \qquad \dot{c} = 0 \tag{4.8}$$

$$\nabla = \mathbf{n}\frac{\mathrm{d}}{\mathrm{d}\nu}, \tag{4.9}$$

where the wave propagates with speed c in the direction of its unit normal vector $\mathbf{n}$ and ν measures distance in the $\mathbf{n}$ direction. Combining (4.7) and (4.9) gives

$$\dot{\psi} = -c\frac{\mathrm{d}\psi}{\mathrm{d}\nu}. \tag{4.10}$$

Application of (4.10) and then (4.9) shows that

$$\nabla \bullet \boldsymbol{\sigma} = \mathbf{n} \bullet \frac{\mathrm{d}\boldsymbol{\sigma}}{\mathrm{d}\nu} = -\frac{1}{c}\mathbf{n} \bullet \dot{\boldsymbol{\sigma}}. \tag{4.11}$$

Using this, linear momentum conservation (4.4) is seen to have the steady-state form:

$$\mathbf{n} \bullet \dot{\boldsymbol{\sigma}} = -\rho c \dot{\mathbf{v}}, \tag{4.12}$$

which is, in incremental form at a fixed material point

$$\boxed{\mathbf{n}\bullet\mathrm{d}\boldsymbol{\sigma}=-\rho c\mathrm{d}\mathbf{v}.} \tag{4.13}$$

Next, choosing $\psi = \mathbf{u}$ in (4.10) shows

$$\mathbf{v}=\dot{\mathbf{u}}=-c\frac{\mathrm{d}\mathbf{u}}{\mathrm{d}\nu}, \tag{4.14}$$

the increment of which at a fixed material point, using (4.8), is

$$\mathrm{d}\mathbf{v}=-c\mathrm{d}\left(\frac{\mathrm{d}\mathbf{u}}{\mathrm{d}\nu}\right). \tag{4.15}$$

Observe that, via (4.9), the infinitesimal strain tensor may be expressed as

$$\boldsymbol{\varepsilon}=\frac{1}{2}\left[\left(\frac{\mathrm{d}\mathbf{u}}{\mathrm{d}\nu}\right)\mathbf{n}+\mathbf{n}\left(\frac{\mathrm{d}\mathbf{u}}{\mathrm{d}\nu}\right)\right]; \tag{4.16}$$

taking the increment of this at a fixed material point, using (4.8) and (4.15), gives

$$\boxed{\mathrm{d}\boldsymbol{\varepsilon}=-\frac{1}{2c}\left[(\mathrm{d}\mathbf{v})\mathbf{n}+\mathbf{n}(\mathrm{d}\mathbf{v})\right].} \tag{4.17}$$

This shows immediately that only three components of total strain can experience incremental change during steady-state smooth wave passage – all components of total strain lying in the plane normal to $\mathbf{n}$ cannot change.

It is forms (4.13) and (4.17), representing the requirements of linear momentum conservation and compatibility, respectively, in incremental form at a fixed material point, that will facilitate – when coupled with constitutive assumptions – derivation of additional restrictions on permissible stress and deformation histories during passage of a steady-state smooth wave (and therefore during passage of a steady-state shock wave). It is interesting to note that (4.13) and (4.17) are identical to the forms *assumed*, on the basis of heuristic arguments, to apply to a material particle during shock wave passage by Drugan and Shen (1987) and Leighton *et al.* (1987), and in the quasi-static case with the right side of (4.13) equaling zero by Drugan and Rice (1984). We have shown, following Drugan (1998), that these forms apply rigorously at a material particle during steady-state smooth wave passage, and that a suitably-chosen smooth wave rigorously represents a steady-state shock wave.

4.2 Weak Constitutive Assumptions Permitting a Broad Class of Rate-Independent Elastic-Plastic Response

Here we describe a rather skeletal set of constitutive assumptions that encompass a wide class of proven constitutive models characterizing many actual structural materials. Nevertheless, these constitutive features are sufficient to lead to some strong conclusions on permissible shock waves in this class of materials, as will be shown.

A total increment of strain $d\boldsymbol{\varepsilon}$ is assumed to decompose additively into elastic $d\boldsymbol{\varepsilon}^e$ and plastic $d\boldsymbol{\varepsilon}^p$ parts, and the elastic part is assumed to be linearly related to an increment in the stress tensor by the elastic compliance tensor $\mathbf{M}$:

$$\boxed{d\boldsymbol{\varepsilon} = d\boldsymbol{\varepsilon}^e + d\boldsymbol{\varepsilon}^p = \mathbf{M} : d\boldsymbol{\sigma} + d\boldsymbol{\varepsilon}^p.} \tag{4.18}$$

The elastic compliance tensor is assumed to be positive-definite and to have the usual index symmetries $M_{ijkl} = M_{jikl} = M_{ijlk} = M_{klij}$ due respectively to symmetry of the stress and strain tensors and to the thermodynamic requirement, for reversible elastic deformations that are either isothermal or adiabatic, of the existence of an elastic strain energy function.

The plastic part of the deformation will be assumed to satisfy the following inequality:

$$\boxed{d\boldsymbol{\sigma} : d\boldsymbol{\varepsilon}^p \geq 0.} \tag{4.19}$$

Physically, this requires that the current plastic strain increment be normal to the material's yield surface at smooth points on the surface, and lie within the cone of limiting normals at any point on the yield surface where a unique normal does not exist, i.e., that is a vertex. We remark that (4.19) is a much weaker restriction than the usual maximum plastic work inequality, which is similar in form to (4.19) except that $d\boldsymbol{\sigma}$ is not restricted to be infinitesimal. Thus, the additional implication of the maximum plastic work inequality – that the material's yield surface be convex – is not demanded by (4.19).

4.3 Restrictions on Steadily Propagating Shock Waves in Rate-Independent Elastic-Plastic Materials

We shall now derive the additional restrictions on steadily propagating shock waves in elastic-plastic materials, beyond those required by the conservation laws, due to the constitutive assumptions outlined in Section 4.2. This will be accomplished by employing the result proved in Section 3: shock waves that propagate under steady-state conditions without rotating are thermodynamically identical to suitably chosen steadily propagating smooth waves, so that restrictions on such shock waves can be derived by analyzing appropriately-selected smooth waves, for which purely mechanical constitutive modeling is sufficient.

Specifically, we will be interested in determining whether, and if so under what conditions, primary field quantities can jump across a steady-state shock wave. Based on what we have proved, this means we must analyze a steadily propagating smooth wave and determine whether the field quantity of interest can accumulate a change across the smooth wave. The minimum necessary condition for such an accumulation is that if one monitors a fixed material point during passage of a smooth wave, the field quantity of interest must experience an incremental change at some time increment during wave passage.

The equations restricting permissible increments at a fixed material point of the fundamental field variables during steady-state smooth wave passage, when the smooth wave is chosen to emulate a shock wave as described previously, are the requirements (4.13) and (4.17) of linear momentum conservation and compatibility, respectively, and the additional restrictions (4.18) and (4.19) of the elastic-plastic constitutive class to be analyzed.

Eliminating $d\boldsymbol{\varepsilon}^p$ in (4.19) via (4.18) gives

$$d\boldsymbol{\sigma} : \left(d\boldsymbol{\varepsilon} - \mathbf{M} : d\boldsymbol{\sigma}\right) \geq 0. \tag{4.20}$$

The implications of this become more clear when (4.17) is used to substitute for $d\boldsymbol{\varepsilon}$:

$$-\frac{1}{2c} d\boldsymbol{\sigma} : \left[\left(d\mathbf{v}\right)\mathbf{n} + \mathbf{n}\left(d\mathbf{v}\right)\right] - d\boldsymbol{\sigma} : \mathbf{M} : d\boldsymbol{\sigma} \geq 0, \tag{4.21}$$

which simplifies further when the requirement (4.5) of angular momentum conservation is enforced:

$$\underline{-\frac{1}{c} \mathbf{n} \bullet d\boldsymbol{\sigma} \bullet d\mathbf{v} - d\boldsymbol{\sigma} : \mathbf{M} : d\boldsymbol{\sigma} \geq 0.} \tag{4.22}$$

The restrictions due to (4.22) will be most clear if we consider different classes of deformations scparatcly.

4.3.1 Restrictions on steadily propagating shock waves in quasi-static deformations. By quasi-static deformations, we refer to those motions that are sufficiently slow that the inertia term in the conservation of linear momentum is negligible. In such situations, the steady-state smooth wave form of linear momentum conservation (4.13) reduces to

$$\mathbf{n} \bullet d\boldsymbol{\sigma} = \mathbf{0}. \tag{4.23}$$

Application of this to (4.22) leads to the requirement

$$-d\boldsymbol{\sigma} : \mathbf{M} : d\boldsymbol{\sigma} \geq 0. \tag{4.24}$$

Since $\mathbf{M}$ is positive-definite, (4.24) immediately shows

$$d\boldsymbol{\sigma} = \mathbf{0}; \tag{4.25}$$

that is, no change in any component of the stress tensor can accumulate during passage of a steady-state smooth wave in a quasi-static deformation. Application of this result to (4.18) shows that no elastic strain changes can accumulate either. Finally, this fact together with the fact that (4.17) rules out changes in total strain components lying in the plane normal to $\mathbf{n}$ means that the only strain components that can accumulate change during passage of a steady-state smooth wave in a quasi-static deformation are $\mathbf{n} \bullet \boldsymbol{\varepsilon}^p = \boldsymbol{\varepsilon}^p \bullet \mathbf{n}$.

Thus, applying the principle proved in Section 3, we conclude that across a steady-state non-rotating shock wave in quasi-static deformations in materials of the class delineated in Section 4.2, the jump restrictions are

$$\boxed{[\![\boldsymbol{\sigma}]\!] = \mathbf{0}, \quad [\![\boldsymbol{\varepsilon}^e]\!] = \mathbf{0}, \quad [\![\boldsymbol{\varepsilon}^p]\!] = \tfrac{1}{2}\left(\mathbf{an} + \mathbf{na}\right)}, \tag{4.26}$$

where $\mathbf{a}$ is an unrestricted vector and $\mathbf{an}$ is a dyad. These results were first derived by using certain (now-validated) ad hoc assumptions and for a smaller class of materials, by Drugan and Rice (1984); the present approach is Drugan's (1998).

4.3.2 Restrictions on steadily propagating shock waves in dynamic deformations. In general dynamic deformations, the full form (4.13) of linear momentum conservation applies. Using this to substitute for $\mathbf{dv}$ in (4.22) gives the general additional restriction

$$\boxed{\left(\mathbf{n} \bullet \mathrm{d}\boldsymbol{\sigma}\right) \bullet \left(\mathbf{n} \bullet \mathrm{d}\boldsymbol{\sigma}\right) - \rho c^2 \mathrm{d}\boldsymbol{\sigma} : \mathbf{M} : \mathrm{d}\boldsymbol{\sigma} \geq 0.} \tag{4.27}$$

This inequality restricts, but does not in general rule out, stress changes across a steadily propagating smooth wave.

However, in addition to (4.27), in order that $\mathrm{d}\boldsymbol{\sigma} \neq \mathbf{0}$ during any part of a steady-state smooth wave passage, when the wave is propagating at any speed other than one of the elastic wave speeds [given by the solutions of (4.31) below], the stress state must remain at yield. This is proved by contradiction: Suppose that $\mathrm{d}\boldsymbol{\sigma}$ takes (or maintains) the stress state below yield. Then, the corresponding strain increment must be purely elastic, so that from (4.18)

$$\mathrm{d}\boldsymbol{\varepsilon} = \mathrm{d}\boldsymbol{\varepsilon}^e = \mathbf{M} : \mathrm{d}\boldsymbol{\sigma} \quad \Rightarrow \quad \mathrm{d}\boldsymbol{\sigma} = \mathbf{C} : \mathrm{d}\boldsymbol{\varepsilon}, \tag{4.28}$$

where $\mathbf{C} = \mathbf{M}^{-1}$ is the elastic modulus tensor. Taking the dot product of (4.28) with $\mathbf{n}$ gives

$$\mathbf{n} \bullet \mathrm{d}\boldsymbol{\sigma} = \mathbf{n} \bullet \mathbf{C} : \mathrm{d}\boldsymbol{\varepsilon}, \tag{4.29}$$

which may be rewritten using the steady-state forms of momentum conservation (4.13) and compatibility (4.17), together with the symmetries of $\mathbf{C}$, as

$$\left(\mathbf{n} \bullet \mathbf{C} \bullet \mathbf{n} - \rho c^2 \mathbf{I}\right) \bullet \mathrm{d}\mathbf{v} = \mathbf{0}, \tag{4.30}$$

where $\mathbf{I}$ is the second-rank identity tensor. This permits $\mathrm{d}\mathbf{v} \neq \mathbf{0}$ only when

$$\det\left(\mathbf{n} \bullet \mathbf{C} \bullet \mathbf{n} - \rho c^2 \mathbf{I}\right) = 0. \tag{4.31}$$

The three solutions of this for c are the elastic wave speeds, and (4.17) and (4.28) show that $\mathrm{d}\mathbf{v} \neq \mathbf{0}$ is needed for $\mathrm{d}\boldsymbol{\sigma} \neq \mathbf{0}$. Thus, we conclude that $\mathrm{d}\boldsymbol{\sigma}$ must be such that $\boldsymbol{\sigma}$ remains at yield for $\mathbf{d\sigma} \neq \mathbf{0}$ during passage of a smooth wave whose propagation speed differs from an elastic wave speed.

We have shown that in order for a steady-state smooth wave to accumulate stress changes during a general dynamic deformation, it must exhibit during at least some portion of the wave a nonzero stress increment, $\mathbf{d\sigma} \neq \mathbf{0}$, which *both*: satisfies (4.27) *and* corresponds to $\boldsymbol{\sigma}$ remaining at yield (unless the wave is propagating at an elastic wave speed).

4.3.3 Restrictions on steadily propagating shock waves in dynamic deformations of ideally plastic materials. The requirements just derived in the previous section result in strong restrictions on shock waves when the material of interest is elastic-ideally plastic (that is, it does not harden nor soften with plastic deformation, so that the yield surface does not change during ongoing deformation). For such materials, the inequality (4.19) becomes a strict equality, which carries through the above analysis so that (4.27) becomes

$$\boxed{(\mathbf{n} \bullet \mathbf{d\sigma}) \bullet (\mathbf{n} \bullet \mathbf{d\sigma}) - \rho c^2 \mathbf{d\sigma} : \mathbf{M} : \mathbf{d\sigma} = 0.} \tag{4.32}$$

This restriction, together with the additional requirement proved in the previous section that the stress state must remain at yield during any portion of a steady-state smooth wave during which stress changes are accumulated, *rule out the possibility of steady-state shock waves propagating at speeds other than the elastic wave speeds in broad classes of dynamic deformations*. Generally speaking, this will be the case for any class of deformation that involves only two independent stress components and a yield condition that contains no linear segments. This is because, for such deformations, (4.32) will require that the increments of the two independent stress components be linearly related, and therefore if the yield condition does not contain any linear segments, both these stress component increments will have to vanish identically, so that no stress changes will be able to occur across any wave not propagating at an elastic wave speed. We now give two specific examples illustrating this. The above results, and the examples to follow, are generalized versions of those originally given by Drugan and Shen (1987).

Example 1. Anti-Plane Shearing

Consider the class of deformations that produce pure anti-plane shearing (and restricting ourselves to materials whose anisotropy permits this), namely, the only nonzero stress components are, in some Cartesian coordinate system, $\sigma_{13}(x_1, x_2) = \sigma_{31}$, $\sigma_{23}(x_1, x_2) = \sigma_{32}$. Without loss of generality, let us choose x_1 to lie in the $\mathbf{n}$-direction. In this situation, (4.32) reduces to

$$(d\sigma_{13})^2 - 4\rho c^2 \left[M_{1313}(d\sigma_{13})^2 + 2M_{1323} d\sigma_{13} d\sigma_{23} + M_{2323}(d\sigma_{23})^2 \right] = 0. \tag{4.33}$$

Regarding this as a quadratic equation for $d\sigma_{23}$, one solves it to obtain:

$$d\sigma_{23} = \frac{-1}{M_{2323}}\left\{M_{1323} \pm \left[M_{1323}^2 + M_{2323}\left(\frac{1}{4\rho c^2} - M_{1313}\right)\right]^{1/2}\right\}d\sigma_{13}. \tag{4.34}$$

This requires σ_{13} to vary linearly with σ_{23} across any steady-state smooth wave, but as shown above these must also satisfy the material's yield condition through any portion of the wave where a stress change may accumulate. Thus, unless the yield condition contains a linear portion with which (4.34) coincides, no stress changes can accumulate unless the wave propagates at an elastic wave speed. [For one simple example, both the Huber-Mises and Tresca yield conditions reduce, in the anti-plane shearing case, to the form $\sigma_{13}^2 + \sigma_{23}^2 = \tau_y^2$, where τ_y is the yield stress in pure shear. This yield condition clearly contains no linear portions.] This means, via (4.13) and (4.17), that no strain nor velocity changes can accumulate either. This shows that no steady-state shock wave is possible in anti-plane shearing deformations for elastic-ideally plastic materials satisfying the constitutive assumptions of Section 4.2 and having a yield condition with no linear portions, unless the shock wave propagates at an elastic wave speed.

Example 2. Plane Strain Deformations of Incompressible Materials

Now consider pure plane strain deformations of a slightly more restricted class of elastic-ideally plastic materials. We assume the material's anisotropy permits plane strain deformation in the x_1, x_2 plane with the only nonzero stress components being $\sigma_{11}(x_1,x_2)$, $\sigma_{12}(x_1,x_2) = \sigma_{21}$, $\sigma_{22}(x_1,x_2)$ and $\sigma_{33}(x_1,x_2)$. Assuming a smooth yield surface, inequality (4.19) requires normality of plastic strain increments to it, and we now assume the material to be elastically and plastically incompressible. These conditions mean that the stress-strain equations (4.18) can now be written as

$$d\boldsymbol{\varepsilon} = \mathbf{M} : d\mathbf{s} + d\Lambda \frac{\partial f(\mathbf{s})}{\partial \mathbf{s}}, \tag{4.35}$$

where $\mathbf{s} = \boldsymbol{\sigma} - \mathbf{I}\,\mathrm{tr}(\boldsymbol{\sigma})/3$ is the deviatoric stress tensor, $f(\mathbf{s}) = 0$ is the yield condition and $d\Lambda \geq 0$ is an undetermined proportionality factor. For maximum clarity of this example, while still retaining a reasonable degree of generality, we will restrict consideration to materials with quadratic yield surfaces whose anisotropy coincides with the material's elastic anisotropy; specifically, we assume:

$$f(\mathbf{s}) = \tfrac{1}{2}\mathbf{s} : (\alpha\mathbf{M}) : \mathbf{s} - \tau_y^2, \tag{4.36}$$

where α is a proportionality factor. Employing (4.36), (4.35) becomes

$$d\boldsymbol{\varepsilon} = \mathbf{M} : d\mathbf{s} + \alpha d\Lambda \mathbf{M} : \mathbf{s}. \tag{4.37}$$

Clearly, isotropic material with a Huber-Mises yield condition is a special case of this, with

$$M_{ijkl} = \frac{1}{2E}\left[-\delta_{ij}\delta_{kl} + \frac{3}{2}\left(\delta_{ik}\delta_{jl} + \delta_{il}\delta_{jk}\right)\right], \quad \alpha = \frac{2E}{3}, \tag{4.38}$$

where E is the Young (tensile) modulus.

First, we observe that the assumption of incompressible material shows from (4.17) that

$$\operatorname{trace}(\mathrm{d}\boldsymbol{\varepsilon}) = -\frac{1}{c}\mathbf{n} \bullet \mathrm{d}\mathbf{v} = 0 \quad \Rightarrow \quad \mathbf{n} \bullet \mathrm{d}\mathbf{v} = 0. \tag{4.39}$$

Employing this in (4.13) then shows

$$\mathbf{n} \bullet \mathrm{d}\boldsymbol{\sigma} \bullet \mathbf{n} = 0. \tag{4.40}$$

Next, the plane strain condition applied to (4.37) requires

$$\mathrm{d}\varepsilon_{33} = M_{33kl}\mathrm{d}s_{kl} + \alpha \mathrm{d}\Lambda M_{33kl}s_{kl} = 0. \tag{4.41}$$

This is a first-order ordinary differential equation for the scalar variable $M_{33kl}s_{kl}$, having the general solution

$$M_{33kl}s_{kl} = Ce^{-\alpha\Lambda}, \tag{4.42}$$

where C is an undetermined constant. To determine this, we consider deformations from an annealed or virgin state, so that initially there are no plastic strains nor plastic strain increments ($\mathrm{d}\Lambda = 0$) but the plane strain condition requires that $M_{33kl}s_{kl} = 0$; application of this initial condition to (4.42) shows that $C = 0$, and thus for all time

$$M_{33kl}s_{kl} = 0. \tag{4.43}$$

Using the fact that $s_{kk} = 0$, (4.43) can be rewritten as:

$$s_{12} = \frac{1}{2M_{3312}}\left[\left(M_{3333} - M_{3311}\right)s_{11} + \left(M_{3333} - M_{3322}\right)s_{22}\right]. \tag{4.44}$$

Let us now choose x_1 to lie in the $\mathbf{n}$-direction. Then (4.40) is

$$\mathrm{d}\sigma_{11} = 0. \tag{4.45}$$

Using (4.45), (4.32) for this case becomes, explicitly:

$$\left(M_{1111}+M_{3333}-2M_{1133}\right)\left(ds_{11}\right)^2+4\left(M_{1112}-M_{1233}\right)ds_{11}ds_{12}$$
$$+2\left(M_{1122}+M_{33333}-M_{1133}-M_{2233}\right)ds_{11}ds_{22}+4\left(M_{1222}-M_{1233}\right)ds_{12}ds_{22} \tag{4.46}$$
$$+\left(M_{2222}+M_{3333}-2M_{2233}\right)\left(ds_{22}\right)^2+\left(4M_{1212}-\frac{1}{\rho c^2}\right)\left(ds_{12}\right)^2=0.$$

Employing the incremental form of (4.44), it is straightforward to show that (4.46) can be written as:

$$A_1\left(ds_{11}\right)^2+A_2ds_{11}ds_{22}+A_3\left(ds_{22}\right)^2=0, \tag{4.47}$$

where A_1, A_2, A_3 are constants whose complicated forms are omitted. The important point is that (4.47) requires s_{11} to be linear in s_{22}, and this fact with (4.44) requires the same of s_{12}, throughout a steady-state smooth wave. However, as shown above, these must also satisfy the material's yield condition through any portion of the wave where a stress change may accumulate. Thus, unless the yield condition contains a linear portion with which (4.47) and (4.44) coincide, no stress changes can accumulate unless the wave propagates at an elastic wave speed. This means, via (4.13) and (4.17), that no strain nor velocity changes can accumulate either. This shows that no steady-state shock wave is possible in plane strain deformations of incompressible elastic-ideally plastic materials satisfying the constitutive assumptions of Section 4.2 (and those introduced at the beginning of this example) and having a yield condition with no linear portions, unless the shock wave propagates at an elastic wave speed.

4.3.4 Restrictions on steadily propagating shock waves in dynamic deformations of hardening elastic-plastic materials with smooth yield surfaces. Here, in addition to the weak constitutive assumptions of Section 4.2, we assume further that the yield surface is always smooth and of the general form

$$f(\boldsymbol{\sigma},\boldsymbol{\kappa})=0, \tag{4.48}$$

where $\boldsymbol{\kappa}$, a second-rank tensor, characterizes the (generally anisotropic) inelastic deformation history. We assume only that $\boldsymbol{\kappa}$ evolves as

$$d\boldsymbol{\kappa}=\mathbf{L}:d\boldsymbol{\varepsilon}^p, \tag{4.49}$$

with the fourth-rank tensor $\mathbf{L}=\mathbf{L}(\boldsymbol{\sigma},\boldsymbol{\varepsilon}^p,\boldsymbol{\kappa})$ being an arbitrary function of the variables indicated and perhaps others. Numerous proposed models of strain hardening are subsumed by this general structure. [The analysis to follow is easily generalized if $\boldsymbol{\kappa}$ represents a *collection* of tensors characterizing the inelastic deformation history.]

Constitutive assumption (4.19) requires normality of plastic strain increments to the yield surface, i.e., an associated flow rule:

$$\mathrm{d}\boldsymbol{\varepsilon}^p = \mathrm{d}\Lambda \frac{\partial f}{\partial \boldsymbol{\sigma}}. \tag{4.50}$$

Prager's consistency condition, that yield (4.48) must continue to be satisfied during ongoing plastic deformation, requires, using the chain rule:

$$\mathrm{d}f = \frac{\partial f}{\partial \boldsymbol{\sigma}} : \mathrm{d}\boldsymbol{\sigma} + \frac{\partial f}{\partial \boldsymbol{\kappa}} : \mathbf{L} : \mathrm{d}\boldsymbol{\varepsilon}^p = 0, \tag{4.51}$$

in light of our smoothness assumption on f and having employed (4.49).

The analysis to follow will be most conveniently performed in terms of the inverted form of the incremental stress-strain equations. First, from (4.18) and (4.50):

$$\mathrm{d}\boldsymbol{\varepsilon} = \mathbf{M} : \mathrm{d}\boldsymbol{\sigma} + \mathrm{d}\Lambda \frac{\partial f(\boldsymbol{\sigma}, \boldsymbol{\kappa})}{\partial \boldsymbol{\sigma}}. \tag{4.52}$$

To invert this, first take the double-dot product of both sides with $\mathbf{C} = \mathbf{M}^{-1}$ and solve for $\mathrm{d}\boldsymbol{\sigma}$:

$$\mathrm{d}\boldsymbol{\sigma} = \mathbf{C} : \mathrm{d}\boldsymbol{\varepsilon} - \mathrm{d}\Lambda \mathbf{C} : \frac{\partial f}{\partial \boldsymbol{\sigma}}. \tag{4.53}$$

To obtain an equation for $\mathrm{d}\Lambda$, we use (4.50) to substitute for $\mathrm{d}\boldsymbol{\varepsilon}^p$ in (4.51) and then the resulting equation to substitute for $\frac{\partial f}{\partial \boldsymbol{\sigma}} : \mathrm{d}\boldsymbol{\sigma}$ obtained by operating on both sides of (4.53) with $\frac{\partial f}{\partial \boldsymbol{\sigma}} :$. Solving this gives

$$\mathrm{d}\Lambda = \frac{\frac{\partial f}{\partial \boldsymbol{\sigma}} : \mathbf{C} : \mathrm{d}\boldsymbol{\varepsilon}}{\frac{\partial f}{\partial \boldsymbol{\sigma}} : \mathbf{C} : \frac{\partial f}{\partial \boldsymbol{\sigma}} - \frac{\partial f}{\partial \boldsymbol{\kappa}} : \mathbf{L} : \frac{\partial f}{\partial \boldsymbol{\sigma}}}, \tag{4.54}$$

so that (4.53) becomes:

$$\mathrm{d}\boldsymbol{\sigma} = \mathbf{C}^{ep} : \mathrm{d}\boldsymbol{\varepsilon}, \tag{4.55}$$

where we have defined the elastic-plastic modulus tensor

$$\mathbf{C}^{ep} \equiv \mathbf{C} - \frac{\left(\mathbf{C} : \frac{\partial f}{\partial \boldsymbol{\sigma}}\right)\left(\frac{\partial f}{\partial \boldsymbol{\sigma}} : \mathbf{C}\right)}{\left(\frac{\partial f}{\partial \boldsymbol{\sigma}} : \mathbf{C} : \frac{\partial f}{\partial \boldsymbol{\sigma}} - \frac{\partial f}{\partial \boldsymbol{\kappa}} : \mathbf{L} : \frac{\partial f}{\partial \boldsymbol{\sigma}}\right)}. \tag{4.56}$$

In the special case of nonhardening material, $\frac{\partial f}{\partial \boldsymbol{\kappa}} = \mathbf{0}$, so that the term containing $\mathbf{L}$ in the denominator of (4.56) vanishes in this case.

Now, to derive a necessary condition for shock wave existence, we apply compatibility (4.17) to substitute for $\mathrm{d}\boldsymbol{\varepsilon}$ in (4.55) and employ the index symmetry of $\mathbf{C}^{ep}$, $C^{ep}_{ijkl} = C^{ep}_{ijlk}$, to obtain

$$\mathrm{d}\boldsymbol{\sigma} = -\frac{1}{c}\left(\mathbf{C}^{ep} \bullet \mathbf{n}\right) \bullet \left(\mathrm{d}\mathbf{v}\right). \tag{4.57}$$

Taking the dot product of both sides of this with $\mathbf{n}$, using (4.13) and rearranging gives:

$$\left(\mathbf{n} \bullet \mathbf{C}^{ep} \bullet \mathbf{n} - \rho c^2 \mathbf{I}\right) \bullet \mathrm{d}\mathbf{v} = \mathbf{0}. \tag{4.58}$$

This shows that in order to accumulate a change in $\mathbf{v}$, we must have

$$\boxed{\det\left(\mathbf{n} \bullet \mathbf{C}^{ep} \bullet \mathbf{n} - \rho c^2 \mathbf{I}\right) = 0}. \tag{4.59}$$

Recall from (4.17) that a change in $\mathbf{v}$ is necessary for a change in $\boldsymbol{\varepsilon}$, which via (4.55) is necessary for a change in $\boldsymbol{\sigma}$. Thus, necessary conditions for existence of a steady-state shock wave propagating at a speed different from an elastic wave speed in elastic-plastic materials of the class described in Section 4.2 with the additional assumptions described at the beginning of this section, are that through at least some portion of a steady-state smooth wave in such material, (4.59) and the yield condition (4.48) [due to the proof in Section 4.3.2], together with the inequality (4.27), are satisfied.

Example: Isotropic material satisfying the Huber-Mises yield condition. Let us illustrate condition (4.59) with a specific example: isotropic material that satisfies the Huber-Mises yield condition. For such a material, the elastic modulus tensor $\mathbf{C}$ adopts the simple form

$$C_{ijkl} = \lambda \delta_{ij}\delta_{kl} + \mu\left(\delta_{ik}\delta_{jl} + \delta_{il}\delta_{jk}\right), \tag{4.60}$$

where λ, μ are the Lamé moduli and δ_{ij} is the Kronecker delta. The Huber-Mises yield condition is

$$f\left(\boldsymbol{\sigma}, \boldsymbol{\kappa}\right) = \frac{1}{2}\mathbf{s} : \mathbf{s} - \tau_Y^2\left(\kappa\right), \tag{4.61}$$

where $\mathbf{s}$ is the deviatoric stress tensor and τ_Y is the yield stress in pure shear, which is a constant for nonhardening material but which changes with plastic deformation for hardening material, and $\kappa = \mathrm{tr}(\boldsymbol{\kappa})/3$. For isotropic hardening,

$$L_{ijkl} = \delta_{ij}\sigma_{kl} \quad \Rightarrow \quad d\kappa = \mathrm{tr}(d\boldsymbol{\kappa})/3 = \boldsymbol{\sigma} : d\boldsymbol{\varepsilon}^{p}. \tag{4.62}$$

From (4.61) it is easily computed that

$$\frac{\partial f}{\partial \boldsymbol{\sigma}} = \mathbf{s}, \quad \frac{\partial f}{\partial \boldsymbol{\kappa}} = -\frac{2}{3}\tau_y(\kappa)\tau_y'(\kappa)\mathbf{I}, \tag{4.63}$$

where $\tau_y'(\kappa) \equiv d\tau_y / d\kappa$. Using (4.63), one calculates

$$\mathbf{C} : \frac{\partial f}{\partial \boldsymbol{\sigma}} = \frac{\partial f}{\partial \boldsymbol{\sigma}} : \mathbf{C} = 2\mu\mathbf{s}, \quad \frac{\partial f}{\partial \boldsymbol{\sigma}} : \mathbf{C} : \frac{\partial f}{\partial \boldsymbol{\sigma}} = 4\mu\tau_y^2,$$
$$\frac{\partial f}{\partial \boldsymbol{\kappa}} : \mathbf{L} : \frac{\partial f}{\partial \boldsymbol{\sigma}} = -4\tau_y^3\tau_y'(\kappa). \tag{4.64}$$

Then, choosing $\mathbf{n}$ to lie in the 1-direction without loss of generality, (4.56) with (4.64) show that

$$\left(\mathbf{n} \bullet \mathbf{C}^{ep} \bullet \mathbf{n}\right)_{ij} = (\lambda + \mu)\delta_{1i}\delta_{1j} + \mu\delta_{ij} - D s_{1i} s_{1j}, \tag{4.65}$$

where we have defined

$$D = \frac{\mu^2}{\tau_y^2\left[\mu + \tau_y\tau_y'(\kappa)\right]}. \tag{4.66}$$

We then calculate from (4.65) that (4.59) takes the explicit form:

$$\left(\mu - \rho c^2\right)\left\{\left(\mu - \rho c^2\right)\left(\lambda + 2\mu - \rho c^2\right) - D\left[\left(\mu - \rho c^2\right)s_{11}^2 + \left(\lambda + 2\mu - \rho c^2\right)\left(s_{12}^2 + s_{13}^2\right)\right]\right\} = 0. \tag{4.67}$$

Thus, for this example, (4.61) and (4.67) must both be satisfied through some portion of a smooth wave in order for changes in $\mathbf{v}$, $\boldsymbol{\varepsilon}$ or $\boldsymbol{\sigma}$ to accumulate. When both can be satisfied, (4.67) is solved to determine the admissible shock wave speeds c; clearly, the non-elastic ones will depend on the stress state and the hardening state.

4.4 Restrictions Continue to Hold for Finite Deviations from Plastic Normality

Restrictions on propagating steady-state shock waves were derived in Section 4.3 for rate-independent elastic-plastic materials within the broad class delineated by the rather skeletal set of constitutive assumptions outlined in Section 4.2. Here we will show that the restrictions derived continue to hold even when one of the key constitutive assumptions made – normality of plastic strain increments to the yield surface – is relaxed by a limited but finite amount. The following is a modification of ideas originally developed by Brannon and Drugan (1993).

Specifically, we shall relax condition (4.19), which was that

$$\mathbf{d}\boldsymbol{\sigma} : \mathbf{d}\boldsymbol{\varepsilon}^{p} \geq 0. \tag{4.68}$$

As noted earlier, this condition requires plastic strain increments to be normal to the yield surface at smooth points. This means that, assuming the yield surface is smooth (the following arguments are easily generalized to permit yield surface vertices), the plastic part of the incremental stress-strain relations (4.18) can be expressed as:

$$\mathrm{d}\boldsymbol{\varepsilon}^{p} = \mathrm{d}\Lambda \frac{\partial f}{\partial \boldsymbol{\sigma}}, \tag{4.69}$$

where $f = 0$ is the yield surface at the current state and $\mathrm{d}\Lambda \geq 0$.

When (4.68) is relaxed, (4.69) must be generalized to permit a component of $\mathrm{d}\boldsymbol{\varepsilon}^{p}$ that is tangent to the yield surface:

$$\mathrm{d}\boldsymbol{\varepsilon}^{p} = \mathrm{d}\Lambda \frac{\partial f}{\partial \boldsymbol{\sigma}} + \mathbf{M}_{1} : \mathrm{d}\boldsymbol{\sigma}, \tag{4.70}$$

where for hardening materials Euler's theorem for homogeneous functions guarantees the existence of the unique fourth-rank tensor $\mathbf{M}_1$, which by definition satisfies $\frac{\partial f}{\partial \boldsymbol{\sigma}} : \left(\mathbf{M}_1 : \mathrm{d}\boldsymbol{\sigma}\right) = 0$ (this simply means that $\mathbf{M}_1 : \mathrm{d}\boldsymbol{\sigma}$ is directed tangent to the yield surface); for nonhardening materials the existence of $\mathbf{M}_1$ is assumed. Thus, (4.70) now replaces (4.69), and (4.68) is replaced by the requirement that the normal component of the plastic strain increment be directed *outward* from the yield surface:

$$\mathrm{d}\Lambda \frac{\partial f}{\partial \boldsymbol{\sigma}} : \mathrm{d}\boldsymbol{\sigma} \geq 0. \tag{4.71}$$

Using (4.70), the full incremental elastic-plastic stress-strain relations (4.18) can be written as

$$\mathrm{d}\boldsymbol{\varepsilon} = \mathrm{d}\boldsymbol{\varepsilon}^{e} + \mathrm{d}\boldsymbol{\varepsilon}^{p} = \left(\mathbf{M} + \mathbf{M}_1\right) : \mathrm{d}\boldsymbol{\sigma} + \mathrm{d}\Lambda \frac{\partial f}{\partial \boldsymbol{\sigma}}. \tag{4.72}$$

Taking the scalar product of this with $\mathrm{d}\boldsymbol{\sigma}$ and then applying (4.71) gives:

$$\mathrm{d}\boldsymbol{\sigma} : \mathrm{d}\boldsymbol{\varepsilon} - \mathrm{d}\boldsymbol{\sigma} : \left(\mathbf{M} + \mathbf{M}_1\right) : \mathrm{d}\boldsymbol{\sigma} \geq 0. \tag{4.73}$$

Use of the steady-state form of compatibility (4.17) to substitute for $\mathrm{d}\boldsymbol{\varepsilon}$, and enforcement of angular momentum conservation (4.5), show that (4.73) can be expressed as:

$$\boxed{-\frac{1}{c}\,\mathbf{n} \bullet \mathrm{d}\boldsymbol{\sigma} \bullet \mathrm{d}\mathbf{v} - \mathrm{d}\boldsymbol{\sigma} : \left(\mathbf{M} + \mathbf{M}_1\right) : \mathrm{d}\boldsymbol{\sigma} \geq 0.} \tag{4.74}$$

This is the generalization of Condition (4.22) for materials whose plastic strain increments deviate from being purely normal to the yield surface. We shall now show that the restrictions derived on steady-state shock waves in Section 4.3 still hold for such materials, provided that the deviations from normality of plastic strain increments are limited (although not merely infinitesimal), in a manner to be made precise.

Restrictions on steadily propagating shock waves in quasi-static deformations for elastic-plastic materials deviating from normality. For quasi-static deformations, we again have the reduced form (4.23) of steady-state linear momentum conservation. Application of this to (4.74) gives:

$$-d\boldsymbol{\sigma} : \left(\mathbf{M} + \mathbf{M}_1\right) : d\boldsymbol{\sigma} \geq 0. \tag{4.75}$$

Recall that we have assumed the elastic compliance tensor $\mathbf{M}$ to be positive-definite. Thus, (4.75) shows that our previously derived restriction (4.25), namely $d\boldsymbol{\sigma} = \mathbf{0}$, meaning that no change in any component of the stress tensor can accumulate during passage of a steady-state smooth wave in a quasi-static deformation, will continue to hold provided that any deviations from plastic normality are such that $(\mathbf{M} + \mathbf{M}_1)$ remains positive-definite. For all such materials, the arguments following (4.25) in Section 4.3.1 still apply, so that the final strong restrictions (4.26) on potential steady-state shock waves still hold. This means that for materials whose elastic compliance tensor $\mathbf{M}$ is strongly positive-definite, quite significant deviations from plastic normality, characterized by $\mathbf{M}_1$, will not alter the strong restrictions (4.26) derived on steady-state shock waves in quasi-static deformations.

5. Application to Steady-State Shock Waves in Elastic-Plastic Media: Finite Deformation Formulation

We shall now apply the principle proved in Section 3 (that restrictions on a steady-state shock wave beyond those required by the general conservation laws can be derived for specific material classes by the analysis of a suitably-chosen smooth wave using a purely mechanical constitutive model) to a very general class of rate-independent elastic-plastic solids within a full finite deformation formulation. This analysis will generalize to large deformations the results derived in Section 4; it is a generalization of the analysis of Drugan and Shen (1990).

5.1 Weak Constitutive Assumptions Permitting a Broad Class of Rate-Independent Elastic-Plastic Response

Here we describe a skeletal set of finite deformation constitutive assumptions that encompass a wide class of constitutive models characterizing many actual structural materials. Nevertheless, these constitutive features are sufficient to lead to some strong restrictions on permissible shock waves, as will be shown.

We consider elastic-plastic materials for which the elastic part of the total response is hyperelastic, while the plastic part is assumed to be rate-independent. As first argued for large deformations by Lee (1969), the total deformation gradient tensor is assumed to admit a multiplicative decomposition into elastic and plastic parts

$$\mathbf{F} = \mathbf{F}^e \bullet \mathbf{F}^p, \tag{5.1}$$

where we employ the interpretations suggested by Willis (1969): $\mathbf{F}^p$ represents the residual (plastic) deformation after a material element has been elastically unloaded and returned to its original orientation, and hence $\mathbf{F}^e$ accounts for the elastic stretch and any rigid-body rotation. Note that these identifications arise naturally in rigorous treatments of crystal plasticity; see, e.g., Rice (1971), Hill and Rice (1972), Havner (1982) and Peirce *et al.* (1982).

The rate-of-deformation tensor $\mathbf{D}$ is expressible as

$$\mathbf{D} = sym\left(\frac{\partial \mathbf{v}}{\partial \mathbf{x}}\right) = sym\left(\dot{\mathbf{F}} \bullet \mathbf{F}^{-1}\right), \tag{5.2}$$

where sym denotes the symmetric part. Substitution of (5.1) into (5.2) shows that $\mathbf{D}$ admits the following additive decomposition:

$$\mathbf{D} = \mathbf{D}^e + \mathbf{D}^p, \tag{5.3}$$

where

$$\mathbf{D}^e = sym\left(\dot{\mathbf{F}}^e \bullet \mathbf{F}^{e^{-1}}\right), \quad \mathbf{D}^p = sym\left(\mathbf{F}^e \bullet \dot{\mathbf{F}}^p \bullet \mathbf{F}^{p^{-1}} \bullet \mathbf{F}^{e^{-1}}\right). \tag{5.4}$$

These expressions show that $\mathbf{D}^e$ and $\mathbf{D}^p$ measure the elastic and plastic parts, respectively, of the rate of stretching with respect to the current lattice orientation in a plastic crystal, as discussed by Peirce *et al.* (1982).

Consistent with this interpretation of the kinematics, let us choose $\mathbf{F} = \mathbf{F}^p$ as the ground state for the elastic response. Thus, we define an elastic Green strain tensor $\mathbf{E}^e$ as

$$\mathbf{E}^e = \frac{1}{2}\left(\mathbf{F}^{e^T} \bullet \mathbf{F}^e - \mathbf{I}\right), \tag{5.5}$$

where $\mathbf{I}$ is the second-rank identity tensor. Our assumption that the elastic part of the response is hyperelastic is then expressed by

$$\mathbf{S} = \frac{\partial W(\mathbf{E}^e)}{\partial \mathbf{E}^e}, \tag{5.6}$$

where we have assumed that the elastic part of the material response is unaffected by plastic shearing (Hill and Rice, 1972), W is the elastic strain energy density function and $\mathbf{S}$ is the

Second Piola-Kirchhoff stress tensor measured from the state $\mathbf{F}=\mathbf{F}^p$, that is

$$\mathbf{S}=\mathbf{F}^{e^{-1}}\bullet\left(J^e\boldsymbol{\sigma}\right)\bullet\mathbf{F}^{e^{-T}}, \tag{5.7}$$

with $J^e=\det(\mathbf{F}^e)$.

We shall restrict attention to materials, such as metals, that exhibit only small elastic stretches, namely

$$\left\|\mathbf{E}^e\right\|=\sqrt{\mathbf{E}^e:\mathbf{E}^e}<<1. \tag{5.8}$$

Given (5.8), it is a sensible assumption that the elastic modulus tensor $\partial^2 W/\partial\mathbf{E}^e\partial\mathbf{E}^e$ is positive-definite, or, equivalently, that $W(\mathbf{E}^e)$ is strictly convex:

$$W(\mathbf{E}_1^e)-W(\mathbf{E}_2^e)-\left.\frac{\partial W(\mathbf{E}^e)}{\partial\mathbf{E}^e}\right|_{\mathbf{E}_1^e}:\left(\mathbf{E}_1^e-\mathbf{E}_2^e\right)\leq 0, \tag{5.9}$$

where $\mathbf{E}_1^e$ and $\mathbf{E}_2^e$ differ in general but the equality in (5.9) holds only when they are equal.

Now, taking increments of (5.6) gives

$$\mathrm{d}\mathbf{S}=\frac{\partial^2 W(\mathbf{E}^e)}{\partial\mathbf{E}^e\partial\mathbf{E}^e}:\mathrm{d}\mathbf{E}^e\equiv\mathbf{C}:\mathrm{d}\mathbf{E}^e, \tag{5.10}$$

so that inverting this gives

$$\mathrm{d}\mathbf{E}^e=\mathbf{M}:\mathrm{d}\mathbf{S},\qquad \mathbf{M}\equiv\mathbf{C}^{-1}. \tag{5.11}$$

Our principal assumption on the inelastic part of the material response is that inelastic deformation increments are normal to the yield surface at smooth points, and lie within the cone of limiting normals at yield surface vertices if such are present. This statement must be expressed in the proper variables for arbitrary finite deformations. First, we observe that the yield surface at a given material state cannot most generally be represented in terms of Cauchy stress, as $f(\boldsymbol{\sigma})=0$ say, because the principle of material frame indifference (i.e., objectivity – invariance to rigid-body rotations) then requires f to satisfy $f(\boldsymbol{\sigma})=f(\mathbf{Q}\bullet\boldsymbol{\sigma}\bullet\mathbf{Q}^T)$ with $\mathbf{Q}$ being an arbitrary proper orthogonal tensor, which is true in general only if f describes isotropic material response. For a more general representation, recall that $\mathbf{F}^e$ contains the (small) elastic stretch and any rigid-body rotation, and employ the polar decomposition theorem to write this as

$$\mathbf{F}^e=\mathbf{V}^e\bullet\mathbf{R}=\left(\mathbf{I}+\boldsymbol{\varepsilon}\right)\bullet\mathbf{R}, \tag{5.12}$$

where $\mathbf{R}^T = \mathbf{R}^{-1}$, $\det(\mathbf{R}) = 1$, $\mathbf{V}^e = \mathbf{V}^{e^T}$, $\boldsymbol{\varepsilon} = \boldsymbol{\varepsilon}^T$, with $\|\boldsymbol{\varepsilon}\| << 1$. Thus, to include materials that experience arbitrarily anisotropic yield, we phrase the yield condition in terms of the rotationless Cauchy stress $\boldsymbol{\tau}$:

$$\boldsymbol{\tau} \equiv \mathbf{R}^T \bullet \boldsymbol{\sigma} \bullet \mathbf{R}. \tag{5.13}$$

Since $\boldsymbol{\tau}$ is invariant to rigid-body rotation, the form of the yield condition expressed in terms of it, $f(\boldsymbol{\tau}) = 0$, need not be restricted further to satisfy objectivity. In terms of this stress measure, the requirement of plastic normality is stated as (using the plastic deformation rate measure that is work-conjugate to $\boldsymbol{\tau}$):

$$d\boldsymbol{\tau} : \left(\mathbf{R}^T \bullet \mathbf{D}^p \bullet \mathbf{R}\right) \geq 0, \tag{5.14}$$

where $d\boldsymbol{\tau}$ is any infinitesimal increment in $\boldsymbol{\tau}$ lying within or on the current yield locus and ending at the current $\boldsymbol{\tau}$ state. In view of our assumption of infinitesimal elastic stretches, and making the additional assumption that plastic compressibility is negligible, (5.14) multiplied by J (noting that $J > 0$ always) may be restated as:

$$d\mathbf{S} : \left(\mathbf{F}^{e^T} \bullet \mathbf{D}^p \bullet \mathbf{F}^e\right) \geq 0. \tag{5.15}$$

Using the fact that $\mathbf{S}$ is symmetric, (5.15) can be rewritten using (5.3) with (5.4) as

$$d\mathbf{S} : \left[\mathbf{F}^{e^T} \bullet \left(\dot{\mathbf{F}} \bullet \mathbf{F}^{-1} - \dot{\mathbf{F}}^e \bullet \mathbf{F}^{e^{-1}}\right) \bullet \mathbf{F}^e\right] = d\mathbf{S} : \left[\mathbf{F}^{e^T} \bullet \dot{\mathbf{F}} \bullet \mathbf{F}^{-1} \bullet \mathbf{F}^e - \mathbf{F}^{e^T} \bullet \dot{\mathbf{F}}^e\right] \geq 0. \tag{5.16}$$

Again using the fact that $\mathbf{S}$ is symmetric, and taking increments of (5.5), (5.16) can be rewritten in incremental form (i.e., having multiplied by dt) as

$$d\mathbf{S} : \left(\mathbf{F}^{e^T} \bullet d\mathbf{F} \bullet \mathbf{F}^{-1} \bullet \mathbf{F}^e\right) - d\mathbf{S} : d\mathbf{E}^e \geq 0. \tag{5.17}$$

Next, using (5.11) and rearranging, (5.17) becomes

$$\left(\mathbf{F}^{-1} \bullet \mathbf{F}^e \bullet d\mathbf{S} \bullet \mathbf{F}^{e^T}\right) : d\mathbf{F} - d\mathbf{S} : \mathbf{M} : d\mathbf{S} \geq 0. \tag{5.18}$$

Because of the presence of the $\mathbf{F}^e$ (or, equivalently within our assumptions, $\mathbf{R}$) terms in (5.18), it does not appear that use of the steady-state finite deformation forms of the conservation laws, derived in Section 2.5, will enable us to derive shock wave restrictions from (5.18) for arbitrarily anisotropic materials that are as explicit as those derived in Section 4.3 within a small-displacement-gradient formulation.

Consequently, we henceforth restrict consideration to *isotropic* materials. As noted above, for such materials the yield condition *can* be expressed simply as an isotropic function of

Cauchy stress, without further restrictions due to objectivity requirements. Then, in terms of this stress measure, the requirement of plastic normality (5.14) simplifies to (using the plastic deformation rate measure that is work-conjugate to $\boldsymbol{\sigma}$):

$$d\boldsymbol{\sigma} : \mathbf{D}^p \geq 0, \tag{5.19}$$

where $d\boldsymbol{\sigma}$ is any infinitesimal increment in $\boldsymbol{\sigma}$ lying within or on the current yield locus and ending at the current $\boldsymbol{\sigma}$ state. Using (5.3) with (5.4) and the symmetry of $\boldsymbol{\sigma}$, (5.19) can be rewritten as, multiplying by dt:

$$d\boldsymbol{\sigma} : \left(d\mathbf{F} \bullet \mathbf{F}^{-1}\right) - d\boldsymbol{\sigma} : \left(d\mathbf{F}^e \bullet \mathbf{F}^{e^{-1}}\right) \geq 0. \tag{5.20}$$

Next, since we assume isotropy, note that the strain energy density function satisfies the condition $W(\mathbf{E}^e) = W(\mathbf{R} \bullet \mathbf{E}^e \bullet \mathbf{R}^T)$; employing this and the chain rule, (5.6) becomes

$$\mathbf{S} = \frac{\partial W(\boldsymbol{\eta}^e)}{\partial \mathbf{E}^e} = \frac{\partial W(\boldsymbol{\eta}^e)}{\partial \boldsymbol{\eta}^e} : \frac{\partial \boldsymbol{\eta}^e}{\partial \mathbf{E}^e} = \mathbf{R}^T \bullet \frac{\partial W(\boldsymbol{\eta}^e)}{\partial \boldsymbol{\eta}^e} \bullet \mathbf{R}, \tag{5.21}$$

where we have defined

$$\boldsymbol{\eta}^e \equiv \mathbf{R} \bullet \mathbf{E}^e \bullet \mathbf{R}^T. \tag{5.22}$$

Thus from (5.21),

$$\boldsymbol{\sigma} \approx \mathbf{R} \bullet \mathbf{S} \bullet \mathbf{R}^T = \frac{\partial W(\boldsymbol{\eta}^e)}{\partial \boldsymbol{\eta}^e} \quad \Rightarrow \quad d\boldsymbol{\sigma} = \frac{\partial^2 W(\boldsymbol{\eta}^e)}{\partial \boldsymbol{\eta}^e \partial \boldsymbol{\eta}^e} : d\boldsymbol{\eta}^e \equiv \mathbf{C} : d\boldsymbol{\eta}^e \quad \Rightarrow \quad d\boldsymbol{\eta}^e = \mathbf{M} : d\boldsymbol{\sigma}. \tag{5.23}$$

Returning to (5.20), note from (5.12) that, since $d\boldsymbol{\sigma}$ is symmetric and $d\mathbf{R} \bullet \mathbf{R}^T$ is antisymmetric, and $\mathbf{V}^e \approx \mathbf{I}$ (although we do not assume $d\mathbf{V}^e \approx \mathbf{0}$)

$$\begin{aligned} d\boldsymbol{\sigma} : \left(d\mathbf{F}^e \bullet \mathbf{F}^{e^{-1}}\right) &= d\boldsymbol{\sigma} : \left[d\left(\mathbf{V}^e \bullet \mathbf{R}\right) \bullet \mathbf{R}^T \bullet \mathbf{V}^{e^{-1}}\right] \\ &= d\boldsymbol{\sigma} : \left[d\mathbf{V}^e \bullet \mathbf{V}^{e^{-1}} + \mathbf{V}^e \bullet d\mathbf{R} \bullet \mathbf{R}^T \bullet \mathbf{V}^{e^{-1}}\right] \\ &\approx d\boldsymbol{\sigma} : \left(d\mathbf{V}^e \bullet \mathbf{V}^{e^{-1}}\right) = d\boldsymbol{\sigma} : d\boldsymbol{\eta}^e. \end{aligned} \tag{5.24}$$

Employing (5.23) and (5.24), and rearranging, (5.20) becomes

$$\boxed{\left(\mathbf{F}^{-1} \bullet \mathrm{d}\boldsymbol{\sigma}\right) : \mathrm{d}\mathbf{F} - \mathrm{d}\boldsymbol{\sigma} : \mathbf{M} : \mathrm{d}\boldsymbol{\sigma} \geq 0.} \tag{5.25}$$

5.2 Restrictions on Steadily Propagating Shock Waves in Isotropic Rate-Independent Elastic-Plastic Materials at Finite Deformation

To employ the constitutive assumptions of the previous section to determine restrictions on steady-state shock waves in isotropic rate-independent elastic-plastic materials within a finite deformation formulation, we will employ our usual procedure of analyzing a steady-state smooth wave. We derived in Section 2.5 the steady-state local forms of the conservation laws that will be applicable here. Writing these in incremental form (by multiplying the Section 2.5 versions by $\mathrm{d}t$), we have from linear momentum conservation (2.27)

$$\boxed{\mathbf{N} \bullet \mathrm{d}\mathbf{T} = \rho_0 c_0^2 \mathrm{d}\mathbf{F} \bullet \mathbf{N},} \tag{5.26}$$

and from compatibility (2.24)

$$\boxed{\mathrm{d}\mathbf{F} = (\mathrm{d}\mathbf{F} \bullet \mathbf{N})\mathbf{N}.} \tag{5.27}$$

We shall also need another implication of (5.27). Recall that the standard formula for the inverse of $\mathbf{F}$ is

$$F_{ij}^{-1} = \frac{C_{ji}}{J}, \tag{5.28}$$

where C_{ij} is the cofactor (signed minor subdeterminant) of the element F_{ij} in $\mathbf{F}$, and $J = \det(\mathbf{F})$. Multiplying (5.28) by J and taking increments gives

$$\mathrm{d}\left(JF_{ij}^{-1}\right) = \mathrm{d}C_{ji}. \tag{5.29}$$

Application of (5.27) to this results in (**Exercise:** Show this.)

$$\boxed{\mathbf{N} \bullet \mathrm{d}\left(J\mathbf{F}^{-1}\right) = \mathbf{0}.} \tag{5.30}$$

We shall assume that the materials under consideration are plastically incompressible; this, together with our assumption of small elastic stretches means that $J \approx 1$. Then application of (5.27), (5.30) and (2.9), respectively, to the first term of (5.25) shows:

$$\begin{aligned}\left(\mathbf{F}^{-1} \bullet \mathrm{d}\boldsymbol{\sigma}\right) : \mathrm{d}\mathbf{F} &= \left(J\mathbf{F}^{-1} \bullet \mathrm{d}\boldsymbol{\sigma}\right) : \left[(\mathrm{d}\mathbf{F} \bullet \mathbf{N})\mathbf{N}\right] = \left[\left(\mathbf{N} \bullet J\mathbf{F}^{-1}\right) \bullet \mathrm{d}\boldsymbol{\sigma}\right] \bullet (\mathrm{d}\mathbf{F} \bullet \mathbf{N}) \\ &= \mathrm{d}\left(J\mathbf{N} \bullet \mathbf{F}^{-1} \bullet \boldsymbol{\sigma}\right) \bullet (\mathrm{d}\mathbf{F} \bullet \mathbf{N}) = (\mathbf{N} \bullet \mathrm{d}\mathbf{T}) \bullet (\mathrm{d}\mathbf{F} \bullet \mathbf{N}).\end{aligned} \tag{5.31}$$

Thus, (5.25) becomes

$$\boxed{\left(\mathbf{N}\bullet\mathrm{d}\mathbf{T}\right)\bullet\left(\mathrm{d}\mathbf{F}\bullet\mathbf{N}\right)-\mathrm{d}\boldsymbol{\sigma}:\mathbf{M}:\mathrm{d}\boldsymbol{\sigma}\geq 0.} \tag{5.32}$$

5.2.1 Restrictions on steadily propagating shock waves in quasi-static deformations. For quasi-static deformations, linear momentum conservation (5.26) simplifies to

$$\mathbf{N}\bullet\mathrm{d}\mathbf{T}=\mathbf{0}. \tag{5.33}$$

Application of this to (5.32) immediately reduces the latter to

$$-\mathrm{d}\boldsymbol{\sigma}:\mathbf{M}:\mathrm{d}\boldsymbol{\sigma}\geq 0. \tag{5.34}$$

Since $\mathbf{M}$ is positive-definite, (5.34) immediately shows

$$\mathrm{d}\boldsymbol{\sigma}=\mathbf{0}; \tag{5.35}$$

that is, no change in any component of the *Cauchy* stress tensor can accumulate during passage of a steady-state smooth wave in a quasi-static deformation. Application of this result to (5.23) shows that no changes in $\boldsymbol{\eta}^e$ can accumulate either, which means that $\mathbf{V}^e$ of (5.12) cannot accumulate change. Finally, using this fact and (5.12) and recalling that (5.27) rules out changes in $\mathbf{F}$ components lying in the plane normal to $\mathbf{N}$, we conclude that the only deformation components that can accumulate change during passage of a wave are $(\mathbf{R}\bullet\mathbf{F}^p)\bullet\mathbf{N}$.

Thus we conclude that across a steady-state non-rotating shock wave in quasi-static deformations in isotropic materials of the class delineated in Section 5.1, the jump restrictions are

$$\boxed{[\![\boldsymbol{\sigma}]\!]=\mathbf{0},\quad [\![\boldsymbol{\eta}^e]\!]=\mathbf{0},\quad [\![\mathbf{F}]\!]=\mathbf{V}^e\bullet[\![\mathbf{R}\bullet\mathbf{F}^p]\!]\bullet\mathbf{N}.} \tag{5.36}$$

5.2.2 Restrictions on steadily propagating shock waves in dynamic deformations. In general dynamic deformations, the full form (5.26) of linear momentum conservation applies. Using this to substitute for $\mathrm{d}\mathbf{F}\bullet\mathbf{N}$ in (5.32) gives the general additional restriction

$$\left(\mathbf{N}\bullet\mathrm{d}\mathbf{T}\right)\bullet\left(\mathbf{N}\bullet\mathrm{d}\mathbf{T}\right)-\rho_0 c_0^2\,\mathrm{d}\boldsymbol{\sigma}:\mathbf{M}:\mathrm{d}\boldsymbol{\sigma}\geq 0. \tag{5.37}$$

Now, from (2.9) and (5.30), note that

$$\mathbf{N}\bullet\mathrm{d}\mathbf{T}=\mathbf{N}\bullet\left(J\mathbf{F}^{-1}\right)\bullet\mathrm{d}\boldsymbol{\sigma}. \tag{5.38}$$

There is no loss in generality if we select $J\mathbf{F}^{-1} = \mathbf{I}$ at the front of the wave; using (5.30), this means that $\mathbf{N} \bullet (J\mathbf{F}^{-1}) = \mathbf{N}$ throughout the wave; thus, (5.38) shows

$$\mathbf{N} \bullet \mathrm{d}\mathbf{T} = \mathbf{N} \bullet \mathrm{d}\boldsymbol{\sigma} \tag{5.39}$$

throughout the wave, so that (5.37) can be written as

$$\boxed{(\mathbf{N} \bullet \mathrm{d}\boldsymbol{\sigma}) \bullet (\mathbf{N} \bullet \mathrm{d}\boldsymbol{\sigma}) - \rho_o c_o^2\, \mathrm{d}\boldsymbol{\sigma} : \mathbf{M} : \mathrm{d}\boldsymbol{\sigma} \geq 0.} \tag{5.40}$$

This inequality restricts, but does not in general rule out, stress changes across a steadily propagating smooth wave. It is the finite deformation version of (4.27), with the important facts that here $\boldsymbol{\sigma}$ is the *Cauchy* stress, and (5.40) only applies to *isotropic* material.

In addition to (5.40), in order that $\mathrm{d}\boldsymbol{\sigma} \neq \mathbf{0}$ during any portion of a steady-state smooth wave passage, when the wave is propagating at any speed other than one of the elastic wave speeds, the stress state must remain at yield. As in the small-displacement-gradient formulation, this is proved by contradiction. Enough subtleties arise in the present finite deformation case to make outlining this version of the proof worthwhile. Suppose that $\mathrm{d}\boldsymbol{\sigma}$ takes (or maintains) the stress state below yield. Then, the corresponding deformation increment, $\mathbf{D}\mathrm{d}t$, must be purely elastic. Thus, from (5.23), (5.24), (5.2) and (5.3), we have, repeatedly using the index symmetry $C_{ijkl} = C_{ijlk}$ of $\mathbf{C}$:

$$\mathrm{d}\boldsymbol{\sigma} = \mathbf{C} : \mathrm{d}\boldsymbol{\eta}^e = \mathbf{C} : \mathbf{D}^e \mathrm{d}t = \mathbf{C} : \mathbf{D}\mathrm{d}t = \mathbf{C} : \left(\mathrm{d}\mathbf{F} \bullet \mathbf{F}^{-1}\right). \tag{5.41}$$

Next, application of compatibility in version (5.27) and then version (5.30), employing the observation made before (5.39), shows

$$\mathbf{C} : \left(\mathrm{d}\mathbf{F} \bullet \mathbf{F}^{-1}\right) = \mathbf{C} : \left[(\mathrm{d}\mathbf{F} \bullet \mathbf{N})\mathbf{N} \bullet \mathbf{F}^{-1}\right] = \mathbf{C} : \left[(\mathrm{d}\mathbf{F} \bullet \mathbf{N})\mathbf{N}\right] = (\mathbf{C} \bullet \mathbf{N}) \bullet (\mathrm{d}\mathbf{F} \bullet \mathbf{N}). \tag{5.42}$$

Taking the dot product of (5.41) with $\mathbf{N}$, then employing (5.39) and (5.42) results in

$$\mathbf{N} \bullet \mathrm{d}\mathbf{T} = (\mathbf{N} \bullet \mathbf{C} \bullet \mathbf{N}) \bullet (\mathrm{d}\mathbf{F} \bullet \mathbf{N}). \tag{5.43}$$

Application of linear momentum conservation (5.26) to this shows finally

$$\left(\mathbf{N} \bullet \mathbf{C} \bullet \mathbf{N} - \rho_0 c_0^2 \mathbf{I}\right) \bullet (\mathrm{d}\mathbf{F} \bullet \mathbf{N}) = \mathbf{0}, \tag{5.44}$$

which permits $\mathrm{d}\mathbf{F} \bullet \mathbf{N} \neq \mathbf{0}$ only when

$$\det\left(\mathbf{N} \bullet \mathbf{C} \bullet \mathbf{N} - \rho_0 c_0^2 \mathbf{I}\right) = 0, \tag{5.45}$$

that is, only when c_0 is one of the elastic wave speeds. Note from (5.27) that $\mathbf{dF} \bullet \mathbf{N} \neq \mathbf{0}$ is needed for nonzero $\mathbf{dF}$, which in turn (5.41) shows is required for nonzero $\mathbf{d}\boldsymbol{\sigma}$. Thus, we have shown that $\mathbf{d}\boldsymbol{\sigma}$ must be such that $\boldsymbol{\sigma}$ remains at yield for $\mathbf{d}\boldsymbol{\sigma} \neq \mathbf{0}$ during passage of a smooth wave not propagating at an elastic wave speed. Therefore, just as in the small-displacement-gradient case, any portion of a smooth wave during which $\mathbf{d}\boldsymbol{\sigma} \neq \mathbf{0}$ must satisfy *both* (5.40) and the yield condition. This means that shock waves propagating at speeds other than elastic wave speeds in nonhardening materials [for which (5.40) becomes an equality] can be ruled out in cases having only two independent stress components and a yield condition not containing a linear portion, as illustrated explicitly in the small-displacement-gradient case.

5.3 Derivation of a Quasi-Thermodynamic Shock Dissipation Condition for Very General Rate-Independent Elastic-Plastic Materials

Here we shall consider a far broader class of rate-independent elastic-plastic materials than in the previous parts of this section, and derive for this class a jump dissipation condition for steady-state shock waves required by the Second Law of thermodynamics. Specifically, the constitutive assumptions will consist of those laid out in Section 5.1 up through (5.7), which permit both elastic and plastic deformations of arbitrary magnitudes; we assume in addition only that the rate of plastic work per unit reference volume at every stage in the deformation history is nonnegative:

$$J\boldsymbol{\sigma} : \mathbf{D}^p \geq 0. \tag{5.46}$$

This is the usually-invoked requirement of nonnegative plastic dissipation.

Rice (1971, 1975) has shown that the Second Law of thermodynamics, in the Clausius-Duhem inequality form, leads for isothermal processes to a weaker restriction than (5.46): namely, that the plastic work rate must exceed the rate of increase of locked-in energy from the material. Examples of constitutive features that involve a *negative* locked-in energy rate include (Rice, 1975): a (severe) Bauschinger effect that commences during unloading in a tensile test while the stress is still acting in tension; and time-dependent strain recovery after reduction of the load on a specimen to a small but still tensile value. Our present constitutive formulation is time-independent, and if we assume that our model excludes other constitutive features that would lead to the release of locked-in energy during deformation, such as a severe Bauschinger effect, (5.46) follows directly from the Second Law.

To derive a jump dissipation inequality implied by (5.46) for steady-state shock waves, we shall again use our approach of analyzing a steady-state smooth wave that emulates a shock. We begin by noting that the integration of (5.46) at a fixed material point just during passage of such a smooth wave, i.e. from the time of the wave's arrival, t^+, to that of its departure, t^-, requires

$$\int_{t^+}^{t^-} J\boldsymbol{\sigma} : \mathbf{D}^p \mathrm{d}t \quad \geq 0. \tag{5.47}$$

Using (5.2) – (5.4), this can be rewritten as, making use of the symmetry of $\boldsymbol{\sigma}$:

$$\int_{\mathbf{F}^+}^{\mathbf{F}^-} J\boldsymbol{\sigma} : \left(\mathrm{d}\mathbf{F} \bullet \mathbf{F}^{-1}\right) - \int_{\mathbf{F}^{e^+}}^{\mathbf{F}^{e^-}} J\boldsymbol{\sigma} : \left(\mathrm{d}\mathbf{F}^e \bullet \mathbf{F}^{e^{-1}}\right) \geq 0. \tag{5.48}$$

To evaluate the second integral in (5.48), rearrange and then apply (5.5) – (5.7):

$$\begin{aligned}\int_{\mathbf{F}^{e^+}}^{\mathbf{F}^{e^-}} J\boldsymbol{\sigma} : \left(\mathrm{d}\mathbf{F}^e \bullet \mathbf{F}^{e^{-1}}\right) &= \int_{\mathbf{F}^{e^+}}^{\mathbf{F}^{e^-}} \left(J\mathbf{F}^{e^{-1}} \bullet \boldsymbol{\sigma}\right) : \left[\left(\mathbf{F}^{e^{-T}} \bullet \mathbf{F}^{e^T}\right) \bullet \mathrm{d}\mathbf{F}^e\right] \\ &= \int_{\mathbf{E}^{e^+}}^{\mathbf{E}^{e^-}} \mathbf{S} : \mathrm{d}\mathbf{E}^e = W\left(\mathbf{E}^{e^-}\right) - W\left(\mathbf{E}^{e^+}\right) = -[\![W(\mathbf{E}^e)]\!].\end{aligned} \tag{5.49}$$

The first integrand in (5.48) can be rewritten by, in order: rearranging, using the stress tensor relation (2.9), employing compatibility (5.27) and rearranging again

$$\begin{aligned}J\boldsymbol{\sigma} : \left(\mathrm{d}\mathbf{F} \bullet \mathbf{F}^{-1}\right) &= \left(J\mathbf{F}^{-1} \bullet \boldsymbol{\sigma}\right) : \mathrm{d}\mathbf{F} = \mathbf{T} : \mathrm{d}\mathbf{F} = \mathbf{T} : \left[\left(\mathrm{d}\mathbf{F} \bullet \mathbf{N}\right)\mathbf{N}\right] \\ &= \left(\mathbf{N} \bullet \mathbf{T}\right) \bullet \left(\mathrm{d}\mathbf{F} \bullet \mathbf{N}\right).\end{aligned} \tag{5.50}$$

5.3.1 Quasi-static deformations. For the case of quasi-static deformations, the conservation of linear momentum simplifies from (5.26) to (5.33), meaning that $\mathbf{N} \bullet \mathbf{T}$ does not vary through the wave. Using this in (5.50), the first integral in (5.48) is thus:

$$\int_{\mathbf{F}^+}^{\mathbf{F}^-} \left(\mathbf{N} \bullet \mathbf{T}\right) \bullet \left(\mathrm{d}\mathbf{F} \bullet \mathbf{N}\right) = \left(\mathbf{N} \bullet \mathbf{T}\right) \bullet \int_{\mathbf{F}^+}^{\mathbf{F}^-} \left(\mathrm{d}\mathbf{F} \bullet \mathbf{N}\right) = -\left(\mathbf{N} \bullet \mathbf{T}\right) \bullet [\![\mathbf{F} \bullet \mathbf{N}]\!]. \tag{5.51}$$

Therefore, from (5.49) and (5.51), for quasi-static deformations (5.47) leads to the following jump dissipation inequality:

$$\boxed{[\![W(\mathbf{E}^e)]\!] - \mathbf{N} \bullet \mathbf{T} \bullet [\![\mathbf{F} \bullet \mathbf{N}]\!] \geq 0.} \tag{5.52}$$

This was derived by Drugan and Shen (1990), and by a completely different method for purely hyperelastic materials by Knowles (1979).

5.3.2 General dynamic deformations. For the general dynamic deformation case, the full form (5.26) of the conservation of linear momentum is employed to substitute for $\mathbf{dF} \bullet \mathbf{N}$ in (5.50); the first integral in (5.48) thus becomes:

$$\begin{aligned}\int_{\mathbf{F}^+}^{\mathbf{F}^-} J\boldsymbol{\sigma} : \left(\mathrm{d}\mathbf{F} \bullet \mathbf{F}^{-1}\right) &= \int_{\mathbf{F}^+}^{\mathbf{F}^-} (\mathbf{N} \bullet \mathbf{T}) \bullet (\mathrm{d}\mathbf{F} \bullet \mathbf{N}) = \frac{1}{\rho_0 c_0^2} \int_{\mathbf{T}^+}^{\mathbf{T}^-} (\mathbf{N} \bullet \mathbf{T}) \bullet (\mathbf{N} \bullet \mathrm{d}\mathbf{T}) \\ &= -\frac{1}{2\rho_0 c_0^2}\left[(\mathbf{N} \bullet \mathbf{T})^+ \bullet (\mathbf{N} \bullet \mathbf{T})^+ - (\mathbf{N} \bullet \mathbf{T})^- \bullet (\mathbf{N} \bullet \mathbf{T})^-\right] \\ &= -\frac{1}{2\rho_0 c_0^2}\left[(\mathbf{N} \bullet \mathbf{T})^+ + (\mathbf{N} \bullet \mathbf{T})^-\right] \bullet [\![\mathbf{N} \bullet \mathbf{T}]\!] \\ &= -\frac{1}{2}\left[(\mathbf{N} \bullet \mathbf{T})^+ + (\mathbf{N} \bullet \mathbf{T})^-\right] \bullet [\![\mathbf{F} \bullet \mathbf{N}]\!],\end{aligned} \tag{5.53}$$

where the last step employs (5.26) integrated across the wave. Therefore, from (5.49) and (5.53), for general dynamic deformations (5.47) leads to the following jump dissipation inequality:

$$\boxed{[\![W(\mathbf{E}^e)]\!] - \mathbf{N} \bullet \overline{\mathbf{T}} \bullet [\![\mathbf{F} \bullet \mathbf{N}]\!] \geq 0.} \tag{5.54}$$

This was originally derived by Drugan and Shen (1990). A comparison of the general dynamic form (5.54) of the jump dissipation condition with the quasi-static special case (5.52) shows that the only difference is that in dynamic deformations, the First Piola-Kirchhoff traction vector can in general jump across the shock, so the average of its values just ahead of and just behind the shock appears in (5.54). Thus, clearly (5.54) reduces to (5.52) in the quasi-static case since then $\mathbf{T}$ has the same value on both sides of the shock wave.

6. Applications of Shock Wave Analysis to Problems in Elastic-Plastic Fracture Mechanics

In this section we shall illustrate the utility of several of the results derived in these notes to some problems in elastic-plastic fracture mechanics. The examples to be described will involve the asymptotic (near-tip) analysis of the elastic-plastic fields near crack tips in nonhardening materials, since these cases permit explicit closed-form analytical treatment. We shall begin with some fundamentals of the asymptotic analysis for nonhardening materials, and then proceed to analyze several specific example problems.

6.1 Asymptotic (Near-Tip) Analysis for Nonhardening Materials

The underlying analysis of this section is based on the papers of Rice (1982) and Drugan (1985). As illustrated in Figure 5, we consider a planar crack (stationary or growing; examples of both situations will be shown) and choose a Cartesian coordinate system oriented such that the crack faces are normal to the x_2-axis and the straight crack front is parallel to the x_3-axis. A polar coordinate system r, θ lies in the x_1, x_2 plane, is centered at the crack tip and, if the crack is growing, moves with the tip through the material. Unit vectors $\mathbf{e}$ and $\mathbf{h}$ point in the radial and circumferential directions, respectively. The crack has length a and if it is propagating, its speed is V. Throughout this section, we will employ a small-displacement-gradient formulation.

It is obvious from Figure 5 (or from the chain rule for partial derivatives) that

$$\frac{\partial r}{\partial x_i} = e_i, \qquad \frac{\partial \theta}{\partial x_i} = \frac{h_i}{r}, \tag{6.1}$$

where

$$e_1 = h_2 = \cos\theta, \qquad e_2 = -h_1 = \sin\theta, \qquad e_3 = h_3 = 0, \tag{6.2}$$

and Latin indices have range $1, 2, 3$ and obey the Einstein summation convention.

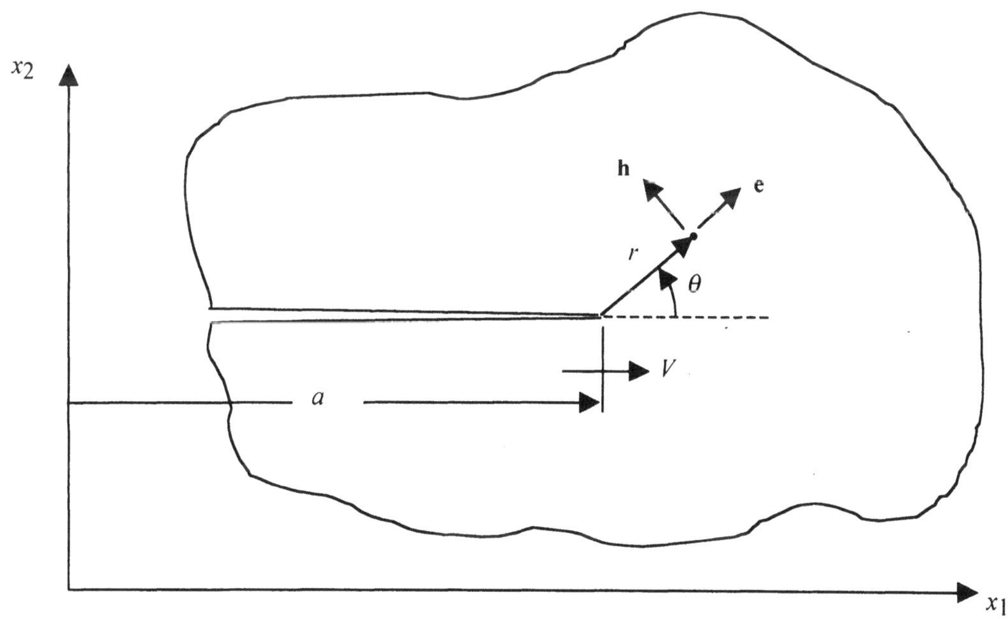

Figure 5. Planar crack, Cartesian and polar coordinate systems.

In index notation, the local form of the conservation of linear momentum, (4.4), is, including body force per unit mass **b**:

$$\frac{\partial \sigma_{ij}}{\partial x_j} + \rho b_i = \rho \dot{v}_i . \tag{6.3}$$

We will consider problems for which the stress and deformation fields do not vary along the crack front, so that they are independent of x_3. It will prove most convenient to express the stress tensor as a function of the polar coordinate system of Figure 5. Then, using the chain rule and (6.1):

$$\frac{\partial \sigma_{ij}}{\partial x_j} = \frac{\partial \sigma_{ij}}{\partial r} e_j + \frac{\partial \sigma_{ij}}{\partial \theta} \frac{h_j}{r} . \tag{6.4}$$

Since our examples will all involve *nonhardening* elastic-plastic materials, stresses cannot become infinite at the crack tip. [Strictly speaking, this is only true for the *deviatoric* components of stress for materials whose yield condition is independent of hydrostatic stress. However, in quasi-static deformations, it can be proved that circumferential equilibrium together with traction-free crack faces requires boundedness of the hydrostatic component of stress also; see Drugan (1985).] This leads to the following important asymptotic result:

$$r \frac{\partial \sigma_{ij}}{\partial r} \to 0 \quad \text{as} \quad r \to 0 . \tag{6.5}$$

A simple proof of this proceeds by contradiction: assuming that $r \frac{\partial \sigma_{ij}}{\partial r}$ exist in the $r \to 0$ limit, we may write

$$r \frac{\partial \sigma_{ij}}{\partial r} \to f_{ij}(\theta, t) \quad \text{as} \quad r \to 0 . \tag{6.6}$$

Treating this as an equality close to the crack tip, dividing both sides by r and integrating gives:

$$\sigma_{ij}(r, \theta, t) = f_{ij}(\theta, t) \ln(r/R) + g_{ij}(\theta, t) \quad \text{as} \quad r \to 0 . \tag{6.7}$$

This shows that stress boundedness in the $r \to 0$ limit demands $f_{ij}(\theta, t) \equiv 0$, which via (6.6) proves (6.5).

To derive an asymptotic form for the material time rate of velocity for a growing crack, let us temporarily introduce another Cartesian coordinate system x, y with the same orientation as the x_i system of Figure 5, except that it is centered at, and moves with, the crack tip when the crack is growing. Then, via the chain rule, writing $\mathbf{v} = \mathbf{v}(x, y, t)$

$$\overset{\bullet}{v}_i = \left.\frac{\partial v_i}{\partial t}\right|_{x,y} - V\frac{\partial v_i}{\partial x}. \tag{6.8}$$

Now, since we anticipate spatial velocity gradients to become infinite as the crack tip is approached, and using the fact that $\partial / \partial x = \partial / \partial x_1$, the asymptotic form of (6.8) is:

$$r\overset{\bullet}{v}_i = -Vr\frac{\partial v_i}{\partial x_1} \quad \text{as} \quad r \to 0. \tag{6.9}$$

Therefore, using (6.5) and (6.9), the asymptotic ($r \to 0$) form of the conservation of linear momentum is, having multiplied (6.4) by r and recognized also that $r\rho b_i \to 0$ as $r \to 0$:

$$\boxed{\frac{\partial \sigma_{ij}}{\partial \theta} h_j = -\rho V r \frac{\partial v_i}{\partial x_1}.} \tag{6.10}$$

For quasi-static deformations (when the crack is stationary, or growing sufficiently slowly that the inertia of near-tip material points can be neglected), (6.10) simplifies to

$$\boxed{\frac{\partial \sigma_{ij}}{\partial \theta} h_j = 0.} \tag{6.11}$$

6.2 Stationary and Quasi-Statically Growing Crack Tip Stress Fields in Plane Strain Deformations of Incompressible Huber-Mises Materials

Let us consider some specific examples of cracks in elastic-ideally plastic Huber-Mises materials, which satisfy the yield criterion

$$\tfrac{1}{2} s_{ij} s_{ij} - \tau_y^2 = 0, \tag{6.12}$$

where τ_y is the (constant) yield stress in pure shear. Taking the θ-derivative of (6.12), which must hold in regions at yield:

$$s_{ij}\frac{\partial s_{ij}}{\partial \theta} = s_{ij}\frac{\partial \sigma_{ij}}{\partial \theta} = 0, \tag{6.13}$$

having used the fact that $s_{kk} = 0$. Noting that the deviatoric stress components can be represented as

$$\begin{aligned} s_{ij} &= s_{rr} e_i e_j + s_{r\theta}\left(e_i h_j + e_j h_i\right) + s_{\theta\theta} h_i h_j \\ &\quad + s_{r3}\left(e_i \delta_{j3} + e_j \delta_{i3}\right) + s_{\theta 3}\left(h_i \delta_{j3} + h_j \delta_{i3}\right) + s_{33}\delta_{i3}\delta_{j3}, \end{aligned} \tag{6.14}$$

use of this and asymptotic equilibrium (6.11) in (6.13) show that in near-tip regions at yield during quasi-static deformations, the stress state must satisfy:

$$s_{rr}\left(e_i\frac{\partial\sigma_{ij}}{\partial\theta}e_j\right)+2s_{r3}e_i\frac{\partial\sigma_{i3}}{\partial\theta}+s_{33}\frac{\partial\sigma_{33}}{\partial\theta}=0. \tag{6.15}$$

The flow rule for this material is of the form (4.37), with (4.38), so that the condition (4.43) for plane strain deformations reduces to $s_{33}=0$, and we have also for such deformations $s_{r3}=0$. Thus, (6.15) reduces to

$$s_{rr}\left(e_i\frac{\partial\sigma_{ij}}{\partial\theta}e_j\right)=0. \tag{6.16}$$

This shows that there are two possible types of near-tip solution sector that are at yield for plane strain deformations:

(*i*) "centered fan plastic sector," in which $s_{rr}=0$.

From (6.11) and (6.12), it is straightforward to show that the most general stress field in such yielded sectors is (**Exercise: Show this.**)

$$\sigma_{r\theta}=\pm\tau_y,\qquad \sigma_{rr}=\sigma_{\theta\theta}=\sigma_{33}=\text{constant}\mp2\tau_y\theta. \tag{6.17}$$

(*ii*) "constant stress plastic sector," in which $\left(e_i\frac{\partial\sigma_{ij}}{\partial\theta}e_j\right)=0.$

From (6.11) and (6.12), it is straightforward to show that the most general stress field in such yielded sectors is (**Exercise: Show this.**)

$$\sigma_{11}=\text{constant},\quad \sigma_{12}=\sigma_{21}=\text{constant},\quad \sigma_{22}=\text{constant}, \tag{6.18}$$

where the constants are chosen to satisfy (6.12).

Now let us consider a stationary plane strain crack in elastic-ideally plastic material and make the plausible assumption that when loading is applied, the material in the vicinity of the crack tip attains yield at all angles about the tip. This means that the near-tip stress field solution must be assembled from some combination of angular sectors drawn from the two permissible types just determined. To determine the continuity conditions on stress that must be satisfied across the borders of such angular sectors, we first realize that since the crack is stationary, these borders will *not* be propagating through the material. Thus, the shock wave conditions derived in Section 4 do *not* apply. However, the general jump condition forms of the conservation laws do apply; in particular, the small-displacement-gradient, quasi-static deformation form of (2.43) applies, which simply requires the *traction* vector to be continuous across inter-sector boundaries. Since such boundaries are asymptotically radial, this means that the stress components $\sigma_{r\theta}$ and $\sigma_{\theta\theta}$ must be continuous. For pure Mode I (tensile) loading, Figure 6 illustrates two permissible near-tip solutions that meet all the requisite boundary (traction-free crack faces, Mode I symmetry) and continuity conditions for a stationary crack.

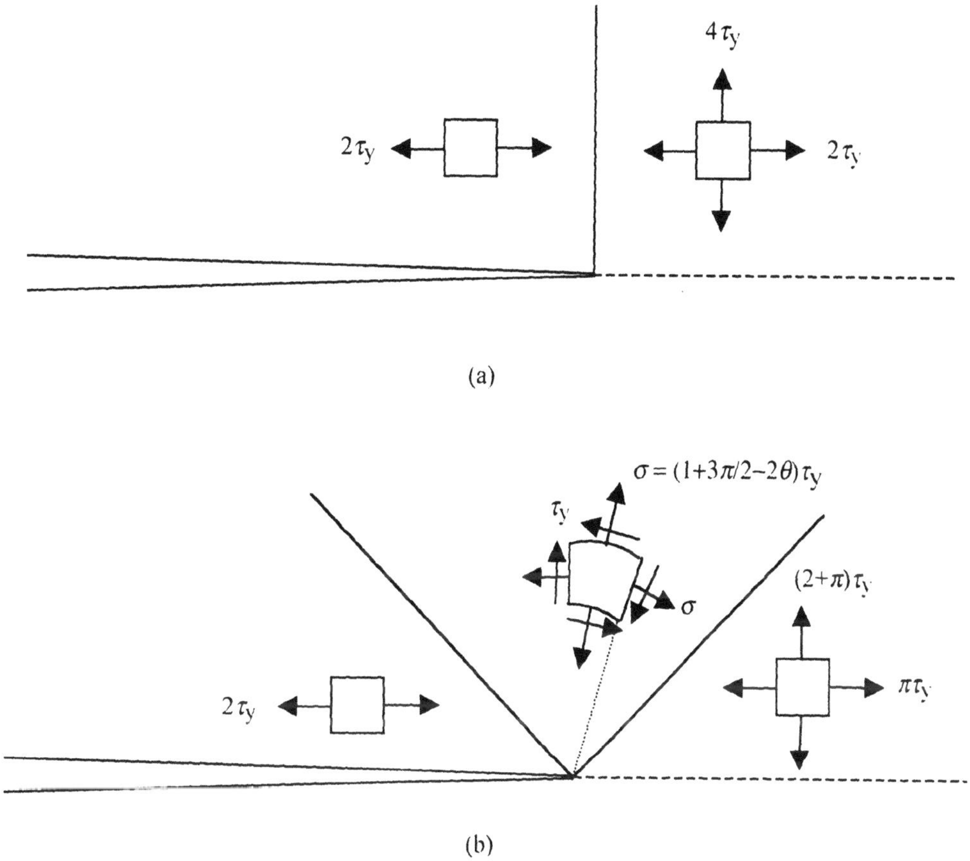

Figure 6. Two admissible plane strain solutions for Pure Mode I loading (symmetric about the crack line) for fully plastic stationary crack near-tip stress fields: (a) two constant stress sectors joined by a discontinuity in σ_{rr} at 90° from the crack line; (b) two constant stress sectors joined at 45° and 135° from the crack line with a centered fan sector, with full stress continuity (Rice, 1967).

In deriving the above results, it was helpful to notice that since $s_{13} = s_{23} = s_{33} = 0$ for incompressible plane strain, the Huber-Mises yield condition (6.12) can be expressed as either:

$$\left(\frac{\sigma_{22}-\sigma_{11}}{2}\right)^2 + \sigma_{12}^2 = \tau_y^2 \quad \text{or} \quad \left(\frac{\sigma_{\theta\theta}-\sigma_{rr}}{2}\right)^2 + \sigma_{r\theta}^2 = \tau_y^2. \tag{6.19}$$

The two fields illustrated in Figure 6 are possible near-tip solutions for a stationary crack. However, for quasi-static crack growth, the results of Section 4 apply to the material considered in this example, and they require that for a propagating shock wave in quasi-static deformations, the *full stress tensor* must be continuous. Thus, we conclude immediately that the solution configuration illustrated in Figure 6(a) *cannot* represent a correct solution near a quasi-statically growing Mode I crack. The configuration illustrated in Figure 6(b) has fully continuous stresses, and thus cannot be ruled out as a growing crack solution on this basis. In fact, interestingly, this configuration is the only possible fully continuous solution that is at yield at all angles about the crack tip and also transmits tension across the crack line, a physical requirement for Mode I loading. However, an analysis of the deformation fields associated with it shows that it cannot be a growing crack solution because the tangential velocity discontinuity necessitated at 135° from the crack line produces negative plastic work (Rice, Drugan and Sham, 1980). Thus, permissible growing crack solutions *must* involve an additional sector of instantaneously purely elastic response; a *family* of Mode I growing crack solutions with this feature, believed to be applicable for a broad range of far-field conditions, is derived by Drugan and Chen (1989).

6.3 Near-Tip Stress Fields for Stationary and Quasi-Statically Growing Anti-Plane Shear Cracks in Ideally Plastic Ductile Single Crystals

Ductile single crystals are characterized by a limited set of slip systems, and yield initiation is accurately predicted by a criterion originally proposed by Schmid, namely that the resolved shear stress acting on a slip plane in the slip direction attain a critical value. Here we consider ideally plastic ductile single crystals, for which the critical value just mentioned is a constant. For such materials, the yield condition can be written as

$$n_i^{(k)}\sigma_{ij}s_j^{(k)} = \tau_y^{(k)}, \tag{6.20}$$

where $n_i^{(k)}$ are components of the unit normal vector to the slip plane, $s_i^{(k)}$ are components of a unit vector parallel to the slip plane and in the slip direction, and $\tau_y^{(k)}$ is the yield stress, all of the kth slip system.

We shall consider here material and crack orientations that permit states of pure anti-plane shearing, so that the only nonzero stress components are $\sigma_{13} = \sigma_{31} \equiv \tau_1$, $\sigma_{23} = \sigma_{32} \equiv \tau_2$. Then, (6.20) requires

$$n_\alpha^{(k)}\tau_\alpha s_3^{(k)} + n_3^{(k)}\tau_\alpha s_\alpha^{(k)} = \tau_y^{(k)}, \tag{6.21}$$

where here and henceforth, Greek subscripts (except θ, which always denotes the polar coordinate) have range 1, 2 and obey the Einstein summation convention. This shows that for each slip system in a given crystal we will have a line in the τ_1, τ_2 plane whose equation is of the form

$$m_1^{(k)}\tau_1 + m_2^{(k)}\tau_2 = \tau_y^{(k)}, \tag{6.22}$$

where $\mathbf{m}^{(k)}$ is a vector in the outer normal direction to a given line. We shall here consider the example situation of a crack in a face-centered cubic (FCC) crystal, with the crack lying on the (0 1 0) plane and having its tip along the $[1\,0\,\bar{1}]$ direction. In this case, Rice and Nikolic (1985) have shown that the yield locus is as illustrated in Figure 7, and the analysis to follow is based on their work.

Let us examine permissible near-tip stress field solutions in regions at yield in this material. In any near-tip sector at yield, (6.22) must be satisfied; that is, the stress state must lie on one of the straight-line segments in Figure 7. First, observe that if the stress state lies at the intersection of any two straight-line segments, i.e. a yield surface vertex, the stress state is fixed – i.e., the solution sector is of the constant stress type. The other possibility is that the stress state lies on one of the slanted lines away from a vertex. In this case, (6.22) applies for a specific line throughout the sector, and hence its θ-derivative must also be satisfied: dropping the superscript (k), we have:

$$m_1 \frac{\partial \tau_1}{\partial \theta} + m_2 \frac{\partial \tau_2}{\partial \theta} = 0. \tag{6.23}$$

To analyze the near-tip stress state permitted in such a sector, observe that equilibrium (6.11) requires that, using (6.2)

$$\frac{\partial \tau_\alpha}{\partial \theta} h_\alpha = 0 \quad \Rightarrow \quad e_2 \frac{\partial \tau_1}{\partial \theta} = e_1 \frac{\partial \tau_2}{\partial \theta}. \tag{6.24}$$

Combining (6.23) and (6.24) gives:

$$\left(m_1 e_1 + m_2 e_2\right) \frac{1}{e_2} \frac{\partial \tau_2}{\partial \theta} = 0. \tag{6.25}$$

This with (6.24) shows that the stresses τ_1, τ_2 must be constant throughout such yielded sectors, except possibly along the single ray from the crack tip for which $\mathbf{e}$ is perpendicular to $\mathbf{m}$, i.e., $\mathbf{m} \cdot \mathbf{e} = 0$. The overall conclusion, then, is that the only admissible yielded solution sector is of constant stress type.

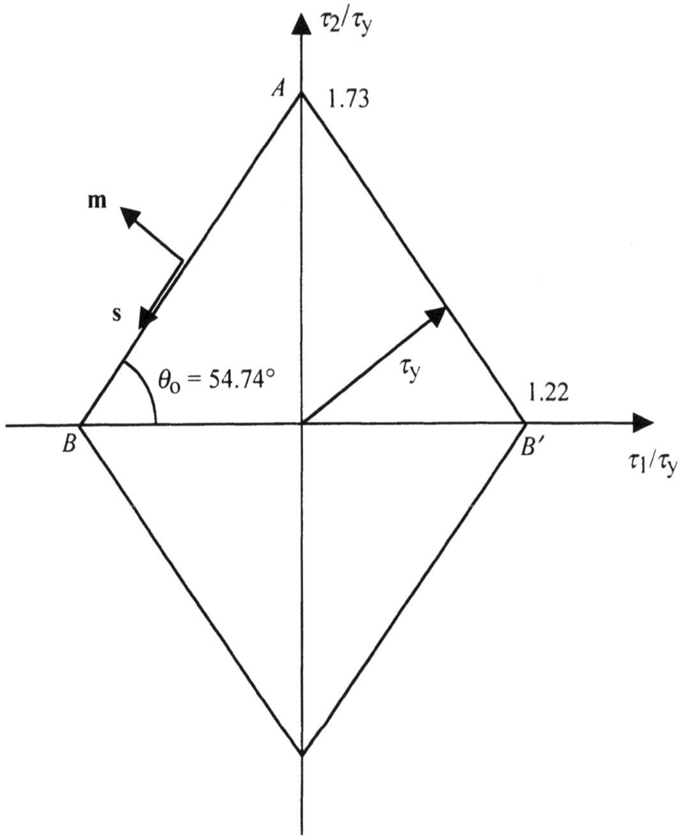

Figure 7. Yield surface for a crack lying on the (0 1 0) plane with its tip along the $[1\,0\,\bar{1}]$ direction, and experiencing anti-plane shear, in an elastic-ideally plastic FCC ductile single crystal.

Thus, if we again make the plausible assumption that the stress state at all angles close to a crack tip has attained yield, solutions must be constructed by assembling sectors of constant stress type that satisfy the yield condition illustrated in Figure 7 together with the requirements of traction continuity between sectors, traction-free crack faces and Mode III antisymmetry. Rice and Nikolic (1985) have shown that the near-tip solution for a crack with the orientation mentioned, and in the FCC material characterized by the yield condition of Figure 7, is as illustrated in Figure 8.

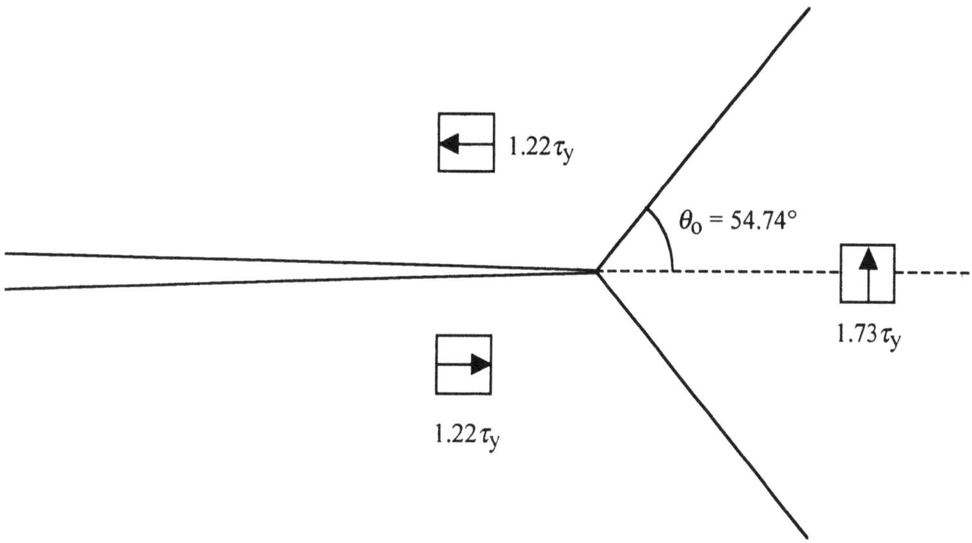

Figure 8. Stationary crack Mode III solution assuming yield at all angles about the crack tip.

As in the isotropic material case, our results from Section 4 show that the solution pictured in Figure 8 *cannot* apply for a quasi-statically growing crack, since it involves stress discontinuities in σ_{r3} across the sector boundaries illustrated. Thus, the quasi-static crack growth solution *must* involve instantaneously elastic sectors, with full stress continuity. The details are given by Rice and Nikolic (1985).

6.4 Near-Tip Stress Fields for Dynamically Growing Anti-Plane Shear Cracks in Ideally Plastic Ductile Single Crystals

Let us consider a very interesting, but far more involved, example: that of a dynamically propagating crack of the same type, and in the same material, as described in Section 6.3. It will still be the case that in near-tip sectors at yield, the stresses must obey relations of the form (6.22), but now asymptotic equilibrium (6.24) is supplanted by the full asymptotic form (6.10) of the conservation of linear momentum, which in the anti-plane shear case becomes

$$\frac{\partial \tau_\alpha}{\partial \theta} h_\alpha = -\rho V r \frac{\partial v_3}{\partial x_1}. \tag{6.26}$$

The analysis of possible near-tip solution sectors that are at yield, and others that are below yield, and the analysis of permissible assembly of such sectors is rather involved, and thus will not be shown here; this is lucidly done by Nikolic and Rice (1988). However, it is extremely interesting to summarize their conclusions, and one may easily verify that their solution satisfies all necessary conditions (**Exercise: Do so.**). They find that the only possible near-tip solution for dynamic Mode III crack growth involves radial shock waves propagating with the growing crack tip. Specifically, their near-tip solution has the following form, half of this antisymmetric field being illustrated in Figure 9: on the upper side of the crack plane, the solution consists of two sectors at yield in which the Cartesian components of stress are constant and no deformation is occurring; these are joined by a radial shock wave, across which stresses and velocity jump and whose angle with respect to the crack line depends on the speed of propagation of the crack. This is because, as discussed in Section 4.3.3, the fields through the shock wave must satisfy both (4.34) *and* the yield condition (here, that of Figure 7 on the line segment connecting A and B), and simultaneous satisfaction of these two conditions specifies the normal speed of propagation, c, of the shock wave that appears in (4.34); this in turn determines the angle the shock wave makes with the crack line, depending on the speed of propagation of the crack. Nikolic and Rice (1988) have determined this angle to be $\theta_0 + \psi$, where θ_0 is as given above in the stationary crack solution and ψ is

$$\psi = \arctan\left[\frac{\sin\theta_0}{\left(\rho V^2 c_{ss}\right)^{-1/2} - \cos\theta_0}\right], \tag{6.27}$$

where c_{ss} is the component of the 2 x 2 elastic compliance matrix in the slip direction s illustrated on the yield surface in Figure 7.

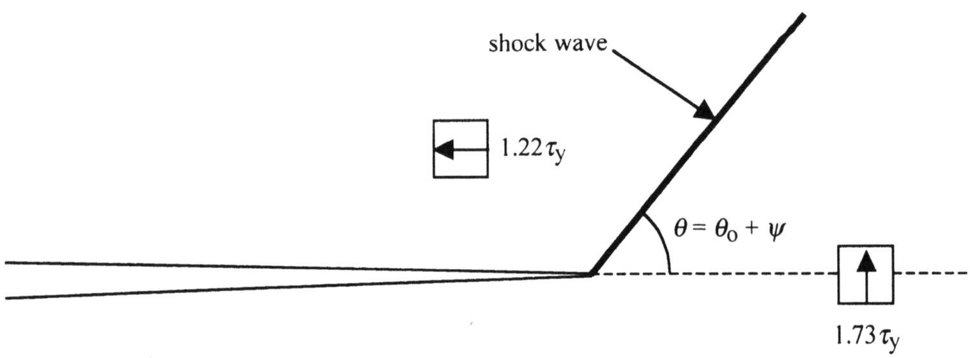

Figure 9. Dynamically propagating Mode III near-tip solution of Nikolic and Rice (1988).

References

Brannon, R. M. and Drugan, W. J. (1993). Influence of non-classical elastic-plastic constitutive features on shock wave existence and spectral solutions. *Journal of the Mechanics and Physics of Solids* 41: 297–330.

Brannon, R. M., Drugan, W. J. and Shen, Y. (1995). Requirements of thermodynamics in the analysis of elastic-plastic shock waves. *Journal of the Mechanics and Physics of Solids* 43: 973–1001.

Courant, R. and Friedrichs, K. O. (1948). *Supersonic Flow and Shock Waves* (Third Printing: 1985). Springer, Berlin.

Drugan, W. J. (1985). On the asymptotic continuum analysis of quasi-static elastic-plastic crack growth and related problems. *Journal of Applied Mechanics* 52: 601–605.

Drugan, W. J. (1986). A more direct and general analysis of moving strong discontinuity surfaces in quasi-statically deforming elastic-plastic solids. *Journal of Applied Mechanics* 53: 224–226.

Drugan, W. J. (1998). Thermodynamic equivalence of steady-state shocks and smooth waves in general media; applications to elastic-plastic shocks and dynamic fracture. *Journal of the Mechanics and Physics of Solids* 46: 313–336.

Drugan, W. J. and Chen, X.-Y. (1989). Plane strain elastic-ideally plastic crack fields for Mode I quasistatic growth at large-scale yielding – I. A new family of analytical solutions. *Journal of the Mechanics and Physics of Solids* 37: 1–26.

Drugan, W. J. and Rice, J. R. (1984). Restrictions on quasi-statically moving surfaces of strong discontinuity in elastic-plastic solids. In Dvorak, G. J. and Shield, R. T., eds., *Mechanics of Material Behavior*. Elsevier Science, Amsterdam. 59–73.

Drugan, W. J. and Shen, Y. (1987). Restrictions on dynamically propagating surfaces of strong discontinuity in elastic-plastic solids. *Journal of the Mechanics and Physics of Solids* 35: 771–787.

Drugan, W. J. and Shen, Y. (1990). Finite deformation analysis of restrictions on moving strong discontinuity surfaces in elastic-plastic materials: Quasi-static and dynamic deformations. *Journal of the Mechanics and Physics of Solids* 38: 553–574.

Hadamard, J. (1903). *Leçons sur la Propagation des Ondes et les Equations de l'Hydrodynamique*. Hermann, Paris.

Havner, K. S. (1982). The theory of finite plastic deformation of crystalline solids. In Hopkins, H. G. and Sewell, M. J., eds., *Mechanics of Solids: The R. Hill 60th Anniversary Volume*. Pergamon Press, Oxford. 265–302.

Hill, R. and Rice, J. R. (1972). Constitutive analysis of elastic-plastic crystals at arbitrary strain. *Journal of the Mechanics and Physics of Solids* 20: 401–413.

Knowles, J. K. (1979). On the dissipation associated with equilibrium shocks in finite elasticity. *Journal of Elasticity* 9: 131–158.

Lee, E. H. (1969). Elastic-plastic deformation at finite strains. *Journal of Applied Mechanics* 36: 1–6.

Leighton, J. T., Champion, C. R. and Freund, L. B. (1987). Asymptotic analysis of steady dynamic crack growth in an elastic-plastic material. *Journal of the Mechanics and Physics of Solids* 35: 541–563.

Nikolic, R. R. and Rice, J. R. (1988). Dynamic growth of anti-plane shear cracks in ideally plastic crystals. *Mechanics of Materials* 7: 163–173.

Peirce, D., Asaro, R. J. and Needleman, A. (1982). An analysis of nonuniform and localized deformation in ductile single crystals. *Acta Metallurgica* 30: 1087–1119.

Rice, J. R. (1967). Mechanics of crack tip deformation and extension by fatigue. In *Fatigue Crack Propagation*, ASTM-STP 415. 247–311.

Rice, J. R. (1971). Inelastic constitutive relations for solids: An internal-variable theory and its application to metal plasticity. *Journal of the Mechanics and Physics of Solids* 19: 433–455.

Rice, J. R. (1975). Continuum mechanics and thermodynamics of plasticity in relation to microscale deformation mechanisms. In Argon, A. S., ed., *Constitutive Equations in Plasticity*. M.I.T. Press. 23–79.

Rice, J. R. (1982). Elastic-plastic crack growth. In Hopkins, H. G. and Sewell, M. J., eds., *Mechanics of Solids: The R. Hill 60th Anniversary Volume*. Pergamon Press, Oxford. 539–562.

Rice, J. R., Drugan, W. J. and Sham, T-L. (1980). Elastic-plastic analysis of growing cracks. In *Fracture Mechanics: 12th Conference*. ASTM STP 700. 189–219.

Rice, J. R. and Nikolic, R. R. (1985). Anti-plane shear cracks in ideally plastic crystals. *Journal of the Mechanics and Physics of Solids* 33: 595–622.

Shen, Y. and Drugan, W. J. (1990). Constraints on moving strong discontinuity surfaces in dynamic plane-stress or plane-strain deformations of stable elastic-ideally plastic materials. *Journal of Applied Mechanics* 57: 569–576.

Willis, J. R. (1969). Some constitutive equations applicable to problems of large dynamic plastic deformation. *Journal of the Mechanics and Physics of Solids* 17: 359–369.

FLUTTER INSTABILITY AND ILL-POSEDNESS IN SOLIDS AND FLUID-SATURATED POROUS MEDIA

B. Loret
Laboratoire Sols, Solides, Structures, Grenoble, France

F.M.F. Simões and J.A.C. Martins
Instituto Superior Técnico, Lisbon, Portogallo

Abstract. For elastoplasticity with locally smooth yield surface and plastic potential, the nature, real or complex, of the squares of the acceleration wave-speeds is analyzed. Emphasis is laid on the effect that some features of the constitutive equations have on that nature. Specifically, our reference problem contemplates an infinite elastic-plastic body endowed with elastic isotropy and deviatoric associativity. In a second step, individual or simultaneous deviations with respect to these reference properties are analyzed. The performed linearized analyses are essentially intended to detect, in the course of a loading process, the onset of wave-speeds whose squares are complex, a phenomenon called flutter. Finite element simulations show how perturbations evolve in such a situation, and, as expected, the plastic loading condition plays a crucial role.
The problem is reconsidered in the context of fluid-saturated porous media whose solid skeleton is elastic-plastic. Unlike for single phase solids, flutter is not excluded in the reference situation where elasticity is isotropic and deviatoric plasticity holds. The propagation of waves of different lengths leads to make a distinction between the two phenomena of flutter instability and flutter ill-posedness. The introdution of a characteristic length in the elastic-plastic behaviour of the solid skeleton may imply the existence of an infimum wavelength below which instability modes do not exist, and well-posedness of the dynamic problem is recovered.

Introduction

The well-posedness of the initial- and boundary-value problem for rate-independent solids requires the dynamical equations of motion to be pointwise hyperbolic. Equivalently the three acceleration wave-speeds must be real and strictly positive, e.g. Truesdell and Noll [1965], Sect. 71.

The squares of the latter are the eigenvalues of the acoustic tensor, which inherits the major symmetry property from the tensor of constitutive moduli. Consequently, the squares of the wave-speeds are real for elastic-plastic solids with associative flow rules, and loss of hyperbolicity occurs when the smaller wave-speed vanishes, Hill [1962], a phenomenon referred to as a *stationary discontinuity*. The quasi-static interpretation of a stationary discontinuity as a strain localization has been exposed by Mandel [1964a,b] and Rice [1976] for materials with both associative and non-associative flow rules. Both Mandel and Rice pointed out the destabilizing influence of non-associativity; Asaro and Rice [1977] investigated the destabilizing effect of various constitutive features that lead to loss of the major symmetry of the tensor of constitutive moduli. These works have been at the root of a number of papers, whose essential aim was to improve upon the somewhat resistant behaviour of classic elastic-plastic solids that yield too late an onset of strain localization as compared with experimental data.

For constitutive equations that lack the major symmetry, the acoustic tensor is not symmetric, and loss of hyperbolicity may occur through a mode other than a stationary discontinuity. This mode has been coined, after a remark of Rice [1976], a *flutter instability* although the terminology *flutter ill-posedness* will appear more appropriate in some contexts, Sect. B.4; it corresponds to the square of a wave-speed becoming complex. It has been the subject of a few investigations in the last decade. Loret et al. [1990] show that it is excluded for a wide range of elastic-plastic solids, namely those whose elasticity is isotropic and whose flow rule is associative with respect to the deviatoric components of the plastic strain rate tensor only, a constitutive feature referred to as *deviatoric associativity*. This proposition will be proved in Sect. A.3. Ottosen and Runesson [1991a] improve slightly upon the former result by showing that, if the unit normals to the plastic potential and to the yield surface are coaxial, with their eigenvalues arranged in the same order, then once again, flutter is precluded.

Deviatoric associativity plays an interesting role in these analyses, although, in a general context, it is neither necessary nor sufficient to exclude flutter. In fact, our plan is precisely to show instances illustrating these statements. These instances concern both single phase solids, Sect. A, and fluid-saturated porous media, Sect. B.

As mentioned above, exclusion of flutter for deviatoric associativity is ensured if the elasticity is isotropic. We are not aware of a result that excludes flutter when the elasticity displays a specific type of anisotropy. On the other hand, one can devise anisotropic elasticity that leads to the onset of flutter even for flow rules that satisfy deviatoric associativity, Sect. A.5.

A route to develop constitutive equations that may lead to the onset of flutter instability has been explored by An and Schaeffer [1992] in two dimensional models and by Loret [1992] in the usual three-dimensional elastic-plastic context. They show that, even for associative flow rules, strict hyperbolicity does not hold and therefore it is possible to define constitutive perturbations involving loss of the major symmetry of the constitutive moduli and leading to the onset of flutter, Sect. A.4. For constitutive perturbations of solids with deviatoric associativity the following results are obtained: any infinitesimal constitutive perturbations from deviatoric associativity, making the normals to the plastic potential $\mathbf{P}$ and to the yield surface $\mathbf{Q}$ non-coaxial, lead to the onset of flutter for a properly chosen plastic modulus; in contrast, a finite amount of deviation with respect to deviatoric associativity is required to reach the onset of flutter when $\mathbf{P}$ and $\mathbf{Q}$ are kept coaxial. This result is in agreement with the work of Ottosen and Runesson [1991a]; that finite amount of deviation from deviatoric associativity is the amount necessary to change the order of the eigenvalues of $\mathbf{P}$ and $\mathbf{Q}$. Incidentally, it is worth mentioning that coaxiality (or non-coaxiality) between specific (and problem dependent) material characteristics has a pervasive influence in all the analyses considered in this review. In all the examples reported here, non-coaxiality has a crucial destabilizing effect and triggers flutter, much like non-associativity does for strain localization.

The former analyses are local, that is, they may be viewed as concerning a point in an infinite solid. Motivated by the strongly destabilizing effects of Coulomb's friction law, another route to the onset of flutter instability may therefore be the incorporation of the effects of boundary conditions on a finite or semi-infinite body, Sect. A.6. Taken independently, neither the constitutive equations nor the boundary conditions we consider would lead to the onset of flutter. However, we shall derive conditions under which their simultaneous influence will indeed lead to the onset of flutter. This analysis can be viewed as an instance where an initial-boundary value problem becomes ill-posed while the rate field equations are still hyperbolic.

The above results concern either the exclusion or provide conditions for the onset of flutter. However, the a priori oscillatory nature of the phenomenon may raise doubts on its real development in initial- and boundary-value problems. The key issue is of course the plastic loading condition. In analytical studies, strict plastic loading is assumed, although a little bit more could be said for neutral loading. An assessment of the occurrence and the consequences of flutter requires the consideration of the fully nonlinear elastic-plastic behaviour of the solid in the analysis of its dynamic response. This is the purpose of the finite element simulations reported in Sect. A.7. These finite element nonlinear dynamic analyses, where both loading and unloading regimes are taken into consideration, indicate in fact that flutter does lead to a oscillatory material response; but this happens in the neighborhood of quasi-static solutions obtained under varying external loadings, and the oscillation amplitude does not explode in time as predicted by linearized stability analyses.

Section B is devoted to the analysis of flutter in fluid-saturated porous media. The elastic-plastic constitutive equations are developed in the framework of the theory of mixtures, B.1. Attention is paid to symmetry conditions, and deviatoric associativity is found to be a useful concept in this context, as it is for solids. The nature of acceleration waves is examined in Sect. B.2. The characteristics of the onset of stationary discontinuities in fluid-saturated porous media are found to be given by the properties of the so-called *underlying drained solid*. On the other hand, in strong contrast with the situation that prevails in solids, deviatoric associativity does not exclude flutter even for isotropic elastic properties. Because the associated algebraic manipulations are easier, the review concerns *incompressible constituents* but complementary studies have shown that they display the essential behaviour of compressible constituents.

For elastic and elastic-plastic porous media with an associative flow-rule, acceleration waves propagate with an amplitude that either decays or remains constant, and the decay coefficient is bounded below and above. For a non-associative flow-rule, the amplitude of acceleration waves may grow for certain plastic moduli when the wave-speeds are real, an unstable behaviour, Sect. B.3. The viscous effects due to Darcy's diffusion of fluid through the solid skeleton does not affect acceleration wave-speeds. This is not so for harmonic waves, where these viscous effects introduce a diffusion length in the governing equations. Crucially, the limit as the wavelength tends to zero of the moduli separating the regions of distinct material behaviours is not continuous, Sect. B.4. This observation leads to make a clear distinction between *flutter instability* and *flutter ill-posedness*.

Despite the presence of characteristic length in the constitutive and field equations, there are material parameters for which there does not exist *an infimum wavelength* for the exponentially growing harmonic solutions, and ill-posedness is not precluded. The existence of such an infimum wavelength requires the introduction of an additional characteristic length in the constitutive equations describing the behaviour of the solid skeleton, Sect. B.5.

Notation: Compact or index tensorial notation will be used throughout this note. Tensor quantities are identified by boldface letters. $\mathbf{I}$ is the second order identity tensor (Kronecker delta). Symbols '$\cdot$' and ':' between tensors of various orders denote their inner product with single and double contraction respectively. The dyadic product of two tensors is indicated with '$\otimes$'. The symbol '$\overline{\underline{\otimes}}$' denotes a symmetrized outer product: componentwise in cartesian axes, for any arbitrary symmetric second order tensors $\mathbf{A}$ and $\mathbf{B}$, $(\mathbf{A}\,\overline{\underline{\otimes}}\,\mathbf{B})_{ijkl}=(A_{ik}B_{jl}+A_{il}B_{jk})/2$. The transpose of an operator is denoted by a superimposed ^{T}. $\|\cdot\|$ denotes the euclidean norm. Repeated latin indices imply summation over the appropriate range.

A Solids

A.1 Constitutive Equations and Deviatoric Associativity

A.1.1 Elasticity, Elastoplasticity and Basic Definitions

We consider elastic-plastic materials which are isotropic in their elastic properties [1] but whose plastic properties may embody any kind of structural or induced anisotropy. The analysis assumes that the materials undergo small perturbations relative to a given unstressed reference state. Therefore all time derivatives coincide with the material derivative denoted by a superimposed dot. The single restriction placed on the tensor of elastic moduli $\mathbb{E}$ defined by the Lamé moduli λ and μ,

$$\mathbb{E} = \lambda\,\mathbf{I}\otimes\mathbf{I} + 2\,\mu\,\mathbf{I}\,\overline{\underline{\otimes}}\,\mathbf{I}\,, \quad \mathbb{E}_{ijkl} = \lambda\, I_{ij}\, I_{kl} + \mu\,(I_{ik}\, I_{jl} + I_{il}\, I_{jk})\,, \tag{A.1.1}$$

is its *positive definiteness* (PD), or equivalently the positive definiteness of the elastic strain energy

$$\text{(PD) of } \mathbb{E} \Leftrightarrow \frac{1}{2}\boldsymbol{\epsilon} : \mathbb{E} : \boldsymbol{\epsilon} = \frac{1}{2}(\lambda + \frac{2}{3}\,\mu)\,\mathrm{tr}^2\boldsymbol{\epsilon} + \mu\,\mathrm{dev}\,\boldsymbol{\epsilon} : \mathrm{dev}\,\boldsymbol{\epsilon} > 0\,, \forall\, \boldsymbol{\epsilon}=\boldsymbol{\epsilon}^{\mathrm{T}} \neq \mathbf{0}\,, \tag{A.1.2}$$

that is,

$$\lambda + \frac{2}{3}\,\mu > 0, \quad \mu > 0\,. \tag{A.1.3}$$

The state of the material is defined by the current value of the stress $\boldsymbol{\sigma}$ and by a set of hidden variables $\mathcal{H}$ which may involve scalars and tensors of various orders. The strain rate $\dot{\boldsymbol{\epsilon}}$ admits an additive decomposition into an elastic part $\dot{\boldsymbol{\epsilon}}_e$ and a plastic part $\dot{\boldsymbol{\epsilon}}_p$,

$$\dot{\boldsymbol{\epsilon}} = \dot{\boldsymbol{\epsilon}}_e + \dot{\boldsymbol{\epsilon}}_p\,. \tag{A.1.4}$$

[1] The only exception to this general framework is Sect. A.5.

Given a plastic modulus h, a smooth yield function $f = f(\sigma, \mathcal{H})$ and a smooth plastic potential $g = g(\sigma, \mathcal{H})$, the elastic-plastic tensor $\mathbb{E}^{ep}$ differs from its elastic counterpart $\mathbb{E}$ by a dyadic term only, namely

$$\mathbb{E}^{ep} = \mathbb{E} - \frac{1}{H}(\mathbb{E} : \mathbf{P}) \otimes (\mathbf{Q} : \mathbb{E}), \tag{A.1.5}$$

where the symmetric second order tensors $\mathbf{P}$ and $\mathbf{Q}$ are the unit outward normals to the plastic potential and to the yield surface respectively. In the following, the modulus H

$$H = h + \mathbf{P} : \mathbb{E} : \mathbf{Q} > 0 \tag{A.1.6}$$

is restricted to be strictly positive to exclude snap-back and locking phenomena. The above expression of the tensor $\mathbb{E}^{ep}$ is obtained by requiring, first, the plastic-strain rate $\dot{\epsilon}_p$ to have the direction of $\mathbf{P}$, and an evolution rule for the set $\mathcal{H}$, namely,

$$\dot{\epsilon}_p = \dot{\Lambda}\mathbf{P}, \quad \dot{\mathcal{H}} = \dot{\Lambda}\mathbf{H}, \quad \dot{\Lambda} \geq 0, \tag{A.1.7}$$

where $\mathbf{H} = \mathbf{H}(\sigma, \mathcal{H})$, and, second, the stress to remain *on* the yield surface, that is f=0 and $\dot{f}=0$, the so-called consistency condition, which, with $h = -\partial f/\partial\mathcal{H} \cdot \mathbf{H}$, provides $\dot{\Lambda} = \mathbf{Q} : \mathbb{E} : \dot{\epsilon}/H$. Thus, at a given state and for a given strain-rate $\dot{\epsilon}$, the material response is plastic if simultaneously $f=0$ and $\mathbf{Q} : \mathbb{E} : \dot{\epsilon} > 0$; the resulting stress rate $\dot{\sigma}$ is then given by $\dot{\sigma} = \mathbb{E}^{ep} : \dot{\epsilon}$; otherwise the material response is elastic and $\dot{\sigma} = \mathbb{E} : \dot{\epsilon}$. So:

$$\dot{\sigma} = \begin{cases} \mathbb{E}^{ep} : \dot{\epsilon} & \text{if } f=0 \text{ and } \mathbf{Q}:\mathbb{E}:\dot{\epsilon} \geq 0; \\ \mathbb{E} : \dot{\epsilon} & \text{otherwise}. \end{cases} \tag{A.1.8}$$

Notice that the relation $\dot{\epsilon} - \dot{\sigma}$ is continuous across *neutral* loadings $\mathbf{Q} : \mathbb{E} : \dot{\epsilon} = 0$. The plastic behaviour is said *associative* or *non-associative* depending on whether the unit tensors $\mathbf{P}$ and $\mathbf{Q}$ are equal or not. Only for an associative behaviour does the elastic-plastic moduli $\mathbb{E}^{ep}$ enjoy the *major* symmetry property $\mathbb{E}^{ep}_{ijkl} = \mathbb{E}^{ep}_{klij}$. The *minor* symmetry property $\mathbb{E}^{ep}_{ijkl} = \mathbb{E}^{ep}_{jikl}$ holds always by necessity (since the stress-rate is symmetric) and the *minor* symmetry property $\mathbb{E}^{ep}_{ijkl} = \mathbb{E}^{ep}_{ijlk}$ holds by choice (since the strain-rate is symmetric).

A.1.2 Deviatoric Associativity

Without loss of generality, one can separate the volumetric and deviatoric contributions to the unit tensors $\mathbf{P}$ and $\mathbf{Q}$ by introducing the angles ψ and χ and the unit deviators $\hat{\mathbf{S}}_Q$ and $\hat{\mathbf{S}}_P$:

$$\mathbf{P} = \cos\chi\, \hat{\mathbf{S}}_P + \sin\chi \frac{\mathbf{I}}{\sqrt{3}}, \quad \mathbf{Q} = \cos\psi\, \hat{\mathbf{S}}_Q + \sin\psi \frac{\mathbf{I}}{\sqrt{3}}. \tag{A.1.9}$$

Observe that these relations are not constitutive relations since any second order tensor can be cast in the above format. For metal plasticity, the influence of the mean-stress on the yield surface and plastic potential are neglected so that both ψ and χ are zero. For geomechanical applications, the dilatancy angle χ is usually smaller than the friction angle, and $(0 \leq)\chi \leq \psi < 90°$. The assumption of *deviatoric associativity* amounts to postulate the unit tensors $\hat{\mathbf{S}}_P$ and $\hat{\mathbf{S}}_Q$ to be colinear. Thus, deviatoric associativity holds if

$$\hat{\mathbf{S}}_P = \hat{\mathbf{S}}_Q = \hat{\mathbf{S}}\,, \tag{A.1.10}$$

and we shall not consider the case $\hat{\mathbf{S}}_P = -\hat{\mathbf{S}}_Q$. In the special case where $\hat{\mathbf{S}}$ is proportional to the stress deviator $\mathrm{s} = \operatorname{dev}\boldsymbol{\sigma}$, the yield function $f = f(\boldsymbol{\sigma}, k)$ and plastic potential $g = g(\boldsymbol{\sigma})$ are of the Drucker-Prager type,

$$f = \cos\psi \;\|\, \mathrm{s} \,\| + \sin\psi\,\frac{\operatorname{tr}\boldsymbol{\sigma}}{\sqrt{3}} - k\,, \quad g = \cos\chi \;\|\, \mathrm{s} \,\| + \sin\chi\,\frac{\operatorname{tr}\boldsymbol{\sigma}}{\sqrt{3}}\,, \tag{A.1.11}$$

and then

$$\mathbf{P} = \frac{\partial g}{\partial \boldsymbol{\sigma}} = \cos\chi\,\hat{\mathbf{S}} + \sin\chi\,\frac{\mathbf{I}}{\sqrt{3}}\,, \quad \mathbf{Q} = \frac{\partial f}{\partial \boldsymbol{\sigma}} = \cos\psi\,\hat{\mathbf{S}} + \sin\psi\,\frac{\mathbf{I}}{\sqrt{3}}\,, \tag{A.1.12}$$

with $\hat{\mathbf{S}} = \mathrm{s}/\|\,\mathrm{s}\,\|$. The angles ψ and χ are related to the friction and dilatancy angles defined by the Coulomb criterion, and the hardening parameter k provides the size of f in the deviatoric plane [2]. The restriction, namely $\hat{\mathbf{S}} \sim \mathrm{s}$, is not required for the present study and any kind of anisotropy can be embodied in $\hat{\mathbf{S}}_P$ and $\hat{\mathbf{S}}_Q$. Nevertheless, the angles ψ and χ will be restricted by the condition $\cos\psi\,\cos\chi > 0$.

For the record, let us note the spectral decomposition of the unit deviator $\hat{\mathbf{S}}$,

$$\hat{\mathbf{S}} = \textstyle\sum_{i=1}^{3} \hat{S}_i\; \mathbf{E}_i \otimes \mathbf{E}_i\,, \tag{A.1.13}$$

where the eigenvectors $\mathbf{E}_i$, $i \in [1,3]$, are of unit norm and the eigenvalues can be defined in terms of a single scalar ℓ, referred to as Lode angle[3],

$$\hat{S}_i = \sqrt{\frac{2}{3}}\,\cos\left(\ell - 2\,(i-1)\frac{\pi}{3}\right), \quad i \in [1,3]\,, \quad \Longrightarrow \; \hat{S}_1 \geq \hat{S}_2 \geq \hat{S}_3\,. \tag{A.1.14}$$

Deviation from deviatoric associativity is most susceptible to observation in the presence of structural or induced anisotropy. However, it can also be displayed by isotropic materials where both the yield surface and plastic potential depend on the three scalar stress-invariants $I_1 = \operatorname{tr}\boldsymbol{\sigma}$, $J_2 = \operatorname{tr}\mathrm{s}^2$, $J_3 = \operatorname{tr}\mathrm{s}^3$. Deviatoric associativity requires a relation to be satisfied by the partial derivatives of f and g with respect to J_2 and J_3, namely, Baker and Desai [1982]:

$$\frac{\partial f}{\partial J_2}\frac{\partial g}{\partial J_3} - \frac{\partial g}{\partial J_2}\frac{\partial f}{\partial J_3} = 0\,. \tag{A.1.15}$$

[2] Let the Coulomb criterion be defined in terms of the principal stresses $\sigma_1 \leq \sigma_2 \leq \sigma_3$ as follows: $-(\sigma_1 - \sigma_3) + (\sigma_1 + \sigma_3)\,\sin\phi - 2\,C\cos\phi = 0$; then the size k is linked to the cohesion C and friction angle ϕ by $k/(\sqrt{3}\,\sin\psi) = C/\tan\phi$ and the cones coinciding with the Coulomb criterion in triaxial compression, e.g. $\sigma_1 < \sigma_2 = \sigma_3$, triaxial extension $\sigma_1 = \sigma_2 < \sigma_3$ and internally tangent are defined by $\tan\psi$ equal respectively to $2\sqrt{2}\,\sin\phi/(3-\sin\phi)$, $2\sqrt{2}\,\sin\phi/(3+\sin\phi)$ and $\sqrt{2}\,\sin\phi/\sqrt{3+\sin^2\phi}$.

[3] Properties of the eigenvalues of the unit deviator $\hat{\mathbf{S}}$ are given in eqns (5.4-7) and in Fig. 2 of LH [1991] and in eqns (A.6) of Loret [1992].

Deviatoric associativity is not a stringent restriction when modeling the behaviour of many geological materials: experimental evidence supporting this assumption is given in the above reference and in Lade and Musante [1977] for both clays and sands. However, these experiments essentially concern states near failure and the experimental validity of deviatoric associativity well inside the failure surface is still an open problem.

A.2 Acceleration Waves

Certain typical phenomena defining the material response, like strain localization, can be analyzed in a strictly static framework. However, they may sometimes advantageously be interpreted in a dynamic context which reveals how the mechanical information provided from the boundary propagates or diffuses through the body. On the other hand, the flutter phenomenon is intrinsically a dynamic phenomenon. Its characteristics will be revealed by the analysis of acceleration waves and harmonic waves.

That is why we turn now to the study of the propagation of acceleration waves, also called waves of second order.

A.2.1 Hadamard Compatibility Relations

At any point $\mathbf{x}$ and at any time t, the velocity $\mathbf{v}(\mathbf{x},t)$, the stress $\boldsymbol{\sigma}(\mathbf{x},t)$, and the mass density $\rho(\mathbf{x},t)$ remain continuous across a second order, or acceleration, wave front with normal $\mathbf{n}(\mathbf{x},t)$ which moves with speed c relative to the material. For any quantity $\phi(\mathbf{x},t)$ continuous across the wave front, Hadamard compatibility relations indicate that there exists a scalar η such that

$$\left[\!\!\left[\frac{\partial \phi}{\partial \mathbf{x}}\right]\!\!\right] = \eta\, \mathbf{n}\,, \quad [\![\dot{\phi}]\!] = -c\eta\,. \tag{A.2.1}$$

These conditions generalize easily to tensors of any order. Hadamard compatibility relations then require that the jump $[\![\partial \mathbf{v}/\partial \mathbf{x}]\!]$ of the velocity gradient and the jump $[\![\dot{\mathbf{v}}]\!]$ of the acceleration across the wave-front are not independent: there exists a discontinuity vector $\boldsymbol{\eta}=\boldsymbol{\eta}(\mathbf{x},t)$ which can be viewed as the jump of the velocity gradient along the normal $\mathbf{n}$, i.e. $\boldsymbol{\eta}=[\![\partial \mathbf{v}/\partial(\mathbf{x}\cdot\mathbf{n})]\!]$, such that

$$\left[\!\!\left[\frac{\partial \mathbf{v}}{\partial \mathbf{x}}\right]\!\!\right] = \boldsymbol{\eta} \otimes \mathbf{n}\,, \quad [\![\dot{\mathbf{v}}]\!] = -c\,\boldsymbol{\eta}\,. \tag{A.2.2}$$

Similarly for the stress with $\boldsymbol{\zeta} = [\![\partial\boldsymbol{\sigma}/\partial(\mathbf{x}\cdot\mathbf{n})]\!]$,

$$\left[\!\!\left[\frac{\partial \boldsymbol{\sigma}}{\partial \mathbf{x}}\right]\!\!\right] = \boldsymbol{\zeta} \otimes \mathbf{n}\,, \quad [\![\dot{\boldsymbol{\sigma}}]\!] = -c\,\boldsymbol{\zeta}\,. \tag{A.2.3}$$

Consequently, if the body force per unit mass $\mathbf{b}$ is continuous, taking the jump of the balance of linear momentum,

$$\operatorname{div}\boldsymbol{\sigma} + \rho\,(\mathbf{b} - \dot{\mathbf{v}}) = \mathbf{0}, \tag{A.2.4}$$

yields $\mathbf{n}\cdot\boldsymbol{\zeta} = -\rho\, c\,\boldsymbol{\eta}$, that is

$$\mathbf{n}\cdot[\![\dot{\boldsymbol{\sigma}}]\!] = \rho\, c^2\, \boldsymbol{\eta}\,. \tag{A.2.5}$$

A.2.2 Elastic and Elastic-Plastic Acoustic Tensors and Wave-Speeds

To pursue the analysis further, we assume the material response to follow the same constitutive cone on both sides of the wave-front, see Rice and Rudnicki [1980] for elaboration. Thus if the material response is plastic (respectively elastic), the disturbances are called plastic waves (respectively elastic waves). Inserting the rate constitutive equations (A.1.8) in (A.2.3) yields an eigenvalue problem for the jump $\boldsymbol{\eta}$:

$$(\mathbf{B}^{(\text{ep})} - c^2\,\mathbf{I}) \cdot \boldsymbol{\eta} = \mathbf{0}\,. \tag{A.2.6}$$

$\mathbf{B} = \mathbf{B}(\mathbf{x}, \mathbf{n})$ denotes the elastic acoustic tensor and $\mathbf{B}^{\text{ep}} = \mathbf{B}^{\text{ep}}(\mathbf{x}, t, \mathbf{n})$ the elastic-plastic acoustic tensor:

$$\mathbf{B}^{(\text{ep})} = \mathbf{n} \cdot \frac{\mathbb{E}^{(\text{ep})}}{\rho} \cdot \mathbf{n}\,, \quad B_{ij}^{(\text{ep})} = \frac{1}{\rho}\, n_k\, \mathbb{E}_{kilj}^{(\text{ep})} n_l\,. \tag{A.2.7}$$

The elastic acoustic tensor $\mathbf{B}$ takes the following form,

$$\mathbf{B} = (c_{\text{l}}^{\text{e}})^2\, \mathbf{n} \otimes \mathbf{n} + (c_{\text{s}}^{\text{e}})^2\, (\mathbf{I} - \mathbf{n} \otimes \mathbf{n})\,, \tag{A.2.8}$$

which shows explicitly that there exists one longitudinal wave with $\boldsymbol{\eta} \sim \mathbf{n}$ and with speed c_{l}^{e}, and two shear-waves with $\boldsymbol{\eta} \sim \mathbf{n}^{\perp}$ with common speed c_{s}^{e}, where

$$(c_{\text{l}}^{\text{e}})^2 = \frac{\lambda + 2\,\mu}{\rho} > (c_{\text{s}}^{\text{e}})^2 = \frac{\mu}{\rho} > 0\,. \tag{A.2.9}$$

The above inequalities are consequences of the positive definiteness of the elastic tensor $\mathbb{E}$, eq (A.1.2). It is useful to record the elastic acoustic tensor $\mathbf{B}$ raised to any power $a \in \mathbb{R}$,

$$(\mathbf{B})^a = (c_{\text{l}}^{\text{e}})^{2a}\, \mathbf{n} \otimes \mathbf{n} + (c_{\text{s}}^{\text{e}})^{2a}\, (\mathbf{I} - \mathbf{n} \otimes \mathbf{n})\,. \tag{A.2.10}$$

Setting $\boldsymbol{\epsilon} = 1/2(\mathbf{m} \otimes \mathbf{n} + \mathbf{n} \otimes \mathbf{m})$ in (A.1.2) shows that positive definiteness of the elastic tensor implies positive definiteness of the elastic acoustic tensor, referred to as *strong ellipticity* (SE), that is, $\forall\, \mathbf{m}$ and $\mathbf{n}$ unit vectors,

$$\text{(SE) of } \mathbb{E} \Leftrightarrow \mathbf{m} \cdot (\mathbf{n} \cdot \mathbb{E} \cdot \mathbf{n}) \cdot \mathbf{m} = (c_{\text{s}}^{\text{e}})^2 + \left((c_{\text{l}}^{\text{e}})^2 - (c_{\text{s}}^{\text{e}})^2\right) (\mathbf{m} \cdot \mathbf{n})^2 > 0\,. \tag{A.2.11}$$

Since $\mathbb{E}$ has the major symmetry, its strong ellipticity is equivalent to its *ellipticity* (E), or, in other words, the positive definiteness of the elastic acoustic tensor implies its invertibility:

$$\text{(E) of } \mathbb{E} \quad \Leftrightarrow \quad \det \mathbf{n} \cdot \frac{\mathbb{E}}{\rho} \cdot \mathbf{n} = \left(\frac{\mu}{\rho}\right)^2 \frac{\lambda + 2\,\mu}{\rho} \neq 0\,, \quad \forall \mathbf{n} \text{ unit vector}\,. \tag{A.2.12}$$

To obtain a tractable expression of the elastic-plastic acoustic tensor, it is instrumental to introduce the two vectors $\mathbf{p} = \mathbf{p}(\mathbf{n})$ and $\mathbf{q} = \mathbf{q}(\mathbf{n})$,

$$\begin{cases} \mathbf{p} = (\mathbb{E} : \mathbf{P}) \cdot \mathbf{n} = \lambda \operatorname{tr} \mathbf{P} \cdot \mathbf{n} + 2\,\mu\, \mathbf{P} \cdot \mathbf{n}\,, \\ \mathbf{q} = \mathbf{n} \cdot (\mathbf{Q} : \mathbb{E}) = \lambda \operatorname{tr} \mathbf{Q} \cdot \mathbf{n} + 2\,\mu\, \mathbf{Q} \cdot \mathbf{n} \end{cases} \tag{A.2.13}$$

the scalars $x = x(\mathbf{n})$, $y = y(\mathbf{n})$,

$$x = -\frac{1}{2\,\mu\rho}(\mathbf{p}\cdot\mathbf{n})(\mathbf{q}\cdot\mathbf{n})\,,\; y = \frac{1}{2\,\mu\rho}\Big(\mathbf{p}\cdot\mathbf{q} - (\mathbf{p}\cdot\mathbf{n})(\mathbf{q}\cdot\mathbf{n})\Big)\,, \tag{A.2.14}$$

and the normalized modulus $\hat{H}$,

$$\hat{H} = \frac{H}{2\,\mu} > 0\,. \tag{A.2.15}$$

Using the decomposition (A.1.9), x and y can be cast in the more explicit format

$$\begin{cases} x = -2(c_s^e)^2 \cos\chi\cos\psi\;(N_P + \mathbf{n}\cdot\hat{\mathbf{S}}_P\cdot\mathbf{n})(N_Q + \mathbf{n}\cdot\hat{\mathbf{S}}_Q\cdot\mathbf{n})\,, \\ y = \;\;2(c_s^e)^2 \cos\chi\cos\psi\;(\mathbf{n}\cdot\hat{\mathbf{S}}_P)\cdot(\mathbf{I}-\mathbf{n}\otimes\mathbf{n})\cdot(\hat{\mathbf{S}}_Q\cdot\mathbf{n})\,, \end{cases} \tag{A.2.16}$$

with

$$N_P = \sqrt{3}\,\tan\chi\,\frac{\lambda + 2\,\mu/3}{2\,\mu}\,,\quad N_Q = \sqrt{3}\,\tan\psi\,\frac{\lambda + 2\,\mu/3}{2\,\mu}\,. \tag{A.2.17}$$

Then

$$\mathbf{B}^{\mathrm{ep}} = \mathbf{B} - \frac{1}{H}\frac{\mathbf{p}\otimes\mathbf{q}}{\rho}\,, \tag{A.2.18}$$

For the jumps η to be non-zero, the wave-speeds c must satisfy the characteristic equation $\det(\mathbf{B}^{\mathrm{ep}} - c^2\,\mathbf{I})=0$, which can be factorized as follows

$$\det(\mathbf{B}^{\mathrm{ep}} - c^2\,\mathbf{I}) = (\,c^2 - (c_s^e)^2\,)\;F(c) \tag{A.2.19}$$

with

$$F(c) = (c^2)^2 - \Big((c_l^e)^2 + (c_s^e)^2 + \frac{x-y}{\hat{H}}\Big)\,c^2 + (c_l^e c_s^e)^2 - (c_l^e)^2\frac{y}{\hat{H}} + (c_s^e)^2\frac{x}{\hat{H}}\,. \tag{A.2.20}$$

The above factorization proves

Proposition A.2.1 Existence of a neutral wave for isotropic elasticity, Mandel [1962].
For any direction of propagation $\mathbf{n}$*, at least one elastic-plastic wave-speed is equal to* c_s^e *and the associated eigenmode* η *is orthogonal to both* $\mathbf{n}$ *and* $\mathbf{q}$*: because of the latter property, the corresponding wave is called* a neutral wave.

If $c=0$ is solution of the characteristic equation, then one speaks of a *stationary discontinuity*: because the mechanical information can no longer propagate in the directions $\mathbf{n}$, the strains accumulate in zones orthogonal to these directions and called *shear-bands*, a phenomenon referred to as *strain localization*. For associative plasticity, the elastic-plastic acoustic tensor is symmetric and the squares of the three wave-speeds are always real, although one of them might be negative (see Sect. A.3.3 below). However for non-associative plasticity, the squares of the wave-speeds are a priori not guaranteed to be real: if the two solutions c^2 of (A.2.20) are complex conjugate, one speaks of *flutter instability*, a term reminiscent of the oscillating behaviour familiar in structural dynamics. Section A.3 shows conditions sufficient to exclude flutter and Sections A.4-A.6 show sufficient conditions for its existence. While, in a general context, neither necessary for its existence nor sufficient to exclude it, deviatoric associativity plays an interesting role in these analyses.

A.3 Deviatoric Associativity with Isotropic Elasticity Excludes Flutter

A.3.1 Conditions for Flutter Instability

For the roots c^2 of the characteristic equation $F(c) = 0$, eqn (A.2.20), to be complex conjugate, its discriminant, or equivalently the function $\Delta(\hat{H})$

$$\Delta(\hat{H}) = \Big(\frac{1}{\hat{H}}\Big)^2 (y-x)^2 + 2\Big(\frac{1}{\hat{H}}\Big)(y+x)\,\tau + \tau^2 < 0\,, \qquad \text{(A.3.1)}$$

should be negative. Now, the problem is to locate the roots $\hat{H}$ of $\Delta(\hat{H}) = 0$. For $\Delta(\hat{H})$ to have positive roots, the following inequalities must hold

$$-\frac{x+y}{\tau} > 0\,, \quad 4\frac{x\,y}{\tau^2} > 0\,, \qquad \text{(A.3.2)}$$

because the scalar τ,

$$\tau = (c_1^{\rm e})^2 - (c_{\rm s}^{\rm e})^2 > 0 \qquad \text{(A.3.3)}$$

is strictly positive as a consequence of the strong ellipticity of $\mathbb{E}$.

A.3.2 Exclusion of Flutter

Let us now assume deviatoric associativity to hold: y, eqn (A.2.14), becomes

$$y = 2\,(c_{\rm s}^{\rm e})^2 \cos\chi \cos\psi \;(\mathbf{n}\cdot\hat{\mathbf{S}})\cdot(\mathbf{I}-\mathbf{n}\otimes\mathbf{n})\cdot(\hat{\mathbf{S}}\cdot\mathbf{n}) \geq 0\,, \qquad \text{(A.3.4)}$$

which is clearly positive or zero because $\mathbf{I}-\mathbf{n}\otimes\mathbf{n}$ is a projection operator. Therefore, it is not possible to satisfy both inequalities (A.3.2) and thus the squares of the wave-speeds can not be complex conjugate. Thus we have proved,

Proposition A.3.1 A case of exclusion of flutter, Loret, Prévost and Harireche [1990].
For an elastic-plastic material with isotropic elasticity and deviatoric associativity, the squares of the body waves are real and therefore flutter instability is excluded.

Finally, let us note that Ottosen and Runesson [1991a] have improved slightly upon Proposition A.3.1 when $0° \leq \chi \leq \psi \leq 90°$: if $\mathbf{P}$ and $\mathbf{Q}$ are coaxial with eigenvalues and eigendirections arranged in the same order, they have shown that flutter instability is also excluded.

A.3.3 Relative Positions of Elastic and Elastic-Plastic Wave-Speeds

For an associative plastic behaviour with arbitrary elastic anisotropy, Mandel [1962] has proved that the squares of the three elastic-plastic wave-speeds are interlaced by their elastic counterparts[4], $(c_3^{\rm P})^2 \leq (c_3^{\rm e})^2 \leq (c_2^{\rm P})^2 \leq (c_2^{\rm e})^2 \leq (c_1^{\rm P})^2 \leq (c_1^{\rm e})^2$. In fact, this interlacing property forces one of the plastic wave-speeds to be equal to the shear wave-speed when elasticity is isotropic, since then the latter has multiplicity two, $c_3^{\rm e}=c_2^{\rm e}=c_{\rm s}^{\rm e}$:

$$(c_3^{\rm P})^2 \leq (c_2^{\rm P})^2 = (c_{\rm s}^{\rm e})^2 \leq (c_1^{\rm P})^2 \leq (c_1^{\rm e})^2\,. \qquad \text{(A.3.5)}$$

[4] This property extends to the four wave-speeds of fluid-saturated porous media, Loret [1990].

Incidentally, notice that these inequalities constitute another proof of Proposition A.2.1.

However, the neutral wave also exists for non-associative plasticity, but then the above interlacing property does not hold. In fact, since

$$F(c_1^e) = -\frac{\mathcal{T}}{\hat{H}}\,x\,, \quad F(c_s^e) = -\frac{\mathcal{T}}{\hat{H}}\,y \leq 0\,, \tag{A.3.6}$$

the squares of the elastic and elastic-plastic wave-speeds can be ordered as follows:

Proposition A.3.2 Relative positions of elastic and elastic-plastic wave-speeds.

$$(c_3^P)^2 \leq (c_2^P)^2 = (c_s^e)^2 \leq \begin{cases} (c_1^P)^2 \leq (c_1^e)^2 & \text{if } x \leq 0\,, \\ (c_1^e)^2 \leq (c_1^P)^2 & \text{if } x \geq 0\,. \end{cases} \tag{A.3.7}$$

For both associative and non-associative plasticity, one wave-speed, namely c_3^P, may be zero or purely imaginary. For associative plasticity, x, eqn (A.2.14), is negative or zero, while x may be positive for non-associative plasticity. Indeed, for definiteness, assume $0° \leq \chi \leq \psi \leq 90°$ so that $N_Q \geq N_P \geq 0$. Therefore $x(\mathbf{n})$ is positive for $\mathbf{n} \cdot \hat{\mathbf{S}} \cdot \mathbf{n}$ belonging to the interval $[-N_Q, -N_P]$. On the other hand, it is easy to see that the range of $\mathbf{n} \cdot \hat{\mathbf{S}} \cdot \mathbf{n}$ is $[\hat{S}_3 < 0, \hat{S}_1 > 0]$ where $\hat{S}_3$ and $\hat{S}_1$ are respectively the smallest and largest eigenvalues of $\hat{\mathbf{S}}$. Thus $x(\mathbf{n})$ remains negative for all directions $\mathbf{n}$ if the intervals $[\hat{S}_3, \hat{S}_1]$ and $[-N_Q, -N_P]$ are disjoint like for associative plasticity. On the other hand, a small dilatancy angle χ and a large friction angle ψ will favor the existence of directions of space $\mathbf{n}$ in which $x(\mathbf{n})$ is positive and therefore the unusual ordering $(A.3.7)_2$ holds.

A.3.4 Wave-Speeds of Higher Multiplicity

In relation with the above ordering of wave-speeds, one can now sharpen Proposition A.2.1. Indeed, one can prove the following

Proposition A.3.3 Non strict hyperbolicity, Loret [1992]. [5]

For the elastic-plastic constitutive equations considered here with deviatoric associativity, the dynamical equations of motion are not strictly hyperbolic.

In fact, for any modulus $\hat{H} > 0$, the wave-speeds in the directions of the eigenvectors of $\hat{\mathbf{S}}$ are of multiplicity at least two and they are equal to the elastic shear wave-speed c_s^e.

The proof of this assertion is given now. For $\mathbf{B}^{ep}$ to have an eigenvalue of multiplicity at least two, a single condition must be satisfied, namely:

$$-y(\mathbf{n}) = 0\,, \tag{A.3.8}$$

[5] The observation that elastic-plastic equations are not strictly hyperbolic has been reported earlier by An and Schaeffer [1992]. However, their example is two-dimensional; therefore, their proof does not carry over to the present framework: their analysis uses a restriction on Poisson's ratio that does no longer appear in the three-dimensional context where there always exists at least one neutral wave traveling at the same speed as the elastic shear-waves.

as results from examination of (A.3.1-A.3.2). This condition can be realized only if $\mathbf{n}$ is one of the eigenvectors of $\hat{\mathbf{S}}$, say $\mathbf{n} = \mathbf{E}_i, i \in [1,3]$; then, by use of the expression of $x(\mathbf{E}_i)$, eqn (A.2.16), with(A.1.10), $\mathbf{B}^{\mathrm{ep}}$ can be recast as:

$$\mathbf{B}^{\mathrm{ep}}(\mathbf{E}_i) = \left((c_1^{\mathrm{e}})^2 + \frac{x(\mathbf{E}_i)}{\hat{H}}\right) \mathbf{E}_i \otimes \mathbf{E}_i + (c_{\mathrm{s}}^{\mathrm{e}})^2\ (\mathbf{I} - \mathbf{E}_i \otimes \mathbf{E}_i) \quad (\text{no sum over i}) . \quad (A.3.9)$$

Therefore, in the eigendirections of $\hat{\mathbf{S}}$, $\mathbf{B}^{\mathrm{ep}}$ is principal and the elastic shear-wave-speed $c_{\mathrm{s}}^{\mathrm{e}}$ has multiplicity at least two. The disturbances traveling along these directions are *neutral*, that is they carry a stress-rate jump that is tangent to the yield surface: this can be shown by computing the strain-rate jump $[\![\dot{\epsilon}]\!]$ through (A.2.2)-(A.2.6) with $c = c_{\mathrm{s}}^{\mathrm{e}}$ and then by realizing that the loading/unloading index $\mathbf{Q} : \mathbb{E} : [\![\dot{\epsilon}]\!]$ vanishes. Finally the elastic shear wave-speed $c_{\mathrm{s}}^{\mathrm{e}}$ may be of multiplicity three in the subset of $\{\mathbf{E}_i, i \in [1,3]\}$ that satisfies the condition $x(\mathbf{E}_i) < 0$ when the modulus $\hat{H}$ takes the *discrete* values:

$$\hat{H} = \frac{-x(\mathbf{E}_i)}{\tau} > 0 . \quad (A.3.10)$$

This subset is never empty. Indeed, if, during a loading process, the modulus $\hat{H}$ decreases monotonously, the other state variables being kept constant, the condition for $c_{\mathrm{s}}^{\mathrm{e}}$ to be a wave-speed of multiplicity three will be met for the critical directions $\mathbf{n}^* \in \{\mathbf{E}_i, i \in [1,3]\}$ that maximize $\hat{H}$ given by (A.3.10), namely:

$$-\text{ for}\quad N_P + N_Q \geq \hat{S}_2 \ :\ \mathbf{n}^* = \begin{cases} \mathbf{E}_1 & \text{if } \hat{S}_1 > \hat{S}_2 \\ \langle \mathbf{E}_1, \mathbf{E}_2 \rangle & \text{if } \hat{S}_1 = \hat{S}_2 \end{cases} , \quad \hat{H}^* = \frac{-x(\mathbf{E}_1)}{\tau} > 0 , \quad (A.3.11)$$

$$-\text{ for}\quad N_P + N_Q \leq \hat{S}_2 \ :\ \mathbf{n}^* = \begin{cases} \mathbf{E}_3 & \text{if } \hat{S}_2 > \hat{S}_3 \\ \langle \mathbf{E}_2, \mathbf{E}_3 \rangle & \text{if } \hat{S}_2 = \hat{S}_3 \end{cases} , \quad \hat{H}^* = \frac{-x(\mathbf{E}_3)}{\tau} > 0 . \quad (A.3.12)$$

In the above formulas, $\langle \mathbf{E}_i, \mathbf{E}_j \rangle$, $i \neq j$, denotes the two-dimensional vector space generated by $\mathbf{E}_i$ and $\mathbf{E}_j$. To derive the above results, one has used once again the fact that, for any unit vector $\mathbf{n}$, $\mathbf{n} \cdot \hat{\mathbf{S}} \cdot \mathbf{n}$ belongs to the interval $[\hat{S}_3 < 0, \hat{S}_1 > 0]$.

A.4 Deviation from Deviatoric Associativity but with Isotropic Elasticity

Proposition A.3.1 shows that non-associative plasticity is a necessary but not sufficient condition for the existence of flutter. On the other hand, since the waves propagating in the eigendirections of $\hat{\mathbf{S}}$ are of multiplicity at least two, one wonders whether or not any infinitesimal deviation from deviatoric associativity is a sufficient condition for the onset of flutter instability to be met. The answer to this problem turns out to be strongly dependent on the anisotropic character and on the coaxial properties of the constitutive equations, Loret [1992].

According to (A.3.1),(A.3.2), flutter instability will occur if there exists a set of unit vectors $\mathcal{N}_{\mathrm{flu}}$ for which both scalars $-x(\mathbf{n})$ and $-y(\mathbf{n})$, eqn (A.2.16), are strictly positive (recall from (A.3.4) that $-y(\mathbf{n})$ is always negative or zero for deviatoric associativity) and if the normalized modulus $\hat{H}$ belongs to the interval $]\hat{H}_-(\mathbf{n}), \hat{H}_+(\mathbf{n})[$ where:

$$\hat{H}_-(\mathbf{n}) = \frac{1}{\tau}\left(\sqrt{-x(\mathbf{n})} - \sqrt{-y(\mathbf{n})}\right)^2 \geq 0, \quad \hat{H}_+(\mathbf{n}) = \frac{1}{\tau}\left(\sqrt{-x(\mathbf{n})} + \sqrt{-y(\mathbf{n})}\right)^2 > 0 , \quad (A.4.1)$$

so that $\Delta(\hat{H})$ is negative. Thus if during a loading process, the modulus $\hat{H}$ decreases monoto-nously while the other state variables are kept constant, the onset of flutter instability will be met for $\hat{H} = \hat{H}^{\text{flu}}$:

$$\hat{H}^{\text{flu}} = \max_{\mathbf{n}\in\mathcal{N}_{\text{flu}}} \frac{1}{\mathcal{T}} \left(\sqrt{-x(\mathbf{n})} + \sqrt{-y(\mathbf{n})}\right)^2 . \tag{A.4.2}$$

Let $\mathbf{n} = \mathbf{n}^{\text{flu}}$ be a direction that realizes this maximum. For $\hat{H} = \hat{H}^{\text{flu}}$, the discriminant $\Delta(\hat{H})$, eqn (A.3.1), is zero and it can be shown that the square c^2_{flu} of the wave-speed of multiplicity 2 associated to a direction $\mathbf{n}^{\text{flu}}$ belongs to the interval $](c^{\text{e}}_{\text{s}})^2, (c^{\text{e}}_{\text{l}})^2[$; indeed by insertion of the expression of $\hat{H}_+(\mathbf{n}^{\text{flu}})$, eqn (A.4.1), in the characteristic equation (A.2.20), one obtains:

$$(c^{\text{e}}_{\text{l}})^2 > c^2_{\text{flu}} = \frac{(c^{\text{e}}_{\text{l}})^2\sqrt{-y(\mathbf{n}^{\text{flu}})} + (c^{\text{e}}_{\text{s}})^2\sqrt{-x(\mathbf{n}^{\text{flu}})}}{\sqrt{-y(\mathbf{n}^{\text{flu}})} + \sqrt{-x(\mathbf{n}^{\text{flu}})}} > (c^{\text{e}}_{\text{s}})^2 . \tag{A.4.3}$$

A.4.1 Generic Constitutive Perturbations from Deviatoric Associativity

Now the key point is to determine whether or not any small deviation from deviatoric associativity ensures the existence of a non empty set $\mathcal{N}_{\text{flu}}$. For that purpose, we first define generic constitutive perturbations from an elastic-plastic model that satisfies deviatoric associativity. Let us consider an elastic-plastic model whose unit normals to the plastic potential and yield surface, $\mathbf{P}$ and $\mathbf{Q}$ respectively, are expressed in the general format (A.1.9). Without loss of generality, the unit deviators $\hat{\mathbf{S}}_P$ and $\hat{\mathbf{S}}_Q$ assumed to be non-colinear can be written in the following form:

$$\hat{\mathbf{S}}_P = \cos\alpha\ \hat{\mathbf{S}} - \sin\alpha\ \Delta\hat{\mathbf{S}} , \quad \hat{\mathbf{S}}_Q = \cos\alpha\ \hat{\mathbf{S}} + \sin\alpha\ \Delta\hat{\mathbf{S}} , \tag{A.4.4}$$

where we introduce the unit deviators $\hat{\mathbf{S}}$ and $\Delta\hat{\mathbf{S}}$ which are orthogonal and the angle $\alpha \in]0, \pi/2[$:

$$\hat{\mathbf{S}} = \frac{\hat{\mathbf{S}}_Q + \hat{\mathbf{S}}_P}{\| \hat{\mathbf{S}}_Q + \hat{\mathbf{S}}_P \|} , \quad \Delta\hat{\mathbf{S}} = \frac{\hat{\mathbf{S}}_Q - \hat{\mathbf{S}}_P}{\| \hat{\mathbf{S}}_Q - \hat{\mathbf{S}}_P \|} , \quad \tan\alpha = \frac{\| \hat{\mathbf{S}}_Q - \hat{\mathbf{S}}_P \|}{\| \hat{\mathbf{S}}_Q + \hat{\mathbf{S}}_P \|} , \tag{A.4.5}$$

and

$$\| \hat{\mathbf{S}} \| = 1, \quad \| \Delta\hat{\mathbf{S}} \| = 1, \quad \hat{\mathbf{S}} : \Delta\hat{\mathbf{S}} = 0 . \tag{A.4.6}$$

A.4.2 Non-Coaxial Perturbations

The question posed above, namely whether or not any infinitesimal constitutive perturbation from deviatoric associativity leads to flutter instability, can now be rephrased as follows: what is the smallest value α that implies the existence of a non-empty set $\mathcal{N}_{\text{flu}}$?

The analysis delineates three cases, namely the case where the eigenspaces of $\hat{\mathbf{S}}_P$ and $\hat{\mathbf{S}}_Q$ are identical, the case where these eigenspaces are disjoint and the case where they share a single vector. Here, we will be content to analyze the first case and to report only the main features of the second case; see Loret [1992] for details. So, let us first announce

Proposition A.4.1 Non-coaxial perturbations from deviatoric associativity, Loret [1992]
Any infinitesimal non-coaxial deviation from deviatoric associativity induces the onset of flutter instability if the modulus $\hat{H}$ is chosen in the interval defined by (A.4.1); the critical directions in which the wave-speeds become complex conjugate are close to the directions in which strict hyperbolicity is lost.

The proof is as follows. In order to determine the set $\mathcal{N}_{\rm flu}$, it is instrumental to rewrite first the scalar $y(\mathbf{n})$, eqn (A.2.16), using the expressions (A.4.4) for $\hat{\mathbf{S}}_P$ and $\hat{\mathbf{S}}_Q$:

$$-y(\mathbf{n}) = 2\,(c_{\rm s}^{\rm e})^2\ \cos\psi\cos\chi\,\big(-\cos^2\alpha\ \Phi(\hat{\mathbf{S}},\mathbf{n}) + \sin^2\alpha\ \Phi(\Delta\hat{\mathbf{S}},\mathbf{n})\big)\,, \tag{A.4.7}$$

where, since $\mathbf{I}-\mathbf{n}\otimes\mathbf{n}$ is a projection operator, Φ is positive or zero:

$$\Phi(\hat{\mathbf{S}},\mathbf{n}) \equiv (\mathbf{n}\cdot\hat{\mathbf{S}})\cdot(\,\mathbf{I}-\mathbf{n}\otimes\mathbf{n}\,)\cdot(\mathbf{n}\cdot\hat{\mathbf{S}}) \geq 0\,. \tag{A.4.8}$$

Therefore, when the eigenspaces of $\hat{\mathbf{S}}$ and $\Delta\hat{\mathbf{S}}$ are disjoint, $\Phi(\hat{\mathbf{S}},\mathbf{E}_i)$ is zero for any eigenvector $\mathbf{E}_i$ of $\hat{\mathbf{S}}$ and $\Phi(\Delta\hat{\mathbf{S}},\mathbf{E}_i)$ is strictly positive. Thus $-y(\mathbf{E}_i)$ is strictly positive for any $\alpha\neq 0$. On the other hand, we have shown that, for deviatoric associativity defined by $\alpha=0$, there exists at least one direction $\mathbf{E}_i$ for which $-x(\mathbf{E}_i)$ is a strictly positive quantity, eqns (A.3.11),(A.3.12). Therefore since $-x$ and $-y$ are both continuous functions of $\mathbf{n}$ and α, there exists a fan of directions $\mathbf{n}$ around at least one $\mathbf{E}_i$ and a range of variation of α, say $\alpha\in\,]0,\alpha_0[\,\subset\,]0,\pi/2[$ for which both $-x(\mathbf{n})$ and $-y(\mathbf{n})$ are strictly positive. Thus $\mathcal{N}_{\rm flu}$ is not empty and the onset of flutter instability is met if the modulus $\hat{H}$ is chosen in the interval defined by (A.4.1).

A.4.3 Coaxial Perturbations

Let us now turn to the case of coaxial perturbations, where the eigenspaces of $\hat{\mathbf{S}}_P$ and $\hat{\mathbf{S}}_Q$ are identical; then so are the eigenspaces of $\hat{\mathbf{S}}$ and $\Delta\hat{\mathbf{S}}$. Consequently, according to (A.4.7),(A.4.8), $-y(\mathbf{n})$ is zero for any $\mathbf{n}=\mathbf{E}_i$ eigenvector of $\hat{\mathbf{S}}$ (and $\Delta\hat{\mathbf{S}}$) and the arguments used above do not apply. Incidentally, it is interesting to observe that, in contrast to the non-coaxial case, the eigenspace of $\hat{\mathbf{S}}$, for which strict hyperbolicity is lost for $\alpha=0$, never undergoes flutter instability; conversely, in the non-coaxial case, these directions are the most likely to undergo flutter instability. To investigate the existence of a non-empty set $\mathcal{N}_{flu}$ is more intricate than in the non-coaxial case. We just give the essential steps and refer to Loret [1992] for details.

Using the spectral properties of the unit deviators $\hat{\mathbf{S}}_P$ and $\hat{\mathbf{S}}_Q$, one first rewrites $-y(\mathbf{n})$ in the format

$$-y(\mathbf{n}) = 2\,(c_{\rm s}^{\rm e})^2\ \cos\psi\cos\chi\,\big(-\Phi(\hat{\mathbf{S}},\mathbf{n}) + \sin^2\alpha\,(\,1-\textstyle\sum_{i=1}^3 n_i^4\,)\big)\,. \tag{A.4.9}$$

One can then show that there exist a value $\alpha_{\rm flu}>0$ of α and directions $\mathbf{n}$ belonging to eigenplanes of $\hat{\mathbf{S}}$, eqn (A.4.5), for which $-y(\mathbf{n})$ is strictly positive:

$$-y(\mathbf{n}) = 2\,(c_{\rm s}^{\rm e})^2\ \cos\psi\cos\chi\ n_i^2\,n_j^2\,\big(-(\hat{S}_i-\hat{S}_j)^2 + 2\,\sin^2\alpha\big)\,,\ \text{ no sum over } i,\,j\,. \tag{A.4.10}$$

In fact one can even show that the maximum of $-y(\mathbf{n})$ occurs for $\mathbf{n}$ belonging to an eigenplane of $\hat{\mathbf{S}}$. The sign of $-x(\mathbf{n})$ is more difficult to study in general but, for sufficiently small N_P and large N_Q, $-x(\mathbf{n})$ takes positive values for any direction $\mathbf{n}$: therefore flutter will occur for $\hat{H} \leq \hat{H}^{\mathrm{flu}}$ defined by (A.4.2). The critical values α_{flu} and $\hat{H}^{\mathrm{flu}}$ depend in a simple way of the Lode angle associated to $\hat{\mathbf{S}}$ (eqns (4.17),(4.18),(4.20) in Loret [1992]). These results are summarized in

Proposition A.4.2 Coaxial perturbations from deviatoric associativity, Loret [1992].
When the normals to the yield surface and to the plastic potential are coaxial, the amount of deviation with respect to deviatoric associativity necessary to induce the onset of flutter instability is found to be finite *and to depend on the triaxiality of the mean deviator* $\hat{\mathbf{S}}$, *eqn (A.4.5). Moreover, the squares of wave-speeds along the eigendirections of* $\hat{\mathbf{S}}_P$ *and* $\hat{\mathbf{S}}_Q$ *are always real.*

A.5 Deviatoric Associativity with Anisotropic Elasticity

We now examine the effects of another type of deviation with respect to deviatoric associativity with elastic isotropy, which was shown to exclude flutter in Sect. A.3. Unlike in the previous section where the perturbation concerned the plastic part of the constitutive equations, we examine now the effects of deviation with respect to elastic isotropy for elastic-plastic models that obey deviatoric associativity. We will show that the two types of deviations share some similarity. Indeed, the analysis will highlight another coaxiality property, now between the fabric tensors $\mathbf{F}$ or $\mathbf{G}$ that define the elastic anisotropy and $\hat{\mathbf{S}}$ which condenses the plastic directional properties. The pervasive influence of noncoaxiality on the onset of flutter will also be noted in the analysis of the effects of a traction free boundary not coaxial with the axes of material orthotropy in Sect. A.6.

A.5.1 Elastic Anisotropy

When the elasticity is isotropic, a key property that facilitates the analysis of elastic-plastic wave-speeds is the fact that the elastic shear wave-speed has double multiplicity, Proposition A.2.1; since the elastic-plastic acoustic tensor is a rank-one modification of its elastic counterpart, the elastic shear wave speed is a neutral elastic-plastic wave speed. This property is lost for general anisotropy. However, the elastic anisotropy developed by Bigoni and Loret [1999] and exposed below is tailored to display the double multiplicity property.

The elastic tensor is obtained by perturbation of isotropy through an arbitrary nondimensional scalar φ and a symmetric and traceless second order fabric tensor $\mathbf{G}$, namely

$$\mathbb{E} = \lambda \mathbf{I} \otimes \mathbf{I} + 2\,\mu\, \mathbf{I} \underline{\overline{\otimes}} \mathbf{I} + (\lambda + \mu)\,(\mathbf{N} \otimes \mathbf{N} - \mathbf{I} \otimes \mathbf{I})\,, \tag{A.5.1}$$

where

$$\mathbf{N} = \mathbf{I} + \varphi\,\mathbf{G}\,. \tag{A.5.2}$$

This type of elasticity can be shown to be a particular case of elastic anisotropy build from a second order tensor using theorems of representation. The Cauchy-Schwarz inequality implies, for all $\boldsymbol{\epsilon}=\boldsymbol{\epsilon}^{\mathrm{T}}\neq\mathbf{0}$,

$$\boldsymbol{\epsilon} : \mathbb{E} : \boldsymbol{\epsilon} \geq \lambda\,(\mathrm{tr}\,\boldsymbol{\epsilon})^2 + 2\,(\mu - \sqrt{3}\,|\lambda + \mu|\;||\varphi\,\mathbf{G}||\,)\,\boldsymbol{\epsilon} : \boldsymbol{\epsilon} + (\lambda + \mu)\,(\varphi\,\mathbf{G} : \boldsymbol{\epsilon})^2\,, \tag{A.5.3}$$

from which the following *sufficient* condition for positive definiteness, and thus strong ellipticity, of $\mathbb{E}$ is obtained:

$$\lambda + \frac{2}{3}\mu > \frac{2}{\sqrt{3}}|\lambda+\mu| \, ||\varphi \mathbf{G}||, \; \mu > \sqrt{3}|\lambda+\mu| \, ||\varphi \mathbf{G}|| \Longrightarrow \mathbb{E} \; (\mathrm{PD}). \tag{A.5.4}$$

Therefore $\lambda+\mu$ and μ are positive, and positive definiteness of $\mathbb{E}$ bounds the anisotropic perturbation,

$$||\varphi \mathbf{G}|| < \frac{\min\{3\lambda/2+\mu, \mu\}}{\sqrt{3}(\lambda+\mu)}. \tag{A.5.5}$$

The elastic acoustic tensor takes the form

$$\mathbf{B}(\mathbf{n}) = \frac{\lambda+\mu}{\rho}\tilde{n}^2 \, \tilde{\mathbf{n}} \otimes \tilde{\mathbf{n}} + \frac{\mu}{\rho}\mathbf{I}, \tag{A.5.6}$$

where $\tilde{\mathbf{n}}$ is the unit vector defined as

$$\tilde{n}\,\tilde{\mathbf{n}} = \mathbf{N}\cdot\mathbf{n}, \qquad \tilde{n} = ||\mathbf{N}\cdot\mathbf{n}|| \;. \tag{A.5.7}$$

Thus a *quasi*-longitudinal mode of propagation along $\tilde{\mathbf{n}}$ with speed c_1^e exists, while all directions in the plane orthogonal to $\tilde{\mathbf{n}}$ are *quasi*-transversal modes with speed c_s^e,

$$(c_1^e)^2 = \frac{\lambda+\mu}{\rho}\tilde{n}^2 + \frac{\mu}{\rho} > (c_s^e)^2 = \frac{\mu}{\rho}. \tag{A.5.8}$$

A.5.2 The Elastic-Plastic Model

When the elastic tensor (A.5.1) is used to build an elastic-plastic model, the resulting elastic-plastic acoustic tensor is given in the format

$$\mathbf{B}^{\mathrm{ep}} = (c_1^e)^2\,(\tilde{n})^2\,\tilde{\mathbf{n}}\otimes\tilde{\mathbf{n}} + (c_s^e)^2\,(\mathbf{I} - (\tilde{n})^2\,\tilde{\mathbf{n}}\otimes\tilde{\mathbf{n}}) - \frac{1}{H}\frac{\mathbf{p}\otimes\mathbf{q}}{\rho}, \tag{A.5.9}$$

with $\mathbf{p} = \mathbf{p}(\mathbf{n}) = (\mathbb{E}:\mathbf{P})\cdot\mathbf{n}$, $\mathbf{q} = \mathbf{q}(\mathbf{n}) = \mathbf{n}\cdot(\mathbf{Q}:\mathbb{E})$ and $H = h + \mathbf{P}:\mathbb{E}:\mathbf{Q}$ exactly as in (A.2.13). Consequently, the eigenmode of propagation in the plane orthogonal to $\tilde{\mathbf{n}}$ and $\mathbf{q}$ is still associated with the quasi-transversal elastic wave-speed c_s^e, and the other two (unordered) elastic-plastic wave-speeds c_1^p and c_2^p are solutions of the equations

$$(c_1^p)^2 + (c_2^p)^2 = (c_1^e)^2 + (c_s^e)^2 + \frac{x-y}{\hat{H}}, \quad (c_1^p)^2\,(c_2^p)^2 = (c_1^e)^2\,(c_s^e)^2 - (c_1^e)^2\frac{y}{\hat{H}} + (c_s^e)^2\frac{x}{\hat{H}}, \tag{A.5.10}$$

exactly as in (A.2.20), but with $x = x(\mathbf{n},\tilde{\mathbf{n}})$ and $y = y(\mathbf{n},\tilde{\mathbf{n}})$ now defined as:

$$x(\mathbf{n},\tilde{\mathbf{n}}) = -\frac{1}{2\,\mu\rho}(\mathbf{p}(\mathbf{n})\cdot\tilde{\mathbf{n}})(\mathbf{q}(\mathbf{n})\cdot\tilde{\mathbf{n}}), \; y(\mathbf{n},\tilde{\mathbf{n}}) = \frac{1}{2\,\mu\rho}\Big(\mathbf{p}(\mathbf{n})\cdot\mathbf{q}(\mathbf{n}) - (\mathbf{p}(\mathbf{n})\cdot\tilde{\mathbf{n}})(\mathbf{q}(\mathbf{n})\cdot\tilde{\mathbf{n}})\Big). \tag{A.5.11}$$

Necessary and sufficient conditions for the squares of the elastic-plastic wave-speeds in the direction $\mathbf{n}$ to be complex conjugates are, as in Sect. A.4, that both $-x(\mathbf{n},\tilde{\mathbf{n}})$ and

$-y(\mathbf{n},\tilde{\mathbf{n}})$ be strictly positive and that the normalized modulus $\hat{H}$ belongs to the interval $]\hat{H}_-, \hat{H}_+[$,

$$\hat{H}_\pm(\mathbf{n}) = \frac{1}{\tau}\left(\sqrt{-x(\mathbf{n},\tilde{\mathbf{n}})} \pm \sqrt{-y(\mathbf{n},\tilde{\mathbf{n}})}\right)^2 \geq 0, \tag{A.5.12}$$

with τ still defined by (A.3.3) but with elastic wave-speeds defined by(A.5.8) above.

The analysis of the sign conditions of $-x$ and $-y$ delineates three cases which highlight the importance of the noncoaxiality of $\mathbf{G}$ and $\hat{\mathbf{S}}$.

A.5.3 Non-Coaxial Fabric

Let us first consider the sign of $-y$. To this purpose, it is useful to introduce the projection operator $\mathbf{J}$ onto the plane orthogonal to $\tilde{\mathbf{n}}$,

$$\mathbf{J} = \mathbf{I} - \tilde{\mathbf{n}} \otimes \tilde{\mathbf{n}}. \tag{A.5.13}$$

Therefore, $-y$ may be recast in a compact form, namely

$$-2\,\mu\,\rho\,y = -\mathbf{p}(\mathbf{n}) \cdot (\mathbf{J} \cdot \mathbf{q}(\mathbf{n})) = -(\mathbf{J} \cdot \mathbf{p}(\mathbf{n})) \cdot (\mathbf{J} \cdot \mathbf{q}(\mathbf{n})). \tag{A.5.14}$$

Since $\mathbf{J} \cdot (\mathbf{N} \cdot \mathbf{n}) = \mathbf{0}$, $\mathbf{J} \cdot \mathbf{p}(\mathbf{n})$ may be simplified to

$$\mathbf{J} \cdot \mathbf{p}(\mathbf{n}) = 2\,\mu\, z(\chi)\, \cos\chi\; \mathbf{J} \cdot (\mathbf{Z}(\chi) \cdot \mathbf{n}), \tag{A.5.15}$$

where the second order symmetric tensor $\mathbf{Z}(\chi)$ has unit norm,

$$z(\chi)\,\mathbf{Z}(\chi) = \hat{\mathbf{S}} - \frac{1}{2\sqrt{3}} \tan\chi\, \mathbf{I}, \quad z(\chi) = \|\, \hat{\mathbf{S}} - \frac{1}{2\sqrt{3}} \tan\chi\, \mathbf{I} \,\|. \tag{A.5.16}$$

Expressions similar to (A.5.15) and (A.5.16) hold for $\mathbf{q}(\mathbf{n})$ with χ replaced by ψ. Since we assume non-associativity, $\chi \neq \psi$, we can introduce the angle $\phi \in]0°, 90°[$ and the two tensors $\mathbf{X}$ and $\mathbf{Y}$,

$$\mathbf{X} = \frac{\mathbf{Z}(\chi) + \mathbf{Z}(\psi)}{2\cos\phi}, \quad \mathbf{Y} = \frac{\mathbf{Z}(\chi) - \mathbf{Z}(\psi)}{2\sin\phi}, \quad \text{with} \quad \cos 2\phi = \mathbf{Z}(\chi) : \mathbf{Z}(\psi). \tag{A.5.17}$$

Notice that $\mathbf{X}$ and $\mathbf{Y}$ are orthogonal and of unit norm,

$$\mathbf{X} : \mathbf{Y} = 0, \quad \|\, \mathbf{X} \,\| = 1, \quad \|\, \mathbf{Y} \,\| = 1. \tag{A.5.18}$$

Expressing $\mathbf{Z}(\chi)$ and $\mathbf{Z}(\psi)$ in terms of $\mathbf{X}$, $\mathbf{Y}$ and ϕ, the quantity $-y$, eqn (A.5.14), is found to be proportional, with a positive proportionality coefficient $2\,(c_s^e)^2\, \cos\chi\, \cos\psi\, z(\chi)\, z(\psi)$, to

$$-\cos^2\phi\, \|\, \mathbf{J}(\mathbf{X} \cdot \mathbf{n}) \,\|^2 + \sin^2\phi\, \|\, \mathbf{J}(\mathbf{Y} \cdot \mathbf{n}) \,\|^2. \tag{A.5.19}$$

Finally, let us denote by $\mathbf{n}_{X^*}$ and $\mathbf{n}_{Y^*}$ the respective eigenvectors of the second order tensors $\mathbf{X}^*$ and $\mathbf{Y}^*$,

$$\mathbf{X}^* = \mathbf{N}^{-\frac{1}{2}} \cdot \mathbf{X} \cdot \mathbf{N}^{-\frac{1}{2}}, \quad \mathbf{Y}^* = \mathbf{N}^{-\frac{1}{2}} \cdot \mathbf{Y} \cdot \mathbf{N}^{-\frac{1}{2}}. \tag{A.5.20}$$

$\mathbf{J}(\mathbf{X}\cdot\mathbf{n})$ and $\mathbf{J}(\mathbf{Y}\cdot\mathbf{n})$ vanish when $\mathbf{n}$ is parallel to $\mathbf{N}^{-\frac{1}{2}}\cdot\mathbf{n}_{X^*}$ and to $\mathbf{N}^{-\frac{1}{2}}\cdot\mathbf{n}_{Y^*}$, respectively. For $\mathbf{J}(\mathbf{X}\cdot\mathbf{n})$ and $\mathbf{J}(\mathbf{Y}\cdot\mathbf{n})$ to vanish for the same direction $\mathbf{n}$, this $\mathbf{n}$ must be an eigenvector of both $\mathbf{G}$ and $\hat{\mathbf{S}}$. Obviously, such an $\mathbf{n}$ does not exist if $\mathbf{G}$ and $\hat{\mathbf{S}}$, and consequently $\mathbf{G}$, $\mathbf{P}$ and $\mathbf{Q}$, do not have any common eigendirection. Then $-y$ is strictly positive along and around the critical directions $\mathbf{n}^{\text{flu}}$ parallel to $\mathbf{N}^{-\frac{1}{2}}\cdot\mathbf{n}_{X^*}$.

We can now consider the sign condition on $-x(\mathbf{n},\tilde{\mathbf{n}})$. First, observe that, when elastic anisotropy decreases, corresponding to small but non zero φ, the three critical directions $\mathbf{n}^{\text{flu}}$ tend to the eigendirections of $\hat{\mathbf{S}}$. Now, we know that, for isotropic elasticity, at least one eigendirection of $\hat{\mathbf{S}}$ exists such that in its neighborhood $-x(\mathbf{n})$ is strictly positive (eqns (3.5-6) in Loret [1992] and eqns (A.3.12) above). Using (A.5.1),(A.5.11)$_1$, it can also be shown that the property still applies to $-x(\mathbf{n},\tilde{\mathbf{n}})$ for small φ. Hence, for sufficiently small φ, at least one direction $\mathbf{n}^{\text{flu}}$ exists in the neighborhood of which both $-x$ and $-y$ are strictly positive and flutter occurs.

This result can be canned in the following

Proposition A.5.1 Non-coaxial elastic perturbations, Bigoni and Loret [1999].

For any elastic-plastic model satisfying deviatoric associativity, there exist infinitesimal anisotropic perturbations from isotropic elasticity which lead to flutter instability if $\hat{H}$ is chosen within the interval defined by (A.5.12).

In the above deductions, the spectra of $\mathbf{X}^*$ and $\mathbf{Y}^*$ are assumed to be separate. However, it may be shown that flutter can also occur when this is not the case.

A.5.4 G and $\hat{\mathbf{S}}$ share one Common Eigendirection

Assume that $\mathbf{G}$ and $\hat{\mathbf{S}}$ have one common eigendirection, say $\mathbf{v}$, corresponding to the intermediate or minimum eigenvalue of $\hat{\mathbf{S}}$ Then $\mathbf{v}$ is also a common eigendirection of $\mathbf{X}$ and $\mathbf{Y}$ and of $\mathbf{X}^*$ and $\mathbf{Y}^*$. In the neighborhood of the directions $\mathbf{N}^{-\frac{1}{2}}\cdot\mathbf{n}_{X^*}$, with $\mathbf{n}_{X^*}$ eigenvector of $\mathbf{X}^*$ orthogonal to $\mathbf{v}$, $-y$ is strictly positive. Repeating the argument employed in the previous subsection, it may be concluded that $-x$ is also strictly positive in the neighborhood of at least one of these directions, and flutter occurs.

A.5.5 Coaxial Fabric

The analysis of the situation where $\mathbf{G}$ and $\hat{\mathbf{S}}$ are coaxial is more intricate. Essentially, the end result is that there may not exist values of φ corresponding to the onset of flutter and satisfying the restriction (A.5.5); if they exist, they should be finite to imply flutter, Bigoni and Loret [1999], Sect. 4.2.3.

A.6 Surface Boundary Effects with Deviatoric Associativity

The analyses of the previous sections are local, that is, they may be viewed as concerning a point in an infinite solid. Another route to the onset of flutter instability may therefore be the incorporation of the effects of boundary conditions on a finite or semi-infinite body. This is the investigation that we envisage in this section. One motivation for this study stems from previous analyses of the strongly destabilizing effects of Coulomb's friction law. Martins et al. [1992] consider the motion of a rigid flat surface over a linear isotropic half-space with frictional contact. If the friction coefficient and the Poisson's ratio are sufficiently large, growing oscillations develop at points just below the contact surface, a situation similar to that of Rayleigh waves with a complex wave-speed, that is a *surface flutter instability*. Similar results are obtained when the deformable half-space is visco-elastic, Martins et al. [1995], or when finite deformations are accounted for by means of a neo-Hookean elastic behaviour, Martins and Faria [1991]. These observations may contribute to an explanation of the origins of the Schallamach waves, which act as 'waves of detachment' that cross the contact from front to rear with rapidly alternating states of contact and loss of contact. Clearly, flutter instability requires the governing equations to possess some kind of non-self-adjointness, although, as pointed out above, non-self-adjointness is not a sufficient condition for the onset of flutter instability in elastic-plastic solids. The lack of symmetry of the tensor of constitutive moduli implies the dynamic equations of motion to be non-self-adjoint, whereas, for frictional contact problems, non-self-adjointness is induced by the boundary conditions. Here, we adopt a somewhat intermediate position by considering an elastic-plastic solid with deviatoric associativity filling a half-space with a traction rate free boundary. So, taken independently, neither the constitutive equations nor the boundary conditions lead to the onset of flutter instability. However, we shall derive conditions under which their simultaneous influence will indeed lead to the onset of flutter instability.

This analysis can be viewed as an instance where an initial-boundary value problem becomes ill-posed while the rate field equations are still hyperbolic. In that sense, contact can be made with analyses of the existence of stationary surface waves in bounded or semi-infinite solids. For example, Benallal et al. [1989] note that failure of the complementing condition is equivalent to the existence of stationary surface waves. Conditions for the existence of these waves are analyzed by Dowaikh and Ogden [1990] for particular pre-stressed incompressible elastic solids and by Needleman and Ortiz [1991] for elastic-plastic solids. The existence of interfacial waves between two perfectly bonded half-spaces is analyzed by Dowaikh and Ogden [1991], and some connection with the conditions for existence of stationary surface waves is established; this problem is also considered by Needleman and Ortiz [1991]. Suo et al. [1992] show that, when the interface between two elastic-plastic half-spaces is compliant and involves a length-scale, the onset of stationary interfacial waves depends on the wavelength and there exists in some circumstances a minimum wavelength for the instability mode.

The analysis, summarizing LMS [1995][6], is organized as follows. We first present the field and boundary conditions, and we obtain a variational formulation for the problem of finding dynamic surface solutions in an elastic-plastic half-space where plastic loading holds pointwise while the traction rates at the boundary are null. The analysis is restricted to orthotropic tensors of constitutive moduli. The variational formulation trivially leads, inter alia, to the expected result that surface flutter instability is ruled out for associative flow rules.

We are naturally led to delineate two cases: the coaxial case, in which one orthotropy direction is normal and the others are tangent to the free surface, and the non-coaxial case, in which only a tangent to the free surface coincides with one of the orthotropy

[6] LMS [1995] stands for Loret, Martins and Simões [1995].

directions. For the non-coaxial case, we show that the onset of flutter instability coincides with the incipience of plasticity, that is the critical plastic modulus at the onset of flutter instability is infinite. This result holds for any type of non-associative flow rules, even satisfying deviatoric associativity. We therefore have an instance where a semi-infinite body may undergo flutter instability whereas the infinite body filled with the same material does not, Sect. A.3. On the other hand, flutter is not excluded in the coaxial case but its onset requires unusual material parameters.

A.6.1 Governing Equations

A.6.1.1 Field equations, boundary and decay conditions

Let us consider a semi-infinite and homogeneous elastic-plastic body that undergoes plastic loading at any point during a fundamental path. Various types of bifurcation modes from this fundamental path are searched for. Therefore, the problems we consider have formally a traction rate free boundary. Let $(\mathbf{e}_1, \mathbf{e}_2, \mathbf{e}_3)$, Fig. A.6.1, be a fixed orthonormal reference frame with the origin on the boundary of the body. Here $\mathbf{e}_1$ is tangent to the free surface, $\mathbf{e}_2$ is the outward normal, and $\mathbf{e}_3$ is the out-of-plane direction. We denote by $\mathbf{x} = (x_1, x_2, x_3)$ the position vector of the particles of the body.

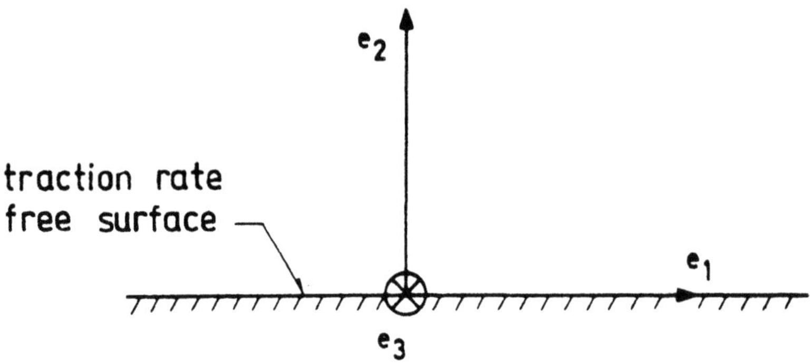

Figure A.6.1 The lower half-space is filled with an elastic-plastic solid and the boundary x_2=0 is traction rate free at any time.

The material is endowed with isotropic elasticity, the elasticity tensor being positive definite, eqn (A.1.3), and strongly elliptic, eqn (A.2.9). Thus one can normalize the Lamé modulus λ, the stress and the mass density ρ by dividing them by μ; formally, μ=1 in the whole Sect. A.6. We are concerned with the propagation of small plane disturbances orthogonal to $\mathbf{e}_3$, which is assumed to be an axis of orthotropy of the elastic-plastic material. The problem that we wish to solve is thus to find a plane velocity field $\mathbf{v}(x_1, x_2, t) = v_1(x_1, x_2, t)\,\mathbf{e}_1 + v_2(x_1, x_2, t)\,\mathbf{e}_2$ in the half-space $x_2 \leq 0$ such that the rate form of the two equations of balance of linear momentum and of the traction free boundary conditions hold for all times t:

$$\dot{\sigma}_{ij,i} = \rho\, \ddot{v}_j \ \text{ for } x_2 < 0\,, \quad \dot{\sigma}_{2j}(x_1, 0, t) = 0 \ \text{ on } x_2 = 0 \quad \text{for } j = 1, 2. \tag{A.6.1}$$

In (A.6.1), and from now on, the latin indices i, j, k, l run from 1 to 2, the summation convention over repeated indices is in force, and a lower comma denotes a partial spatial derivative. Insertion of the constitutive equations (A.1.8) in the above equations provides a system of two partial differential equations with two boundary conditions: for $j = 1, 2$,

$$\mathbb{E}^{ep}_{ijkl}\, v_{k,li} = \rho \ddot{v}_j \ \text{ for } x_2 < 0\,, \quad \mathbb{E}^{ep}_{2jk1} v_{k,1}(x_1, 0, t) + \mathbb{E}^{ep}_{2jk2} v_{k,2}(x_1, 0, t) = 0 \text{ on } x_2 = 0\,. \tag{A.6.2}$$

We seek surface solutions $\mathbf{v}$ to these equations in the form [7]

$$v_k = V_k(n_1\, x_2)\, \exp(\mathrm{i} n_1 (x_1 - c\, t))\,, \quad \text{for } k = 1, 2. \tag{A.6.3}$$

Here n_1 is a real number that represents the angular frequency of the solutions along the axis $\mathbf{e}_1$; $V_k(n_1\, x_2)$ are sufficiently smooth functions defined for $x_2 \leq 0$ with an appropriate decay as $x_2 \to -\infty$, namely

$$\lim_{x_2 \to -\infty} V_k(n_1\, x_2) = 0\,, \quad \text{for } k = 1, 2; \tag{A.6.4}$$

c is a complex number such that $\mathcal{R}e\, c$ represents the speed at which the solution propagates along the axis $\mathbf{e}_1$ and $n_1 \mathcal{I}m\, c$ represents the rate of exponential growth or decay of that solution in time. Note that the Rayleigh waves in an elastic half-space are of the form (A.6.3),(A.6.4) with c^2 real and positive. Here we are particularly interested in detecting the situations in which, by influence of the non-associativity, c^2 ceases to be a real positive number. Although not needed for the variational formulation, we shall later specialize the solutions to the form

$$v_k = U_k\, \exp(\mathrm{i} n_1 (x_1 + N\, x_2 - c\, t))\,, \quad \text{for } k = 1, 2, \tag{A.6.5}$$

that is,

$$V_k(n_1\, x_2) = U_k\, \exp(\mathrm{i}\, N\, n_1\, x_2)\,, \quad \text{for } k = 1, 2, \tag{A.6.6}$$

where U_k is an undetermined amplitude and N is a complex number that satisfies

$$\mathcal{I}m\, N < 0\,. \tag{A.6.7}$$

Insertion of (A.6.3) in the field equations (A.6.2) leads to a system of two linear and homogeneous second order ordinary differential equations for the vector $\mathbf{V}$ of components V_k,

$$\mathbf{A}\mathbf{V}'' + \mathbf{B}\mathbf{V}' + (\mathbf{C} - X^2\, \mathbf{I})\mathbf{V} = \mathbf{0}\,, \tag{A.6.8}$$

with boundary conditions

$$\mathrm{i}\, \mathbb{E}^{ep}_{2jk1}\, V_k(0) + \mathbb{E}^{ep}_{2jk2}\, V_k'(0) = 0 \quad \text{for } j = 1, 2. \tag{A.6.9}$$

Here $\mathbf{A}$, $\mathbf{B}$ and $\mathbf{C}$ are 2×2 matrices,

$$A_{jk} = -\mathbb{E}^{ep}_{2jk2}\,, \quad B_{jk} = -\mathrm{i}(\mathbb{E}^{ep}_{1jk2} + \mathbb{E}^{ep}_{2jk1})\,, \quad C_{jk} = \mathbb{E}^{ep}_{1jk1}\,, \tag{A.6.10}$$

[7] The usual notation for complex numbers is used: $\mathrm{i} = \sqrt{-1}$, $\mathcal{R}e\, c$ denotes the real part of the complex number c, $\mathcal{I}m\, c$ its imaginary part, $\bar{c}$ its conjugate.

a superimposed $'$ denotes a derivative with respect to $n_1\, x_2$, and

$$X = \frac{c}{c_s^e}\,. \tag{A.6.11}$$

A.6.1.2 **The nature of the solutions**

For solutions that satisfy the field and boundary equations, we delineate four situations depending on the nature of the square of the wave-speed c^2, i.e. of X^2:

- if X^2 is real positive, the situation is similar to that of Rayleigh waves for an elastic solid, the solution being periodic in time (a *surface wave*);
- if X^2 is zero, we have a *stationary surface wave*; notice that the onset of such a wave is a priori different from the onset of a stationary body wave since, for the latter, satisfaction of boundary conditions is not required;
- if X^2 is real strictly negative, the solution explodes as time elapses, and we have a *surface divergence instability*;
- if X^2 is complex, the solution grows while oscillating, a situation termed *surface flutter instability*.

A.6.1.3 **Sesquilinear form and variational formulation**

The above eigenproblem can be given a compact variational formulation. Note that the eigenfunctions V_k may be complex: an oscillatory behaviour along the axis $\mathbf{e}_2$ is admissible provided it satisfies the decay condition (A.6.4). On the other hand, the non-symmetry of $\mathbb{E}^{ep}$ may lead to complex eigenvalues X^2. Hence, instead of working with bilinear forms on spaces of real functions, we have to work with sesquilinear forms on spaces of complex functions.

Formally multiplying (A.6.8) by the complex conjugates of sufficiently smooth functions $W_k(n_1\, x_2)$, $k = 1, 2$, with compact support in $]-\infty, 0[$, integrating the product from $-\infty$ to 0, and performing an integration by parts of the terms involving $\mathbb{E}^{ep}_{2jk1}$ and $\mathbb{E}^{ep}_{2jk2}$, we obtain, for all such vectors $\mathbf{W}$:

$$\int_{-\infty}^{0} \overline{W_j'}\, \mathbb{E}^{ep}_{2jk2}\, V_k' + \mathrm{i}\,(\overline{W_j'}\, \mathbb{E}^{ep}_{2jk1}\, V_k - \overline{W_j} \mathbb{E}^{ep}_{1jk2}\, V_k') + \overline{W_j}\, \mathbb{E}^{ep}_{1jk1}\, V_k \; dx_2 \tag{A.6.12}$$
$$+\overline{W_j}(0)\,\Big(\mathrm{i}\,\mathbb{E}^{ep}_{2jk1}\, V_k(0) + \mathbb{E}^{ep}_{2jk2}\, V_k'(0)\Big) = X^2 \int_{-\infty}^{0} \overline{W_k}\, V_k\, dx_2\,.$$

Taking into account the boundary conditions (A.6.9), we obtain:

Proposition A.6.1 Variational formulation through sesquilinear forms, LMS [1995].

The boundary value problem over the half-space defined in the previous section reduces[8] *to finding the squares of the speed of propagation-like scalars* $X \in \mathbb{C}$ *and the non zero functions* $\mathbf{V} \in V \equiv H_1(I) \times H_1(I)$, *with* $I \equiv]-\infty, 0[$, *such that*

$$a(\mathbf{V}, \mathbf{W}) = X^2\, (\mathbf{V}, \mathbf{W})\,, \quad \forall\, \mathbf{W} \in V\,, \tag{A.6.13}$$

[8] Conversely, solutions of the variational equations can be shown to be solutions of the boundary value problem (A.6.8),(A.6.9),(A.6.4)

where we have introduced the symmetric sesquilinear form that represents the inner product over $L^2(I) \times L^2(I)$,

$$(\mathbf{V}, \mathbf{W}) = \overline{(\mathbf{W}, \mathbf{V})} = \int_{-\infty}^{0} \overline{W_k}\, V_k \, dx_2 \,, \tag{A.6.14}$$

and the sesquilinear form $a(\mathbf{V}, \mathbf{W})$ defined by the integral in the left handside of (A.6.12), which can be recast in the form

$$a(\mathbf{V}, \mathbf{W}) = \int_{-\infty}^{0} \overline{D_i\, W_j}\, \mathbb{E}^{\text{ep}}_{ijkl}\, D_k\, W_l \, dx_2 \quad \text{with} \quad D_i(\cdot) \equiv \mathrm{i}\, I_{1i}(\cdot) + I_{2i}(\cdot)' \,. \tag{A.6.15}$$

For associative plasticity, the elastic-plastic tensor is symmetric, that is it has the major symmetry, and the associated sesquilinear form (A.6.15) is *symmetric*,

$$a(\mathbf{V}, \mathbf{W}) = \overline{a(\mathbf{W}, \mathbf{V})} \,, \quad \forall\, \mathbf{V}, \mathbf{W} \in V \,. \tag{A.6.16}$$

As a consequence of this symmetry, we have

Proposition A.6.2 Variational formulation for associative plasticity, LMS [1995].
For associative plasticity, the scalars X^2 that solve the variational equation (A.6.13) are real.

A.6.2 The Non-Coaxial Case

Non-coaxiality of the orthotropy axes and of the traction rate free boundary (Fig. A.6.2) will be shown to give rise to flutter instability for non-associative flow rules right at the incipience of plasticity, that is for $H^{-1} \approx 0$. The variational formulation of the previous

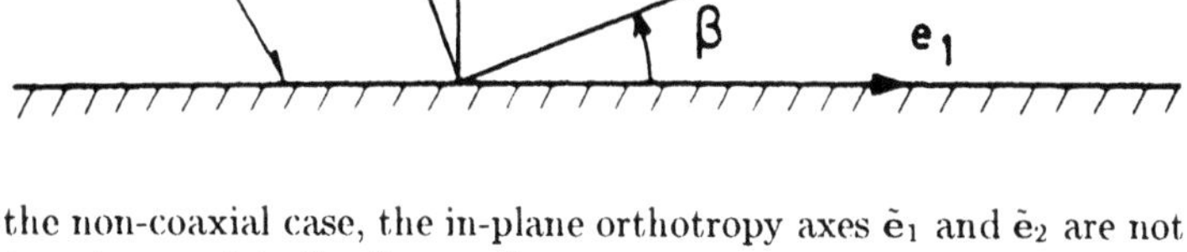

Figure A.6.2 For the non-coaxial case, the in-plane orthotropy axes $\tilde{\mathbf{e}}_1$ and $\tilde{\mathbf{e}}_2$ are not aligned with the tangent and normal to the free surface.

section will be used to establish that the derivative dX^2/dH^{-1} at H^{-1}=0 is complex if the normal to the boundary is not an axis of orthotropy of the constitutive tensor of plastic moduli, Sect. A.6.2.2. A linearized analysis of the field equations will also give additional information on the two mode shapes, that is $dN^{(1)}/dH^{-1}$ and $dN^{(2)}/dH^{-1}$ at H^{-1}=0, Sect. A.6.2.4. The evolution, with H^{-1} increasing monotonously from 0, of the triplet $(X^2, N^{(1)}, N^{(2)})$ is analyzed quantitatively in Sect. A.6.2.3: this evolution displays the respective influences of the features that trigger flutter instability, namely, deviation with respect to associativity and non-coaxiality with the boundary.

The results of this section do not require the constitutive equations to satisfy deviatoric associativity.

A.6.2.1 **Non-coaxial constitutive equations**

We first rewrite the constitutive equations in a format adequate to our present purpose. Assuming that the horizontal direction $\mathbf{e}_3$ is an orthotropy direction for the solid under consideration, the rate constitutive equations may be expressed in the $(\mathbf{e}_1, \mathbf{e}_2, \mathbf{e}_3)$-axes in the form:

$$\left\{\begin{aligned} \dot\sigma_{11} &= L_{11}\,\dot\epsilon_{11} + L_{12}\,\dot\epsilon_{22} + 2\,L_{13}\,\dot\epsilon_{12}\,, \\ \dot\sigma_{22} &= L_{21}\,\dot\epsilon_{11} + L_{22}\,\dot\epsilon_{22} + 2\,L_{23}\,\dot\epsilon_{12}\,, \\ \dot\sigma_{12} &= L_{31}\,\dot\epsilon_{11} + L_{32}\,\dot\epsilon_{22} + 2\,L_{33}\,\dot\epsilon_{12}\,. \end{aligned}\right. \tag{A.6.17}$$

Although the results obtained in this section do not require deviatoric associativity, one may identify the coefficients L_{ij} appearing above for an elastic-plastic solid obeying deviatoric associativity (in the formulas below, i and j vary from 1 to 2, there is no sum over repeated indices, and the terms in brackets hold for deviatoric associativity only, Sect. A.1.2):

$$\left\{\begin{aligned} L_{ii} &= \mathbb{E}^{\text{ep}}_{iiii}\ [= \lambda + 2 - r(N_P + \hat S_{ii})(N_Q + \hat S_{ii})], \quad L_{33} = \mathbb{E}^{\text{ep}}_{1212}\ [= -r(\hat S_{12})^2]\,, \\ L_{ij} &= \mathbb{E}^{\text{ep}}_{iijj}\ [= \lambda \quad\ - r(N_P + \hat S_{ii})(N_Q + \hat S_{jj})], \\ L_{i3} &= \mathbb{E}^{\text{ep}}_{ii12}\ [= -r(N_P + \hat S_{ii})\hat S_{12}]\,,\ L_{3i} = \mathbb{E}^{\text{ep}}_{12ii}\ [= -r(N_Q + \hat S_{ii})\hat S_{12}]\,, \end{aligned}\right. \tag{A.6.18}$$

with $r = 4\cos\chi\,\cos\psi/H$ and N_P, N_Q defined by (A.2.17).

A.6.2.2 **Surface flutter instability: a first general result**

The constitutive tensor moduli $\mathbb{E}^{\text{ep}}$, eqn (A.1.5), may be split into a symmetric part and an antisymmetric part as follows (accounting for the major symmetry of the elastic tensor moduli $\mathbb{E}$),

$$\mathbb{E}^{\text{ep}} = \mathbb{E} + H^{-1}\,\mathbb{E}^{\text{p}}, \quad \mathbb{E}^{\text{p}} = \mathbb{E}^{\text{sp}} + \mathbb{E}^{\text{ap}}, \quad \mathbb{E}^{\text{sp}} = \frac{1}{2}(\mathbb{E}^{\text{p}} + (\mathbb{E}^{\text{p}})^{\text{T}}), \quad \mathbb{E}^{\text{ap}} = \frac{1}{2}(\mathbb{E}^{\text{p}} - (\mathbb{E}^{\text{p}})^{\text{T}})\,, \tag{A.6.19}$$

where the transpose operates on two first and the two last indices, that is $(\mathbb{E}^{\text{p}})^{\text{T}}_{ijkl} = \mathbb{E}^{\text{p}}_{klij}$. Accordingly, the sesquilinear form (A.6.15) may be split into three sesquilinear forms, *symmetric* in the sense of (A.6.16) and independent of H,

$$a(\mathbf{V},\mathbf{W}) = a^{\text{e}}(\mathbf{V},\mathbf{W}) + H^{-1}\,a^{\text{sp}}(\mathbf{V},\mathbf{W}) + \mathrm{i}\,H^{-1}\,a^{\text{ap}}(\mathbf{V},\mathbf{W})\,, \tag{A.6.20}$$

defined by *symmetric* moduli through the following correspondences:

$$a^{\text{e}} \longleftrightarrow \mathbb{E}\,, \quad a^{\text{sp}} \longleftrightarrow \mathbb{E}^{\text{sp}}\,, \quad a^{\text{ap}} \longleftrightarrow \frac{1}{\mathrm{i}}\mathbb{E}^{\text{ap}}\,. \tag{A.6.21}$$

The eigenproblem (A.6.13) becomes:

$$a^{\text{e}}(\mathbf{V},\mathbf{W}) + H^{-1}\,a^{\text{sp}}(\mathbf{V},\mathbf{W}) + \mathrm{i}\,H^{-1}\,a^{\text{ap}}(\mathbf{V},\mathbf{W}) = X^2\,(\mathbf{V},\mathbf{W})\,, \quad \forall\ \mathbf{W} \in V\,. \tag{A.6.22}$$

The elastic situation $H^{-1}=0$ reduces to the symmetric variational statement,

$$a^{\mathrm{e}}(\mathbf{V}_R,\mathbf{W}) = X_R^2\,(\mathbf{V}_R,\mathbf{W})\,, \quad \forall\,\mathbf{W} \in V\,, \tag{A.6.23}$$

in terms of the real (from Proposition A.6.2) Rayleigh wave-speed X_R and of the eigenmode $\mathbf{V}_R$. Exploiting further the symmetry of the sesquilinear form a^{e} and of the inner product (A.6.14) provides an equivalent statement for (A.6.23),

$$a^{\mathrm{e}}(\mathbf{W},\mathbf{V}_R) = X_R^2\,(\mathbf{W},\mathbf{V}_R)\,, \quad \forall\,\mathbf{W} \in V\,. \tag{A.6.24}$$

There exists a single elastic eigenmode $\mathbf{V}_R$ which is solution to this problem, e.g. Achenbach [1975], p.187: it is the sum of two terms of the type (A.6.6), with

$$\begin{cases} V_{1R}(n_1\,x_2)/U &= \exp(\mathrm{i}\,N_R^{(1)} n_1\,x_2) + (X_R^2/2-1) \quad \exp(\mathrm{i}\,N_R^{(2)} n_1\,x_2)\,, \\ V_{2R}(n_1\,x_2)/U N_R^{(1)} &= \exp(\mathrm{i}\,N_R^{(1)} n_1\,x_2) + (X_R^2/2-1)^{-1} \exp(\mathrm{i}\,N_R^{(2)} n_1\,x_2)\,, \end{cases} \tag{A.6.25}$$

where n_1 is a real number and the arbitrary amplitude U may be chosen to be real as well, $N_R^{(1)}$ and $N_R^{(2)}$ are purely imaginary numbers,

$$N_R^{(1)} = -\mathrm{i}\left(1-\frac{X_R^2}{\lambda+2}\right)^{\frac{1}{2}}, \quad N_R^{(2)} = -\mathrm{i}\,(1-X_R^2)^{\frac{1}{2}}\,, \tag{A.6.26}$$

and X_R^2 is the single root smaller than 1 of the cubic equation,

$$-(X^2)^3 + 8\,(X^2)^2 - 24\,\frac{\lambda+4/3}{\lambda+2}X^2 + 16\,\frac{\lambda+1}{\lambda+2} = 0\,. \tag{A.6.27}$$

In order to study the eigenproblem (A.6.22) in the neighbourhood of the known elastic case, we compute the first variation of that equation at $H^{-1}=0$, holding fixed all other constitutive parameters:

$$\begin{aligned} a^{\mathrm{e}}(\delta\mathbf{V},\mathbf{W}) + \delta H^{-1}\,a^{\mathrm{sp}}(\mathbf{V}_R,\mathbf{W}) + \mathrm{i}\,\delta H^{-1}\,a^{\mathrm{ap}}(\mathbf{V}_R,\mathbf{W}) &= \\ X_R^2\,(\delta\mathbf{V},\mathbf{W}) + \delta X^2\,(\mathbf{V},\mathbf{W})\,,\ \forall\,\mathbf{W} \in V\,. \end{aligned} \tag{A.6.28}$$

Setting $\mathbf{W}=\mathbf{V}_R$ and using (A.6.24) which eliminates the first terms in the left and right handsides of (A.6.28), we obtain,

$$\left(\frac{dX^2}{dH^{-1}}\right)_{|\,H^{-1}=0} = \frac{a^{\mathrm{sp}}(\mathbf{V}_R,\mathbf{V}_R)}{(\mathbf{V}_R,\mathbf{V}_R)} + \mathrm{i}\,\frac{a^{\mathrm{ap}}(\mathbf{V}_R,\mathbf{V}_R)}{(\mathbf{V}_R,\mathbf{V}_R)}\,. \tag{A.6.29}$$

For an orthotropic elastic-plastic solid with the direction $\mathbf{e}_3$ as a principal direction of orthotropy, there are two independent constitutive moduli involved in $a^{\mathrm{ap}}(\mathbf{V}_R,\mathbf{V}_R)$:

$$\begin{cases} \mathcal{F} \equiv \mathbb{E}^{\mathrm{ap}}_{1122} = \frac{1}{2}(L_{12}-L_{21})[=-4\,\cos\chi\,\cos\psi(N_P-N_Q)(\hat{S}_{22}-\hat{S}_{11})]\,, \\ \mathcal{E} \equiv \mathbb{E}^{\mathrm{ap}}_{1112} = \frac{1}{2}(L_{13}-L_{31}) = \mathbb{E}^{\mathrm{ap}}_{2212} = \frac{1}{2}(L_{23}-L_{32})[=-2\,\cos\chi\,\cos\psi(N_P-N_Q)\,\hat{S}_{12}]\,. \end{cases} \tag{A.6.30}$$

In fact, V_{1R} is real and V'_{2R} is imaginary, (A.6.25), and this implies, after some algebra, that the coefficient of $\mathcal{F}$ in $a^{\text{ap}}(\mathbf{V}_R, \mathbf{V}_R)$ vanishes, that is the in-axis components of the tensor moduli do not contribute to that expression either. Using the expressions (A.6.25) of the elastic eigenmode and performing the integration in x_2 from $-\infty$ to 0 yields:

$$a^{\text{ap}}(\mathbf{V}_R, \mathbf{V}_R) = 2\,\mathcal{E}\frac{U^2}{n_1}\left(1 + (N_R^{(1)})^2\right)\frac{N_R^{(1)} - N_R^{(2)}}{N_R^{(1)} + N_R^{(2)}} \tag{A.6.31}$$

which is clearly different from zero in the non-coaxial case [9], $\mathcal{E} \neq 0$. Consequently $\mathcal{I}m\,(\delta X^2) \neq 0$ and therefore we have proved

Proposition A.6.3 Surface flutter in the non-coaxial case, LMS [1995].

For an orthotropic elastic-plastic half-space with one principal direction of orthotropy parallel to its boundary, namely $\mathbf{e}_3$*, the imaginary part of the derivative of the square of the wave-speed with respect to* H^{-1} *estimated at* H^{-1}=0 *is different from zero iff the normal to the boundary of the half-space is not a principal direction of orthotropy and the flow rule is non-associative.*

A.6.2.3 **Linearized eigenmodes around** $H^{-1} = 0$

Satisfaction of the field equations (A.6.2) results in a system of two linear equations for the components U_1 and U_2 of the solution vector (A.6.6):

$$\begin{bmatrix} L_{11} + (L_{13} + L_{31})\,N + L_{33}\,N^2 - X^2 & L_{13} + (L_{12} + L_{33})\,N + L_{32}\,N^2 \\ L_{31} + (L_{21} + L_{33})\,N + L_{23}\,N^2 & L_{22}\,N^2 + (L_{23} + L_{32})\,N + L_{33} - X^2 \end{bmatrix}\begin{bmatrix} U_1 \\ U_2 \end{bmatrix} = \begin{bmatrix} 0 \\ 0 \end{bmatrix}. \tag{A.6.32}$$

The vanishing of the determinant of this system yields a relation linking N, X^2 and the constitutive coefficients, say:

$$\phi(N\,, X^2\,, H^{-1}) = 0\,. \tag{A.6.33}$$

Linearization of this equation about each of the two elastic solutions $N_R^{(\alpha)}$, $\alpha = 1, 2$, and X_R^2 yields an approximate value for $N^{(\alpha)}$,

$$\frac{dN^{(\alpha)}}{dH^{-1}} = -\left(\frac{\partial \phi}{\partial N^{(\alpha)}}\right)^{-1}\left(\frac{\partial \phi}{\partial H^{-1}} + \frac{\partial \phi}{\partial X^2}\,\frac{dX^2}{dH^{-1}}\right), \tag{A.6.34}$$

with dX^2/dH^{-1} given by (A.6.29), and all derivatives estimated at H^{-1}=0.

A.6.2.4 **Eigenmodes for** H **finite**

The boundary conditions (A.6.8) have been incorporated in the variational formulation. However, in order to perform a parameter analysis, one combines two solutions $U_k^{(\alpha)}$, $k = 1, 2$, indexed by $\alpha = 1, 2$, of the form (A.6.6), each of them satisfying the field equations, (A.6.8) or (A.6.32), that is

$$w^{(\alpha)} \equiv \frac{U_2^{(\alpha)}}{U_1^{(\alpha)}} = -\frac{L_{11} + (L_{13} + L_{31})\,N^{(\alpha)} + L_{33}\,(N^{(\alpha)})^2 - X^2}{L_{13} + (L_{12} + L_{33})\,N^{(\alpha)} + L_{32}\,(N^{(\alpha)})^2}\,, \quad \text{for } \alpha = 1, 2. \tag{A.6.35}$$

[9] For $\hat{\mathbf{S}}$ coaxial with the surface, $\mathcal{E}$ vanishes and therefore $\mathcal{I}m\,(\delta X^2) = 0$, i.e. flutter does not occur at H^{-1}=0.

The two available modes are combined in order to satisfy the boundary conditions:

$$\begin{cases} \sum_{\alpha=1}^{2} \left(L_{33}\,(N^{(\alpha)} + w^{(\alpha)}) + L_{32}\,N^{(\alpha)}\,w^{(\alpha)} + L_{31}\right) U_1^{(\alpha)} = 0\,, \\ \sum_{\alpha=1}^{2} \left(L_{21} + L_{23}\,N^{(\alpha)} + (L_{22}\,N^{(\alpha)} + L_{23})\,w^{(\alpha)}\right)\; U_1^{(\alpha)} = 0\,. \end{cases} \tag{A.6.36}$$

Indeterminacy of this system yields a relation linking X^2 and two symmetric functions of $N^{(1)}$ and $N^{(2)}$, namely their sum and product. This relation is of the form

$$\gamma(N^{(1)},\, N^{(2)},\, X^2,\, H^{-1}) = a\,(X^2)^2 + b\,X^2 + c = 0\,. \tag{A.6.37}$$

To obtain numerical results, we use a continuation method, starting from the known elastic solution $(N_R^{(1)},\, N_R^{(2)},\, X_R^2)$ and increment H^{-1}, solving at each step the system of three non-linear algebraic equations for the triplet $(N^{(1)},\, N^{(2)},\, X^2)$, with $\mathcal{I}m\, N^{(1)} < 0$, $\mathcal{I}m\, N^{(2)} < 0$,

$$\begin{cases} \phi(N^{(1)}\,, X^2\,, H^{-1}) = 0\,, & \text{field equation for mode } 1\,, \\ \phi(N^{(2)}\,, X^2\,, H^{-1}) = 0\,, & \text{field equation for mode } 2\,, \\ \gamma(N^{(1)},\, N^{(2)},\, X^2,\, H^{-1}) = 0\,, & \text{boundary condition}\,. \end{cases} \tag{A.6.38}$$

Figures A.6.3 and 4 display the influence of the non-coaxiality parameter $\hat{S}_{12}$ on the evolution in the complex plane of X^2, $N^{(1)}$ and $N^{(2)}$ while the modulus H decreases but the regime remains hyperbolic, Fig. A.6.6. In contrast to the coaxial case ($\hat{S}_{12} = 0$), X^2 becomes complex right at the incipience of plasticity; the value of the imaginary part of X^2 near $H^{-1} = 0$ is roughly proportional to the parameter $\hat{S}_{12}$. In the non-coaxial case, the values of $N^{(1)}$ and $N^{(2)}$ acquire real parts at the incipience of plasticity, whereas in the coaxial case, this happens later at point C that does not correspond to the onset of flutter; in the latter case, the onset of flutter occurs much later, at the point F at which the equalities $\mathcal{R}e\, N^{(1)} = -\mathcal{R}e\, N^{(2)}$ and $\mathcal{I}m\, N^{(1)} = \mathcal{I}m\, N^{(2)}$ cease to hold. Figures 9 and 10 of LMS [1995] display the influence of the non-associativity parameter $\tan\psi - \tan\chi$ on the evolution in the complex plane of $(N^{(1)},\, N^{(2)},\, X^2)$. Clearly, flutter instability does not occur in the associative case and, in the non-associative non-coaxial cases, the onset of flutter instability coincides with the incipience of plasticity. The value of the imaginary part of X^2 near $H^{-1} = 0$ is proportional to the difference $\tan\psi - \tan\chi$. Moreover, the values of $N^{(1)}$ and $N^{(2)}$ acquire a real part right at the incipience of plasticity even in the associative case, since, in this example, $\hat{S}_{12}$ is different from zero. A comparison betwen the linearized solution, eqns (A.6.29),(A.6.34), and the solution of the non-linear system (A.6.38) is shown in Figs. 11 and 12 of LMS [1995].

Figures A.6.3 and 4 display clearly the difference between the coaxial and the non-coaxial situations: in the coaxial case, flutter, if it occurs, appears well inside the plastic regime, whereas, in the non-coaxial case, it occurs right at the incipience of plasticity with an intensity roughly proportional to the non-coaxiality and non-associativity parameters.

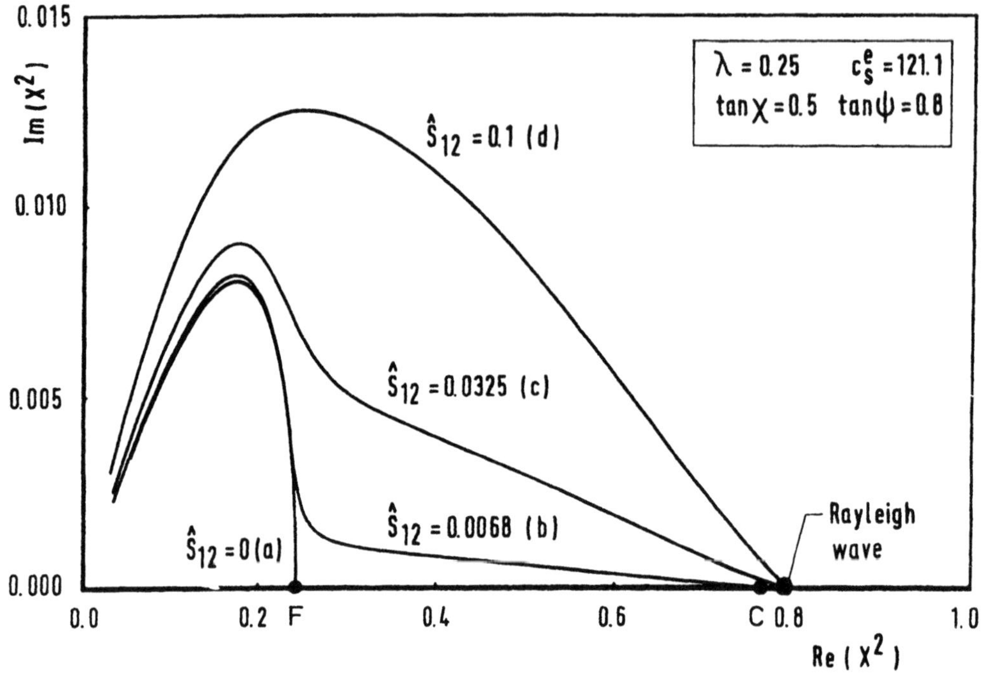

Figure A.6.3 Evolution in the complex plane of the square of the normalized wave-speed X^2 while H^{-1} varies from 0 (elasticity) to 0.458 for several values of $\hat{S}_{12}$.

	$\hat{S}_{11}$	$\hat{S}_{22}$	$\hat{S}_{33}$	$\hat{S}_{12}$
(a)	0.8078	−0.3012	−0.5066	0
(b)	0.807	−0.301	−0.506	0.0068
(c)	0.81	−0.30	−0.51	0.0325
(d)	0.8	−0.3	−0.5	0.1

A.6.3 The Coaxial Case

A.6.3.1 Field equations and boundary conditions

In the coaxial case, certain constitutive coefficients vanish, namely from (A.6.18),

$$L_{13} = L_{23} = L_{31} = L_{32} = 0\,, \tag{A.6.39}$$

and the field equations for each mode (A.6.32) simplify to

$$\begin{bmatrix} L_{11} + N^2 - X^2 & (L_{12} + 1)\,N \\ (1 + L_{21})\,N & L_{22}\,N^2 + 1 - X^2 \end{bmatrix} \begin{bmatrix} U_1 \\ U_2 \end{bmatrix} = \begin{bmatrix} 0 \\ 0 \end{bmatrix}. \tag{A.6.40}$$

Therefore satisfaction of the rate field equations (A.6.40) for a non-trivial mode **V** reduces to

$$(N^2)^2 - S\,N^2 + P = 0\,, \tag{A.6.41}$$

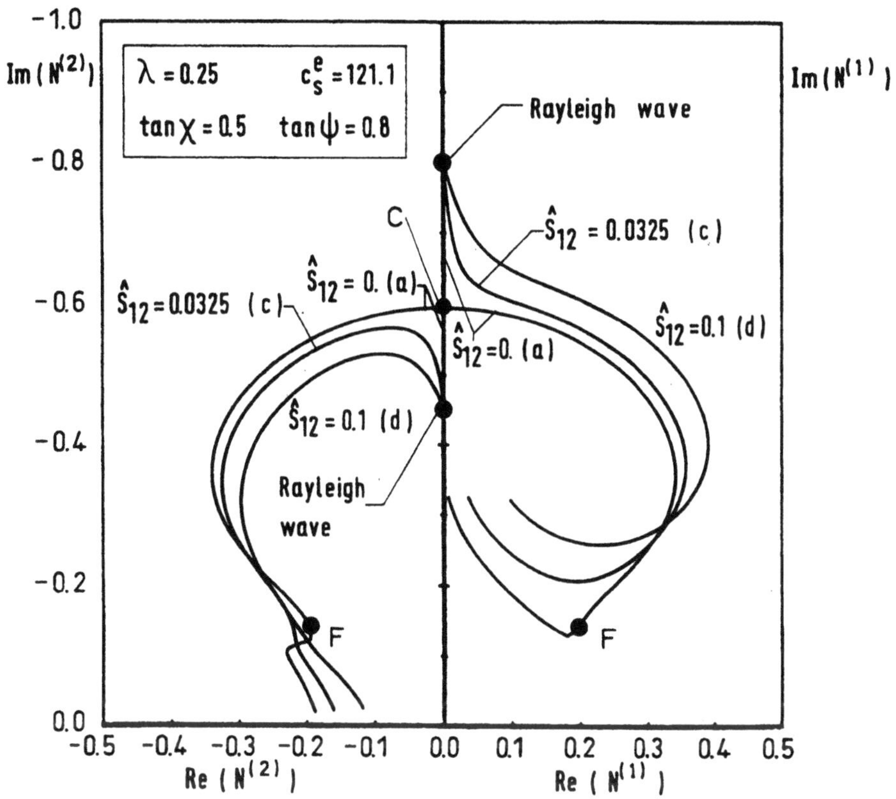

Figure A.6.4 Evolution in the complex plane of $N^{(1)}$ and $N^{(2)}$ while H^{-1} varies from 0 (elasticity) to 0.458; $\tan\chi$=0.5 $\neq$ $\tan\psi$=0.8. The components of $\hat{\mathbf{S}}$ are given in the legend of Fig. A.6.3.

where, assuming $L_{22} \neq 0$,

$$S = \frac{1}{L_{22}}\big(- \Delta + (L_{12} + L_{21}) + (1 + L_{22})\, X^2 \big)\,, \quad P = \frac{1}{L_{22}}(1 - X^2)(L_{11} - X^2)\,. \tag{A.6.42}$$

The scalar Δ involved in S,

$$\Delta = L_{11}\, L_{22} - L_{12}\, L_{21}\,, \tag{A.6.43}$$

is affine in terms of H^{-1}. Since (A.6.41) is quadratic in N, it has at most two solutions satisfying the decay condition. Hence we can construct $\mathbf{V}(n_1\, x_2)$ by superposition of two modes of the form (A.6.6),

$$\mathbf{V}(n_1\, x_2) = \mathbf{U}^{(1)} \exp(\mathrm{i}\, N^{(1)}\, n_1\, x_2) \; + \; \mathbf{U}^{(2)} \exp(\mathrm{i}\, N^{(2)}\, n_1\, x_2)\,, \tag{A.6.44}$$

where each of the two $\mathbf{U}^{(\alpha)}$ satisfies the field equations (A.6.40), and hence from the first and second of (A.6.40), for $\alpha = 1, 2$,

$$w^{(\alpha)} \equiv \frac{U_2^{(\alpha)}}{U_1^{(\alpha)}} = -\frac{L_{11} + (N^{(\alpha)})^2 - X^2}{(L_{12}+1)\, N^{(\alpha)}} = -\frac{(1+L_{21})\, N^{(\alpha)}}{L_{22}\,(N^{(\alpha)})^2 + 1 - X^2}\,. \tag{A.6.45}$$

To close the problem, we require the boundary conditions at $x_2 = 0$ to be satisfied, namely from (A.6.36),

$$\begin{bmatrix} N^{(1)} + w^{(1)} & N^{(2)} + w^{(2)} \\ L_{21} + L_{22}\, N^{(1)}\, w^{(1)} & L_{21} + L_{22}\, N^{(2)}\, w^{(2)} \end{bmatrix} \begin{bmatrix} U_1^{(1)} \\ U_2^{(2)} \end{bmatrix} = \begin{bmatrix} 0 \\ 0 \end{bmatrix}. \tag{A.6.46}$$

Indeterminacy of this system yields, using (A.6.45),

$$N^{(1)}\, N^{(2)} = \frac{X^2\,(X^2 - L_{11})}{\Delta - L_{22}\, X^2} = \frac{(X^2-1)(\Delta - L_{22}\, X^2)}{L_{22}\, X^2}. \tag{A.6.47}$$

Thus the square of the speed of propagation-like scalar X satisfies the equation:

$$F(X) = c_0\,(X^2)^3 + c_1\,(X^2)^2 + c_2\, X^2 + c_3 = 0\,, \tag{A.6.48}$$

with coefficients

$$c_0 = L_{22}(1-L_{22})\,,\ c_1 = L_{22}(L_{22}-L_{11}+2\,\Delta)\,,\ c_2 = -\Delta\,(2\,L_{22}+\Delta)\,,\ c_3 = \Delta^2\,. \tag{A.6.49}$$

A.6.3.2 **Surface flutter**

Now that we have established the equations for the main unknowns, namely, the propagation direction $\mathbf{n} = (n_1, n_2)$ and the wave-speed c, we can analyze the nature of the solutions.

Given the constitutive parameters, we first calculate the three squares of the wave-speeds by solving (A.6.48); for each solution, we calculate by (A.6.41) the four associated directions of propagation, out of which we select, when they are available, those two $N^{(1)}$ and $N^{(2)}$ that satisfy the decay condition (A.6.4). The key difficulty is that these two solutions may not satisfy the boundary conditions at the free surface represented by (A.6.47). In fact, while the roots $N^{(1)}$ and $N^{(2)}$ of (A.6.41) always satisfy (A.6.47) in absolute value, their signs might be in contradiction with the decay condition.

We assume that during deformation processes H^{-1} increases monotonously from 0 to ∞. At $H^{-1} = 0$, the behaviour is elastic and the cubic equation (A.6.48) yields the elastic Rayleigh wave $c_R < c_s^e$, and the corresponding mode (A.6.25) is a surface wave. As the control parameter H^{-1} increases, we look for situations in which some X^2 solution of (A.6.48) ceases to be a real positive number and the appropriate sign on $N^{(1)}\, N^{(2)}$ is satisfied: the onset of a dynamic instability. Surface instabilities are of interest when they precede possible body instabilities: conditions for hyperbolicity of the field equations are considered in Sect. A.6.3.4.

The analysis of surface instability is described in LMS [1995]. It is observed that, whenever the onset of flutter instability is reached, the (real) speed of the prolongation

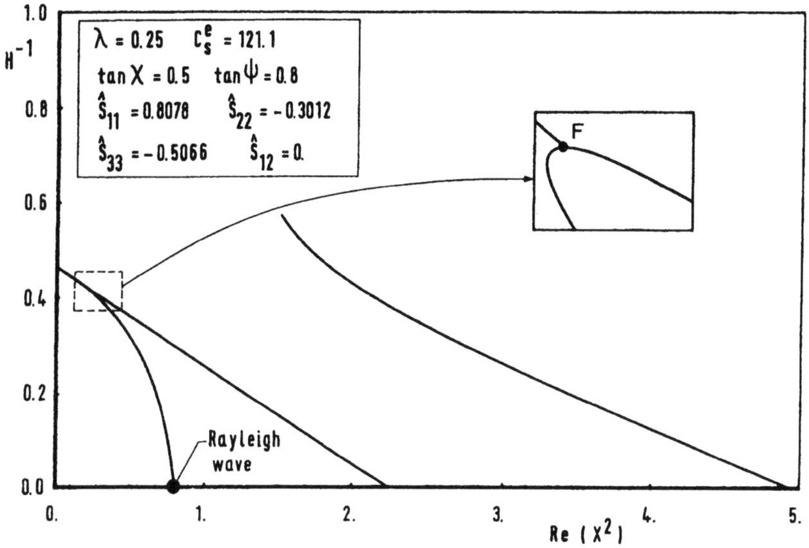

Figure A.6.5 Evolution of the three roots of the cubic equation (A.6.48) in the coaxial case, $\hat{S}_{12}=0$. Flutter occurs when the prolongation in the plastic range of the elastic Rayleigh wave-speed coalesces with the real intermediate root of the cubic equation (A.6.48).

in the plastic range of the elastic Rayleigh wave coalesces with the (real) intermediate root of the cubic equation (A.6.48). At this stage both solutions have a surface character, even if, along some phase of the incrementation process between $H^{-1}=0$ and the critical value of H^{-1}, one or both of them do not have such a surface character. The evolution of the three roots of the cubic equation with H^{-1} shown in Fig. A.6.5 is typical. At the onset of flutter instability, the two smaller roots of (A.6.48) coalesce and, similarly to what happens in finite dimensional systems, Huseyin [1978], the partial derivative of the control parameter H^{-1} with respect to X^2 vanishes at the onset of flutter. The evolution in the complex plane of $(X^2, N^{(1)} N^{(2)})$ is shown in Figs. A.6.3 and 4, case (a) $\hat{S}_{12}=0$.

A.6.3.3 **Regimes of field equations**

For elastic-plastic solids with deviatoric associativity, loss of hyperbolicity coincides with the onset of a stationary body wave, that is c or $X=0$, while at least one N^2 should be non-negative real, because, in an infinite body, only plane waves need to be considered. Thus from (A.6.41), hyperbolicity is retained if the inequalities below hold[10],

$$\begin{cases} \text{if } S_0 \geq 0, \text{ then } SB_0 \equiv P_0 - S_0^2/4 > 0; \\ \text{if } S_0 \leq 0, \text{ then } SB_0 \equiv P_0 > 0. \end{cases} \tag{A.6.50}$$

[10] SB is an acronym for Stationary Body wave.

The onset of a stationary wave is then detected by the criterion, see Fig. A.6.6-(a),

$$\begin{cases} \text{if } S_0 \geq 0, \text{ then } SB_0 \equiv P_0 - S_0^2/4 = 0; \\ \text{if } S_0 \leq 0, \text{ then } SB_0 \equiv P_0 = 0. \end{cases} \tag{A.6.51}$$

In the above, S_0 and P_0 denote the values of S and P, eqn (A.6.42), at $X = 0$. Note that at the beginning of the deformation process, $H^{-1} = 0$, $\Delta = 4(\Lambda + 1)$, $S_0 = -2$ and $P_0 = 1$.

Fig. A.6.6-(b) shows also the characteristic plastic moduli

- h_S at which $S_0 = 0$;
- h_{SB} at which $SB_0 = 0$, more precisely $h_{SB} = h_{SB}^{(+)}$ if $SB_0 = 0$ while $S_0 \geq 0$ and $h_{SB} = h_{SB}^{(-)}$ if $SB_0 = 0$ while $S_0 \leq 0$: hyperbolicity is retained when $h > h_{SB}$.

In addition, it can be proved that, in the hyperbolic regime, *stationary surface waves* can occur only in the HI regime, for non-associative plasticity, whereas, for associative plasticity, they can occur only along the parabola $SB_0 = 0$, $S_0 \leq 0$. Moreover, surface divergence instabilities cannot occur before the stationary surface waves. Body divergence can be reached only after stationary body discontinuity at $X = 0$ since there is no body flutter instability due to deviatoric associativity.

As can be checked from Fig. A.6.6-(a), the path in the plane (S_0, SB_0) corresponding to the example shown in Fig. A.6.3 and 4 remains in the hyperbolic regime, and thus we have an instance where surface flutter instability precedes any other kind of body or surface instability.

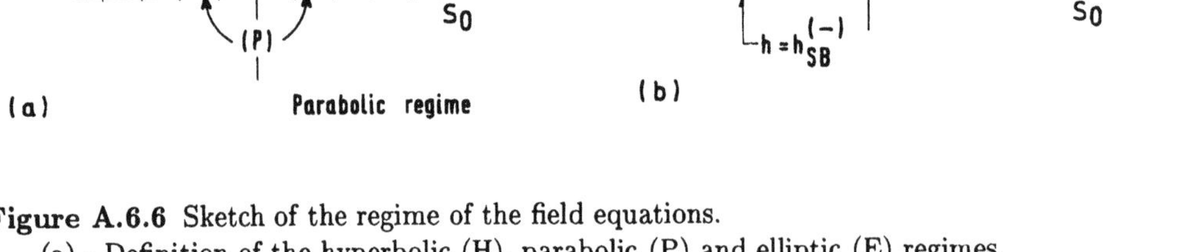

Figure A.6.6 Sketch of the regime of the field equations.

(a) - Definition of the hyperbolic (H), parabolic (P) and elliptic (E) regimes.

(H): the two N^2 solutions of (A.6.41) are complex conjugate if $SB_0 = P_0 - S_0^2/4 > 0$ (HC) or real negative if $SB_0 \leq 0$ (HI);

(P): one N^2 is real positive, the other is real negative;

(E): the two N^2 are real positive.

(b) - Characteristic moduli.

A.7 Finite Element Analysis

The onset of flutter instability and ill-posedness in evolution problems involving rate-independent solids has been the subject of numerous studies summarized in the previous sections. However the physical and structural consequences of flutter in elastic-plastic solids remain mostly unknown. Even the proper conclusions that can be extracted from an indication of flutter in a linearized stability analysis are not clear: in the above stability analyses the tangent constitutive moduli corresponding to plastic loading were always considered, but the growing oscillatory nature of the flutter solutions implies that, in the true elastic-plastic solid, successive periods of loading and unloading would occur immediately or shortly after the initiation of such dynamic solutions. Without further clarification, this raises serious doubts on the validity of the adopted linearization procedure and on the attempts to extract meaningful conclusions from such linearized analyses. An assessment of the occurrence and the consequences of flutter requires the consideration of the fully nonlinear elastic-plastic behaviour of the solid in the analysis of its dynamic response. These are the purpose and the approach adopted in finite element studies reported in the present section. Numerical predictions of flutter based on linearized dynamic stability analyses (plastic loading assumed in all the plastified region) are confronted here with the results of finite element nonlinear dynamic analyses, where both loading and unloading regimes are taken into consideration. The numerical results show that the nonlinear dynamic behaviour of the system is quite different in the cases of constant or variable applied forces. Only in the second case a correlation is found between the predictions of the above linearized analyses and the nonlinear dynamic solutions. These numerical results, together with some developments in Martins, Simões, Gastaldi and Marques [1995] and in Martins, Barbarin, Raous and Costa [1999], will motivate the interpretation for flutter proposed in Sect. A.8.

A.7.1 Description of the Problem

We study a problem in which conditions are created that are reminiscent of those considered in Sect. A.6 for the occurrence of surface flutter with a non-associative elastic-plastic body. We consider a homogeneous body in a plane state of deformation, with unit dimensions in the plane 1 m × 1 m and with an elastic-perfectly plastic behaviour, i.e. $h=0$, satisfying the Drucker-Prager yield criterion, with a non-associative flow rule. The elastic properties of the solid are Young's modulus E=2200 MPa, Poisson's ratio $\nu=0.1$ and the plastic properties are $\tan\chi=0.5$, $\tan\psi=0.7$, k=1 MPa, see Sect. A.1.2 for notation. The density ρ is 2500 kg/m^3. The body is fixed on the surface Γ_1, lateral displacements are restricted but the vertical displacements are free on the surface Γ_2, the surface Γ_3 is free and normal and tangential distributed forces are applied simultaneously on the surface Γ_4, Fig. A.7.1. Eight-noded isoparametric finite elements are used in the three uniform meshes employed in this study: mesh 1 has 7×7 elements and the meshes 2 and 3 have 10×10 and 14×14 elements, respectively.

The finite element program developed and tested in Simões [1997] is used to solve the quasi-static and the dynamic problems discussed next, as well as the corresponding linearized stability eigenproblems. The elastic-plastic behaviour is numerically treated

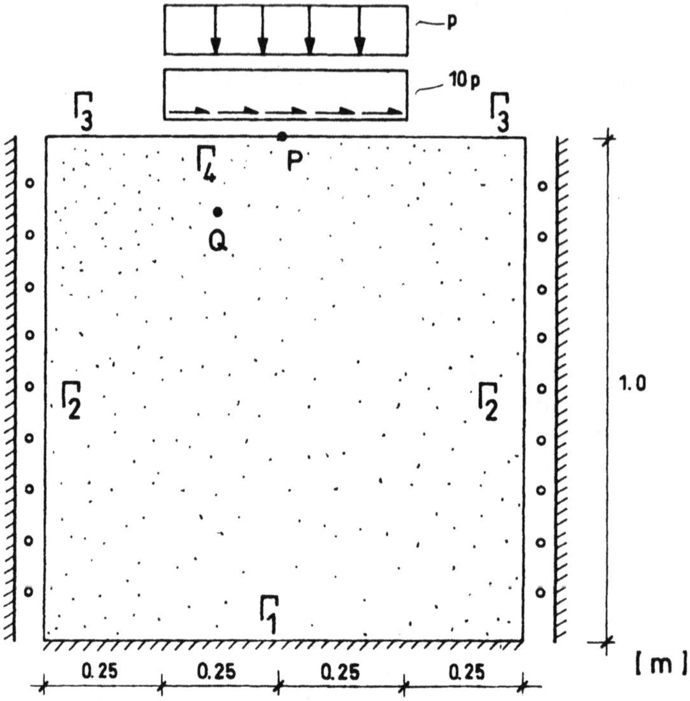

Figure A.7.1 Non-associative elastic-plastic body in a plane state of deformation and loaded on the surface Γ_4. Geometry and boundary conditions of the problem.

with the return mapping algorithm, as adapted to the non-associative model of Drucker-Prager with isotropic hardening; the resulting non-linear problem is solved with Newton-Raphson iterations that employ the tangent stiffness matrix or the consistent tangent stiffness matrix, Simo and Taylor [1985]. The time integration of the dynamic equations is performed either explicitly, by central differences, or implicitly by the α-method, Hilber at al. [1977]. The generalized stability eigenproblem, that involves a non-symmetric tangent stiffness matrix, is solved by the method of Arnoldi (software package ARPACK, http://www.netlib.no:80/netlib/scalapack). For the two first kinds of problems, comparisons with the results of the commercial finite element program ABAQUS were frequently performed.

A.7.2 Quasi-Static Solutions

The quasi-static trajectories are obtained by incrementing the total load $p = 250\,\mathrm{kN/m}$ in 100 steps. The quasi-static vertical displacement of point P, Fig. A.7.1, obtained with mesh 3, is plotted in Fig. A.7.2 against the load parameter p. The plastic region of the body at the load level $p_5 = 125\,\mathrm{kN/m}$ is shown in Fig. A.7.3.

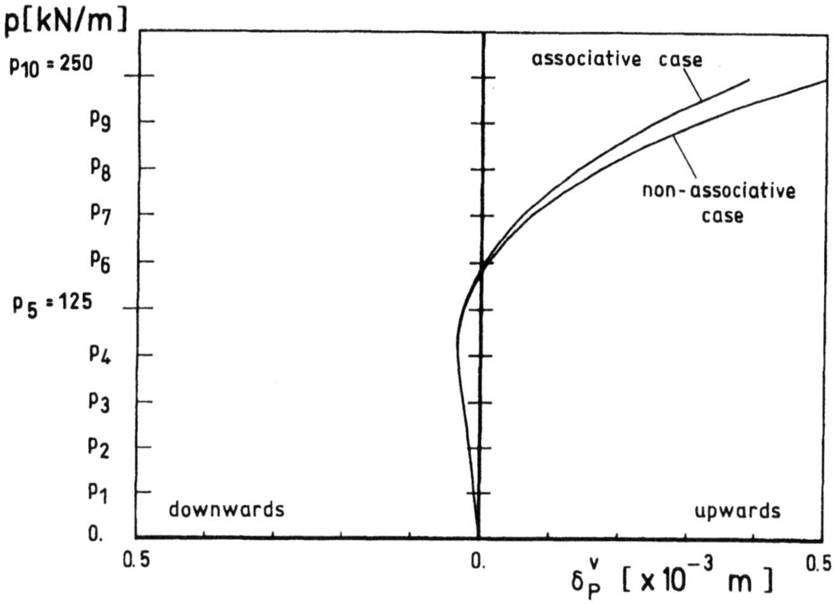

Figure A.7.2 Evolution of the vertical displacement of point P, see Fig. A.7.1, when the load p is applied quasi-statically for the associative material with tan ψ=tan χ=0.5, and for the non-associative material with tan ψ=0.7>tan χ=0.5.

A.7.3 Linearized Stability Analysis

In the neighborhood of an equilibrium configuration along some quasi-static path, a formal linearization of the dynamic equations (and of the constitutive equations) can be performed, in analogy to those used in the analytical studies of Sect. A.6. We are led to a system of second order homogeneous ordinary differential equations for the displacements relative to the equilibrium configuration $\Delta\,\mathbf{q}(t)$, namely

$$\mathbf{M}\,\Delta\,\ddot{\mathbf{q}}(t) + \mathbf{K}\,\Delta\,\mathbf{q}(t) = \mathbf{0}\,, \tag{A.7.1}$$

where $\mathbf{M}$ is the mass matrix and the tangent stiffness matrix $\mathbf{K}$ is constructed with the elastic-plastic tensor that corresponds to plastic loading in the element integration points where the stress state is on the yield surface, and with the elastic constitutive tensor at the integration points where the stress state is inside the yield surface. Letting $\Delta\,\mathbf{q}(t)$ be of the form

$$\Delta\,\mathbf{q}(t) = \mathbf{V}\,\exp(-i\,\omega\,t), \tag{A.7.2}$$

the following eigenproblem of size N is obtained:

$$\text{Find } \omega \in \mathbb{C} \text{ and } \mathbf{V} \in \mathbb{C}^N\,,\ \mathbf{V} \neq \mathbf{0}\,,\ \text{such that } (\mathbf{K} - \omega^2\,\mathbf{M})\mathbf{V} = \mathbf{0}\,. \tag{A.7.3}$$

Note that, as it is well-known, the non-associativity of the flow rule originates a non-symmetric tangent stiffness matrix, which may lead to angular frequencies ω with complex squares and consequent flutter (in the linearized problem (A.7.1)).

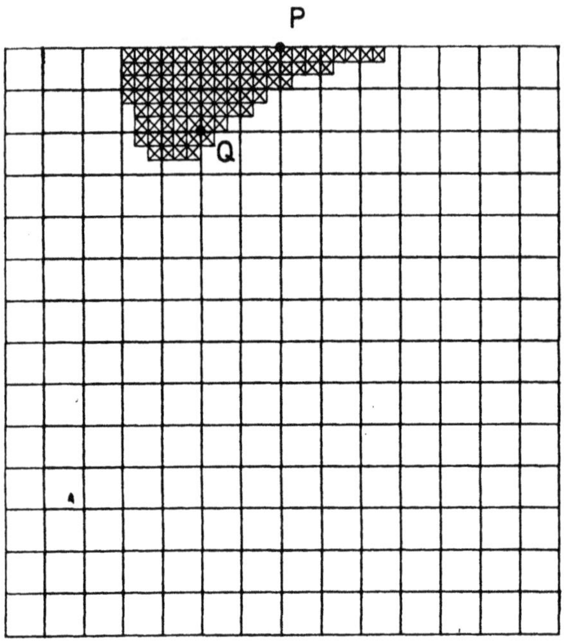

Figure A.7.3 Plastic region in the body at the load level p_5 on mesh 3.

The eigenproblem (A.7.3) is solved for successive equilibrium configurations of the body at increasing values of the load parameter p. As a reference, let us note that the smallest elastic eigenvalue is 1490 rad/s. It is observed that complex eigenvalues ω with positive imaginary parts occur as soon as plastic deformation initiates in the body, $p \geq p_2$. The real and imaginary parts of the complex eigenvalues with positive imaginary part are plotted in the Argand planes of Figs. A.7.4-(a), (b) and (c), for three load levels p_6, p_8, and p_{10}. From these figures and from the whole set of results obtained, it is possible to conclude that:

- the number of complex eigenvalues with positive imaginary part increases as the load parameter and the size of the plastified region increase,
- for each load level and as the mesh is refined, the number of complex eigenvalues ω with positive imaginary parts increases, as well as the number of those ω that have larger positive imaginary parts.

This increase of the number of instability modes and of their rates of growth in time as the mesh is refined is believed to be a manifestation, in the approximate finite dimensional problems, that the corresponding limit continuum problem is *linearly ill-posed.*

A.7.4 Nonlinear Dynamic Analyses

For the load level p_5, various nonlinear dynamic analyses are performed. We group them in the following sets: 1) non-associative case with constant applied forces; 2) non-associative case with variable applied forces; 3) associative case with variable applied forces. In all these analyses the mesh 3 is used.

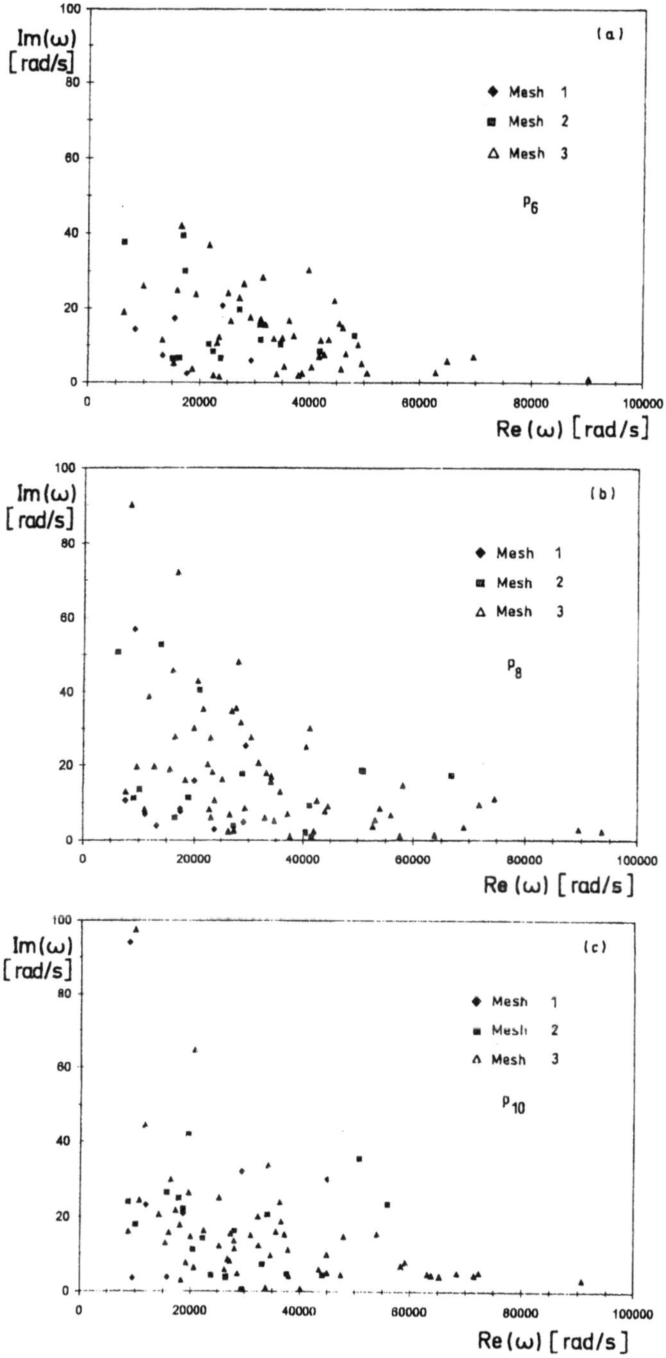

Figure A.7.4 Real and imaginary parts of the complex eigenvalues with positive imaginary part for three load levels p: (a): p=p_6, (b): p=p_8, and (c): p=p_{10}. The number of complex eigenvalues ω with positive imaginary part $\mathcal{I}m\,\omega > 0$ increases as the load plevel p and the size of the plastified region increase. For each load level and as the mesh is refined, the number of complex eigenvalues with $\mathcal{I}m\,\omega > 0$ increases, as well as the number of those ω that have larger positive imaginary parts.

A.7.4.1 Non-associative flow rule with constant applied forces

The loads on the surface of the body are kept constant at the level p_5, and the initial conditions are the following: the initial displacements are the previously obtained static equilibrium displacements at the load level p_5, plus a small perturbation along the real components of the eigenvector that corresponds to the complex frequency with smaller real part; the initial velocity is the real velocity that is consistent with the modal perturbation considered in the initial displacement. Phase plane plots for the vertical displacement and velocity components q_2 and $\dot{q}_2$ of node Q of the finite element mesh, Fig. A.7.1, located inside the plastified region are shown in Fig. A.7.5-(a)-(b) for some initial perturbation of amplitude $\epsilon=10^{-5}$, and in Fig. A.7.5-(c)-(d) for a ten times smaller perturbation $\epsilon=10^{-6}$. Similar tests were performed with initial perturbations given along other instability modes, and it was also checked that, in the absence of any initial perturbation, the dynamic solution remains in the initial equilibrium configuration.

The results in Fig. A.7.5 and the whole set of results obtained suggest that the equilibrium configuration is *stable*: the dynamic solution deviates as little as desired from that equilibrium state provided sufficiently small initial perturbations are considered.

A.7.4.2 Non-associative flow rule with variable applied forces

The load parameter p has the initial value of p_5, but now it is increased at a constant rate $\dot{p}$. The initial displacements are the previously obtained static equilibrium displacements at the load level p_5, and the initial velocities are taken equal to the velocities of the quasi-static solution at the load level p_5, assuming that the loading in the quasi-static and the dynamic phases is performed at the same rate. Similarly to what was done in the previous subsection, additional initial perturbations could also be explicitly considered, but actually that is not necessary: the neglected inertia forces in the computed quasi-static solution and the finite precision of all numerical computations and results will be sufficient perturbations for the problem now at hand. The phase plane evolution of the vertical displacement and velocity components of node Q, Fig. A.7.1, situated inside the plastified region are shown in Fig. A.7.6-(a)-(b) for two load rates $\dot{p}$=294.1 kN/m/s and 2941 kN/m/s.

In these plots and in all the other obtained in the analyses of this type, it can be seen that the dynamic response deviates oscillating from the corresponding quasi-static trajectory, even in the absence of any (explicit) initial perturbation. The curves of load parameter versus vertical displacement of point P, for the quasi-static solution and for the dynamic solutions are represented in Fig. A.7.7. It is clear that the oscillations shown in Fig. A.7.6 have no significant consequences on the load-displacement behaviour of the system.

A.7.4.3 Associative flow rule with variable applied forces

The dynamic behaviour obtained in the previous non-associative case is now compared with the one that is obtained when the flow rule is associated. The elastic-plastic body considered now has the same geometry, it is subjected to the same kinematic restrictions and to the same loading histories as the one of the previous case. It has also the same material properties as those of the previous case, with the exception that now $\tan\chi = \tan\psi = 0.7$, i.e. the friction and the dilation angles are equal, so that the flow rule is associated. The load-displacement diagrams of Fig. A.7.2 show that, in terms of the involved stiffness, the associated and the non-associated problems are not significantly

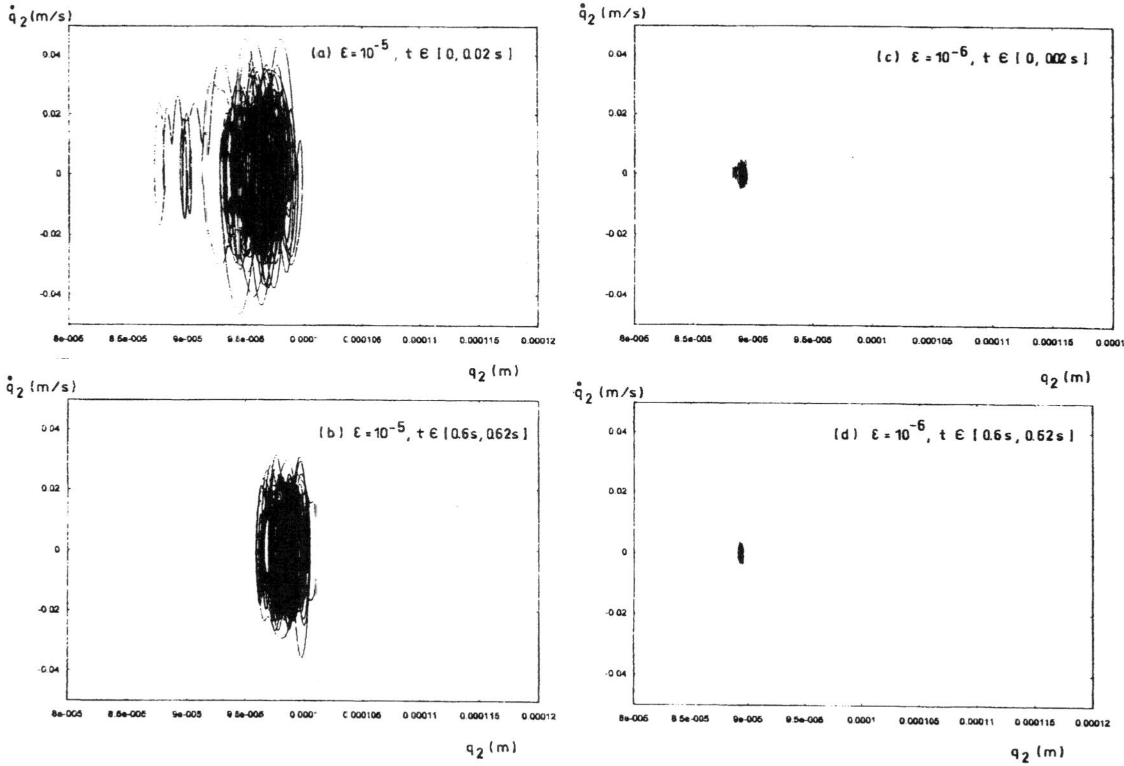

Figure A.7.5 Phase plane showing the vertical displacement q_2 and velocity $\dot{q}_2$ of the point Q, see Fig. A.7.1, when, at load level $p=p_5$, the initial displacements are perturbed in a dynamic analysis by the real part of the eigenmode whose complex eigenvalue has the smallest real part and the body is left to evolve at constant load. The perturbation of amplitude ϵ affects the material response at early times, while later the state tends to rest.

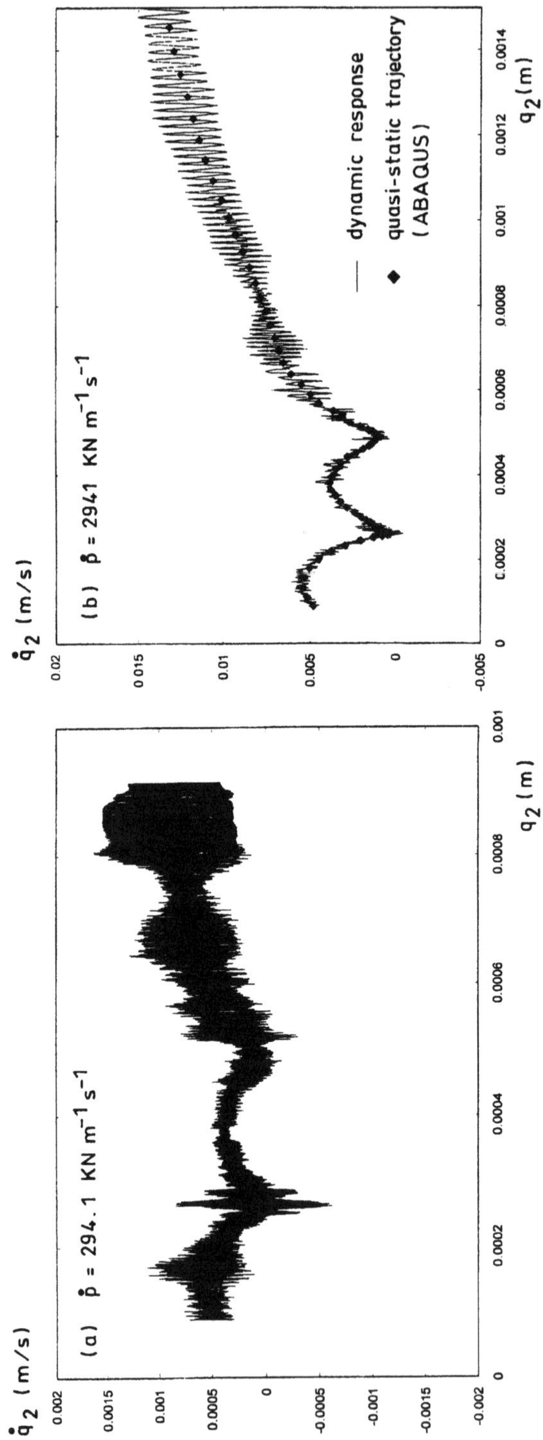

Figure A.7.6 Phase plane showing the vertical displacement q_2 and velocity $\dot{q}_2$ of point Q, see Fig. A.7.1, when, from load level $p{=}p_5$, the load p is applied quasi-statically and dynamically with two load rates $\dot{p}$. Non-associative behaviour.
(a): $\dot{p}$=294.1 kN/m/s, (b): $\dot{p}$= 2941 kN/m/s

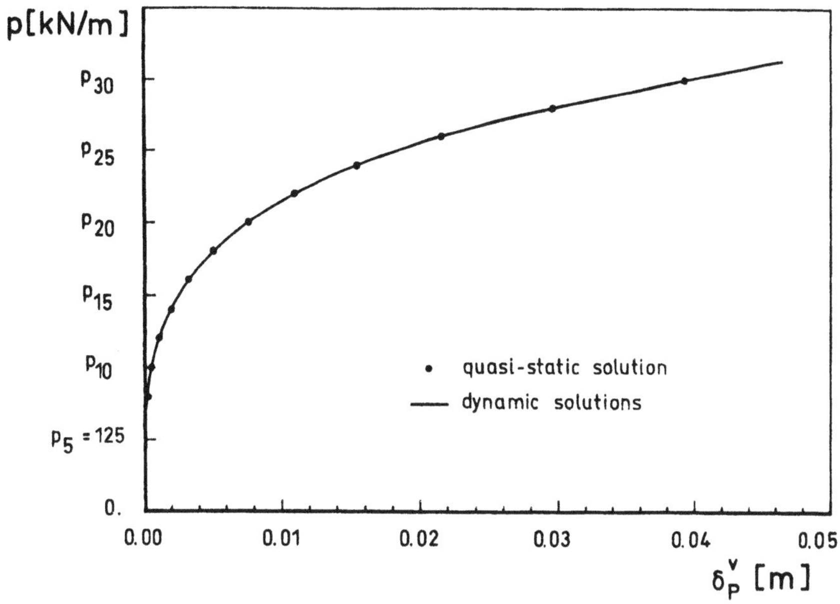

Figure A.7.7 Evolution of the vertical displacement of point P, see Fig. A.7.1, when the load p is applied quasi-statically and dynamically from load level $p=p_5$.

different in the loading range of interest. The main results of the present analysis are phase plane plots of the type shown in Fig. A.7.8-(a)-(b). These plots are to be compared with those of Fig. A.7.6-(a)-(b) for non-associative behaviour. It is clear that the deviation from the smooth quasi-static solution and the oscillations present in the dynamic solution in the associative case are significantly smaller than in the non-associative case.

A.7.5 Concluding Remarks

The numerical solutions obtained suggest thus that, when the linearized problem (A.7.3) has complex eigenvalues ω, one cannot conclude that the corresponding equilibrium configuration (at constant applied forces) is unstable. On the contrary, the numerical results suggest that the equilibrium configuration is stable. However, if, in the same circumstances, the applied forces are varied, the numerical results show that growing flutter type dynamic solutions exist in the neighborhood of the corresponding quasi-static trajectory. The non-linearity of the elastic-plastic constitutive laws has the consequence that the growth of the perturbed solutions does not continue at the exponential rate predicted by the linearized stability analyses. In the problems studied in this chapter, this has the consequence that the load-displacement curves are not significantly disturbed by flutter.

A.8 Dynamic Behaviour in the Neighbourhood of a Quasi-Static Path

The numerical results of the previous section suggest that the occurrence of growing *flutter type* dynamic solutions for the equations that govern the behaviour of non-associative

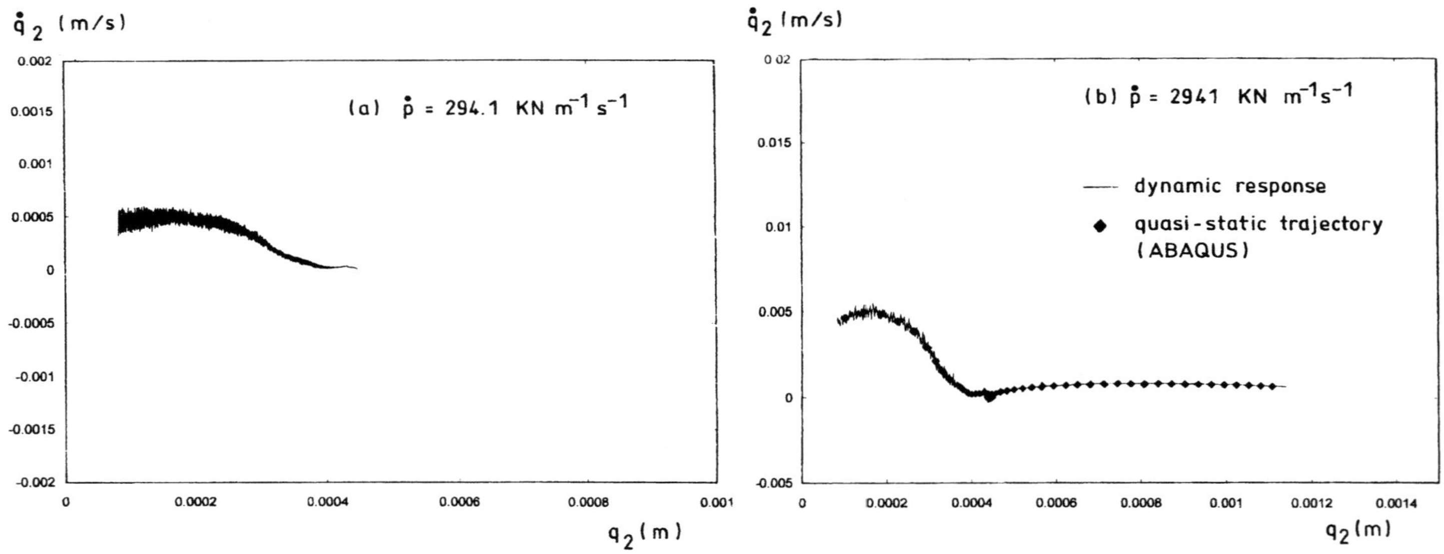

Figure A.7.8 Phase plane showing the vertical displacement q_2 and velocity $\dot{q}_2$ of point Q, see Fig. A.7.1, when, from load level $p=p_5$, the load p is applied dynamically with two load rates $\dot{p}$. Associative behaviour (compare with Figs. A.7.6).

elastic-plastic solids is closely related with the existence of some external time-varying applied forces or displacements that provide a permanent source of energy and originate a quasi-static solution, about which the growing oscillation develops. This should not be surprising for various reasons. On one hand the classical mechanical examples of flutter have similar characteristics, e.g. Holmes [1977]: it is the wind that provides the permanent source of energy for the *panel flutter*; it is the fluid flow that provides the permanent source of energy in the case of the flutter of a *fluid conveying pipe*. On the other hand, in the quite similar situation of the flutter of an elastic system in contact with an obstacle along which the non-associative Coulomb friction law holds, the flutter phenomenon is characterized in Martins et al. [1999] as a growing oscillation in the neighbourhood of a portion of a quasi-static solution, or as the instability of the steady sliding of the system, when a permanent energy source is provided by an imposed tangential motion of the system support or of the obstacle.

Similarly to Martins et al. [1999], the theoretical concepts that should be applied to the characterisation of flutter in elastic-plastic solids can be loosely classified as concepts of *dynamic instability of quasi-static paths*. However, to the authors' knowledge, such concepts have not been consistently defined yet. In fact, the concept of Lyapunov stability has been for long applied to the study of the dynamic stability of equilibrium configurations of elastic or elastic-plastic systems, under constant applied forces. However, in what concerns quasi-static paths under varying external loads, any attempt to apply the classical concepts of Lyapunov stability is immediately confronted with the following difficulty: the quasi-static solutions whose *stability* is sought, are not, in general, true solutions of the corresponding dynamic problem. Although supposedly small and accordingly neglected, the accelerations of the quasi-static solutions are not null in general. In the context of frictional contact problems, this difficulty was avoided in Martins et al. [1999] by considering only *straight* quasi-static paths, that is with constant speed and null acceleration, that are also solutions of the dynamic equations of motion. In this manner, if the portion of the quasi-static and dynamic solution under consideration extends in time up to infinity, the question addressed by Martins et al. [1999], and that is also the object of our interest here, is the dynamic (Lyapunov) instability of that solution. But, while coinciding as much as possible with the concept of Lyapunov instability in these particular circumstances, a sufficiently general concept of dynamic instability of a quasi-static path must be valid also in cases where the quasi-static paths are not straight.

Some proposals exist in the literature for the characterisation of the *stability of quasi-static paths*. In the definition of Petryk [1993], a quasi-static deformation process is said to be unstable in the energy sense, at some stage of the deformation, if the respective velocity solution does not minimize the value of the second order rate of change of the energy functional on the space of the admissible velocities. This energy functional is the sum of the work of deformation in the body with the potential energy of the loading device. Because of the first of these contributions, it depends on the deformation history, and because of its physical interpretation, it is also known, Nguyen [1994], as the *energy expense* provided by the external world to the system and to the loading device. Petryk relates his energetic criterion to a definition of instability involving finite deviations from the unperturbed path, produced by infinitesimal perturbing forces acting in any finite time interval. However, in the context of *stability of paths*, inertia effects are not taken into account: the relationship with some concept of *dynamic instability* is not apparent. A similar energetic criterion for the stability of quasi-static paths is proposed by Bažant and Cedolin [1991]: the criterion is presented as a result of thermodynamic considerations and, again, inertia effects are disregarded. Because of the *energetic* character of the proposed criterion, *its validity is carefully restricted* by Petryk [1993] *to the associative case.*

In the next section A.8.1, we present the linearized form of the equations that govern the evolution of the difference between the dynamic and the quasi-static solution of an elastic-plastic solid subjected to external forces or displacements varying linearly in time, assuming that *plastic loading holds on the quasi-static path.* Those equations are discussed and interpreted in section A.8.2, where we explore their analogies with the equations that govern the dynamic stability of steady frictional sliding states, and with the equations that govern the dynamic behaviour of a visco-elastic model of Maxwell. An equation of balance of energy that involves the difference between the dynamic and the quasi-static solutions and that applies to the associative case is established in section A.8.3. Dynamic instability of a quasi-static path is discussed in section A.8.4 in the particular case of straight quasi-static paths along flat portions of the yield surface and of the plastic potential; some general observations on ill-posedness will be provided also in the same section. A singular perturbation framework for the generalization of the concept of dynamic instability of a quasi-static path to the general case of curved paths and curved yield surfaces and plastic potentials is proposed in A.8.5.

A.8.1 Linearized Equations Governing the Difference between Dynamic and Quasi-Static Solutions in the Plastic Loading Regime

In order to derive these linearized governing equations, we essentially follow the procedure suggested in Simões [1997, p. 220] to establish the equations needed for a *second order stability analysis* incorporating inertia effects. Along this derivation, we provide an interpretation of the governing variables and of various aspects of the governing equations that differs significantly from second order bifurcation analyses. We consider the elastic-plastic body subjected to an external loading, which is assumed

(H1) to be controlled by a single non-dimensional load parameter λ with a constant non-vanishing time rate of change γ: the body forces are of the form $\mathbf{b}(\mathbf{x},t)=\lambda(t)\,\bar{\mathbf{b}}(\mathbf{x})$, and, similarly, the prescribed tractions or displacements on the boundary of the body are given by $\mathbf{t}(\mathbf{x},t)=\lambda(t)\,\bar{\mathbf{t}}(\mathbf{x})$, and $\mathbf{u}(\mathbf{x},t)=\lambda(t)\,\bar{\mathbf{u}}(\mathbf{x})$, respectively.

The subscript $(\cdot)_{qs}$ and superscript $(\cdot)^{qs}$ will denote fields of the *quasi-static* solution. The symbol $\delta(\cdot)$ is used to denote the difference between the dynamic and the quasi-static fields. e.g. $\delta\mathbf{v}(\mathbf{x},t)=\mathbf{v}(\mathbf{x},t)-\mathbf{v}^{qs}(\mathbf{x},t)$. Since the equations of balance of linear momentum in the dynamic and quasi-static regimes are, respectively,

$$\begin{aligned} \rho\,\dot{\mathbf{v}} - \operatorname{div}\boldsymbol{\sigma} &= \mathbf{b}\,, \\ -\operatorname{div}\boldsymbol{\sigma}^{qs} &= \mathbf{b}\,, \end{aligned} \tag{A.8.1}$$

then the equations that govern the difference fields $\delta\mathbf{v}$ and $\delta\boldsymbol{\sigma}$ are

$$\rho\,\delta\dot{\mathbf{v}} - \operatorname{div}(\delta\boldsymbol{\sigma}) = -\rho\,\dot{\mathbf{v}}^{qs}\,, \tag{A.8.2}$$

where $\delta\dot{\mathbf{v}} \equiv \dot{\mathbf{v}} - \dot{\mathbf{v}}^{\mathrm{qs}}$ and $\dot{\mathbf{v}}^{\mathrm{qs}}$ is the acceleration of the quasi-static solution, which, by definition, is neglected when computing the quasi-static solution $(\mathrm{A.8.1})_2$.

As for the elastic-plastic behaviour of the solid, the following is assumed:

(H2) the tensor of the elasticity moduli $\mathbb{E}$ is constant;
(H3) the plastic modulus h is constant;
(H4) the unit normals to the yield surface $\mathbf{Q}$ and plastic potential $\mathbf{P}$ depend only on the stress $\boldsymbol{\sigma}$:

$$\mathbf{Q} = \mathbf{Q}(\boldsymbol{\sigma}), \quad \mathbf{P} = \mathbf{P}(\boldsymbol{\sigma}), \tag{A.8.3}$$

and, as further consequence of the previous assumptions,

$$H = H(\boldsymbol{\sigma}) = h + \mathbf{Q}(\boldsymbol{\sigma}) : \mathbb{E} : \mathbf{P}(\boldsymbol{\sigma}); \tag{A.8.4}$$

(H5) plastic loading holds in the quasi-static solution during the time interval of interest (that may extend up to infinity), i.e. there exists a positive constant η such that

$$\mathbf{Q}^{\mathrm{qs}} : \mathbb{E} : \dot{\boldsymbol{\epsilon}}^{\mathrm{qs}} \geq \eta > 0, \quad \dot{\boldsymbol{\sigma}}^{\mathrm{qs}} = \mathbb{E}^{\mathrm{ep}}_{\mathrm{qs}} : \dot{\boldsymbol{\epsilon}}^{\mathrm{qs}}, \tag{A.8.5}$$

where $\mathbb{E}^{\mathrm{ep}}_{\mathrm{qs}} = \mathbb{E}^{\mathrm{ep}}(\boldsymbol{\sigma}^{\mathrm{qs}}) = \mathbb{E} - 1/H^{\mathrm{qs}}(\mathbb{E} : \mathbf{P}^{\mathrm{qs}}) \otimes (\mathbf{Q}^{\mathrm{qs}} : \mathbb{E})$ with $\mathbf{P}^{\mathrm{qs}} = \mathbf{P}(\boldsymbol{\sigma}^{\mathrm{qs}})$, $\mathbf{Q} = \mathbf{Q}(\boldsymbol{\sigma}^{\mathrm{qs}})$, $H^{\mathrm{qs}} = H(\boldsymbol{\sigma}^{\mathrm{qs}}) = h + \mathbf{Q}^{\mathrm{qs}} : \mathbb{E} : \mathbf{P}^{\mathrm{qs}}$;
(H6) plastic loading also holds at the initial time t_0 for the perturbed dynamic solution, i.e. the perturbed stress state at the initial time $\boldsymbol{\sigma}(t_0)$ is *on* the yield surface and the differences $\delta\boldsymbol{\sigma}(t_0)$ and $\delta\dot{\boldsymbol{\epsilon}}(t_0)$ are sufficiently small that

$$\mathbf{Q}(\boldsymbol{\sigma}(t_0)) : \mathbb{E} : \dot{\boldsymbol{\epsilon}}(t_0) > 0, \tag{A.8.6}$$

By continuity, it follows that, while the difference between the quasi-static and the dynamic solution is sufficiently small, plastic loading also holds for the perturbed dynamic solution:

$$\mathbf{Q} : \mathbb{E} : \dot{\boldsymbol{\epsilon}} > 0, \quad \dot{\boldsymbol{\sigma}} = \mathbb{E}^{\mathrm{ep}} : \dot{\boldsymbol{\epsilon}}, \tag{A.8.7}$$

where $\mathbb{E}^{\mathrm{ep}}$ is given by (A.1.5), with $\mathbf{P}$, $\mathbf{Q}$ and H depending on the stress $\boldsymbol{\sigma}$ of the dynamic solution. In view of all the above, the linearized form of

$$\delta\dot{\boldsymbol{\sigma}} = \dot{\boldsymbol{\sigma}} - \dot{\boldsymbol{\sigma}}^{\mathrm{qs}} = \left(\mathbb{E}^{\mathrm{ep}}_{\mathrm{qs}} + \delta\mathbb{E}^{\mathrm{ep}}\right) : \left(\dot{\boldsymbol{\epsilon}}^{\mathrm{qs}} + \delta\dot{\boldsymbol{\epsilon}}\right) - \mathbb{E}^{\mathrm{ep}}_{\mathrm{qs}} : \dot{\boldsymbol{\epsilon}}^{\mathrm{qs}}, \tag{A.8.8}$$

is

$$\delta\dot{\boldsymbol{\sigma}} = \delta\mathbb{E}^{\mathrm{ep}} : \dot{\boldsymbol{\epsilon}}^{\mathrm{qs}} + \mathbb{E}^{\mathrm{ep}}_{\mathrm{qs}} : \delta\dot{\boldsymbol{\epsilon}}. \tag{A.8.9}$$

The linearized form of $\delta\mathbb{E}^{\mathrm{ep}}$ is in turn defined by

$$H^{\mathrm{qs}}\,\delta\mathbb{E}^{\mathrm{ep}} = \frac{\delta H}{H^{\mathrm{qs}}}(\mathbb{E} : \mathbf{P}^{\mathrm{qs}}) \otimes (\mathbf{Q}^{\mathrm{qs}} : \mathbb{E}) - (\mathbb{E} : \delta\mathbf{P}) \otimes (\mathbf{Q}^{\mathrm{qs}} : \mathbb{E}) - (\mathbb{E} : \mathbf{P}^{\mathrm{qs}}) \otimes (\delta\mathbf{Q} : \mathbb{E}), \tag{A.8.10}$$

where, since h is constant,

$$\delta H = \delta\mathbf{Q} : \mathbb{E} : \mathbf{P}^{\mathrm{qs}} + \mathbf{Q}^{\mathrm{qs}} : \mathbb{E} : \delta\mathbf{P}, \tag{A.8.11}$$

and then

$$\delta\mathbb{E}^{\rm ep} = -\frac{1}{H^{\rm qs}}\Big((\mathbb{E}^{\rm ep}_{\rm qs} : \delta\mathbf{P}) \otimes (\mathbf{Q}^{\rm qs} : \mathbb{E}) + (\mathbb{E} : \mathbf{P}^{\rm qs}) \otimes (\delta\mathbf{Q} : \mathbb{E}^{\rm ep}_{\rm qs})\Big)\,. \tag{A.8.12}$$

In view of (A.8.9) and (A.8.12), the difference of the stress rates can also be written as

$$\delta\dot{\boldsymbol{\sigma}} = \mathbf{A}^{\rm qs} : \delta\boldsymbol{\sigma} + \mathbb{E}^{\rm ep}_{\rm qs} : \delta\dot{\boldsymbol{\epsilon}}\,, \tag{A.8.13}$$

where, with the subscript symbol $_{,\sigma}$ denoting a partial derivative with respect to $\boldsymbol{\sigma}$, $\mathbf{A}^{\rm qs} = \mathbf{A}[\boldsymbol{\sigma}^{\rm qs}, \dot{\boldsymbol{\epsilon}}^{\rm qs}] \equiv (\mathbb{E}^{\rm ep}_{,\sigma})^{\rm qs} : \dot{\boldsymbol{\epsilon}}^{\rm qs}$ is defined as

$$\mathbf{A}^{\rm qs} = -\frac{1}{H^{\rm qs}}\Big((\mathbf{Q}^{\rm qs} : \mathbb{E} : \dot{\boldsymbol{\epsilon}}^{\rm qs})\, \mathbb{E}^{\rm ep}_{\rm qs} : \mathbf{P}^{\rm qs}_{,\sigma} + (\mathbb{E} : \mathbf{P}^{\rm qs}) \otimes (\dot{\boldsymbol{\sigma}}^{\rm qs} : \mathbf{Q}^{\rm qs}_{,\sigma}) \Big)\,. \tag{A.8.14}$$

and

$$\mathbf{P}^{\rm qs}_{,\sigma} = g^{\rm qs}_{,\sigma\sigma}\,, \quad \mathbf{Q}^{\rm qs}_{,\sigma} = f^{\rm qs}_{,\sigma\sigma}\,. \tag{A.8.15}$$

In this manner, the partial differential equations that govern the evolution of the fields of the linearized differences in velocities and stresses ($\delta\mathbf{v}$ and $\delta\boldsymbol{\sigma}$) can be written as a *linear, non-homogeneous, first order system with variable coefficients*:

$$\rho\,\delta\dot{\mathbf{v}} = \operatorname{div}(\delta\boldsymbol{\sigma}) - \rho\,\dot{\mathbf{v}}^{\rm qs}\,, \tag{A.8.16}$$

$$\delta\dot{\boldsymbol{\sigma}} = \mathbb{E}^{\rm ep}_{\rm qs} : \boldsymbol{\epsilon}(\delta\mathbf{v}) + \mathbf{A}^{\rm qs} : \delta\boldsymbol{\sigma}\,, \tag{A.8.17}$$

where

$$\boldsymbol{\epsilon}(\delta\mathbf{v}) = \delta\dot{\boldsymbol{\epsilon}} = \frac{1}{2}(\nabla\mathbf{v} + \nabla^{\rm T}\mathbf{v})\,. \tag{A.8.18}$$

Assuming further that, along the considered quasi-static path,

(H7) the tensor of the elastic-plastic moduli $\mathbb{E}^{\rm ep}_{\rm qs}$ is invertible, with inverse

$$\mathbb{F}^{\rm ep}_{\rm qs} = (\mathbb{E}^{\rm ep}_{\rm qs})^{-1} = \mathbb{E}^{-1} + \frac{1}{h}\mathbf{P}^{\rm qs} \otimes \mathbf{Q}^{\rm qs}\,, \tag{A.8.19}$$

and pre-multiplying (A.8.13) by $\mathbb{F}^{\rm ep}_{\rm qs}$, we get the following alternative form to (A.8.17),

$$\mathbb{F}^{\rm ep}_{\rm qs} : \delta\dot{\boldsymbol{\sigma}} = \boldsymbol{\epsilon}(\delta\mathbf{v}) + \mathbf{D}^{\rm qs} : \delta\boldsymbol{\sigma}\,, \tag{A.8.20}$$

where $\mathbf{D}^{\rm qs} = \mathbf{D}[\boldsymbol{\sigma}^{\rm qs}, \dot{\boldsymbol{\sigma}}^{\rm qs}]$ is obtained from $\mathbf{A}^{\rm qs}$,

$$\mathbf{D}^{\rm qs} = \mathbb{F}^{\rm ep}_{\rm qs} : \mathbf{A}^{\rm qs} = -\frac{1}{h}\Big((\dot{\boldsymbol{\sigma}}^{\rm qs} : \mathbf{Q}^{\rm qs})\, \mathbf{P}^{\rm qs}_{,\sigma} + \mathbf{P}^{\rm qs} \otimes (\dot{\boldsymbol{\sigma}}^{\rm qs} : \mathbf{Q}^{\rm qs}_{,\sigma})\Big)\,. \tag{A.8.21}$$

Note that the system of equations (A.8.16) and (A.8.17), or (A.8.20), is non-homogeneous because of the term $\rho\,\dot{\mathbf{v}}^{\rm qs}$ in (A.8.16) associated with the acceleration of the quasi-static solution. Note also that this system has variable (time dependent) coefficients because $\mathbb{E}^{\rm ep}_{\rm qs} = \mathbb{E}^{\rm ep}(\boldsymbol{\sigma}^{\rm qs})$ and $\mathbf{A}^{\rm qs} = \mathbf{A}[\boldsymbol{\sigma}^{\rm qs}, \dot{\boldsymbol{\epsilon}}^{\rm qs}]$, or $\mathbb{F}^{\rm ep}_{\rm qs} = \mathbb{F}^{\rm ep}(\boldsymbol{\sigma}^{\rm qs})$ and $\mathbf{D}^{\rm qs} = \mathbf{D}[\boldsymbol{\sigma}^{\rm qs}, \dot{\boldsymbol{\sigma}}^{\rm qs}]$, vary with the time dependent quasi-static stresses and strains.

A.8.2 Two Important Analogies and Some Remarks and Interpretations

The governing equations (A.8.16)-(A.8.17) obtained above have a form that may seem somewhat unusual, so that to stress their analogy with the equations that govern other mechanical problems may clarify their content.

We consider first the related three degree-of-freedom problem of the dynamic stability of the steady sliding of an elastic system in frictional contact with a flat obstacle that moves with a prescribed non-vanishing and constant tangential velocity. This problem was discussed by Moirot [1998] and by Cho and Barber [1999]. The system has one particle of mass m, and its displacement vector $\mathbf{u}(t) \in \mathbb{R}^3$ has two components (assembled in $\mathbf{u}_T(t) \in \mathbb{R}^2$ parallel to the plane of the obstacle, and one component $u_N(t)$ normal to the obstacle; $\mathbf{r}(t) \in \mathbb{R}^3$, $\mathbf{r}_T(t) \in \mathbb{R}^2$ and $r_N(t) \in \mathbb{R}$ denote the corresponding vectors and components of the reaction of the obstacle on the particle. The notations $\mathbf{f}$, $\mathbf{f}_T$, f_N are used for the corresponding constant external applied forces. The prescribed tangential velocity of the obstacle is denoted by $\mathbf{v}_T(t) \in \mathbb{R}^2$, $\mathbf{v}_T(t) \neq \mathbf{0}$, and the unit vector that defines the direction and orientation of that velocity is the vector $\mathbf{i}_T = \mathbf{v}_T/|\mathbf{v}_T|$. In view of the decomposition of the displacement vector into tangential and normal components, the 3×3 mass and stiffness matrices of the system, $\mathbf{M}$ and $\mathbf{K}$ respectively, have the structure indicated next

$$\mathbf{M} = m \begin{bmatrix} \mathbf{I} & \mathbf{0}^{\mathrm{T}} \\ \mathbf{0} & 1 \end{bmatrix}, \quad \mathbf{K} = \begin{bmatrix} \mathbf{K}_{TT} & \mathbf{K}_{NT}^{\mathrm{T}} \\ \mathbf{K}_{NT} & K_{NN} \end{bmatrix}, \tag{A.8.22}$$

where $\mathbf{I}$ and $\mathbf{K}_{TT}$ are 2 x 2 matrices, $\mathbf{0}$ (zero) and $\mathbf{K}_{NT}$ are 1 x 2 matrices, and K_{NN} is a real number. It is assumed that, for the data in presence, contact with the obstacle is preserved: the normal displacement is null, $u_N = 0$, both in the steady-sliding configuration and in the dynamic evolution of the system. It is further assumed that in both of these situations the normal reaction is strictly compressive, $r_N < 0$, and the particle slides on the obstacle: the reaction force is on the Coulomb friction cone (the friction coefficient is denoted by μ in this sub-section), that is the *yield surface* $f(\mathbf{r})$,

$$f(\mathbf{r}) = |\mathbf{r}_T| + \mu\, r_N = 0\,, \tag{A.8.23}$$

and the non-associated *flow rule* gives the direction of the sliding speed $\dot{\mathbf{u}}_T - \mathbf{v}_T$,

$$\dot{\mathbf{u}}_T - \mathbf{v}_T = |\dot{\mathbf{u}}_T - \mathbf{v}_T| \frac{\partial g}{\partial \mathbf{r}} = |\dot{\mathbf{u}}_T - \mathbf{v}_T| \frac{\mathbf{r}_T}{|\mathbf{r}_T|}, \tag{A.8.24}$$

where the plastic potential g is defined by

$$g(\mathbf{r}) = |\mathbf{r}_T|\,. \tag{A.8.25}$$

In these circumstances, the equations that govern the steady-sliding equilibrium and the dynamic evolution of the particle are the following:

$$\begin{bmatrix} \mathbf{K}_{TT} \\ \mathbf{K}_{NT} \end{bmatrix} \mathbf{u}_T^{\mathrm{qs}} = \begin{bmatrix} \mathbf{f}_T \\ f_N \end{bmatrix} + \begin{bmatrix} \mathbf{r}_T^{\mathrm{qs}} \\ r_N^{\mathrm{qs}} \end{bmatrix} = \begin{bmatrix} \mathbf{f}_T \\ f_N \end{bmatrix} + r_N^{\mathrm{qs}} \begin{bmatrix} -\mu\, \mathbf{i}_T \\ 1 \end{bmatrix}, \tag{A.8.26}$$

$$m \begin{bmatrix} \mathbf{I} \\ \mathbf{0} \end{bmatrix} \ddot{\mathbf{u}}_T(t) + \begin{bmatrix} \mathbf{K}_{TT} \\ \mathbf{K}_{NT} \end{bmatrix} \mathbf{u}_T(t) = \begin{bmatrix} \mathbf{f}_T \\ f_N \end{bmatrix} + \begin{bmatrix} \mathbf{r}_T(t) \\ r_N(t) \end{bmatrix} = \begin{bmatrix} \mathbf{f}_T \\ f_N \end{bmatrix} + r_N(t) \begin{bmatrix} \mu \mathbf{j}_T \\ 1 \end{bmatrix}, \quad \text{(A.8.27)}$$

where $\mathbf{j}_T \equiv (\dot{\mathbf{u}}_T(t) - \mathbf{v}_T)/|\dot{\mathbf{u}}_T(t) - \mathbf{v}_T|$. By subtraction of these equations followed by linearization, one gets

$$m \begin{bmatrix} \mathbf{I} \\ \mathbf{0} \end{bmatrix} \delta\ddot{\mathbf{u}}_T(t) + \begin{bmatrix} \mathbf{K}_{TT} \\ \mathbf{K}_{NT} \end{bmatrix} \delta\mathbf{u}_T(t) = \begin{bmatrix} \delta\mathbf{r}_T(t) \\ \delta r_N(t) \end{bmatrix} = \begin{bmatrix} \mu\, \delta(\mathbf{r}_N(t)\, \mathbf{j}_T) \\ \delta(r_N(t)) \end{bmatrix}. \quad \text{(A.8.28)}$$

where, similarly to the previous subsection, for any time-dependent function $a(t)$, $\delta a(t) \equiv a(t) - a^{\text{qs}}$. Since, in the first rows,

$$\delta\mathbf{r}_T(t) \approx \mu\, r_N^{\text{qs}}\, \delta\mathbf{j}_T - \mu\, \delta r_N\, \mathbf{i}_T = \mu\, (\mathbf{I} - \mathbf{i}_T \otimes \mathbf{i}_T) \frac{r_N^{\text{qs}}}{|\mathbf{v}_T|}\, \delta\dot{\mathbf{u}}_T - \mu\, \delta r_N\, \mathbf{i}_T\,, \quad \text{(A.8.29)}$$

and, from the last row,

$$\delta r_N = \mathbf{K}_{NT}\, \delta\mathbf{u}_T\,, \quad \text{(A.8.30)}$$

it follows that, after elimination of δr_N, the two linearized equations that govern the difference between the tangential displacements of the dynamic solution and the steady sliding configuration are:

$$\mathbf{M}^*_{TT}\, \delta\ddot{\mathbf{u}}_T(t) + \mathbf{C}^*_{TT}\, \delta\dot{\mathbf{u}}_T(t) + \mathbf{K}^*_{TT}\, \delta\mathbf{u}_T(t) = \mathbf{0}\,, \quad \text{(A.8.31)}$$

where

$$\mathbf{M}^*_{TT} = m\,\mathbf{I}\,, \quad \mathbf{C}^*_{TT} = \mu \frac{r_N^{\text{qs}}}{|\mathbf{v}_T|} (\mathbf{I} - \mathbf{i}_T \otimes \mathbf{i}_T)\,, \quad \mathbf{K}^*_{TT} = \mathbf{K}_{TT} + \mu\, \mathbf{i}_T \otimes \mathbf{K}_{NT}\,. \quad \text{(A.8.32)}$$

This example is discussed here because it shows, in a simple finite dimensional context, that the rate independent friction law of Coulomb (a prototype for non-associative plasticity) leads to rate dependent effects in the equations that govern the variation about a steady-sliding solution. It is important to observe that these effects appear because of the curvature of the yield surface f (A.8.23) and the plastic potential g (A.8.25) when a variation of the three-dimensional friction law is computed. Note that, although in this example the steady sliding corresponds to a configuration of static equilibrium of the system, $\dot{\mathbf{u}}^{\text{qs}}=\mathbf{0}$, $\ddot{\mathbf{u}}^{\text{qs}}=\mathbf{0}$, the situation is indeed of the same type as the quasi-static solutions with strict plastic loading considered in Sect. A.8.1: in the steady sliding configuration, the modulus of the sliding speed (the equivalent plastic strain rate) is positive $|\dot{\mathbf{u}}_T^{\text{qs}} - \mathbf{v}_T| = |\mathbf{0} - \mathbf{v}_T| > 0$. The example would even look more similar to the problems treated in the previous subsection if we considered the equivalent situation of a fixed flat obstacle and a tangential velocity $-\mathbf{v}_T$ prescribed to the spring supports, instead of the presented situation of fixed spring supports and an obstacle with a prescribed tangential velocity $\mathbf{v}_T$.

It is also important to notice that the elastic and the plastic elements in this example are associated *in parallel*: the forces in the plastic element (the contact reactions) and the forces in the elastic elements are not the same, and indeed they are summed in the balance equations (A.8.26), (A.8.27). This is the reason why, in the variational equations (A.8.31), the viscous damping term $\mathbf{C}^*_{TT}\delta\dot{\mathbf{u}}_T$ resulting from the variation of the plastic element is summed to the stiffness term $\mathbf{K}^*_{TT}\delta\mathbf{u}_T$, which in fact combines elastic and plastic contributions, eqn (A.8.32). The variational equations (A.8.31) have thus the structure of the balance equations for a Kelvin-Voigt visco-elastic model,

$$M\ddot{X} + C\dot{X} + KX = 0\,. \tag{A.8.33}$$

In what concerns the elastic-plastic solids considered in the previous sections, it is clear that the elastic and the plastic elements are associated *in series*. It is then to be expected that the viscous effects resulting from the variation of the plastic element will combine with the stiffness effects (which indeed have elastic and plastic contributions) in the same manner as in a Maxwell visco-elastic model. Denoting by Σ the common traction in the elastic and the viscous elements, the sum of the deformation rates of both elements for this model reads:

$$\dot{X} = \frac{\dot{\Sigma}}{K} + \frac{\Sigma}{C}\,, \tag{A.8.34}$$

while the balance of linear momentum is

$$M\ddot{X} = -\dot{\Sigma}\,. \tag{A.8.35}$$

Denoting $\dot{X}$ by V, the following system of first order equations is obtained:

$$M\dot{V} = -\Sigma\,, \tag{A.8.36}$$

$$\frac{\dot{\Sigma}}{K} = V - \frac{\Sigma}{C}. \tag{A.8.37}$$

The physical interpretation of the various quantities present in the governing equations (A.8.16) and (A.8.17) becomes apparent when compared to (A.8.36) and (A.8.37). In particular, it is important to notice that the fourth order tensor $\mathbf{D}^{\mathrm{qs}}$, eqn (A.8.21), plays the role of the inverse of a tensor of viscous damping moduli. The motivation for the assumption (H7) and the pre-multiplication of (A.8.13) by $\mathbb{F}^{\mathrm{ep}}_{\mathrm{qs}}$ to get (A.8.20) becomes apparent if we establish the energy balance of the Maxwell model. Multiplying (A.8.36) by V, (A.8.37) by Σ and adding the resulting equations, we get

$$M\dot{V}V + \frac{\dot{\Sigma}}{K}\Sigma = -\frac{\Sigma^2}{C}\,, \tag{A.8.38}$$

so that the obvious result is obtained that the time rate of change of the total, kinetic plus elastic, mechanical energy is equal to the power dissipated in the viscous element:

$$\frac{d}{dt}\left(\frac{1}{2}MV^2 + \frac{1}{2}\frac{\Sigma^2}{K}\right) = -\frac{\Sigma^2}{C}\,. \tag{A.8.39}$$

The corresponding energy balance equation for the system of (A.8.16) and (A.8.17), in the case of associative elastic-plastic behaviour, will be established in Sect. A.8.3.

The assumptions (H1) to (H6) used in the previous section deserve also some comments, since they are not equally important or necessary. The assumptions (H2) to (H4) are not at all essential for the presented theoretical developments: they are introduced only to keep the theory sufficiently simple and easy to interpret, although having already all the typical ingredients. The assumptions (H5) and (H6) are essential for the developments in this section, since they guarantee that plastic loading, and the corresponding constitutive moduli, hold in the vicinity of the quasi-static solution: this is essential to give some sense to the present study on flutter instability. Finally, the importance of assumption (H1) results from the fact that a definition of instability of a quasi-static path, particularly for rate-independent materials, requires an appropriate choice of a time-like parameter to parameterize the quasi-static solutions; the load parameter λ will be used in Sect. A.8.5 with that purpose, but it is likely that more general choices will be needed in some circumstances.

A final remark on the governing equations (A.8.16) and (A.8.17) is due. In order to avoid the mixed formulation in those equations, it would seem desirable to write $\delta\boldsymbol{\sigma}$ as a function of the kinematic variables. A simple but erroneous guess would be to take $\delta\boldsymbol{\sigma}= \mathbb{E}^{\text{ep}}_{\text{qs}} : \delta\boldsymbol{\epsilon}$ in (A.8.16) and (A.8.17). This in general is not correct because the essential assumption needed to derive it, the flow rule, is valid for true, dynamic or quasi-static, evolutions of the system but not for differences between true evolutions: $\dot{\boldsymbol{\epsilon}}_p^{\text{qs}}= \dot{\Lambda}^{\text{qs}}\,\mathbf{P}^{\text{qs}}$, $\dot{\boldsymbol{\epsilon}}_p= \dot{\Lambda}\,\mathbf{P}$, but $\delta\dot{\boldsymbol{\epsilon}}_p= \dot{\boldsymbol{\epsilon}}_p - \dot{\boldsymbol{\epsilon}}_p^{\text{qs}}$ does not have to have the direction of any normal to the plastic potential, since it does not correspond to any true evolution of the system. On the other hand, $\delta\dot{\boldsymbol{\epsilon}}_p \approx \delta\dot{\Lambda}\,\mathbf{P}^{\text{qs}} + \dot{\Lambda}^{\text{qs}}\,\delta\mathbf{P}$, a result that is consistent with all the derivations presented in Sect. A.8.1.

A.8.3 Energy Balance in the Associative Case

An energy balance statement can be obtained by integrating over the body Ω the inner products of (A.8.16) by $\delta\mathbf{v}$ and of (A.8.17) by $\delta\boldsymbol{\sigma}$,

$$\int_\Omega \rho\,\delta\mathbf{v}\cdot\delta\dot{\mathbf{v}}\,d\Omega + \int_\Omega \delta\boldsymbol{\sigma} : \mathbb{F}^{\text{ep}}_{\text{qs}} : \delta\dot{\boldsymbol{\sigma}}\,d\Omega = \tag{A.8.40}$$

$$= \int_\Omega \delta\mathbf{v}\cdot\operatorname{div}(\delta\boldsymbol{\sigma})\,d\Omega - \int_\Omega \delta\mathbf{v}\cdot\rho\,\dot{\mathbf{v}}^{\text{qs}}\,d\Omega + \int_\Omega \delta\boldsymbol{\sigma} : \boldsymbol{\epsilon}(\delta\mathbf{v})\,d\Omega + \int_\Omega \delta\boldsymbol{\sigma} : \mathbf{D}^{\text{qs}} : \delta\boldsymbol{\sigma}\,d\Omega\,.$$

It can be observed that

$$\int_\Omega \delta\mathbf{v}\cdot\operatorname{div}(\delta\boldsymbol{\sigma})\,d\Omega = \int_\Gamma \delta\mathbf{v}\cdot\delta\boldsymbol{\sigma}\cdot\mathbf{n}\,d\Gamma - \int_\Omega \boldsymbol{\epsilon}(\delta\mathbf{v}) : \delta\boldsymbol{\sigma}\,d\Omega = -\int_\Omega \delta\boldsymbol{\sigma} : \boldsymbol{\epsilon}(\delta\mathbf{v})\,d\Omega$$

because, on the surface Γ of the body, either the velocities or the tractions are prescribed, so that $\delta\mathbf{v}=\mathbf{0}$ or $\delta\boldsymbol{\sigma}\cdot\mathbf{n}=\mathbf{0}$ on Γ. On the other hand,

$$\rho\,\delta\mathbf{v}\cdot\delta\dot{\mathbf{v}}=\frac{d}{dt}\Big(\frac{1}{2}\rho\,\delta\mathbf{v}\cdot\delta\mathbf{v}\Big)\,,\quad \delta\boldsymbol{\sigma} : \mathbb{F}^{\text{ep}}_{\text{qs}} : \delta\dot{\boldsymbol{\sigma}}=\frac{d}{dt}\Big(\frac{1}{2}\,\delta\boldsymbol{\sigma} : \mathbb{F}^{\text{ep}}_{\text{qs}} : \delta\boldsymbol{\sigma}\Big) - \frac{1}{2}\,\delta\boldsymbol{\sigma} : \dot{\mathbb{F}}^{\text{ep}}_{\text{qs}} : \delta\boldsymbol{\sigma}\,,$$

and, in the associative case, using (A.8.15) and (A.8.3), one has for the rate of change of $\big(\mathbb{F}^{\text{ep}}_{,\sigma}\big)^{\text{qs}}$, namely $\dot{\mathbb{F}}^{\text{ep}}_{\text{qs}}=\big(\mathbb{F}^{\text{ep}}_{,\sigma}\big)^{\text{qs}} : \dot{\boldsymbol{\sigma}}^{\text{qs}}$, the following relations,

$$\dot{\mathbb{F}}^{\text{ep}}_{\text{qs}}=-\frac{1}{h}\Big(\big(\mathbf{Q}^{\text{qs}}_{,\sigma} : \dot{\boldsymbol{\sigma}}^{\text{qs}}\big)\otimes\mathbf{Q}^{\text{qs}} + \mathbf{Q}^{\text{qs}}\otimes\big(\mathbf{Q}^{\text{qs}}_{,\sigma} : \dot{\boldsymbol{\sigma}}^{\text{qs}}\big)\Big)\,,\ \ \mathbf{D}^{\text{qs}} - \frac{1}{2}\,\dot{\mathbb{F}}^{\text{ep}}_{\text{qs}}=-\frac{1}{2}\,\dot{\Lambda}^{\text{qs}}\,f^{\text{qs}}_{,\sigma\sigma}\,.$$

Collecting now all these results, the energy balance statement that involves the differences between the dynamic and the quasi-static solutions can be recast in the following form, in the associative case:

$$\begin{aligned}\frac{d}{dt}\Big(\frac{1}{2}\int_\Omega \rho\,\delta\mathbf{v}\cdot\delta\mathbf{v}\,d\Omega + \frac{1}{2}\int_\Omega \delta\boldsymbol{\sigma} : \mathbb{F}^{\mathrm{ep}}_{\mathrm{qs}} : \delta\boldsymbol{\sigma}\,d\Omega\Big) = \\ = -\frac{1}{2}\int_\Omega \dot{\Lambda}^{\mathrm{qs}}\,\delta\boldsymbol{\sigma} : \boldsymbol{f}^{\mathrm{qs}}_{,\sigma\sigma} : \delta\boldsymbol{\sigma}\,d\Omega - \int_\Omega \rho\,\delta\mathbf{v}\cdot\dot{\mathbf{v}}^{\mathrm{qs}}\,d\Omega\,.\end{aligned} \tag{A.8.41}$$

Since the convexity of the yield surface implies

$$\delta\boldsymbol{\sigma} : \boldsymbol{f}^{\mathrm{qs}}_{,\sigma\sigma} : \delta\boldsymbol{\sigma} \geq 0, \quad \forall\,\delta\boldsymbol{\sigma}\,, \tag{A.8.42}$$

and since $\dot{\Lambda}^{\mathrm{qs}} > 0$, the first term of the right-hand side of (A.8.41), which is associated with the curvature of the yield surface, is non-positive, i.e. it cannot provide energy to the system.

A.8.4 Dynamic Instability in the Particular Case of *Straight* Quasi-Static Paths. General Observations on Ill-Posedness

We assume that

(H8) the quasi-static solution is straight, i.e. the quasi-static acceleration is null,

$$\dot{\mathbf{v}}^{\mathrm{qs}} = \mathbf{0}\,; \tag{A.8.43}$$

(H9) for all particles of the body and for the whole time interval of interest, the quasi-static solution evolves on flat portions of the yield surface and of the plastic potential, i.e.

$$\mathbf{P}^{\mathrm{qs}}_{,\sigma} = g^{\mathrm{qs}}_{,\sigma\sigma} = \mathbf{0}\,, \quad \mathbf{Q}^{\mathrm{qs}}_{,\sigma} = f^{\mathrm{qs}}_{,\sigma\sigma} = \mathbf{0}\,, \quad \mathbf{D}^{\mathrm{qs}} = \mathbf{0}\,. \tag{A.8.44}$$

In these circumstances, the governing equations (A.8.16), (A.8.17) reduce to

$$\rho\,\delta\dot{\mathbf{v}} = \mathrm{div}(\delta\boldsymbol{\sigma})\,, \tag{A.8.45}$$

$$\delta\dot{\boldsymbol{\sigma}} = \mathbb{E}^{\mathrm{ep}}_{\mathrm{qs}} : \boldsymbol{\epsilon}(\delta\mathbf{v})\,, \tag{A.8.46}$$

where now $\mathbb{E}^{\mathrm{ep}}_{\mathrm{qs}}$ is independent of time and position. Then, differentiating (A.8.45) with respect to time and substituting (A.8.46) in the resulting equation, we get the second order, homogeneous system of partial differential equations with constant coefficients:

$$\rho\,\delta\ddot{\mathbf{v}} - \mathrm{div}(\,\mathbb{E}^{\mathrm{ep}}_{\mathrm{qs}} : \boldsymbol{\epsilon}(\delta\mathbf{v})\,) = \mathbf{0}\,. \tag{A.8.47}$$

These equations have the same structure as the dynamic rate equations but the unknown field $\delta\mathbf{v}$ in (A.8.47) is the velocity difference between a perturbed dynamic solution and a quasi-static solution in plastic loading that is also an unperturbed dynamic solution. As a result of the assumption (H6), plastic loading also holds for the perturbed dynamic solution, while $\delta\mathbf{v}$ is sufficiently small. This is the reason for the use of the loading constitutive moduli $\mathbb{E}^{\mathrm{ep}}_{\mathrm{qs}}$ in the stability analyses. This procedure has, of course, important limitations:

- the effect of initial perturbations leading to immediate unloading is not studied;
- the solutions of (A.8.47) cease to be valid at the time t_1 when

$$\mathbf{Q}\big(\sigma^{qs}(t_1) + \mathbb{E}^{ep}_{qs} : \int_{t_0}^{t_1} \epsilon(\delta\mathbf{v}(\tau))\, d\tau\,\big) : \mathbb{E} : \big(\dot{\epsilon}^{qs}(t_1) + \epsilon(\delta\mathbf{v}(t_1))\big) = 0\,, \qquad (A.8.48)$$

first occurs at some point of the body. But, since the sufficiently smooth solutions $\delta\mathbf{v}(t)$ of (A.8.47) are such that in the meantime $\mathbf{v}(t) = \mathbf{v}^{qs}(t) + \delta\mathbf{v}(t)$ is a plastic-loading solution of the original dynamic elastic-plastic problem, it is possible to conclude:

Proposition A.8.1 *Let the assumptions (H1) to (H6), (H8) and (H9) hold for a quasi-static solution that extends in time up to infinity. If the zero solution $\delta\mathbf{v}(t) = \mathbf{0}$ of the linear system (A.8.47) is dynamically unstable, then the corresponding quasi-static and dynamic solution of the original elastic-plastic system is also dynamically unstable.*

Note that in this proposition the same space norm is used to assess dynamic instability of the zero solution of (A.8.47) and of the quasi-static and dynamic solution of the original elastic-plastic system. Note also that the variations of strain rate $\epsilon(\delta\mathbf{v})$ and the plastic loading condition $(A.8.5)_1$ are assumed to have well defined pointwise values. The (otherwise arbitrary) space norm used in the characterization of the $\epsilon - \delta$ neighborhoods involved in Liapunov instability must then be strong enough to guarantee small pointwise variations of strain rate (hence plastic loading) inside those small neighborhoods.

Analytic studies of dynamic instability involving surface and plane harmonic wave mode perturbations in the plastic loading regime are provided in Sects. A.6 and B.4. The conditions on the imaginary parts of the wave-speeds that imply dynamic instability are then mentioned and will not be discussed further in this section. We only wish to stress here that it is the interpretation of the variable $\delta\mathbf{v}$ as a velocity perturbation relative to a quasi-static solution, in equations of the type (A.8.47), that makes possible the discussion of instability, namely flutter instability, of straight quasi-static paths of an elastic-plastic system. These ideas were initially advanced in Simões [1997], p. 213, where it was observed that, when the quasi-static strain-rate is in the interior of the plastic loading constitutive cone and the plastic loading constitutive moduli $\mathbb{E}^{ep}_{qs}$ are constant in a neighbourhood of the quasi-static path, a *linearized stability analysis*, based on the dynamic rate equations, would coincide with a *second order stability analysis*, i.e. an analysis based on equations governing the variation relative to the quasi-static solution like the present (A.8.47). When the governing equations do not have characteristic lengths or characteristic times, i.e. for non-regularized rate-independent materials, it is known that the existence of growing dynamic solutions (dynamic instability) actually leads to (linear) ill-posedness, since those growing solutions can occur for arbitrarily small wave lengths, and their rate of exponential growth in time is unbounded. This situation is also detected by the existence of non-real speeds of propagation of acceleration waves. When characteristic lengths or rate effects exist, the onsets of instability and ill-posedness may not coincide. We make these observations at this point to stress the fact that (linear) ill-posedness in the plastic loading regime is equally well identified by using the rate form of the balance of linear momentum (A.2.4), namely $\operatorname{div}\dot{\sigma} - \rho\,\ddot{\mathbf{v}} = \mathbf{0}$ with $\dot{\sigma} = \mathbb{E}^{ep}_{qs} : \epsilon(\mathbf{v})$, or by using the linearized equations (A.8.16)-(A.8.17). In fact the analysis of the propagation of acceleration waves with the system (A.8.16)-(A.8.17) leads to the same results of Sects. (A.2)-(A.5), if the quasi-static acceleration, the term $-\rho\dot{\mathbf{v}}^{qs}$, and the stress difference in the term $\mathbf{D}^{qs} : \delta\boldsymbol{\sigma}$ are assumed to be continuous across the wave front. In other words, the non-homogeneous and the viscous terms in (A.8.16)-(A.8.17) that are distinct from all those considered in Sects. (A.2)-(A.5), do not have any contribution to the acceleration wave problem; hence they have no contribution for the problem ill-posedness. The possible contribution of those terms to the dynamic instability of the quasi-static solution will be discussed in the next sub-section.

A.8.5 A Singular Perturbation Framework for the Study of General Quasi-Static Paths

A starting point for a general concept of *dynamic instability of a quasi-static path* is an understanding on how a quasi-static solution can be obtained in the truly dynamic world of our daily life. A quasi-static solution can be obtained if the external loading varies sufficiently slowly that *there is time enough* for any available damping mechanism to damp out the oscillations induced by those changes in loading and by all unavoidable perturbations to the system. But it is clear that, in some circumstances, no matter how slowly we change the external loading and no matter how carefully we try to avoid external perturbations, some existing quasi-static paths cannot be observed, unless some features of the system are changed, like its stiffness or damping properties. These are the situations that we wish to detect, and these are the quasi-static solutions that we wish to classify as *dynamically unstable.*

As observed in the final part of Sect. A.8.2, the load parameter λ is used here to parameterize the quasi-static solutions; in order to study the process of approximation of the dynamic equations and their solutions to the quasi-static ones, it is convenient to re-write the governing dynamic equations (A.8.16)-(A.8.17) in terms of the adopted quasi-static time-like parameter λ. The corresponding transformation of variables is given by

$$\frac{d\lambda}{dt} = \gamma = \text{constant} > 0,$$

so that for any time-dependent quantity $a = a(t)$,

$$\dot{a} = \frac{d\,a}{dt} = \gamma\,\frac{d\,a}{d\lambda} = \gamma\,a'\,.$$

For simplicity, each function of time t is denoted by the same letter as the corresponding transformed function of λ. In this manner (A.8.16) and (A.8.17) become

$$\gamma^2\,\rho\,\delta\mathbf{u}'' = \operatorname{div}(\delta\boldsymbol{\sigma}) - \gamma^2\,\rho\,(\mathbf{u}'')^{qs}\,, \tag{A.8.49}$$

and

$$\delta\boldsymbol{\sigma}' = \mathbb{E}^{ep}_{qs} : \boldsymbol{\epsilon}(\delta\mathbf{u}') + \mathbf{A}[\boldsymbol{\sigma}^{qs}, (\boldsymbol{\epsilon}')^{qs}] : \delta\boldsymbol{\sigma}\,, \tag{A.8.50}$$

or

$$\mathbb{F}^{ep}_{qs} : \delta\boldsymbol{\sigma}' = \boldsymbol{\epsilon}(\delta\mathbf{u}') + \mathbf{D}[\boldsymbol{\sigma}^{qs}, (\boldsymbol{\sigma}')^{qs}] : \delta\boldsymbol{\sigma}\,. \tag{A.8.51}$$

A motivation for this change of variables and formulation can be found in Martins et al. [1995], where the process of approximation (or not!) of a quasi-static solution by dynamic solutions, as a parameter proportional to the loading rate decreases, is illustrated in a finite dimensional frictional contact example that has a divergence instability.

The partial differential equations in (A.8.49) and (A.8.50) define, in an infinite dimensional context, a problem of a class that has been systematically studied for ordinary differential equations: the class of singular perturbation problems. Similarly to the mechanical problem that is of our interest here, those problems involve at least two distinct time scales, a fast (dynamic) time scale t and a slow (quasi-static) time-scale λ. Similarly to what happens above, these time scales are related by a small parameter ϵ that multiplies the time derivative of highest order of (one of) the governing equations of the system. In the more standard form, those singular perturbation problems are governed by a system of ordinary equations of the type:

$$\begin{cases} \epsilon\, \mathbf{y}'(\lambda) = \boldsymbol{\psi}(\mathbf{x},\mathbf{y},\epsilon,\lambda)\,, \\ \mathbf{x}'(\lambda) = \boldsymbol{\xi}(\mathbf{x},\mathbf{y},\epsilon,\lambda)\,, \end{cases} \tag{A.8.52}$$

where $\mathbf{x}=\mathbf{x}(\lambda)$, $\mathbf{y}=\mathbf{y}(\lambda)$ are vector functions of appropriate dimensions. Similarly to our mechanical problem, a reduced (quasi-static) problem can be defined that governs the slow evolution $\mathbf{x}^{qs}=\mathbf{x}^{qs}(\lambda)$, $\mathbf{y}^{qs}=\mathbf{y}^{qs}(\lambda)$ of the system:

$$\begin{cases} \mathbf{0} = \boldsymbol{\psi}(\mathbf{x}^{qs},\mathbf{y}^{qs},0,\lambda)\,, \\ (\mathbf{x}^{qs})'(\lambda) = \boldsymbol{\xi}(\mathbf{x}^{qs},\mathbf{y}^{qs},0,\lambda)\,. \end{cases} \tag{A.8.53}$$

For some given finite or infinite interval of slow time (load parameter), the first important question to answer is in what circumstances the fast (dynamic) solutions approach (or not!) the solutions of the reduced (quasi-static) problem when the small parameter ϵ is decreased to zero. Other more advanced mathematical issues have also been object of study and, probably, will also prove to be relevant for the mechanical problems that concern us here. But the extension to infinite dimensional non-smooth (elastic-plastic or visco-plastic) dynamic systems of some of the more elementary results and procedures of the Tikhonov-Levinson theory, e.g. O'Malley [1991], already appears to be a huge but fruitful research effort to be undertaken.

B Fluid-Saturated Porous Media

We consider fluid-saturated porous media in the framework of the theory of mixtures. The basic notions and the necessary extensions to include an irreversible behaviour are first recalled. Next, based on the nature of the acceleration wave-speeds, precise results concerning the onsets of stationary discontinuity and flutter are derived. The propagation of acceleration waves and harmonic waves reveals interesting features that permits to distinguish flutter instability and flutter ill-posedness. In fact, the diffusion length is not sufficient for the existence of an infimum wavelength below which flutter and ill-posedness are excluded: for that, an intrinsic length should be present in the constititive equations describing the behaviour of the solid skeleton.

B.1 Field and Constitutive Equations

B.1.1 Basics of Mixture Theory

The constitutive equations are derived within the general framework of mixture theory, see e.g. Truesdell and Toupin [1960], Green and Naghdi [1969], Bowen [1976]. The presentation below taps heavily on Rizzi and Loret [1999] which extend the constitutive equations of LH [1991] [11], first by incorporating anisotropic elasticity and second by using a general effective stress as a state variable.

We consider here porous media which are made up of two constituents, a solid and a fluid, namely water, which are viewed as two independent overlapping continua. Phases (solid (s) and fluid (w)) represent the constituents when viewed as part of the mixture. The solid phase is usually referred to as *solid skeleton*. Each constituent has a mass M_α and a volume V_α, $\alpha{=}s,w$, which make up the total mass $M{=}M_s{+}M_w$ and the total volume $V{=}V_s{+}V_w$ of the mixture.

At each point of each phase, we define *intrinsic* or *true* quantities, labeled by subscripts, and *apparent* or *partial* quantities, labeled by superscripts. For example, the intrinsic mass density is defined as $\rho_\alpha{=}M_\alpha/V_\alpha$, whereas the apparent mass density is defined by $\rho^\alpha{=}M_\alpha/V$; hence $\rho^\alpha{=}\mathrm{n}^\alpha\,\rho_\alpha$ where $\mathrm{n}^\alpha = V_\alpha/V$ is the volume fraction of phase α. Since we consider fluid-saturated porous media, the volume fractions satisfy the constraint

$$\mathrm{n}^{\mathrm{s}} + \mathrm{n}^{\mathrm{w}} = 1\,. \tag{B.1.1}$$

The equations of balance of momentum of the phases involve two (symmetric) *partial stress tensors* $\boldsymbol{\sigma}^s$ and $\boldsymbol{\sigma}^w$, which make up the total stress tensor of the mixture

$$\boldsymbol{\sigma} = \boldsymbol{\sigma}^s + \boldsymbol{\sigma}^w \tag{B.1.2}$$

when the contribution due to diffusion is neglected. To each partial stress tensor $\boldsymbol{\sigma}^\alpha$ is associated an *intrinsic stress tensor* $\boldsymbol{\sigma}_\alpha$ defined as $\boldsymbol{\sigma}_\alpha{=}\boldsymbol{\sigma}^\alpha/\mathrm{n}^\alpha$. The partial stress tensor of the fluid phase, associated to a linear compressible barotropic fluid, is defined in terms of its partial p^{w} or intrinsic p_{w} pressure as

$$\boldsymbol{\sigma}^w = -\mathrm{p}^{\mathrm{w}}\,\mathbf{I} = -\mathrm{n}^{\mathrm{w}}\mathrm{p}_{\mathrm{w}}\,\mathbf{I}\,. \tag{B.1.3}$$

Each phase α is endowed with its own (infinitesimal) apparent strain tensor $\boldsymbol{\epsilon}^\alpha$ defined as the symmetrized gradient of its macroscopic displacement field. As for constituents, they are generally compressible. Incompressibility of one or both constituents can be viewed as limit cases. If both solid and fluid constituents are incompressible, the equations of mass balance imply that the rates of volume change must satisfy the constraint $\mathrm{div}\,(\mathrm{n}^{\mathrm{s}}\,\mathbf{v}_s + \mathrm{n}^{\mathrm{w}}\,\mathbf{v}_w) = 0$ which, upon linearization around a natural state takes the form,

$$\mathrm{n}^{\mathrm{s}}\,\mathrm{div}\,\mathbf{v}_s + \mathrm{n}^{\mathrm{w}}\,\mathrm{div}\,\mathbf{v}_w = \mathrm{n}^{\mathrm{s}}\,\mathbf{I} : \dot{\boldsymbol{\epsilon}}^s + \mathrm{n}^{\mathrm{w}}\,\mathbf{I} : \dot{\boldsymbol{\epsilon}}^w = 0\,. \tag{B.1.4}$$

For each phase, $\alpha = s,\ w$, of the porous medium, the balance of momentum

$$\mathrm{div}\,\boldsymbol{\sigma}^\alpha + \hat{\mathbf{p}}_\alpha + \rho^\alpha\,(\mathbf{b}_\alpha - \mathbf{a}_\alpha) = \mathbf{0}\,, \qquad (\text{no sum over } \alpha)\,, \tag{B.1.5}$$

[11] LH [1991] stands for Loret and Harireche [1991].

involves, in addition to the usual terms present in single phase solids, namely the divergence of the stress tensor, the body force per unit mass $\mathbf{b}_\alpha$ and the acceleration $\mathbf{a}_\alpha$, the apparent mass density ρ^α and the momentum supply $\hat{\mathbf{p}}_\alpha$ to the phase α by the rest of the mixture. Momentum supplies are subject to the constraint

$$\hat{\mathbf{p}}_s + \hat{\mathbf{p}}_w = \mathbf{0}. \tag{B.1.6}$$

The balance of momentum equation holds pointwise, except along a singular surface of order 1, e.g. Eringen and Ingram [1965]. Singular surfaces of order 1 display a velocity discontinuity, while, across singular surfaces of order 2, the velocity is continuous but the acceleration and the velocity gradient are not. Here, only the existence and propagation of the latter are considered.

Since we envisage a linearized analysis around an equilibrium state, the mass density ρ^α in (B.1.5) is fixed to its reference value and the acceleration $\mathbf{a}_\alpha$ is approximated by $\dot{\mathbf{v}}_\alpha$; thus we shall make no difference between the speed of displacement W and the two speeds of propagation of an acceleration wave with respect to the particles of each of the two phases α, $c_\alpha = W - \mathbf{v}_\alpha \cdot \mathbf{n}$. To calculate the acceleration wave-speeds, we need constitutive assumptions for the momentum supplies. They will be assumed continuous across singular surfaces of order 2, that is across the front of the acceleration wave. We neglect the buoyancy term since it would disappear anyway in this linearized analysis. We assume that the momentum supplies describe diffusion of the fluid through the solid skeleton, for which we simply adopt an isotropic Darcy's law that introduces a single constant material parameter $\xi > 0$, proportional to the inverse of permeability, Bowen [1976]:

$$\hat{\mathbf{p}}_s = -\hat{\mathbf{p}}_w = -\xi\,(\mathbf{v}_s - \mathbf{v}_w)\,. \tag{B.1.7}$$

The rates of the body forces $\mathbf{b}_\alpha$, $\alpha = s, w$, are also assumed to be continuous across the wave-front. Application of Hadamard's compatibility conditions (A.2.5) to the equations of balance of momentum of the two phases then yields

$$\mathbf{n} \cdot [\![\dot{\sigma}^\alpha]\!] = \rho^\alpha\, c^2\, \eta_\alpha\,, \quad \alpha = s, w\,, \tag{B.1.8}$$

where we have used the approximations $c \equiv W \approx c_\alpha$, $\alpha = s, w$.

B.1.2 Elastic Constitutive Equations

The constitutive relations for the elastic mixture with an anisotropic solid phase are postulated in a form that enjoys the major symmetry property, so that they may be derived from a quadratic strain energy function $\mathcal{W}(\boldsymbol{\epsilon}^s, \boldsymbol{\epsilon}^w)$:

$$\begin{cases} \boldsymbol{\sigma}^s = \mathbb{E}^{ss} : \boldsymbol{\epsilon}^s + \mathbb{E}^{sw} : \boldsymbol{\epsilon}^w = \mathbb{E}^{ss} : \boldsymbol{\epsilon}^s + \lambda^{\mathrm{sw}}\, \mathbf{e}^{sw}\, (\mathbf{I} : \boldsymbol{\epsilon}^w) \\ \boldsymbol{\sigma}^w = \mathbb{E}^{ws} : \boldsymbol{\epsilon}^s + \mathbb{E}^{ww} : \boldsymbol{\epsilon}^w = \left(\lambda^{\mathrm{sw}}\, \mathbf{e}^{sw} : \boldsymbol{\epsilon}^s + \lambda^{\mathrm{ww}}\, (\mathbf{I} : \boldsymbol{\epsilon}^w)\right) \mathbf{I}\,. \end{cases} \tag{B.1.9}$$

The symmetry property is ensured by the major symmetry of the diagonal moduli $\mathbb{E}^{ss}$ and $\mathbb{E}^{ww} = \lambda^{\mathrm{ww}}\, \mathbf{I} \otimes \mathbf{I}$ and by the fact that the coupling moduli are the transpose of one another: $\mathbb{E}^{sw} = \lambda^{\mathrm{sw}}\, \mathbf{e}^{sw} \otimes \mathbf{I} = (\mathbb{E}^{ws})^{\mathrm{T}}$. The diagonal moduli $\mathbb{E}^{ss}$ are also assumed to display the minor symmetries. Since second order tensor $\mathbf{e}^{sw}$ is symmetric, the coupling moduli display the minor symmetries as well. The tensor $\mathbf{e}^{sw}$ will not be spherical if the material behaviour is not isotropic: indeed, even if the fluid acts through its volume change, its contribution to $\boldsymbol{\sigma}^s$ may not be spherical.

The apparent constitutive parameters introduced in (B.1.9), namely the symmetric fourth order tensor $\mathbb{E}^{ss}$, the (nondimensional) symmetric second order tensor $\mathbf{e}^{sw}$, normalized to $\| \mathbf{e}^{sw} \| = \sqrt{3}$, and the scalars λ^{sw}, λ^{ww} are unspecified at the present stage but they may be identified from intrinsic material parameters measured in appropriate experimental tests as shown by Biot and Willis [1957] (see Note 2.3 in Rizzi and Loret [1999]).

To any (nondimensional) symmetric second order tensor $\boldsymbol{\alpha}$, we associate *a generalized effective stress* through the following formula (in the framework of a continuum mechanics approach, tensile stresses are positive):

$$\bar{\boldsymbol{\sigma}} \equiv \boldsymbol{\sigma} + \mathrm{p_w}\, \boldsymbol{\alpha} \,. \tag{B.1.10}$$

The stress tensor just defined is said to be effective in the sense that later it will be assumed to characterize completely the elastic-plastic behaviour of the solid skeleton. In fact, later on, the yield function and other plastic properties will be based on this effective stress space. While this *plastic* effective stress is a constitutive feature, the *elastic* effective stress, also called the Nur-Byerlee effective stress, is uniquely defined by the value $\boldsymbol{\alpha}=\boldsymbol{\alpha}^{\mathrm{NB}}$ which allows decoupling of the elastic response (see below). The tensor $\boldsymbol{\alpha}^{\mathrm{NB}}$ may be identified from measurable material parameters for both isotropic, Nur and Byerlee [1971], and anisotropic solid skeletons, Carroll [1979], Rizzi and Loret [1999]. When $\boldsymbol{\alpha}=\mathbf{I}$, Terzaghi's effective stress is recovered, which may be interpreted as a particular case of the Nur-Byerlee effective stress when the solid constituent is much stiffer than the drained solid skeleton.

The relation between $\bar{\boldsymbol{\sigma}}$ and the strains $\boldsymbol{\epsilon}^s$ and $\boldsymbol{\epsilon}^w$ will be needed later for characterizing the elastic-plastic response. For that purpose, the effective stress definition (B.1.10) may be rephrased, with the help of (B.1.2), in terms of the partial stress tensor $\boldsymbol{\sigma}^s$ and of the intrinsic fluid pressure $\mathrm{p_w}$:

$$\bar{\boldsymbol{\sigma}} = \boldsymbol{\sigma}^s + \mathrm{p_w}\, (\boldsymbol{\alpha} - \mathrm{n^w} \mathbf{I}) \,. \tag{B.1.11}$$

Then, the first of (B.1.9) may be replaced by

$$\bar{\boldsymbol{\sigma}} = \bar{\mathbb{E}}^{ss} : \boldsymbol{\epsilon}^s + \bar{\mathbb{E}}^{sw} : \boldsymbol{\epsilon}^w \tag{B.1.12}$$

with fourth order tensors $\bar{\mathbb{E}}^{ss}$ and $\bar{\mathbb{E}}^{sw}$ defined as

$$\bar{\mathbb{E}}^{ss} = \mathbb{E}^{ss} + \boldsymbol{\beta}_{\mathrm{s}} \otimes \lambda^{\mathrm{sw}}\, \mathbf{e}^{sw} \,, \qquad \bar{\mathbb{E}}^{sw} = \mathbb{E}^{sw} + \lambda^{\mathrm{ww}}\, \boldsymbol{\beta}_{\mathrm{s}} \otimes \mathbf{I} = \lambda^{\mathrm{sw}}\, \bar{\mathbf{e}}^{sw}_{\mathrm{s}} \otimes \mathbf{I} \tag{B.1.13}$$

where

$$\boldsymbol{\beta}_{\mathrm{s}} = \boldsymbol{\beta}_{\mathrm{s}}(\boldsymbol{\alpha}) = \mathbf{I} - \frac{\boldsymbol{\alpha}}{\mathrm{n^w}} \tag{B.1.14}$$

and the block $\lambda^{\mathrm{sw}}\, \bar{\mathbf{e}}^{sw}_{\mathrm{s}}$ is formally defined as

$$\lambda^{\mathrm{sw}}\, \bar{\mathbf{e}}^{sw}_{\mathrm{s}} = \lambda^{\mathrm{sw}}\, \mathbf{e}^{sw} + \lambda^{\mathrm{ww}}\, \boldsymbol{\beta}_{\mathrm{s}} \,. \tag{B.1.15}$$

Definition (B.1.15) is made in this form for the sake of keeping in (B.1.12) and $(B.1.13)_2$ a notation formally analogous to $(B.1.9)_1$. If $\lambda^{sw}=0$, $\lambda^{sw}\,\bar{\mathbf{e}}_S^{sw}$ in (B.1.15) should be substituted by $\lambda^{ww}\,\boldsymbol{\beta}_S$.

Finally, let us define the constitutive moduli $\mathbb{E}^{DS}$ of the *underlying drained elastic solid* obtained by assuming a zero fluid pressure, $\boldsymbol{\sigma}^w=\mathbf{0}$. Elimination of the fluid strain $\mathbf{I}:\boldsymbol{\epsilon}^w$ from the elastic constitutive equations (B.1.9) yields:

$$\bar{\boldsymbol{\sigma}} = \boldsymbol{\sigma}^s = \boldsymbol{\sigma} = \mathbb{E}^{DS} : \boldsymbol{\epsilon}^s \qquad \text{where} \qquad \mathbb{E}^{DS} \equiv \mathbb{E}^{ss} - \frac{(\lambda^{sw})^2}{\lambda^{ww}}\, \mathbf{e}^{sw} \otimes \mathbf{e}^{sw} \,. \tag{B.1.16}$$

The strain energy function $\mathcal{W}(\boldsymbol{\epsilon}^s, \boldsymbol{\epsilon}^w)$ is assumed to be positive definite so that the stress-strain relations are invertible. Since $\mathcal{W}$ may be cast in the form

$$\mathcal{W}(\boldsymbol{\epsilon}^s, \boldsymbol{\epsilon}^w) = \frac{1}{2}\, \boldsymbol{\epsilon}^s : \mathbb{E}^{DS} : \boldsymbol{\epsilon}^s + \frac{1}{2\lambda^{ww}}\, \left(\lambda^{sw}\, \mathbf{e}^{sw} : \boldsymbol{\epsilon}^s + \lambda^{ww}\, \mathbf{I} : \boldsymbol{\epsilon}^w\right)^2 , \tag{B.1.17}$$

positive definiteness of the strain energy function is equivalent to

$$\mathbb{E}^{DS} \text{ positive definite}\,, \quad \lambda^{ww} > 0\,. \tag{B.1.18}$$

The elastic constitutive equations (B.1.12) decouple when $\bar{\mathbb{E}}^{sw}=\mathbf{0}$, namely for $\lambda^{sw}\,\bar{\mathbf{e}}_S^{sw}=\mathbf{0}$, eqn (B.1.15), which happens for the choice $\boldsymbol{\beta}_S=\boldsymbol{\beta}_S(\boldsymbol{\alpha}^{NB})=\boldsymbol{\beta}_S^{NB}$ that corresponds to take $\boldsymbol{\alpha}=\boldsymbol{\alpha}^{NB}$ in the effective stress definition (B.1.10):

$$\boldsymbol{\alpha}^{NB} = n^w \left(\mathbf{I} + \frac{\lambda^{sw}}{\lambda^{ww}}\, \mathbf{e}^{sw}\right), \qquad \boldsymbol{\beta}_S^{NB} = -\frac{\lambda^{sw}}{\lambda^{ww}}\, \mathbf{e}^{sw} \,. \tag{B.1.19}$$

Then $\bar{\mathbb{E}}^{ss}=\mathbb{E}^{DS}$, and (B.1.12) simplifies to

$$\bar{\boldsymbol{\sigma}} = \mathbb{E}^{DS} : \boldsymbol{\epsilon}^s . \tag{B.1.20}$$

Isotropic mixture

For a porous medium defined by isotropic solid constituent moduli $\mathbb{E}_s$ and isotropic drained moduli $\mathbb{E}^{DS}$, with respective bulk moduli $1/K_s=\mathbf{I}:\mathbb{E}_s^{-1}:\mathbf{I}$, $1/K^{DS}=\mathbf{I}:(\mathbb{E}^{DS})^{-1}:\mathbf{I}$ and necessarily equal shear moduli $\mu_s=\mu^{DS}=\mu$, both $\boldsymbol{\alpha}^{NB}$ and $\mathbf{e}^{sw}$ are isotropic and $\boldsymbol{\alpha}^{NB}=(1-K^{DS}/K_s)\,\mathbf{I}$, $\mathbf{e}^{sw}=\mathbf{I}$, and $\lambda^{sw}/\lambda^{ww}= n^s/n^w-K^{DS}/(n^w K_s)$. The equations (B.1.9) simplify to

$$\begin{cases} \boldsymbol{\sigma}^s = \mathbb{E}^{ss} : \boldsymbol{\epsilon}^s + \lambda^{sw}\, \mathbf{I}\, (\mathbf{I} : \boldsymbol{\epsilon}^w) \\ \boldsymbol{\sigma}^w = \left(\lambda^{sw}\, \mathbf{I} : \boldsymbol{\epsilon}^s + \lambda^{ww}\, \mathbf{I} : \boldsymbol{\epsilon}^w\right) \mathbf{I}\,. \end{cases} \tag{B.1.21}$$

On the basis $(\mathcal{I}, \mathcal{J})$, with

$$\mathcal{I} = \mathbf{I} \otimes \mathbf{I}\,, \quad \mathcal{J} = \mathbf{I} \underline{\overline{\otimes}} \mathbf{I}\,, \tag{B.1.22}$$

the tensor moduli $\mathbb{E}^{ss}$ and $\mathbb{E}^{DS}$ have components $(\lambda^{ss}, 2\,\mu)$ and $(\lambda^{DS}, 2\,\mu)$ respectively with

$$\lambda^{DS} \equiv \lambda^{ss} - \frac{(\lambda^{sw})^2}{\lambda^{ww}}\,. \tag{B.1.23}$$

The positive definiteness of the strain energy function, eqn (B.1.18), takes then the usual explicit form

$$K^{\rm DS} = \lambda^{\rm DS} + \frac{2}{3}\mu > 0, \quad \mu > 0 \text{ and } \lambda^{\rm ww} > 0. \qquad \text{(B.1.24)}$$

Incompressible constituents

If both constituents become incompressible, the fluid stress will be indeterminate if the following asymptotic conditions are satisfied:

$$\lambda^{\rm ww} \to \infty\,, \quad \lambda^{\rm sw} \to \infty\,, \quad \frac{\lambda^{\rm sw}}{\lambda^{\rm ww}} \mathbf{e}^{sw} \to \frac{\rm n^s}{\rm n^w}\mathbf{I}\,. \qquad \text{(B.1.25)}$$

The fluid volume change is given in terms of the volume change of the solid skeleton, eqn (B.1.4), and, due to (B.1.19),(B.1.20),(B.1.25), the constitutive equations reduce to

$$\bar{\boldsymbol{\sigma}} = \mathbb{E}^{\rm DS} : \boldsymbol{\epsilon}^s\,, \qquad \text{(B.1.26)}$$

where the elastic effective stress $\bar{\boldsymbol{\sigma}}$ now collapses to Terzaghi's effective stress,

$$\bar{\boldsymbol{\sigma}} = \boldsymbol{\sigma} + {\rm p_w}\,\mathbf{I}\,. \qquad \text{(B.1.27)}$$

B.1.3 Elastic-Plastic Constitutive Equations

The plastic properties are defined in terms of the generalized effective stress $\bar{\boldsymbol{\sigma}}$ instead of the more classical Terzaghi's effective stress as e.g. in Rice [1975]. Plastic behaviour emanates from the solid phase whereas the fluid constituent is taken as a linear compressible barotropic fluid. However, both solid and fluid *phases* will develop irreversible strains as pointed out by Bažant and Krizek [1975] and Rice [1975]. The plastic strain rate of the fluid phase will be assumed to be proportional to the plastic strain rate of the solid skeleton, in such a way that the incompressibility constraint (B.1.4) be satisfied, for both elastic unloading and plastic loading, in the limit case of incompressibility of both constituents.

B.1.3.1 **Basic assumptions**
The basic assumptions of the present elastic-plastic formulation are the following:
(a) The yield condition f=0 is defined in the space of effective stresses $\bar{\boldsymbol{\sigma}}$, eqn (B.1.10): $f=f(\bar{\boldsymbol{\sigma}},\mathcal{H})=0$, where $\mathcal{H}$ is a set of hardening variables. The yield function f is assumed to be smooth at the point of interest and its outward normal indicated by $\mathbf{Q}=\partial f/\partial\bar{\boldsymbol{\sigma}}$ is assumed to be of unit norm.
(b) The (infinitesimal) strain rates of the two phases are additively decomposed into an elastic part and a plastic part, namely

$$\dot{\boldsymbol{\epsilon}}^\alpha = \dot{\boldsymbol{\epsilon}}^\alpha_e + \dot{\boldsymbol{\epsilon}}^\alpha_p\,, \qquad \alpha = s, w\,. \qquad \text{(B.1.28)}$$

The plastic strain rate of the solid skeleton is given by the following flow rule:

$$\dot{\boldsymbol{\epsilon}}^s_p = \dot{\Lambda}\,\mathbf{P}\,, \qquad \dot{\Lambda} \geq 0\,, \qquad \text{(B.1.29)}$$

where $\dot{\Lambda}$ is the plastic multiplier and $\mathbf{P}$ is a unit symmetric tensor indicating the direction of flow.

(c) The plastic strain rate of the fluid phase is linked to the plastic strain rate of the solid skeleton by the following *"plastic coupling equation"*:

$$\mathbf{I} : \dot{\boldsymbol{\epsilon}}_p^w = \boldsymbol{\beta} : \dot{\boldsymbol{\epsilon}}_p^s \,, \tag{B.1.30}$$

where $\boldsymbol{\beta}$ is a (nondimensional) symmetric second order tensor. Eqn (B.1.30) makes it possible for the constitutive equations to be consistent with the kinematic constraint (B.1.4) when the two constituents are incompressible. Indeed then, in order this constraint to be satisfied for *both* elastic and plastic material responses, the following asymptotic relation should be satisfied

$$\boldsymbol{\beta} \to \boldsymbol{\beta}^{\mathrm{I}} = -\frac{\mathrm{n}^{\mathrm{s}}}{\mathrm{n}^{\mathrm{w}}} \mathbf{I} \,. \tag{B.1.31}$$

Eqn (B.1.30) generalizes the parallel hypothesis made by LH [1991] in the isotropic case, where they assumed $\boldsymbol{\beta}=\boldsymbol{\beta}^{\mathrm{I}}$. Note that, generally, the tensor $\boldsymbol{\beta}$ needs not to be related to the tensor $\boldsymbol{\alpha}$. However, the particular choice $\boldsymbol{\beta}=\boldsymbol{\beta}_{\mathrm{S}}(\boldsymbol{\alpha})$, eqn (B.1.14), brings an interesting structure in the elastic-plastic equations, namely the symmetry of the tangent operators when $\mathbf{P}=\mathbf{Q}$, as discussed later in Sect. B.1.3.3.
(d) The evolution of the hardening variables $\mathcal{H}$ together with the hardening modulus h are defined in the following form:

$$\dot{\mathcal{H}} = \dot{\Lambda}\, \mathbf{H}(\bar{\boldsymbol{\sigma}}, \mathcal{H}) \; ; \qquad h = -\frac{\partial f}{\partial \mathcal{H}} \cdot \mathbf{H} \,, \tag{B.1.32}$$

where just here '$\cdot$' means inner product between tensorial quantities of the same nature.
(e) The elastic constitutive laws in rate form are derived from (B.1.12) and from the second of (B.1.9):

$$\begin{cases} \dot{\bar{\boldsymbol{\sigma}}} = \bar{\mathbb{E}}^{ss} : \dot{\boldsymbol{\epsilon}}_e^s + \bar{\mathbb{E}}^{sw} : \dot{\boldsymbol{\epsilon}}_e^w = \bar{\mathbb{E}}^{ss} : \dot{\boldsymbol{\epsilon}}_e^s + \lambda^{\mathrm{sw}}\, \bar{\mathbf{e}}_{\mathrm{S}}^{sw}\, (\mathbf{I} : \dot{\boldsymbol{\epsilon}}_e^w) \\ \dot{\boldsymbol{\sigma}}^w = \mathbb{E}^{ws} : \dot{\boldsymbol{\epsilon}}_e^s + \mathbb{E}^{ww} : \dot{\boldsymbol{\epsilon}}_e^w = \left(\lambda^{\mathrm{sw}}\, \mathbf{e}^{sw} : \dot{\boldsymbol{\epsilon}}_e^s + \lambda^{\mathrm{ww}}\, (\mathbf{I} : \dot{\boldsymbol{\epsilon}}_e^w)\right) \mathbf{I} \,. \end{cases} \tag{B.1.33}$$

Notice that in this linearized analysis, (B.1.10) and (B.1.11) remain true in terms of rates, namely

$$\dot{\bar{\boldsymbol{\sigma}}} = \dot{\boldsymbol{\sigma}} + \dot{p}_w\, \boldsymbol{\alpha} = \dot{\boldsymbol{\sigma}}^s + \dot{p}_w\, (\boldsymbol{\alpha} - \mathrm{n}^{\mathrm{w}} \mathbf{I}) \;. \tag{B.1.34}$$

Hypotheses (a) and (c) are central to the present derivation. They clearly imply that the elastic-plastic response of the mixture is directly tied to the choices made for the second order tensors $\boldsymbol{\alpha}$ in (B.1.10) and $\boldsymbol{\beta}$ in (B.1.30). However, we will show that the underlying drained solid, and consequently the onset of strain localization, is independent of these constitutive tensors.

B.1.3.2 **Tangent stiffness**

When plastic loading occurs, solving the consistency condition $\dot{f}=0$ for the plastic multiplier $\dot{\Lambda}$, we get

$$\dot{\Lambda} = \frac{1}{H} \left(\mathbf{Q} : \bar{\mathbb{E}}^{ss} : \dot{\boldsymbol{\epsilon}}^s + \mathbf{Q} : \bar{\mathbb{E}}^{sw} : \dot{\boldsymbol{\epsilon}}^w\right) , \quad H \equiv h + \mathbf{Q} : (\bar{\mathbb{E}}^{ss} + \lambda^{\mathrm{sw}}\, \bar{\mathbf{e}}_{\mathrm{S}}^{sw} \otimes \boldsymbol{\beta}) : \mathbf{P} > 0 \tag{B.1.35}$$

where the modulus H is assumed to be strictly positive. Substituting (B.1.35) into the elastic constitutive laws (B.1.33) and in (B.1.34), we may recast the rate stress-strain relations in the form

$$\begin{cases} \dot{\boldsymbol\sigma}^s = \mathbb{E}^{ss}_{\rm ep} : \dot{\boldsymbol\epsilon}^s + \mathbb{E}^{sw}_{\rm ep} : \dot{\boldsymbol\epsilon}^w = \mathbb{E}^{ss}_{\rm ep} : \dot{\boldsymbol\epsilon}^s + \lambda^{\rm sw}\, \mathbf{e}^{sw}_{\rm ep}\, (\mathbf{I} : \dot{\boldsymbol\epsilon}^w) \\ \dot{\boldsymbol\sigma}^w = \mathbb{E}^{ws}_{\rm ep} : \dot{\boldsymbol\epsilon}^s + \mathbb{E}^{ww}_{\rm ep} : \dot{\boldsymbol\epsilon}^w = \left(\lambda^{\rm sw}\, \mathbf{e}^{ws}_{\rm ep} : \dot{\boldsymbol\epsilon}^s + \lambda^{\rm ww}\, e^{ww}_{\rm ep}\, (\mathbf{I} : \dot{\boldsymbol\epsilon}^w)\right) \mathbf{I} \end{cases} \tag{B.1.36}$$

where the fourth order tensor moduli $\mathbb{E}^{ss}_{\rm ep}$, $\mathbb{E}^{sw}_{\rm ep}$, $\mathbb{E}^{ws}_{\rm ep}$, $\mathbb{E}^{ww}_{\rm ep}$ are given by

$$\mathbb{E}^{ss}_{\rm ep} = \mathbb{E}^{ss} - \frac{1}{H}\,(\mathbb{E}^{ss} + \lambda^{\rm sw}\, \mathbf{e}^{sw} \otimes \boldsymbol\beta) : \mathbf{P} \otimes \mathbf{Q} : (\mathbb{E}^{ss} + \boldsymbol\beta_{\rm S} \otimes \lambda^{\rm sw}\, \mathbf{e}^{sw})\,, \tag{B.1.37}$$

$$\mathbb{E}^{sw}_{\rm ep} = \lambda^{\rm sw}\, \mathbf{e}^{sw}_{\rm ep} \otimes \mathbf{I}\,, \qquad \mathbb{E}^{ws}_{\rm ep} = \mathbf{I} \otimes \lambda^{\rm sw}\, \mathbf{e}^{ws}_{\rm ep}\,, \qquad \mathbb{E}^{ww}_{\rm ep} = \lambda^{\rm ww}\, e^{ww}_{\rm ep}\, \mathbf{I} \otimes \mathbf{I}\,, \tag{B.1.38}$$

with

$$\begin{aligned} \lambda^{\rm sw}\, \mathbf{e}^{sw}_{\rm ep} &= \lambda^{\rm sw}\, \mathbf{e}^{sw} - \frac{1}{H}(\mathbf{Q} : \lambda^{\rm sw}\, \bar{\mathbf{e}}^{sw}_{\rm S})\ \ (\mathbb{E}^{ss} + \lambda^{\rm sw}\, \mathbf{e}^{sw} \otimes \boldsymbol\beta) : \mathbf{P}\,, \\ \lambda^{\rm sw}\, \mathbf{e}^{ws}_{\rm ep} &= \lambda^{\rm sw}\, \mathbf{e}^{sw} - \frac{1}{H}(\lambda^{\rm sw}\, \bar{\mathbf{e}}^{sw} : \mathbf{P})\ \ \mathbf{Q} : (\mathbb{E}^{ss} + \boldsymbol\beta_{\rm S} \otimes \lambda^{\rm sw}\, \mathbf{e}^{sw})\,, \\ \lambda^{\rm ww}\, e^{ww}_{\rm ep} &= \lambda^{\rm ww} - \frac{1}{H}(\lambda^{\rm sw}\, \bar{\mathbf{e}}^{sw} : \mathbf{P})\,(\mathbf{Q} : \lambda^{\rm sw}\, \bar{\mathbf{e}}^{sw}_{\rm S}) = \lambda^{\rm ww}\, \frac{H^{\rm DS}}{H} > 0\,. \end{aligned} \tag{B.1.39}$$

In the last expression of $\lambda^{\rm ww}\, e^{ww}_{\rm ep}$, use has been made of (B.1.46) to be established later. Analogously to $\lambda^{\rm sw}\, \bar{\mathbf{e}}^{sw}_{\rm S}$, eqn (B.1.15), we have posed

$$\lambda^{\rm sw}\, \bar{\mathbf{e}}^{sw} = \lambda^{\rm sw}\, \mathbf{e}^{sw} + \lambda^{\rm ww}\, \boldsymbol\beta\,. \tag{B.1.40}$$

B.1.3.3 **Symmetry and coupling issues**

The tensor moduli in (B.1.36) depend on the choices made for second order tensors $\boldsymbol\alpha$ and $\boldsymbol\beta$ which enter in our basic assumptions (B.1.10) and (B.1.30). To reduce the arbitrariness of these tensors, one may think to require the tangent stiffness to display the *major symmetry* property

$$\mathbb{E}^{ss}_{\rm ep} = (\mathbb{E}^{ss}_{\rm ep})^{\rm T}\,, \qquad \mathbb{E}^{ww}_{\rm ep} = (\mathbb{E}^{ww}_{\rm ep})^{\rm T}\,, \qquad \mathbb{E}^{ws}_{\rm ep} = (\mathbb{E}^{sw}_{\rm ep})^{\rm T} \tag{B.1.41}$$

when $\mathbf{P}=\mathbf{Q}$. Inspection of (B.1.37)-(B.1.39) reveals that the major symmetry property holds for $\mathbf{P}=\mathbf{Q}$ if $\boldsymbol\beta=\boldsymbol\beta_{\rm S}$, $\boldsymbol\beta_{\rm S}$ already defined in (B.1.14):

$$\boldsymbol\beta = \boldsymbol\beta_{\rm S}(\boldsymbol\alpha) = \mathbf{I} - \frac{\boldsymbol\alpha}{\rm n^w} \qquad \Longleftrightarrow \qquad \text{“major symmetry”, eqn (B.1.41)}\,. \tag{B.1.42}$$

Moreover, the tangent stiffness simplifies when the Nur-Byerlee effective stress is used, that is for $\boldsymbol\alpha=\boldsymbol\alpha^{\rm NB}$. Indeed if $\boldsymbol\beta_{\rm S}=\boldsymbol\beta^{\rm NB}_{\rm S}$, eqn (B.1.19)$_2$, we get, similarly to the elastic response (B.1.20), $\dot{\bar{\boldsymbol\sigma}}=\mathbb{E}^{\rm DS}_{\rm ep}: \dot{\boldsymbol\epsilon}^s$; furthermore, the plastic contributions to the tensor moduli $\mathbb{E}^{sw}_{\rm ep}$ and $\mathbb{E}^{ww}_{\rm ep}$ vanish; finally, if $\boldsymbol\beta=\boldsymbol\beta_{\rm S}=\boldsymbol\beta^{\rm NB}_{\rm S}$, $\mathbb{E}^{ws}_{\rm ep}$ becomes purely elastic as well, and then

$$\mathbb{E}^{ss}_{\rm ep} = \mathbb{E}^{\rm DS}_{\rm ep} + \frac{(\lambda^{\rm sw})^2}{\lambda^{\rm ww}}\, \mathbf{e}^{sw} \otimes \mathbf{e}^{sw}\,, \quad \mathbb{E}^{sw}_{\rm ep} = \mathbb{E}^{sw}\,, \quad \mathbb{E}^{ws}_{\rm ep} = \mathbb{E}^{ws}\,, \quad \mathbb{E}^{ww}_{\rm ep} = \mathbb{E}^{ww}\,, \tag{B.1.43}$$

where $\mathbb{E}^{\rm DS}_{\rm ep}$ is the elastic-plastic stiffness of the underlying drained solid as shown in the next section.

B.1.3.4 **The underlying drained solid**

Submitting the porous medium to the local drainage condition $\dot{\boldsymbol{\sigma}}^w=\mathbf{0}$ (which collapses to $\boldsymbol{\sigma}^w=\mathbf{0}$ in the elastic case), we obtain a relation between the rate of partial stress and the rate of strain of the solid skeleton which has exactly the structure of a single-phase elastic-plastic material which we refer to as the underlying drained solid.

LH [1991] realized that the onset of strain localization of fluid-saturated isotropic elastic-plastic porous media is simultaneous with that of the underlying drained solid. Rizzi and Loret [1999] generalized this remarkable property in three respects: (i) by using anisotropic elastic constitutive (B.1.9), (ii) by allowing an arbitrary tensor $\boldsymbol{\alpha}$ in the definition (B.1.10) of the effective stress, and (iii) an arbitrary tensor $\boldsymbol{\beta}$ in the plastic coupling equation (B.1.30) for the plastic strain rates of the two phases. It turns out that the underlying drained solid does not depend on the tensors $\boldsymbol{\alpha}$ and $\boldsymbol{\beta}$, which enter in the definitions (B.1.10) and (B.1.30). The consequences for the onset of strain localization will be drawn in the next section.

Rizzi and Loret [1999] provide two proofs of the fact that the elastic-plastic stiffness $\mathbb{E}^{\rm DS}_{\rm ep}$ of the underlying drained solid is independent of $\boldsymbol{\alpha}$ and $\boldsymbol{\beta}$: (I) a qualitative proof by imposing directly the consistency condition for $\dot{\boldsymbol{\sigma}}^w=\mathbf{0}$; (II) an algebraic check by imposing the constraint $\dot{\boldsymbol{\sigma}}^w=\mathbf{0}$ in the rate equations (B.1.36).

Derivation (I)

Since $\dot{\boldsymbol{\sigma}}^w=\mathbf{0}$, it follows, from the rate forms of (B.1.2), (B.1.16) and from (B.1.34), that

$$\dot{\bar{\boldsymbol{\sigma}}} = \dot{\boldsymbol{\sigma}}^s = \dot{\boldsymbol{\sigma}} = \mathbb{E}^{\rm DS} : \dot{\boldsymbol{\epsilon}}^s_e \,. \tag{B.1.44}$$

When plastic loading occurs, the consistency condition $\dot{f}=0$ yields the plastic multiplier $\dot{\Lambda}$,

$$\dot{\Lambda} = \frac{1}{H^{\rm DS}}\, \mathbf{Q} : \mathbb{E}^{\rm DS} : \dot{\boldsymbol{\epsilon}}^s \,, \qquad H^{\rm DS} = h + \mathbf{Q} : \mathbb{E}^{\rm DS} : \mathbf{P} > 0 \tag{B.1.45}$$

where, like its counterpart H, eqn (B.1.35)$_2$, the modulus $H^{\rm DS}$ is supposed to be strictly positive. Note that $H^{\rm DS}$ differs from H only in its elastic part, namely $\mathbf{Q}$:$\mathbb{E}^{\rm DS}$:$\mathbf{P}$, now built with the drained elastic stiffness:

$$H = H^{\rm DS} + \frac{1}{\lambda^{\rm sw}}\, (\lambda^{\rm sw}\, \bar{\mathbf{e}}^{sw} : \mathbf{P})\, (\mathbf{Q} : \lambda^{\rm sw}\, \bar{\mathbf{e}}^{sw}_{\rm S}) \,. \tag{B.1.46}$$

However, when the anisotropic Nur-Byerlee effective stress is used ($\boldsymbol{\alpha}=\boldsymbol{\alpha}^{\rm NB}$), H and $H^{\rm DS}$ turn out to coincide since $\lambda^{\rm sw}\, \bar{\mathbf{e}}^{sw}_{\rm S}=\mathbf{0}$. Upon substitution of $\dot{\Lambda}$, eqn (B.1.45), in the rate equation (B.1.44), the elastic-plastic stiffness $\mathbb{E}^{\rm DS}_{\rm ep}$ emerges as

$$\dot{\bar{\boldsymbol{\sigma}}} = \dot{\boldsymbol{\sigma}}^s = \dot{\boldsymbol{\sigma}} = \mathbb{E}^{\rm DS}_{\rm ep} : \dot{\boldsymbol{\epsilon}}^s \,, \qquad \mathbb{E}^{\rm DS}_{\rm ep} = \mathbb{E}^{\rm DS} - \frac{1}{H^{\rm DS}}\, \mathbb{E}^{\rm DS} : \mathbf{P} \otimes \mathbf{Q} : \mathbb{E}^{\rm DS} \,. \tag{B.1.47}$$

Derivation (II)

Since $\lambda^{\rm ww}$, eqn (B.1.18), H and $H^{\rm DS}$ are assumed to be strictly positive, the scalar $e_{\rm ep}^{ww}$, eqn (B.1.39), is strictly positive as well. Then $\mathbf{I}:\dot{\boldsymbol\epsilon}^{w}$ can be eliminated from the constitutive equations (B.1.36) when $\dot{\boldsymbol\sigma}^{w}=\mathbf{0}$ yielding:

$$\dot{\boldsymbol\sigma}^{s} = \left(\mathbb{E}_{\rm ep}^{ss} - \frac{\lambda^{\rm sw}\,\mathbf{e}_{\rm ep}^{sw}\otimes\lambda^{\rm sw}\,\mathbf{e}_{\rm ep}^{ws}}{\lambda^{\rm ww}\,e_{\rm ep}^{ww}}\right):\dot{\boldsymbol\epsilon}^{s}\,. \tag{B.1.48}$$

Finally, using (B.1.37), (B.1.39) and (B.1.46), the operator appearing above can be shown to reduce to $\mathbb{E}_{\rm ep}^{\rm DS}$, eqn (B.1.47):

$$\mathbb{E}_{\rm ep}^{ss} - \frac{\lambda^{\rm sw}\,\mathbf{e}_{\rm ep}^{sw}\otimes\lambda^{\rm sw}\,\mathbf{e}_{\rm ep}^{ws}}{\lambda^{\rm ww}\,e_{\rm ep}^{ww}} = \mathbb{E}_{\rm ep}^{\rm DS}\,. \tag{B.1.49}$$

Incidentally, from (B.1.37)-(B.1.39), note that the isotropic constitutive equations used in LH [1991], Sect. 3.2, and LSM [1997], Sect. 2, correspond to $\boldsymbol\beta=-{\rm n}^{\rm s}/{\rm n}^{\rm w}\,\mathbf{I}$, eqn (B.1.31), and $\boldsymbol\alpha=\mathbf{I}$ in agreement with (B.1.42) since, on the basis $(\mathcal{I},\mathcal{J})$, $\mathbb{E}^{ss}+\lambda^{\rm sw}\,\mathbf{e}^{sw}\otimes\boldsymbol\beta=\mathbb{E}^{ss}+\boldsymbol\beta_{\rm S}\otimes\lambda^{\rm sw}\,\mathbf{e}^{sw}$ has components $(\lambda^{\rm ss}-{\rm n}^{\rm s}/{\rm n}^{\rm w}\,\lambda^{\rm sw},\,2\,\mu)$ and $\lambda^{\rm sw}\,\bar{\mathbf{e}}^{sw}=\lambda^{\rm sw}\,\bar{\mathbf{e}}_{\rm S}^{sw}$ is equal to $(\lambda^{\rm sw}-{\rm n}^{\rm s}/{\rm n}^{\rm w}\,\lambda^{\rm ww})\,\mathbf{I}$.

B.1.3.5 **Incompressible constituents**

In order the incompressibility constraint, eqn (B.1.4), to be satisfied, $\boldsymbol\beta$ should be equal to $-{\rm n}^{\rm s}/{\rm n}^{\rm w}\,\mathbf{I}$, eqn (B.1.31). Hence, if (B.1.42) holds, $\boldsymbol\alpha$ is equal to $\mathbf{I}$, that is the elastic and plastic effective stresses are equal to Terzaghi's effective stress. Consequently, both the elastic and plastic constitutive moduli associated to the effective stress are equal to their drained counterparts: the elastic equations are given by (B.1.26) and the elastic-plastic rate equations have the form

$$\dot{\bar{\boldsymbol\sigma}} = \mathbb{E}_{\rm ep}^{\rm DS}:\dot{\boldsymbol\epsilon}^{s}\,, \tag{B.1.50}$$

with $\mathbb{E}_{\rm ep}^{\rm DS}$ defined by (B.1.47).

B.2 Acceleration Waves, Stationary Discontinuities and Flutter Instabilities

B.2.1 Acceleration Waves

Starting from the general constitutive equations including elastic anisotropy and arbitrary effective stress (B.1.36), we first establish the characteristic equation that provides the four elastic-plastic wave-speeds. The elastic wave-speeds are obtained as a particular case, simply by setting the plastic modulus h equal to ∞. For incompressible constituents, the number of finite wave-speeds is reduced to 3 due to the constraint (B.1.4). Next we examine the particularly important situation where one wave-speed vanishes, a phenomenon referred to as strain localization in a quasi-static context. Finally, we show that, in contrast to the single phase case, Sect. A.3, deviatoric associativity with isotropic elasticity does not preclude the existence of body flutter.

B.2.1.1 **The characteristic equation for the wave-speeds**

We begin by

Proposition B.2.1 Kinematics of the acceleration wave in the fluid phase.

The normal jump of the velocity gradient in the fluid phase $\boldsymbol{\eta}_w$, *eqn (A.2.2), is aligned with the wave normal* $\mathbf{n}$,

$$\boldsymbol{\eta}_w = \eta_w \, \mathbf{n} \, . \tag{B.2.1}$$

The proof is trivial: simply write Hadamard's equation for the fluid phase, (B.1.8), accounting for the fact that the rate of fluid stress is isotropic, that is

$$\mathbf{n} \cdot [\![-\dot{p}^w \, \mathbf{I}]\!] = \rho^w \, c^2 \, \boldsymbol{\eta}_w \, . \tag{B.2.2}$$

Consequently, insertion of the constitutive equations (B.1.36) in the discontinuity relations (B.1.8) provides the 4×4 generalized eigenvalue system for the wave-speeds c^2 and the jumps $[\boldsymbol{\eta}] = [\boldsymbol{\eta}_s, \eta_w]^{\mathrm{T}}$,

$$\begin{bmatrix} \mathbf{n} \cdot \mathbb{E}^{ss}_{\mathbf{ep}} \cdot \mathbf{n} - c^2 \, \rho^s \mathbf{I} & \mathbf{n} \cdot \lambda^{\mathrm{sw}} \, \mathbf{e}^{sw}_{\mathbf{ep}} \\ \lambda^{\mathrm{sw}} \, \mathbf{e}^{ws}_{\mathbf{ep}} \cdot \mathbf{n} & \lambda^{\mathrm{ww}} \, e^{ww}_{\mathbf{ep}} - c^2 \rho^w \end{bmatrix} \begin{bmatrix} \boldsymbol{\eta}_s \\ \eta_w \end{bmatrix} = \begin{bmatrix} \mathbf{0} \\ 0 \end{bmatrix} . \tag{B.2.3}$$

To establish the above generalized eigenvalue problem, one has assumed the two sides of the wave-front to follow the plastic branch of the constitutive cone. Through simple manipulations one can transform the generalized eigenvalue problem into a classic eigenvalue problem, $[\mathbf{A}^{\#} - c^2 \mathbf{I}_4][\boldsymbol{\eta}^{\#}] = 0$, or,

$$\begin{bmatrix} \mathbf{n} \cdot \mathbb{E}^{ss}_{\mathbf{ep}} \cdot \mathbf{n} / \rho^s - c^2 \, \mathbf{I} & \mathbf{n} \cdot \lambda^{\mathrm{sw}} \, \mathbf{e}^{sw}_{\mathbf{ep}} / \sqrt{\rho^s \, \rho^w} \\ \lambda^{\mathrm{sw}} \, \mathbf{e}^{ws}_{\mathbf{ep}} \cdot \mathbf{n} / \sqrt{\rho^s \, \rho^w} & \lambda^{\mathrm{ww}} \, e^{ww}_{\mathbf{ep}} / \rho^w - c^2 \end{bmatrix} \begin{bmatrix} \sqrt{\rho^s} \, \boldsymbol{\eta}_s \\ \sqrt{\rho^w} \, \eta_w \end{bmatrix} = \begin{bmatrix} \mathbf{0} \\ 0 \end{bmatrix} , \tag{B.2.4}$$

where $[\mathbf{I}_4]$ is the identity matrix over the set of 4×4 matrices. Consequently, there are at most four possible modes of propagation, each one associated with a solution c^2 of (B.2.3). In addition, (B.2.4) shows clearly that, if it exists, the major symmetry of the constitutive equations is inherited by the normalized pseudo-acoustic matrix $[\mathbf{A}^{\#}]$. Thus, for an elastic-plastic skeleton obeying an associative flow rule, the eigenvalue problem is symmetric and, therefore, the squares of the wave-speeds and the eigenvectors are real. For isotropic mixtures with compressible constituents, LH [1991] present a detailed analysis of the nature, real or complex, of the elastic-plastic wave-speeds. They also order the four elastic wave-speeds: they correspond to a shear wave-speed of double multiplicity which affects only the solid phase, and to two longitudinal wave-speeds, so-called Biot's first and second waves, which affect both the solid and fluid phases.

B.2.1.2 **Incompressible constituents**

When the constituents are both incompressible, the discontinuities $\boldsymbol{\eta}_\alpha$ are constrained by the incompressibility condition (B.1.4), namely:

$$\mathrm{n}^{\mathrm{s}} \, (\boldsymbol{\eta}_s \cdot \mathbf{n}) \, + \, \mathrm{n}^{\mathrm{w}} \, (\boldsymbol{\eta}_w \cdot \mathbf{n}) \, = \, 0 \, . \tag{B.2.5}$$

Therefore the number of modes of propagation is now at most three. The associated eigenvalue problem can be obtained using the limit procedure (B.1.25), or directly following LH [1991]: using (B.1.2),(B.1.3) and (B.1.26), $\boldsymbol{\sigma}^s = \bar{\boldsymbol{\sigma}} - \mathrm{n}^{\mathrm{s}} \, \mathrm{p_w} \, \mathbf{I}$, one eliminates first the jump of the rate of fluid pressure from the two equations (B.1.8) and second the jump of velocity gradient $\boldsymbol{\eta}_w$ with (B.2.1),(B.2.5). The rate constitutive equations $\dot{\boldsymbol{\sigma}}=\mathbb{E}^{\rm DS}_{\rm ep}:\dot{\boldsymbol{\epsilon}}^s$ then yield

$$\Big(\mathbf{B}^{\rm DS}_{\rm ep} - c^2\,\mathbf{I}(\mathrm{r})\Big)\cdot\boldsymbol{\eta}_s = \mathbf{0}\,. \tag{B.2.6}$$

The elastic and elastic-plastic acoustic tensors of the drained solid, $\mathbf{B}^{\rm DS}$ and $\mathbf{B}^{\rm DS}_{\rm ep}$ respectively, are defined as for a single phase solid,

$$\mathbf{B}^{\rm DS} = \mathbf{n}\cdot\frac{\mathbb{E}^{\rm DS}}{\rho^s}\cdot\mathbf{n}\,,\quad \mathbf{B}^{\rm DS}_{\rm ep} = \mathbf{n}\cdot\frac{\mathbb{E}^{\rm DS}_{\rm ep}}{\rho^s}\cdot\mathbf{n} = \mathbf{B}^{\rm DS} - \frac{\mathbf{p}\otimes\mathbf{q}}{\rho^s\,H^{\rm DS}}\,, \tag{B.2.7}$$

with $\mathbf{p}$ and $\mathbf{q}$ defined similarly to (A.2.13) but in relation to $\mathbb{E}^{\rm DS}$,

$$\mathbf{p} = (\mathbb{E}^{\rm DS}:\mathbf{P})\cdot\mathbf{n}\,,\quad \mathbf{q} = \mathbf{n}\cdot(\mathbf{Q}:\mathbb{E}^{\rm DS})\,. \tag{B.2.8}$$

$\mathbf{I}(\mathrm{r})$ differs from the identity $\mathbf{I}=\mathbf{I}(1)$ due to the scalar r greater than 1:

$$\mathbf{I}(\mathrm{r}) = \mathbf{I} + (r-1)\,\mathbf{n}\otimes\mathbf{n}\,,\quad r = 1 + \Big(\frac{\mathrm{n}^{\rm s}}{\mathrm{n}^{\rm w}}\Big)^2\frac{\rho^w}{\rho^s} > 1. \tag{B.2.9}$$

Due to the presence of r, the characteristic equation $\det\big(\mathbf{B}^{\rm DS}_{\rm ep} - c^2\,\mathbf{I}(\mathrm{r})\big)=0$ differs from its counterpart for single phase solids. Indeed, for isotropic elasticity,

$$\det\big(\mathbf{B}^{\rm DS}_{\rm ep} - c^2\,\mathbf{I}(\mathrm{r})\big) = (c^2 - (c^{\rm e}_{\rm s})^2)\,F(c)\quad\text{with}\quad F(c) = u_4\,(c^2)^2 + u_2\,c^2 + u_0\,, \tag{B.2.10}$$

and the coefficients u_k involve both an elastic and a plastic part, namely,

$$u_k = u_{{\rm e}k} + \frac{u_{{\rm p}k}}{\hat{H}}\,,\quad k = 0,2,4\,, \tag{B.2.11}$$

with $\hat{H}$ the normalized modulus,

$$\hat{H} = \frac{H^{\rm DS}}{2\,\mu}\,, \tag{B.2.12}$$

and

$$\left\{\begin{array}{lll} u_{\rm e4}=r, & u_{\rm e2}=-r\,\big((c^{\rm e}_{\rm s})^2+(c^{\rm e}_{\rm l})^2\big), & u_{\rm e0}=r\,(c^{\rm e}_{\rm s})^2\,(c^{\rm e}_{\rm l})^2, \\ u_{\rm p4}=0, & u_{\rm p2}=r\,y-x, & u_{\rm p0}=-r\,y\,(c^{\rm e}_{\rm l})^2+x\,(c^{\rm e}_{\rm s})^2\,, \end{array}\right. \tag{B.2.13}$$

in which we have anticipated the fact that there exist an elastic shear-wave with wave-speed $c^{\rm e}_{\rm s}$ of double multiplicity and an elastic longitudinal wave with speed $c^{\rm e}_{\rm l}$,

$$(c^{\rm e}_{\rm s})^2 = \frac{\mu}{\rho^s}\,,\quad (c^{\rm e}_{\rm l})^2 = \frac{1}{r}\frac{\lambda^{\rm DS}+2\,\mu}{\rho^s}\,. \tag{B.2.14}$$

The scalars $x=x(\mathbf{n})$ and $y=y(\mathbf{n})$ are still given by (A.2.16) but ρ^s in place of ρ and with $\lambda^{\rm DS}$ in place of λ in (A.2.17). The nature of the elastic-plastic wave-speeds will be analyzed in Sects. B.2.2, B.2.3. In contrast to the case of single phase solids, eqn (A.2.9), and despite the inequalities (B.1.24), the presence of the coefficient $r>1$ makes it possible for the elastic shear wave-speed to be larger than the elastic longitudinal wave-speed, a circumstance which will be crucial in the analysis of flutter instability.

B.2.2 Stationary Discontinuity

Now that we have established the equations for the acceleration wave-speeds, we have the tools to study the strain localization phenomenon that, in this dynamic analysis, corresponds to the existence of a stationary discontinuity: $c = 0$ is solution of the generalized eigenvalue problem (B.2.3).

According to (B.1.8), the jump of traction rate of each partial stress $\boldsymbol{\sigma}^\alpha$, $\alpha=s,w$, vanishes across a stationary discontinuity, namely $\mathbf{n}\cdot[\![\dot{\boldsymbol{\sigma}}^\alpha]\!]=0$. While the tangent components of the rate of partial stress $\dot{\boldsymbol{\sigma}}^s$ may be discontinuous, the fact that the fluid stress is isotropic, eqn (B.1.3) or (B.2.2), implies trivially,

Proposition B.2.2 Jump of rate of fluid pressure.
The jump of the rate of fluid pressure across a stationary discontinuity vanishes:

$$[\![-\dot{p}^w]\!] = 0\,. \tag{B.2.15}$$

This simple observation indicates that, across a stationary discontinuity, the material behaves as if it were drained. In fact, setting c=0 in (B.2.4), the condition for non zero jumps η_α, $\alpha = s, w$, is that the determinant of the 4×4 matrix $[\mathbf{A}^\#]$ vanishes,

$$\det[\mathbf{A}^\#] = \det\begin{bmatrix} \mathbf{n}\cdot\mathbb{E}^{ss}_{\rm ep}\cdot\mathbf{n}/\rho^s & \mathbf{n}\cdot\lambda^{\rm sw}\,\mathbf{e}^{sw}_{\rm ep}/\sqrt{\rho^s\,\rho^w} \\ \lambda^{\rm sw}\,\mathbf{e}^{ws}_{\rm ep}\cdot\mathbf{n}/\sqrt{\rho^s\,\rho^w} & \lambda^{\rm ww}\,e^{ww}_{\rm ep}/\rho^w \end{bmatrix} = 0\,. \tag{B.2.16}$$

Now, a 4×4 matrix $[\mathbf{Z}]$ can be decomposed multiplicatively in the following format:

$$[\mathbf{Z}] = \begin{bmatrix}[\mathbf{Z}^{ss}] & [\mathbf{Z}^{sw}] \\ [\mathbf{Z}^{ws}] & Z^{ww}\end{bmatrix} = \begin{bmatrix}[\mathbf{I}_3] & [\mathbf{Z}^{sw}]/Z^{ww} \\ [\mathbf{0}] & 1\end{bmatrix}\begin{bmatrix}[\mathbf{Z}^{ss}] - [\mathbf{Z}^{sw}]\,[\mathbf{Z}^{ws}]/Z^{ww} & [\mathbf{0}] \\ [\mathbf{Z}^{ws}] & Z^{ww}\end{bmatrix}. \tag{B.2.17}$$

where $[\mathbf{Z}^{ss}]$, $[\mathbf{Z}^{sw}]$, $[\mathbf{Z}^{ws}]$ have respective sizes 3×3, 3×1, 1×3, Z^{ww} is a scalar different from 0 and $[\mathbf{I}_3]$ is the identity matrix over the set of 3×3 matrices. Therefore,

$$\det[\mathbf{Z}] = Z^{ww}\det\left([\mathbf{Z}^{ss}] - \frac{1}{Z^{ww}}[\mathbf{Z}^{sw}]\,[\mathbf{Z}^{ws}]\right). \tag{B.2.18}$$

Consequently, since $\lambda^{\rm ww}\,e^{ww}_{\rm ep}>0$, eqn (B.1.39)$_3$, the singularity of $[\mathbf{A}^\#]$ reduces to the singularity of the Schur complement of $\lambda^{\rm ww}\,e^{ww}_{\rm ep}/\rho^w$ in $[\mathbf{A}^\#]$, which, due to (B.1.49), is equal to the acoustic tensor $\mathbf{B}^{\rm DS}_{\rm ep}$ of the drained solid, eqn (B.2.7). In fact

$$\det[\mathbf{A}^\#] = \lambda^{\rm ww}\,e^{ww}_{\rm ep}\;\det\left[\mathbf{n}\cdot\left(\frac{\mathbb{E}^{ss}_{\rm ep}}{\rho^s} - \frac{1}{\rho^s}\frac{\lambda^{\rm sw}\,\mathbf{e}^{sw}_{\rm ep}\otimes\lambda^{\rm sw}\,\mathbf{e}^{ws}_{\rm ep}}{\lambda^{\rm ww}\,e^{ww}_{\rm ep}}\right)\cdot\mathbf{n}\right] = \lambda^{\rm ww}\,e^{ww}_{\rm ep}\;\det\mathbf{B}^{\rm DS}_{\rm ep}\,. \tag{B.2.19}$$

Therefore we have proved

Proposition B.2.3 Onset of strain localization in fluid-saturated porous media, LH [1991], Rizzi and Loret [1999].
The onsets of strain localization, or equivalently of a stationary discontinuity, in the fluid-saturated porous medium and in its underlying drained solid are simultaneous. This

result holds for the materials that enter the general elastic-plastic constitutive equations exposed in Sect. B.1.3, independently of
- *their elastic or plastic anisotropic characters,*
- *the compressible or incompressible behaviour of the constituents,*
- *the effective stress that enters the plastic criterion, that is of the tensor* $\boldsymbol{\alpha}$, *eqn (B.1.10),*
- *the plastic coupling equation that links the plastic strain in the fluid phase to the plastic strain in the solid skeleton, that is of the tensor* $\boldsymbol{\beta}$, *eqn (B.1.30).*

Since the drained solid has the structure of an elastic-plastic single-phase solid, the tools developed for single-phase materials may be applied as they are to the stiffness of the drained solid, see for example Rudnicki and Rice [1975] and more recent developments for isotropic solids in Bigoni and Hueckel [1991], Ottosen and Runesson [1991b], Neilsen and Schreyer [1993], and for anisotropic solids in Rizzi and Loret [1997], Loret and Rizzi [1997], Bigoni and Loret [1999].

In sum, not all the features of the constitutive equations transpire in the condition for strain localization. This result applies only to the onset of strain localization and it should not be misunderstood. Indeed, it is clear that, during general deformation processes, all the details of the constitutive equations matter and influence the material state and, hence, indirectly, the occurrence of strain localization and, later, the post-localization regime, where the growth of the shear band is highly influenced by the local permeability properties, see e.g. Loret and Prévost [1991].

Assuming that during a deformation process the modulus H^{DS} (or equivalently the plastic modulus h) decreases continuously, the onset of a stationary discontinuity c=0 is obtained, from (B.2.10)-(B.2.13), as (cf Sect. 6.3 in LH [1991]),

$$\hat{H}^{\mathrm{sta}} = \max_{\mathbf{n}\in\mathcal{N}} \frac{y}{(c_{\mathrm{s}}^{\mathrm{e}})^2} - \frac{x}{r\,(c_{\mathrm{l}}^{\mathrm{e}})^2}\,, \tag{B.2.20}$$

with $(c_{\mathrm{s}}^{\mathrm{e}})^2$ and $(c_{\mathrm{l}}^{\mathrm{e}})^2$ given by (B.2.14) and where the maximization is to be performed over the set of directions $\mathcal{N}$ such that H^{DS} or $\hat{H}$ is strictly positive.

B.2.3 Onset of Flutter Instability

If (B.1.42) holds, the constitutive equations for fluid-saturated porous media are such that the tangent stiffness has the major symmetry when $\mathbf{P}$=$\mathbf{Q}$, in which case the behaviour can be called associative. Consequently, akin to the situation that prevails for single-phase solids, Sect. A.2.2, the squares of the elastic-plastic wave-speeds are real for fluid-saturated porous media that have an associative behaviour. Also, interlacing properties between the four squares of the elastic wave-speeds and the four squares of the elastic-plastic wave-speeds, in the vein of Sect. A.3.3, have been obtained by Loret [1990]. On the other hand, unlike the situation that prevails for single phase solids, Sect. A.3, flutter instability is not excluded even for fluid-saturated porous media with isotropic elastic properties and a plastic behaviour obeying deviatoric associativity. We consider below only the case of incompressible constituents which is much more tractable than the compressible case; in fact, besides additional algebraic difficulties, the compressible case does not show features qualitatively different from the incompressible case, LH [1991].

The problem that we address is the following: assuming that during a deformation process the plastic modulus h is decreasing, what is the largest plastic modulus h such that there exists at least one direction $\mathbf{n}=\mathbf{n}^{\mathrm{flu}}$ along which the characteristic equation (B.2.10) has two complex conjugate roots $c^2 = c_{\mathrm{flu}}^2$? These roots will be complex conjugate if the discriminant $\Delta(\hat{H})$ of (B.2.10) is negative

$$\Delta(\hat{H}) = \Big(\frac{1}{\hat{H}}\Big)^2 (r\,y - x)^2 + 2\Big(\frac{1}{\hat{H}}\Big)(r\,y + x)\,\tau(\mathrm{r}) + \tau(\mathrm{r})^2 < 0. \tag{B.2.21}$$

Here, unlike in (A.3.3), the scalar $\tau(\mathrm{r})$,

$$\tau(\mathrm{r}) = r\,\left((c_1^{\mathrm{e}})^2 - (c_{\mathrm{s}}^{\mathrm{e}})^2\right)\,, \tag{B.2.22}$$

has not a definite sign since c_1^{e} depends itself on r, eqn (B.2.14). Notice that the expressions (A.3.1) and (B.2.21) are identical for $r = 1$, that is for a single phase solid. For $\tau(\mathrm{r})$=0, $\Delta(\hat{H})$ can not be negative and therefore flutter instability is precluded. For $\tau(\mathrm{r})\neq 0$, if a direction $\mathbf{n}$ exists for which the inequality (B.2.21) is satisfied, then in that direction the following inequalities must hold:

$$-\frac{x + r\,y}{\tau(\mathrm{r})} > 0\,,\quad 4\,\frac{x\,y}{\tau(\mathrm{r})^2} > 0\,. \tag{B.2.23}$$

For deviatoric associativity, $y(\mathbf{n})$ is strictly positive unless $\mathbf{n}$ is an eigenvector of $\hat{\mathbf{S}}$ and then y vanishes. Consequently for flutter instability to occur, $x(\mathbf{n})$ must be positive according to $(\mathrm{B.2.23})_2$ and $\tau(\mathrm{r})$ must be negative according to $(\mathrm{B.2.23})_1$. If there exists a set of directions $\mathcal{N}_{\mathrm{flu}}$ for which these sign conditions are satisfied, flutter will occur if the normalized modulus $\hat{H}$ belongs to the interval $]\hat{H}_-(\mathbf{n}), \hat{H}_+(\mathbf{n})[$ where

$$\hat{H}_-(\mathbf{n}) = -\frac{1}{\tau(\mathrm{r})}\left(\sqrt{x(\mathbf{n})} - \sqrt{r\,y(\mathbf{n})}\right)^2 \geq 0,\quad \hat{H}_+(\mathbf{n}) = -\frac{1}{\tau(\mathrm{r})}\left(\sqrt{x(\mathbf{n})} + \sqrt{r\,y(\mathbf{n})}\right)^2 > 0\,. \tag{B.2.24}$$

Thus if during a loading process, the modulus $\hat{H}$ decreases monotonously while the other state variables are kept constant, the onset of flutter instability will be met for $\hat{H} = \hat{H}^{\mathrm{flu}}$:

$$\hat{H}^{\mathrm{flu}} = \max_{\mathbf{n}\in\mathcal{N}_{\mathrm{flu}}}\; -\frac{1}{\tau(\mathrm{r})}\left(\sqrt{x(\mathbf{n})} + \sqrt{r\,y(\mathbf{n})}\right)^2 \tag{B.2.25}$$

Let $\mathbf{n} = \mathbf{n}^{\mathrm{flu}}$ be a direction that realizes this maximum. For $\hat{H} = \hat{H}^{\mathrm{flu}}$, the discriminant $\Delta(\hat{H})$, eqn (B.2.21), is zero and it can be shown that the square c_{flu}^2 of the wave-speed of multiplicity 2 associated to a direction $\mathbf{n}^{\mathrm{flu}}$ belongs to the interval $](c_1^{\mathrm{e}})^2, (c_{\mathrm{s}}^{\mathrm{e}})^2[$; indeed by insertion of the expression of $\hat{H}_+(\mathbf{n}^{\mathrm{flu}})$, eqn (B.2.24), in the characteristic equation (B.2.10), one obtains:

$$(c_{\mathrm{s}}^{\mathrm{e}})^2 > c_{\mathrm{flu}}^2 = \frac{(c_1^{\mathrm{e}})^2\sqrt{r\,y(\mathbf{n}^{\mathrm{flu}})} + (c_{\mathrm{s}}^{\mathrm{e}})^2\sqrt{x(\mathbf{n}^{\mathrm{flu}})}}{\sqrt{r\,y(\mathbf{n}^{\mathrm{flu}})} + \sqrt{x(\mathbf{n}^{\mathrm{flu}})}} > (c_1^{\mathrm{e}})^2\,. \tag{B.2.26}$$

Now, for deviatoric associativity, the range of the scalar $\mathbf{n}\cdot\hat{\mathbf{S}}\cdot\mathbf{n}$ involved in the definition of $x(\mathbf{n})$ is $[\hat{S}_3 < 0, \hat{S}_1 > 0]$ and $x(\mathbf{n})$ remains negative for all directions $\mathbf{n}$ if the intervals $[\hat{S}_3, \hat{S}_1]$ and $[-N_Q, -N_P]$ are disjoint like for associative plasticity, see Sect. A.3.3.

We summarize these results in

Proposition B.2.4 Onset of flutter for incompressible fluid-saturated porous media, LH [1991].

For incompressible porous media with isotropic elasticity and a plastic flow rule that satisfies deviatoric associativity, flutter instability is excluded if any of the conditions below holds:

$$(\mathcal{E}_1): \mathbf{P} = \mathbf{Q}; \quad (\mathcal{E}_2): \tau(\mathrm{r}) \geq 0; \quad (\mathcal{E}_3): -N_P < \hat{S}_3 . \tag{B.2.27}$$

If none of these conditions hold during a deformation process in which the plastic modulus decreases monotonously, then flutter instability will occur when the normalized modulus $\hat{H}$ *reaches the critical value* $\hat{H}^{\mathrm{flu}}$ *defined by (B.2.25), where the maximization is to be performed over the set* $\mathcal{N}_{\mathrm{flu}}$ *:*

$$\mathcal{N}_{\mathrm{flu}} = \{\mathbf{n} \;/\; \| \mathbf{n} \| = 1 ;\; \mathbf{n} \text{ not eigenvector of } \hat{\mathbf{S}} ;\; -N_Q < \mathbf{n} \cdot \hat{\mathbf{S}} \cdot \mathbf{n} < -N_P\} . \tag{B.2.28}$$

B.3 Growth or Decay of Propagating Acceleration Waves

In Sect. B.2.1, we have formally defined the modes under which plane acceleration waves propagate. We shall now study how their amplitudes vary in time. This analysis requires knowledge of the evolution of the wave-speed on the wave front . In contrast to linear elasticity where the wave-speeds in a given direction of propagation depend on material constants only, the elastic-plastic wave-speeds depend on the material state, that is the stresses and the hardening parameters. In a general initial- and boundary-value problem, the latter are functions of point. To be able to carry the analysis further, we consider a hypothetical situation where the material state *on the wave-front* is constant; then, so remains the wave-speed c.

We begin by establishing the differential equation governing the evolution in time of the amplitude of the propagation mode η_s. Next, we will analyze the coefficients of growth or decay depending on the details of the constitutive equations, with special emphasis on the consequences of the nature of the wave-speeds.

To simplify the presentation, we consider only incompressible constituents, and direct the reader to LSM [1997] [12] for the compressible case.

B.3.1 Equations of Propagation

The smoothness assumptions on the body forces and momentum supplies enunciated in Sect. B.2.1 are still valid. Thus, upon derivation with respect to time, the linearized equations of balance of momentum (B.1.5) yield the discontinuity equations:

$$\rho^{\alpha}[\![\ddot{\mathbf{v}}_{\alpha}]\!] = [\![\operatorname{div} \dot{\sigma}^{\alpha}]\!] - \epsilon_{\alpha}\,\xi\,([\![\dot{\mathbf{v}}_s]\!] - [\![\dot{\mathbf{v}}_w]\!]) , \tag{B.3.1}$$

with $\epsilon_{\alpha} = 1$ for $\alpha = s$, $\epsilon_{\alpha} = -1$ for $\alpha = w$. The evolution of the amplitude of the acceleration wave-front is expressed through a differential equation in terms of the *displacement derivative*, Truesdell and Toupin [1960], eqn (179.8):

$$\frac{\delta}{\delta t}[\![\dot{\mathbf{v}}_{\alpha}]\!] = \frac{\partial}{\partial t}[\![\dot{\mathbf{v}}_{\alpha}]\!] + \frac{\partial}{\partial \mathbf{x}}[\![\dot{\mathbf{v}}_{\alpha}]\!] \cdot (c\,\mathbf{n}) \tag{B.3.2}$$

[12] LSM [1997] stands for Loret, Simões and Martins [1997].

Now, since the velocity is continuous across the front that travels at constant speed c in a given direction $\mathbf{n}$, the *iterated kinematical compatibility equation*, Truesdell and Toupin [1960], eqn (181.8), simplifies to

$$2\,\frac{\delta}{\delta t}[\![\dot{\mathbf{v}}_\alpha]\!] = [\![\ddot{\mathbf{v}}_\alpha]\!] - c^2\,\boldsymbol{\Delta}_\alpha \tag{B.3.3}$$

where the vector $\boldsymbol{\Delta}_\alpha$ is called *induced discontinuity* :

$$\boldsymbol{\Delta}_\alpha = \mathbf{n}\cdot\left[\!\!\left[\frac{\partial^2\mathbf{v}_\alpha}{\partial\mathbf{x}^2}\right]\!\!\right]\cdot\mathbf{n}, \quad \Delta_{\alpha\,k} = \left[\!\!\left[\frac{\partial^2 v_{\alpha\,k}}{\partial x_i\partial x_j}\right]\!\!\right] n_i n_j, \quad \text{for } i,\,j,\,k \in [1,3]. \tag{B.3.4}$$

Due to the *iterated geometrical condition of compatibility*, Truesdell and Toupin [1960], eqn (176.8), the discontinuity of the second spatial derivative of the velocity can be cast into a dyadic form that also involves the induced discontinuity:

$$\left[\!\!\left[\frac{\partial^2\mathbf{v}_\alpha}{\partial\mathbf{x}^2}\right]\!\!\right] = \boldsymbol{\Delta}_\alpha\otimes\mathbf{n}\otimes\mathbf{n}, \quad \left[\!\!\left[\frac{\partial^2 v_{\alpha\,k}}{\partial x_i\partial x_j}\right]\!\!\right] = \Delta_{\alpha\,k}\,n_i n_j, \quad \text{for } i,\,j,\,k \in [1,3]. \tag{B.3.5}$$

Eliminating the discontinuity $[\![\ddot{\mathbf{v}}_\alpha]\!]$, $\alpha{=}s,\,w$, from (B.3.1),(B.3.3), and using the rate constitutive equations (B.1.50) in terms of the effective stress $\bar{\boldsymbol{\sigma}}{=}\boldsymbol{\sigma}^s - \mathrm{n^s/n^w}\,\boldsymbol{\sigma}^w$, eqn(B.1.27), one obtains the evolution equations for the discontinuities of acceleration $[\![\dot{\mathbf{v}}_\alpha]\!]$, $\alpha{=}s,\,w$:

$$2\,\rho^s\frac{\delta}{\delta t}[\![\dot{\mathbf{v}}_s]\!] = \rho^s\,(\mathbf{B}^{\mathrm{DS}}_{\mathrm{ep}} - c^2\,\mathbf{I})\cdot\boldsymbol{\Delta}_s + \frac{\mathrm{n^s}}{\mathrm{n^w}}[\![\mathrm{div}\,\dot{\boldsymbol{\sigma}}^w]\!] - \xi\,([\![\dot{\mathbf{v}}_s]\!] - [\![\dot{\mathbf{v}}_w]\!])\,, \tag{B.3.6}$$

$$2\,\rho^w\frac{\delta}{\delta t}[\![\dot{\mathbf{v}}_w]\!] = [\![\mathrm{div}\,\dot{\boldsymbol{\sigma}}^w]\!] - \rho^w c^2\boldsymbol{\Delta}_w + \xi\,([\![\dot{\mathbf{v}}_s]\!] - [\![\dot{\mathbf{v}}_w]\!])\,. \tag{B.3.7}$$

Due to (A.2.2) and (B.2.1) and to the incompressibility condition (B.1.4), the discontinuity of acceleration in the fluid phase is aligned with the wave normal $\mathbf{n}$,

$$[\![\dot{\mathbf{v}}_w]\!] = -\frac{\mathrm{n^s}}{\mathrm{n^w}}\Big([\![\dot{\mathbf{v}}_s]\!]\cdot\mathbf{n}\Big)\,\mathbf{n}\,. \tag{B.3.8}$$

Similarly, applying (A.2.1) to $\partial\dot{p}^w/\partial\mathbf{x}$ shows that $[\![\mathrm{div}\,\dot{\boldsymbol{\sigma}}^w]\!]$ is aligned with $\mathbf{n}$. Also due to the incompressibility condition (B.1.4) and to (B.3.5), the induced discontinuities are balanced by the condition:

$$\mathrm{n^s}(\boldsymbol{\Delta}_s\cdot\mathbf{n}) + \mathrm{n^w}(\boldsymbol{\Delta}_w\cdot\mathbf{n}) = 0\,. \tag{B.3.9}$$

Therefore, the projection of (B.3.7) along $\mathbf{n}$ can be written as

$$-2\,\rho^w\frac{\delta}{\delta t}\Big(\frac{\mathrm{n^s}}{\mathrm{n^w}}([\![\dot{\mathbf{v}}_s]\!]\cdot\mathbf{n})\,\mathbf{n}\Big) = [\![\mathrm{div}\,\dot{\boldsymbol{\sigma}}^w]\!] + \rho^w c^2\frac{\mathrm{n^s}}{\mathrm{n^w}}(\boldsymbol{\Delta}_s\cdot\mathbf{n})\,\mathbf{n} + \frac{\xi}{\mathrm{n^w}}([\![\dot{\mathbf{v}}_s]\!]\cdot\mathbf{n})\mathbf{n}\,. \tag{B.3.10}$$

Subtraction of (B.3.10) multiplied by $\mathrm{n^s/n^w}$ from (B.3.6) yields the evolution equation for the 3-component vector $\mathbf{V} = [\![\dot{\mathbf{v}}_s]\!]$ in the solid phase

$$2\,\mathbf{I}(\mathrm{r})\cdot\frac{\delta\mathbf{V}}{\delta t} = \Big(\mathbf{B}^{\mathrm{DS}}_{\mathrm{ep}} - c^2\,\mathbf{I}(\mathrm{r})\Big)\cdot\boldsymbol{\Delta}_s - \xi\,\frac{\mathbf{E}}{\rho^s}\cdot\mathbf{V}\,, \tag{B.3.11}$$

where the 3×3 matrix $\mathbf{I}(\mathrm{r})$ has been defined by (B.2.9) and the 3x3 symmetric matrix $\mathbf{E}$ takes a similar form:

$$\mathbf{E} = \Big(\mathbf{I} + \frac{\mathrm{n^s}}{\mathrm{n^w}}\mathbf{n}\otimes\mathbf{n}\Big)^2 = \mathbf{I} + \Big(\frac{1}{(\mathrm{n^w})^2} - 1\Big)\mathbf{n}\otimes\mathbf{n}\,. \tag{B.3.12}$$

The wave-front in the fluid-phase is then given by (B.3.8). Since we assume the state on the wave-front to be fixed, the unit left and right eigenvectors of the generalized eigenvalue problem (B.2.6), namely $\mathbf{e}^L$ and $\mathbf{e}^R$, are constant. Therefore, multiplication of the evolution equation (B.3.11) by the unit left eigenvector $\mathbf{e}^L$ yields a differential equation for the amplitude $|\mathbf{V}|$ that is integrated to

$$\frac{|\mathbf{V}|(t)}{|\mathbf{V}|(0)} = \exp\Big(-\frac{\xi}{2\rho^s}\zeta\, t\Big), \quad \zeta = \frac{\mathbf{e}^L\cdot\mathbf{E}\cdot\mathbf{e}^R}{\mathbf{e}^L\cdot\mathbf{I}(\mathrm{r})\cdot\mathbf{e}^R}\,. \tag{B.3.13}$$

Since the coefficient ξ is a real positive number, growth or decay of the wave-front is governed by the sign of the nondimensional *decay* coefficient ζ

For an elastic or an elastic-plastic skeleton obeying an associative flow rule, the eigenvalue problem, eqn (B.2.6), is symmetric, the wave-speeds are real and the left and right eigenvectors are one and the same. Since the matrices $\mathbf{I}(\mathrm{r})$, eqn (B.2.9), and $\mathbf{E}$, eqn (B.3.12), are symmetric positive definite, the coefficient ζ is then positive and propagation is accompanied by a wave-front whose amplitude either strictly decays or remains constant: the fact that the two constituents are incompressible excludes the possibility that the fastest wave can propagate without dissipation, a situation referred to as *dynamic compatibility*, Biot [1956]. In fact, one can sharpen this remark. Indeed, if one transforms the generalized eigenvalue problem (B.2.6) into a classic eigenvalue problem

$$\big((\mathbf{B}^{\mathrm{DS}}_{\mathrm{ep}})^{\#} - c^2\,\mathbf{I}\big)\cdot\eta_s^{\#} = \mathbf{0}\,, \tag{B.3.14}$$

with

$$(\mathbf{B}^{\mathrm{DS}}_{\mathrm{ep}})^{\#} = \mathbf{I}(\mathrm{r})^{-\frac{1}{2}}\cdot\mathbf{B}^{\mathrm{DS}}_{\mathrm{ep}}\cdot\mathbf{I}(\mathrm{r})^{-\frac{1}{2}}, \quad \eta_s^{\#} = \mathbf{I}(\mathrm{r})^{\frac{1}{2}}\cdot\eta_s\,, \quad \mathbf{I}(\mathrm{r})^a = \mathbf{I} + (r^a - 1)\,\mathbf{n}\otimes\mathbf{n}\,, \tag{B.3.15}$$

for any real a, and if one denotes by $\mathbf{e}^{L\#} = \mathbf{I}(\mathrm{r})^{\frac{1}{2}}\cdot\mathbf{e}^L$ and $\mathbf{e}^{R\#} = \mathbf{I}(\mathrm{r})^{\frac{1}{2}}\cdot\mathbf{e}^R$ the new eigenvectors, then one obtains

Proposition B.3.1 The decay coefficient as a Rayleigh quotient, LSM [1997].

For elastic or elastic-plastic mixtures with an associative flow rule, the decay coefficient ζ appears in the form of a Rayleigh quotient associated to the symmetric positive definite matrix $\mathbf{E}^{\#} = \mathbf{I}(\mathrm{r})^{-\frac{1}{2}}\cdot\mathbf{E}\cdot\mathbf{I}(\mathrm{r})^{-\frac{1}{2}}$:

$$\zeta = \frac{\mathbf{e}^{R\#}\cdot\mathbf{E}^{\#}\cdot\mathbf{e}^{R\#}}{\mathbf{e}^{R\#}\cdot\mathbf{e}^{R\#}}\,. \tag{B.3.16}$$

Consequently, it is bounded below and above by the smallest and largest eigenvalues of this matrix:

$$\text{for elasticity or associative elastoplasticity} \quad 1 \le \zeta \le \frac{1}{r(\mathrm{n^w})^2}\,. \tag{B.3.17}$$

The sense of the inequality is due to the definition of r, eqn (B.2.9),

$$\frac{1}{r(\mathrm{n}^{\mathrm{w}})^2} > 1\,. \tag{B.3.18}$$

This proposition does not hold a priori if the solid skeleton obeys a non-associative flow rule, $\mathbf{P}\neq\mathbf{Q}$.

The evolution equation (B.3.11) yields not only the decay coefficient but also the part of the induced discontinuity which is orthogonal to the eigenmode $\mathbf{e}^R$: in fact, when the wave-speed has single multiplicity, one can show that this part is uniquely defined.

On a direct evaluation of the decay coefficient ζ

To obtain the decay coefficient for a given wave-speed c, the natural approach is to calculate first the associated left and right eigenvectors $\mathbf{e}^R$ and $\mathbf{e}^L$. However, if c is a wave-speed of single multiplicity, considerable algebraic simplifications result from the alternative procedure below.

Let $c^2(0)=c^2$ be an eigenvalue with single multiplicity of the generalized eigenproblem (B.2.6) and let $\mathbf{e}^R(0)=\mathbf{e}^R$ and $\mathbf{e}^L(0)=\mathbf{e}^L$ be the associated right and left eigenvectors. Consider now the *extended* generalized eigenvalue problem:

$$\Big(\mathbf{B}_{\mathrm{ep}}^{\mathrm{DS}} + \varphi\,\mathbf{E} - c^2(\varphi)\,\mathbf{I}(\mathrm{r})\Big)\cdot\mathbf{e}^R(\varphi) = \mathbf{0}\,. \tag{B.3.19}$$

For φ a sufficiently small real number, the eigenvalue $c^2(\varphi)$ is an analytic function of φ, Stoer and Bulirsch [1980], p.389. Therefore scalar multiplication of (B.3.19) by the left eigenvector $\mathbf{e}^L(0)=\mathbf{e}^L$, differentiation with respect to φ of the resulting product and estimation of the derivative at $\varphi=0$ leads to

Proposition B.3.2 Direct evaluation of the decay coefficient, Stoer and Bulirsch[1980].

Let $c(0)$ be a solution of single multiplicity of the eigenvalue problem (B.3.19), with $\mathbf{e}^R$ and $\mathbf{e}^L$ the respective right and left eigenvectors. Then the decay coefficient associated to $c(0)$ is given by

$$\frac{dc^2}{d\varphi}(\varphi=0) = \zeta = \frac{\mathbf{e}^L\cdot\mathbf{E}\cdot\mathbf{e}^R}{\mathbf{e}^L\cdot\mathbf{I}(\mathrm{r})\cdot\mathbf{e}^R}\,. \tag{B.3.20}$$

Notice that the products by $\mathbf{e}^L$ are not inner products in the usual sense, that is, they remain unchanged even if $\mathbf{e}^L$ is complex, Wilkinson [1965], eqn (3.10), p.4.

The procedure breaks down at an eigenvalue of multiplicity greater than one, where the nature of the eigenspace may be more complex. Let us simply consider the case of a double eigenvalue, for $\mathbf{A}$ a non-symmetric matrix. Then, the eigenvectors may not be unique or, if the eigenspace is defective, the left and right eigenvectors may be well-defined and then they are orthogonal with respect to the matrix $\mathbf{I}(\mathrm{r})$, that is, Wilkinson [1965], eqn (7.5), p.10:

$$\mathbf{e}^L\cdot\mathbf{I}(\mathrm{r})\cdot\mathbf{e}^R = 0 \tag{B.3.21}$$

which is in strong contrast with the case of matrices with distinct eigenvalues, where this equality occurs only for $\mathbf{e}^L$ and $\mathbf{e}^R$ associated to distinct eigenvalues. This result suggests that, in the vicinity of a wave-speed of multiplicity greater than 1, the decay coefficient for non-associative flow rules may become unbounded: this phenomenon will actually be observed for double wave-speeds different from elastic wave-speeds.

B.3.2 Evolution of the Amplitude of the Wave-Front

For a plastic eigenvalue c^2 of single multiplicity, the decay coefficient ζ can be calculated using the procedure described above. One computes first the eigenvalues $c^2(\varphi)$ that make singular the matrix $\mathbf{Z}$ [13]:

$$\mathbf{Z} = \mathbf{B}_{\mathbf{ep}}^{\mathrm{DS}} + \varphi\,\mathbf{E} - c^2(\varphi)\,\mathbf{I}(\mathrm{r})\,. \tag{B.3.22}$$

The determinant of $\mathbf{Z}$ can be cast in the form:

$$\det \mathbf{Z} = \Big((c_{\mathrm{s}}^{\mathrm{e}})^2 + \varphi - c^2\Big)\,F(c) \quad \text{with} \quad F(c) = u_4\,(c^2)^2 + u_2(\varphi)\,c^2 + u_0(\varphi)\,. \tag{B.3.23}$$

The coefficients $u_k, k = 0$ to 2, involve an elastic and a plastic contribution as in (B.2.11). For each $k \in [0,2]$, the elastic contribution $u_{\mathrm{e}k}$ is a polynomial of degree k in φ while the plastic contribution $u_{\mathrm{p}k}$ is a polynomial of degree $\max(k-1,0)$, with the term of order zero given by (B.2.13):

$$\begin{cases} u_{\mathrm{e}4}(\varphi) = u_{\mathrm{e}4}(0)\,, \quad u_{\mathrm{p}4}(\varphi) = u_{\mathrm{p}4}(0) = 0\,, \\ u_{\mathrm{e}2}(\varphi) = u_{\mathrm{e}2}(0) \;-\; \varphi\, r\left(1 + \dfrac{1}{r\,(\mathrm{n}^{\mathrm{w}})^2}\right), \quad u_{\mathrm{p}2}(\varphi) = u_{\mathrm{p}2}(0)\,, \\ u_{\mathrm{e}0}(\varphi) = u_{\mathrm{e}0}(0) + \varphi\left(r\,(c_{\mathrm{l}}^{\mathrm{e}})^2 + \dfrac{(c_{\mathrm{s}}^{\mathrm{e}})^2}{(\mathrm{n}^{\mathrm{w}})^2}\right) + \dfrac{\varphi^2}{(\mathrm{n}^{\mathrm{w}})^2}\,, \quad u_{\mathrm{p}0}(\varphi) = u_{\mathrm{p}0}(0) + \varphi\left(x - \dfrac{y}{(\mathrm{n}^{\mathrm{w}})^2}\right). \end{cases} \tag{B.3.24}$$

Notice first that, if the shear wave-speed $c_{\mathrm{s}}^{\mathrm{e}}$ has multiplicity one, then its associated wave-speed in the extended eigenvalue problem is defined by $c^2(\varphi) = (c_{\mathrm{s}}^{\mathrm{e}})^2 + \varphi$ and therefore (B.3.20) implies that the associated decay coefficient is equal to one like for an elastic behaviour, Biot [1956].

Let us now turn to the study of the decay coefficients associated to the plastic wave-speeds. If the solutions $c_{\epsilon_c}^2(\varphi), \epsilon_c = \pm 1$, to $F(c) = 0$ are real, namely

$$c_{\epsilon_c}^2(\varphi) = \frac{1}{2u_4}\Big(-u_2(\varphi) + \epsilon_c\sqrt{\Delta(\hat{H},\varphi)}\Big), \quad \Delta(\hat{H},\varphi) = u_2^2(\varphi) - 4\,u_0\,u_4(\varphi) \geq 0\,, \tag{B.3.25}$$

then the coefficients $\zeta_{\epsilon_c} = (dc_{\epsilon_c}^2/d\varphi)(\varphi = 0)$ associated to $c_{\epsilon_c}^2$ are simply:

$$\zeta_{\epsilon_c} = -\frac{1}{2\,u_4}\left(\frac{du_2}{d\varphi} + \frac{\epsilon_c}{\sqrt{\Delta}}\Big(-u_2\frac{du_2}{d\varphi} + 2u_4\frac{du_0}{d\varphi}\Big)\right)(\varphi = 0)\,. \tag{B.3.26}$$

When applied to the coefficients u_k, $k = 0$ to 2, (B.3.24), the above formulas become, outside the flutter region,

$$\zeta_{\epsilon_c} = \frac{1}{2}\Big(1 + \frac{1}{r\,(\mathrm{n}^{\mathrm{w}})^2}\Big) - \frac{\epsilon_c}{2\sqrt{\Delta(\hat{H},\varphi{=}0)}}\Big(1 - \frac{1}{r\,(\mathrm{n}^{\mathrm{w}})^2}\Big)\Big(1 + \frac{1}{\hat{H}}\frac{x + r\,y}{\tau(\mathrm{r})}\Big)\,. \tag{B.3.27}$$

[13] For $\varphi = 0$, the right and left eigenvectors are given in Appendix A.2 of LMS [1997] and the eigenspaces are analyzed in detail in their Table 1.

Key parameters that decide of the nature of the wave-speeds are, besides the normalized plastic modulus $\hat{H}$, eqn (B.2.12), the scalar $\tau(\mathrm{r})$, eqn (B.2.22), and the couple $(r\,y,\, x)$, with r defined by (B.2.9) and the scalars $x = x(\mathbf{n})$ and $y = y(\mathbf{n})$ are still given by (A.2.16) but with λ^{DS} in place of λ in (A.2.17). Indeed, the discriminant $\Delta(\hat{H},0)$, eqn (B.3.25), is given by (B.2.21) and it is expressed in terms of these parameters. In the elastic limit, $\hat{H} \to \infty$, the decay coefficients (B.3.27) simplify due to (B.2.21):

$$\zeta_{\epsilon_c} = \frac{1}{2}\left(1 + \frac{1}{r\left(\mathrm{n}^{\mathrm{w}}\right)^2}\right) - \frac{\epsilon}{2}\left(1 - \frac{1}{r\left(\mathrm{n}^{\mathrm{w}}\right)^2}\right) = \begin{cases} 1 & \text{for } c = c_s^e \\ \dfrac{1}{r\left(\mathrm{n}^{\mathrm{w}}\right)^2} & \text{for } c = c_l^e \end{cases} \tag{B.3.28}$$

where $\epsilon = \epsilon_c\,\mathrm{sign}(\tau(\mathrm{r}))$ is equal to -1 for the shear wave-speed and 1 for the longitudinal wave-speed. Notice that, due to the inequality (B.3.18), the elastic longitudinal wave is always more diffusive than the shear wave, irrespective of their relative order. In fact, the bounds (B.3.17) that hold for both elastic and associative elastic-plastic materials are realized by the elastic waves.

The bounds (B.3.17) do not hold for non-associative flow rules. To elaborate further on that point, we assume deviatoric associativity to hold, that is, the deviatoric parts of the unit normals to the plastic potential $\mathbf{P}$ and to the yield surface $\mathbf{Q}$ are equal. Then, the scalar $y = y(\mathbf{n})$, eqn (A.3.4), is positive or zero; since $x = x(\mathbf{n})$ is negative for an associative flow rule, a necessary but not sufficient condition for $x = x(\mathbf{n})$ to be positive is that the volumetric parts of $\mathbf{P}$ and $\mathbf{Q}$ are not equal. Moreover if $\tau(\mathrm{r})$ is negative, complex wave-speeds are available for the directions $\mathbf{n}$ such that $x = x(\mathbf{n}) > 0$ and for normalized plastic moduli $\hat{H}$ within a certain range that renders the discriminant $\Delta(\hat{H},0)$, eqn (B.2.21), negative (see Sect. B.2.3).

A qualitative analysis of the shape and sign of the decay coefficients ζ associated to the two plastic wave-speeds involves the normalized plastic modulus $\hat{H}$, the scalar $\tau(\mathrm{r})$ and the directions of propagation $\mathbf{n}$ contained implicitly in the couple $(r\,y,\, x)$. This analysis delineates four cases as follows.

<u>Case a</u>: $\tau(\mathrm{r}) < 0,\ x(\mathbf{n}) > 0,\ r\,y(\mathbf{n}) > 0$.

Along the directions $\mathbf{n}$ that ensure both $x(\mathbf{n})$ and $r\,y(\mathbf{n})$ to be strictly positive, flutter may occur for a certain range of plastic moduli. Indeed, the discriminant $\Delta(\hat{H},0)$ is zero when the normalized plastic modulus is equal to $\hat{H}^{\mathrm{flu}}_{\epsilon_H}$, $\epsilon_H = \pm 1$, where

$$\hat{H}^{\mathrm{flu}}_{\epsilon_H} = -\frac{1}{\tau(\mathrm{r})}\left(\sqrt{r\,y} + \epsilon_H\sqrt{x}\right)^2 \tag{B.3.29}$$

and $\Delta(\hat{H},0)$ is negative for $\hat{H} \in [\hat{H}^{\mathrm{flu}}_{\epsilon=-1},\, \hat{H}^{\mathrm{flu}}_{\epsilon=1}]$. Near the zeros of Δ, ζ_{ϵ_c} behaves essentially like $1/\sqrt{\Delta}$, to within a multiplicative coefficient whose sign is that of $-\epsilon_H\epsilon_c$.

The decay coefficient ζ_{ϵ_c} is zero when the normalized plastic modulus is equal to $\hat{H}^{\mathrm{gr}}_{\epsilon=\epsilon_c}$ where

$$\hat{H}^{\mathrm{gr}}_{\epsilon} = -\frac{1}{\tau(\mathrm{r})}\left(\sqrt{r\,y} + \epsilon\frac{\sqrt{x}}{\sqrt{r}\mathrm{n}^{\mathrm{w}}}\right)\left(\sqrt{r\,y} + \epsilon\sqrt{x}\sqrt{r}\mathrm{n}^{\mathrm{w}}\right) \tag{B.3.30}$$

should be positive, which is always true for the largest plastic wave-speed $\epsilon_c = 1$; for the smallest plastic wave-speed $\epsilon_c = -1$, the sign of this modulus depends on the propagation direction $\mathbf{n}$. Indeed, the following inequalities can be proved:

$$\hat{H}^{\text{gr}}_{\epsilon=1} \geq \hat{H}^{\text{flu}}_{\epsilon=1} \geq \hat{H}^{\text{flu}}_{\epsilon=-1} \geq \begin{cases} 0 \geq \hat{H}^{\text{gr}}_{\epsilon=-1} \text{ for } \sqrt{r}\,\text{n}^{\text{w}} < \dfrac{\sqrt{r\,y}}{\sqrt{x}} < \dfrac{1}{\sqrt{r}\,\text{n}^{\text{w}}}\,; \\ \hat{H}^{\text{gr}}_{\epsilon=-1} \geq 0 \text{ otherwise.} \end{cases} \tag{B.3.31}$$

The qualitative variations of the decay coefficients ζ as functions of the normalized modulus $\hat{H}$ are shown on Figs. B.3.1-(a),(b). Notice the behaviour of ζ at the extremities of the flutter region where the plastic wave-speed has multiplicity two: the eigenspace of $\mathbf{B}^{\text{DS}}_{\text{ep}}$ is defective by one dimension; the left and right eigenvectors are $\mathbf{I}(\text{r})$-orthogonal, eqn (B.3.21), so that ζ is unbounded.

Case b: $x(\mathbf{n}) < 0,\ r\,y(\mathbf{n}) > 0$.
Now if x is negative, it is clear that none of the moduli that make zero the discriminant $\Delta(\hat{H}, 0)$ and the decay coefficients is real positive. Consequently, the decay coefficients remain real positive and finite, Fig. B.3.1-(c).

Case c: $\tau(\text{r}) > 0,\ x(\mathbf{n}) > 0,\ r\,y(\mathbf{n}) > 0$.
The four moduli that make zero the discriminant $\Delta(\hat{H}, 0)$ and the decay coefficients can be ordered as follows:

$$\left.\begin{array}{ll} \text{for } \sqrt{r}\,\text{n}^{\text{w}} < \dfrac{\sqrt{r\,y}}{\sqrt{x}} < \dfrac{1}{\sqrt{r}\,\text{n}^{\text{w}}}, & \hat{H}^{\text{gr}}_{\epsilon=-1} \geq 0 \\ \text{otherwise,} & 0 \geq \hat{H}^{\text{gr}}_{\epsilon=-1} \end{array}\right\} \geq \hat{H}^{\text{flu}}_{\epsilon=-1} \geq \hat{H}^{\text{flu}}_{\epsilon=1} \geq \hat{H}^{\text{gr}}_{\epsilon=1}\,. \tag{B.3.32}$$

Consequently the discriminant $\Delta(\hat{H}, 0)$ is strictly positive so that the decay coefficients are finite. On the other hand, the coefficient $\zeta_{\epsilon_c=-1}$ associated to the smallest plastic wave-speed may be zero for some directions when $\hat{H}^{\text{gr}}_{\epsilon=-1}$ is positive, Fig. B.3.1-(d).

Case d: $x(\mathbf{n}) = 0$ or $r\,y(\mathbf{n}) = 0$.
When x or $r\,y$ are zero, at least one plastic wave-speed is equal to an elastic wave-speed, the so-called neutral wave: Table 1 of LSM [1997] summarizes the different possibilities and gives the associated eigenspaces.

B.3.3 Examples

In summary, whether elastic or elastic-plastic with an associative flow rule, incompressible constituents are characterized by a strictly positive and bounded decay coefficient, eqn (B.3.17). In strong contrast, for non-associative flow rules, there may exist, depending on material parameters, directions along which acceleration waves grow exponentially in time.

Figures B.3.2-(a),(b) are intended to quantify these qualitative features displayed on Figs. B.3.1. For that purpose, the material termed *material 1* in LH [1991] and LSM [1997] is considered. Its volume fraction n^{w} is 0.02, the apparent densities are related by $\rho^s/\rho^w = 2.5\,\text{n}^{\text{s}}/\text{n}^{\text{w}}$ and $\lambda^{\text{DS}}/\mu = 15$ so that $\tau(\text{r}) < 0$. The material obeys deviatoric associativity, with $\psi = 30°$ and $\chi = 0°$ and the eigenvalues of the unit deviator

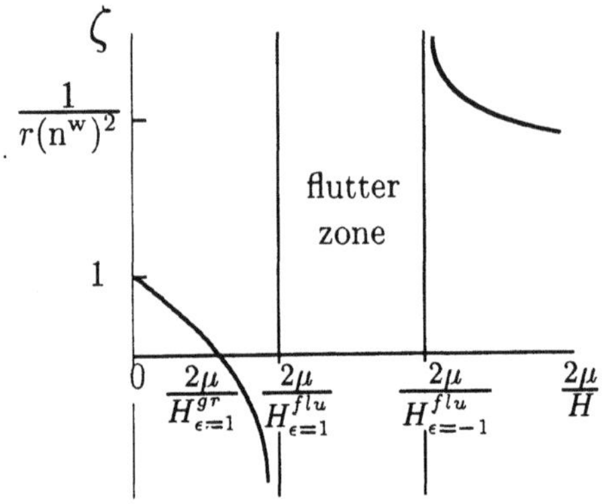

(a) $\tau(\mathrm{r}) < 0$, $x(\mathbf{n}) > 0$, $r\,y(\mathbf{n}) > 0$.
Largest plastic wave-speed.

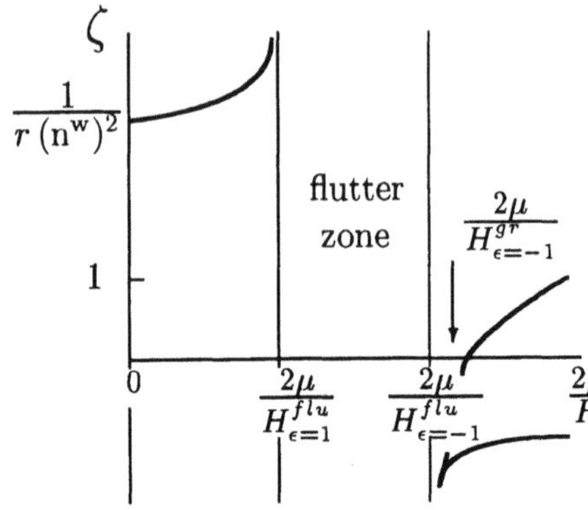

(b) $\tau(\mathrm{r}) < 0$, $x(\mathbf{n}) > 0$, $r\,y(\mathbf{n}) > 0$.
Smallest plastic wave-speed.
For small normalized moduli $\frac{2\mu}{H}$,
the sign of ζ depends on the direction $\mathbf{n}$.

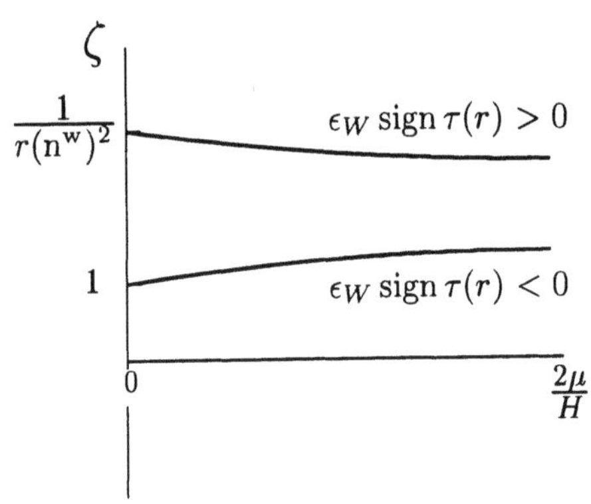

(c) $x(\mathbf{n}) < 0$, $r\,y(\mathbf{n}) > 0$.
This case is typical of associative flow rules,
$1 \leq \zeta \leq 1/r(\mathrm{n}^{\mathrm{w}})^2$.

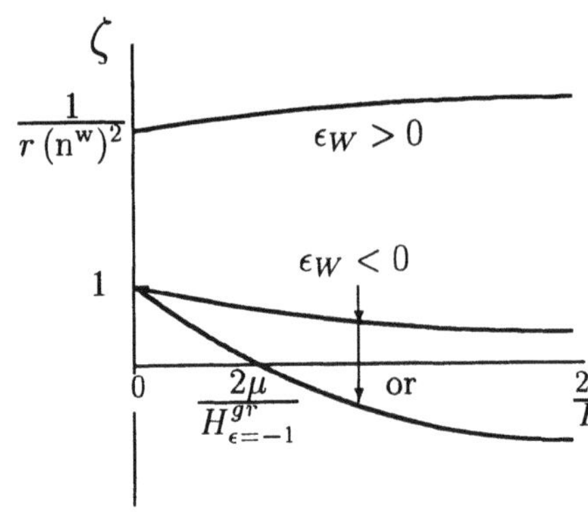

(d) $\tau(\mathrm{r}) < 0$, $x(\mathbf{n}) > 0$, $r\,y(\mathbf{n}) > 0$.
The sign of the decay coefficient associated to the wave-speed which is equal to c_s^e at the elastic limit depends on the direction $\mathbf{n}$.

Figure B.3.1 Sketch of the variations of the decay coefficient ζ in a direction of propagation $\mathbf{n}$. The decay coefficients associated to the elastic shear and longitudinal waves are equal to 1 and $1/r(\mathrm{n}^{\mathrm{w}})^2$, respectively.

$\hat{\mathbf{S}}$ are completely defined by the Lode angle $\ell = 20°$, eqn (A.1.14). Figures B.3.2 display the decay coefficient ζ associated to the largest and smallest plastic wave-speed for appropriate ranges of the plastic moduli starting at the elastic limit $1/\hat{H}=0$.

The directions of propagation $\mathbf{n}$ belong to the plane defined by the eigenvectors associated to the major and minor eigenvalues of $\hat{\mathbf{S}}$; the angle made by the normals $\mathbf{n}$ and the direction of the major eigenvector is denoted by θ; recall that the direction of propagation affects the sign of the scalar $x(\mathbf{n})$. When $\tau(\mathrm{r})$ is negative, flutter is possible for a certain range of plastic moduli and in certain directions of propagation, see figures 5 and 8.b of LH [1991]. In contrast, when $\tau(\mathrm{r})$ is positive, flutter is excluded for any plastic modulus and any direction of propagation as illustrated by Fig. B.3.1-(d). Figures B.3.2 are in agreement with these observations. The important point here is to note the quantitative influence of the plastic moduli and of the directions of propagation on the values of the decay coefficients.

When in a range of the normalized modulus a wave-speed c is not real, either due to flutter, c complex, or to divergence, c imaginary, we have not plotted the growth coefficient because boundary-value problems in that range will be shown to be ill-posed in Sect. B.4. On the other hand, the range where the amplitude of the acceleration waves grows will be associated to an instability.

B.4 Harmonic Instabilities

A linearized analysis of the stability of the flow in elastic-plastic fluid-saturated porous media with incompressible constituents is performed. As in the previous section, the analysis is restricted to solid skeletons whose plastic dilatant behaviour obeys a flow rule that preserves deviatoric associativity. The dynamic growth or decay of a small plane harmonic perturbation is analyzed, and the results are compared with those of the analysis based on acceleration waves performed in the previous section. We aim at providing a link between these two kinds of analyses and, especially, at clarifying:

(i) the reasons for the growth of acceleration waves in the non-associative case, while their speeds are real;
(ii) the relationship between the nature and the values of the speeds of propagation of these two kinds of waves;
(iii) the relationship between the rates of growth in time of these two kinds of waves, as well as between the parameter ranges where such growth occurs.

As observed in previous sections, the exclusive consideration of the plastic loading regime in the present linearized stability analysis precludes an assessment of the full consequences of the detected flutter instabilities, since the growing oscillatory nature of the flutter solutions leads to situations of local elastic unloading that are quite distinct from the linearized loading regime considered in this stability analysis. Nevertheless, it makes sense to compare the results of the present harmonic wave analysis with those obtained in Sect. B.3 and in LH [1991] and LSM [1997], because only the linearized constitutive equations for the plastic loading regime were also used there. The characteristic equation that yields the harmonic wave-speeds is obtained in subsection B.4.1. In subsection B.4.2, quasi-explicit expressions for the plastic modulus at the onset of divergence or flutter are

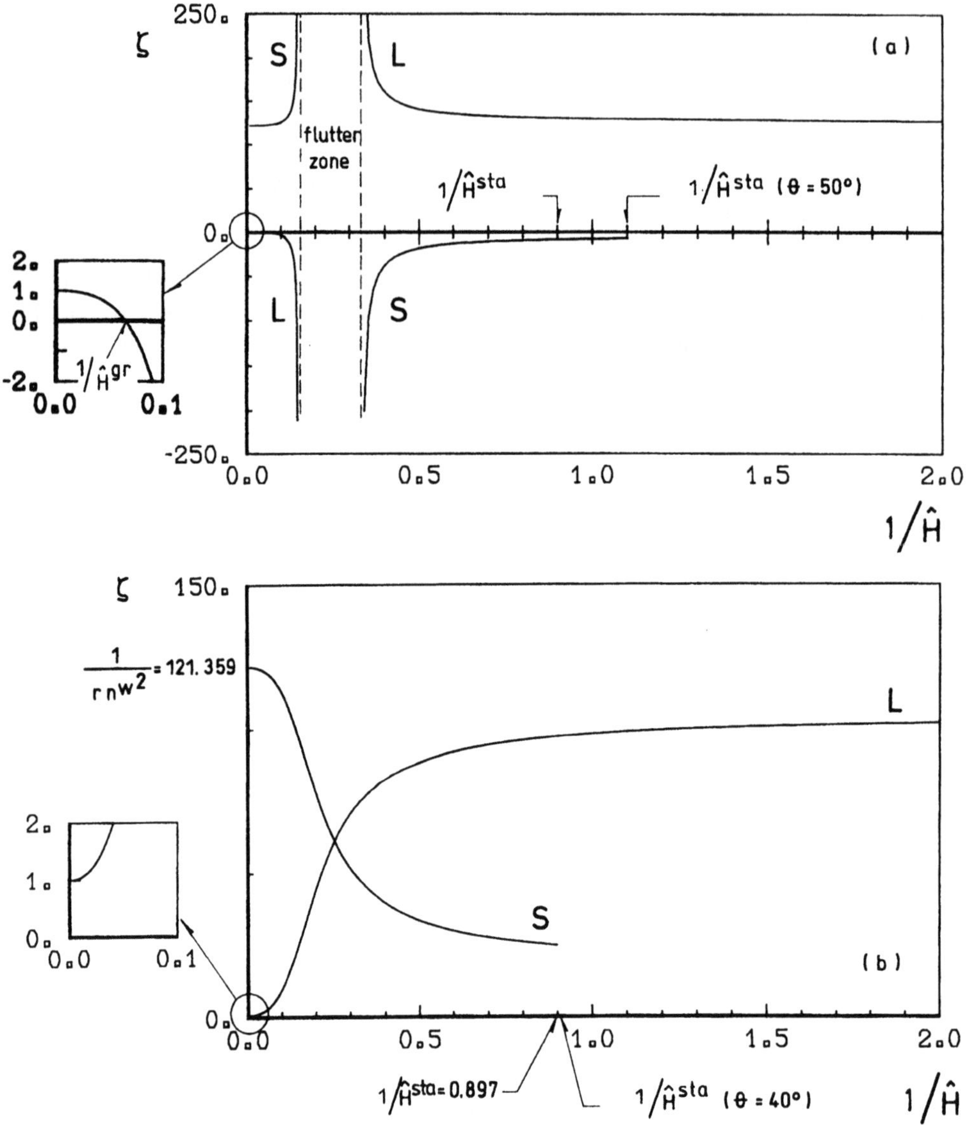

Figure B.3.2 Porous material with incompressible constituents and non-associative plasticity. Variations of the decay coefficient associated to the largest (L) and to the smallest (S) wave-speeds as a function of the inverse normalized modulus $1/\hat{H}$. (a) For the direction of propagation $\mathbf{n}$ defined by $\mathbf{n} \in [\mathbf{E}_1, \mathbf{E}_3]$, angle $\theta=(\mathbf{n}, \mathbf{E}_1)=50°$, where $\mathbf{E}_1$ and $\mathbf{E}_3$ are the major and minor eigendirections of $\hat{\mathbf{S}}$, eqn (A.1.13), the coefficient $x=x(\mathbf{n})$ is positive and flutter is available in a range of normalized moduli. (b) For $\theta=40°$, x is negative and flutter is excluded along that direction of propagation.
Material parameters: $\lambda^{DS}/\mu=15$ ($\Rightarrow \tau(\mathrm{r}) < 0$), $\psi = 30°$, $\chi = 0°$, $\ell=20°$.

obtained, and a relationship between the results of the analyses based on acceleration waves and on harmonic waves is established. The analysis is illustrated and commented with the help of examples shown in subsection B.4.3.

Following SML [1999] [14], we consider only incompressible constituents, but the numerical results described in Simões [1997] suggest that the major qualitative features of the results presented in this section also hold true when both the fluid and the solid constituents are compressible.

B.4.1 Characteristic Equations for Harmonic Waves

Assuming that the body forces $\mathbf{b}_\alpha, \alpha = s, w$, are constant in time, we seek solutions in terms of velocity fields

- that satisfy the constitutive equations (B.1.50), together with the incompressibility constraint (B.1.4) and the rate form of the equations of linear momentum balance for each phase (B.1.5),
- and that represent plane harmonic waves of assigned wavenumber k propagating with the (possibly complex) speed of propagation c along the direction $\mathbf{n}$, namely:

$$\begin{aligned} \text{for } \alpha = \mathrm{s,w}, \quad \mathbf{v}_\alpha(\mathbf{x},t) &= \hat{\mathbf{v}}_\alpha \exp\left(\mathrm{i}\,k\,(\mathbf{n}\cdot\mathbf{x} - c\,t)\right), \\ &= \hat{\mathbf{v}}_\alpha \exp(k\,\mathcal{I}m\,c\;t)\exp\left(\mathrm{i}\,k\,(\mathbf{n}\cdot\mathbf{x} - \mathcal{R}e\,c\;t)\right), \end{aligned} \tag{B.4.1}$$

where the $\hat{\mathbf{v}}_\alpha$'s are eigenvectors to be defined. The wavenumber k is linked to wavelength L by the usual relation,

$$k = \frac{2\pi}{L}. \tag{B.4.2}$$

Notice that Darcy's law introduces a characteristic length-scale L_{dif} in the field equations, namely

$$\frac{1}{L_{\text{dif}}} = \frac{\xi}{c_s^e}\left(\frac{1}{\rho^s} + \frac{1}{\rho^w}\right). \tag{B.4.3}$$

We also assume that the indeterminate rate of the partial pressure in the fluid p^w is a periodic quantity:

$$\dot{\mathrm{p}}^\mathrm{w} = \hat{\mathrm{p}}^\mathrm{w}\,\mathrm{n}^\mathrm{w}\,k\exp\left(\mathrm{i}\,k\,(\mathbf{n}\cdot\mathbf{x} - c\,t)\right). \tag{B.4.4}$$

Introduction of the rate constitutive equations (B.1.50) with $\bar{\sigma}$ defined by (B.1.27) in the rate form of the balance of momentum of the solid phase provides one set of three equations for the eigenmode $(\hat{\mathbf{v}}_s, \hat{\mathbf{v}}_w, \hat{\mathrm{p}}^\mathrm{w})$. The rate form of the balance of momentum in the fluid phase provides another set of three equations while the incompressibility constraint (B.1.4) yields the seventh relation. The resulting algebraic system can be cast in the form

$$\begin{bmatrix} \rho^s\,\mathbf{B}^{\mathrm{DS}}_{\mathrm{ep}} - (\mathrm{i}\,c\,\xi k^{-1} + \rho^s\,c^2)\mathbf{I} & \mathrm{i}\,c\,\xi k^{-1}\mathbf{I} & \mathrm{i}\,\mathrm{n}^\mathrm{s}\,\mathbf{n} \\ \mathrm{i}\,c\,\xi k^{-1}\mathbf{I} & -(\mathrm{i}\,c\,\xi k^{-1} + \rho^w\,c^2)\,\mathbf{I} & \mathrm{i}\,\mathrm{n}^\mathrm{w}\,\mathbf{n} \\ \mathrm{i}\,\mathrm{n}^\mathrm{s}\,\mathbf{n}^\mathrm{T} & \mathrm{i}\,\mathrm{n}^\mathrm{w}\,\mathbf{n}^\mathrm{T} & 0 \end{bmatrix} \begin{bmatrix} \hat{\mathbf{v}}_s \\ \hat{\mathbf{v}}_w \\ \hat{\mathrm{p}}^\mathrm{w} \end{bmatrix} = \begin{bmatrix} 0 \\ 0 \\ 0 \end{bmatrix}. \tag{B.4.5}$$

[14] SML [1999] stands for Simões, Martins and Loret [1999].

The eigenmodes $\hat{\mathrm{p}}^{\mathrm{w}}$ and $\hat{\mathbf{v}}_w$ can be expressed in terms of $\hat{\mathbf{v}}_s$, namely,

$$\hat{\mathrm{p}}^{\mathrm{w}} = \left(\mathrm{i}\, c\, \xi k^{-1} + \mathrm{n}^{\mathrm{s}}\, \rho^w\, c^2\right) \frac{\mathrm{i}\, \hat{\mathbf{v}}_s \cdot \mathbf{n}}{(\mathrm{n}^{\mathrm{w}})^2}\,, \tag{B.4.6}$$

and

$$\hat{\mathbf{v}}_w = \frac{\mathrm{i}\, c\, \xi k^{-1}}{\mathrm{i}\, c\, \xi k^{-1} + \rho^w\, c^2}\, \hat{\mathbf{v}}_s - \frac{\mathrm{i}\, c\, \xi k^{-1} + \mathrm{n}^{\mathrm{s}}\, \rho^w\, c^2}{\mathrm{i}\, c\, \xi k^{-1} + \quad \rho^w\, c^2}\, \frac{\hat{\mathbf{v}}_s \cdot \mathbf{n}}{\mathrm{n}^{\mathrm{w}}}\, \mathbf{n}\,, \tag{B.4.7}$$

while $\hat{\mathbf{v}}_s$ satisfies the relation $\mathbf{L} \cdot \hat{\mathbf{v}}_s = \mathbf{0}$, with

$$\mathbf{L} = (\mathrm{i}\, c\, \xi k^{-1} + \mathrm{n}^{\mathrm{s}}\, \rho^w\, c^2)\, \rho^s\, \mathbf{B}^{\mathrm{DS}}_{\mathrm{ep}} - (\mathrm{i}\, c\, \xi k^{-1}\, c^2 + \rho^s \rho^w\, c^4)\, \mathbf{I} - (\mathrm{i}\, c\, \xi k^{-1} + \mathrm{n}^{\mathrm{s}}\, \rho^w\, c^2)^2\, \frac{\mathbf{n} \otimes \mathbf{n}}{(\mathrm{n}^{\mathrm{w}})^2}\,.$$

Defining a cartesian coordinate system $(\mathbf{e}_1, \mathbf{e}_2, \mathbf{e}_3)$ in such a way that $\mathbf{e}_1 = \mathbf{n}$ and $\mathbf{e}_3 \cdot \mathbf{n} = 0$, the characteristic equation for the wave-speeds provided by the singularity of the coefficient matrix $\mathbf{L}$ can be factorized as a product of polynomials in the normalized wave-speed $X = c/c^{\mathrm{e}}_{\mathrm{s}}$:

$$X^3 \;\; F(\mathrm{i}\, X)\, G(\mathrm{i}\, X) = 0\,, \tag{B.4.8}$$

with

$$G(\mathrm{i}\, X) = (\mathrm{i}\, X)^3 - \alpha\, (\mathrm{i}\, X)^2 + (\mathrm{i}\, X) - \alpha\, \beta\,, \tag{B.4.9}$$

and

$$F(\mathrm{i}\, X) = u_5\, (\mathrm{i}\, X)^5 + u_4\, (\mathrm{i}\, X)^4 + u_3\, (\mathrm{i}\, X)^3 + u_2\, (\mathrm{i}\, X)^2 + u_1\, (\mathrm{i}\, X) + u_0\,. \tag{B.4.10}$$

The real coefficients u_k, $k \in [0, 5]$, depend on the nondimensional wavelength α,

$$\alpha = \frac{1}{2\, \pi}\, \frac{L}{L_{\mathrm{dif}}}\,, \tag{B.4.11}$$

on the quantities $x(\mathbf{n})$ and $y(\mathbf{n})$ defined earlier by (A.2.16), (A.2.17) but with λ^{DS} in place of λ, and they depend also on the scalar r defined by (B.2.9) and on the scalars β, γ and δ below,

$$\beta = \frac{\rho^s}{\rho^s + \rho^w}\,, \quad \gamma = \frac{(c^{\mathrm{e}}_1)^2}{(c^{\mathrm{e}}_{\mathrm{s}})^2}\,, \quad \delta = \frac{1 - \beta}{r\, (\mathrm{n}^{\mathrm{w}})^2}\,. \tag{B.4.12}$$

They can be split into an elastic and plastic part as in (B.2.11) and then, SML [1999],

$$\begin{cases} u_{\mathrm{e}5} = r\,, \quad u_{\mathrm{p}5} = 0\,, & u_{\mathrm{e}4} = -\alpha\, r\, (1 + \delta)\,, \quad u_{\mathrm{p}4} = 0\,, \\ u_{\mathrm{e}3} = r\, (1 + \alpha^2\, \delta + \gamma)\,, & u_{\mathrm{p}3} = (c^{\mathrm{e}}_{\mathrm{s}})^{-2} (-r\, y + x)\,, \\ u_{\mathrm{e}2} = -r\, \alpha\, (\beta + \gamma + \delta)\,, & u_{\mathrm{p}2} = (c^{\mathrm{e}}_{\mathrm{s}})^{-2} \alpha\, (r\, y\, (\beta + \delta) - x)\,, \\ u_{\mathrm{e}1} = r\, (\alpha^2\, \beta\, \delta + \gamma), & u_{\mathrm{p}1} = (c^{\mathrm{e}}_{\mathrm{s}})^{-2} (-r\, y\, (\alpha^2\, \beta\, \delta + \gamma) + x)\,, \\ u_{\mathrm{e}0} = -r\, \alpha\, \beta\, \gamma\,, & u_{\mathrm{p}0} = (c^{\mathrm{e}}_{\mathrm{s}})^{-2} \alpha\, \beta\, (r\, y\, \gamma - x)\,. \end{cases} \tag{B.4.13}$$

The following observations may shed some light on these algebraic expressions:

- Eqn (B.4.8) has always the root $X = 0$, but it corresponds to the trivial solution $\hat{\mathbf{v}}_s = \hat{\mathbf{v}}_w = 0$, $\hat{\mathrm{p}}^{\mathrm{w}} = 0$, and it is not related to a stationary discontinuity.

- The roots X of $G(\mathrm{i}\,X) = 0$ in (B.4.8) have, according to Routh's criterion, negative imaginary part or are zero if $\alpha = 0$. Besides no plastic characteristics enter in G.
- The characteristic equation for an elastic porous medium with incompressible constituents is recovered if H^{DS}, hidden in the u_k's through (B.2.11),(B.2.12), is set equal to ∞ in (B.4.8): then the polynomial equation for X is still given by (B.4.8) but now with

$$F(\mathrm{i}\,X) = r\,G(\mathrm{i}\,X)\,\left((\mathrm{i}\,X)^2 - \alpha\,\delta\,(\mathrm{i}\,X) + \gamma\right)\,, \tag{B.4.14}$$

which, for $\alpha \neq 0$, has complex roots X with strictly negative imaginary parts.
- As $\alpha \to 0$, the coefficients u_k, $k \in [0,5]$, tend to 0 for k even, and, for $k = 2\,k'+1$ odd, they tend to the corresponding coefficients $u_{k'}/s_{k'}$, $k' \in [0,2]$, of the characteristic equation (B.2.13) for the acceleration waves [15].
- The plastic characteristics enter only in F. Moreover, since the coefficients u_k are real, the polynomial of degree five $F(\mathrm{i}\,X)$ will have in general one purely imaginary root X, and two pairs of roots X symmetric with respect to the imaginary axis, i.e. if $X = \mathcal{R}e\,X + \mathcal{I}m\,X$ is a root, so is $-\mathcal{R}e\,X + \mathcal{I}m\,X$.

We summarize these observations in

Proposition B.4.1 Acceleration and harmonic wave-speeds in incompressible porous media, SML [1999].

As $\alpha = 1/(2\pi)\,L/L_{\mathrm{dif}}$ tends to zero, i.e. for infinitely small wavelengths or for infinitely small viscous damping associated to Darcy's law, the harmonic wave-speeds tend to the acceleration wave-speeds provided by (B.2.10).

B.4.2 Dynamic Instabilities

Assuming that during a deformation process the modulus H^{DS} (or equivalently the plastic modulus h) decreases continuously starting from $H^{\mathrm{DS}} = \infty$ (elasticity), we look for situations in which divergence or flutter-type growth occurs along at least one direction of space $\mathbf{n}$.

B.4.2.1 Divergence instability and stationary waves

Setting X equal to zero in (B.4.10), we obtain the condition for the onset of a stationary wave given by (B.2.20). As we already know, the viscous effects due to Darcy's law play no role on the determination of the onset of stationary waves. Since the solutions X of (B.4.10) are in general complex, transitions to divergence ($\mathcal{I}m\,X > 0, \mathcal{R}e\,X = 0$) can occur either when $\mathcal{R}e\,X \to 0$ and $\mathcal{I}m\,X \to 0$ (a stationary wave) or when $\mathcal{I}m\,X > 0$ and $\mathcal{R}e\,X \neq 0$, Fig. B.4.1. However in the last case divergence occurs necessarily after the occurrence of flutter ($\mathcal{I}m\,X > 0, \mathcal{R}e\,X \neq 0$). We therefore do not investigate these situations in more detail.

B.4.2.2 Flutter

First of all, notice that, for elastic mixtures ($H^{\mathrm{DS}} = \infty$) or for elastic-plastic mixtures with associative plasticity ($\mathbf{P} = \mathbf{Q}$), the matrix in (B.4.5) is symmetric, and so does $\mathbf{L}$.

---

[15] Up to a scaling coefficient $s_{k'} = ((c_s^e)^2)^{2-k'}$ due to the fact that the characteristic equation for the acceleration waves is written in terms of c, not in terms of the normalized velocity c/c_s^e.

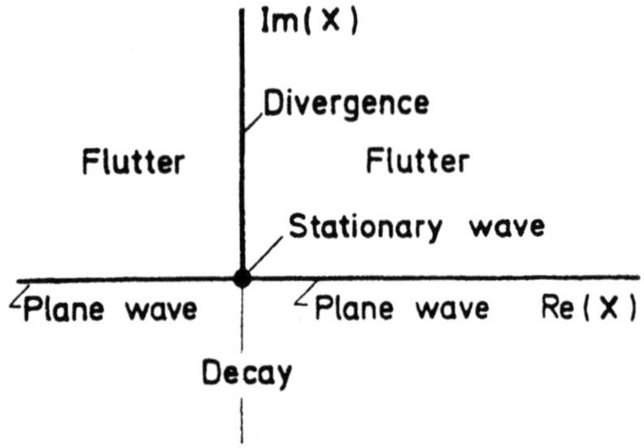

Figure B.4.1 Sketch of the five possibilities for the nature of the normalized wave-speed X. $\mathcal{I}m\,X$ negative, zero or positive implies motions with an amplitude which is, respectively, exponentially decaying, constant or exponentially growing in time.

Therefore the normalized speeds X are real or complex with negative imaginary parts, so that flutter is excluded. For non-associative plasticity ($\mathbf{P} \neq \mathbf{Q}$), one of the roots of (B.4.8) may become complex with positive imaginary part, leading to an harmonic solution with flutter-type growth. As already observed, only the plastic wave-speeds X that are roots of $F(\mathrm{i}\,X)=0$ may become complex leading to flutter. Looking for transitions to flutter, at which a root of $F(\mathrm{i}\,X)=0$ corresponds to a real wave-speed X, we obtain, by separating the real and imaginary parts of (B.4.10),

$$\hat{H}^{\mathrm{flu}}(\alpha) = r\,(c_{\mathrm{s}}^{\mathrm{e}})^2 \max_{\mathbf{n}\in\mathcal{N}} \frac{-X^2(r\,y - x) + r\,y\,(\alpha^2\beta\,\delta + \gamma) - x}{X^4 - X^2(1 + \alpha^2\delta + \gamma) + \alpha^2\beta\,\delta + \gamma}\,, \tag{B.4.15}$$

where the X^2 are chosen among the real roots of the cubic equation for X^2,

$$a_0\,X^6 + a_2\,X^4 + a_4\,X^2 + a_6 = 0\,, \tag{B.4.16}$$

where $a_0 = r\,y - \tilde{x}$, $a_2 = (\alpha^2\delta^2 - 2\,\gamma)\,r\,y - (\alpha^2 - 2)\,\tilde{x}$, $a_4 = \gamma^2 r\,y - (1 - 2\,\alpha^2\beta)\,\tilde{x}$, $a_6 = -\alpha^2\beta^2\,\tilde{x}$ and $\tilde{x}=x/r\,(\mathrm{n}^{\mathrm{w}})^2$. A study of the behaviour of these solutions in the long wavelength limit can be found in SML [1999]. Here we shall recall only the results for short wavelengths ($\alpha \ll 1$), since this is the case that provides a relationship between the present analysis and the acceleration wave analysis of the previous section. For $\alpha \ll 1$, the transition boundary (B.4.15) simplifies to $\hat{H}_\epsilon^{\mathrm{gr}}$ given by (B.3.30). Along this boundary, we have

$$X^2 \approx \frac{\gamma\,\sqrt{r\,y} + \epsilon\sqrt{\tilde{x}}/(\sqrt{r}\,\mathrm{n}^{\mathrm{w}})}{\sqrt{r\,y} + \epsilon\sqrt{\tilde{x}}/(\sqrt{r}\,\mathrm{n}^{\mathrm{w}})}\,. \tag{B.4.17}$$

This transition boundary exists only when $x \geq 0$, which may happen only if the flow rule is non-associative. When $\tau(r) < 0$, eqn (B.2.22), inequalities (B.3.31) hold, and therefore, there exists always at least one transition boundary given by (B.3.30) with $\epsilon = +1$ and it coincides with the boundary that defines the transition from decay to growth in time of the amplitude of the largest acceleration wave in the non-associative elastic-plastic fluid-saturated mixture with incompressible constituents, Fig. B.3.1-(a). When the transition boundary given by (B.3.30) with $\epsilon=-1$ exists, it coincides with the boundary that defines the transition from growth to decay in time of the amplitude of the smallest acceleration wave, Fig. B.3.1-(b). When $\tau(r)>0$, inequalities (B.3.32) hold, and therefore, it may exist one transition boundary given by (B.3.30) with $\epsilon=-1$: if it exists, it coincides with the boundary that defines the transition from growth to decay in time of the amplitude of the smallest acceleration wave in the non-associative elastic-plastic fluid-saturated mixture with incompressible constituents, Fig. 4 in LSM [1997].

Furthermore, when the acceleration wave-speeds are real, then, according to Proposition (B.4.1), $\lim_{\alpha\to 0} \mathcal{I}m\, X(\alpha)=0$. The limit as $\alpha \to 0$ of the coefficient that controls the growing time behaviour of the harmonic waves, namely $k\, \mathcal{I}m\, c = c_s^e / L_{\text{dif}}\, \mathcal{I}m\, X/\alpha$, can be estimated by l'Hôpital's rule and the continuity of $d\,\mathcal{I}m\, X(\alpha)/d\alpha$ at $\alpha=0$. Differentiating (B.4.10), estimating the result at $\alpha=0$, and using the fact that the acceleration wave-speeds are roots of (B.2.10), this limit is found to be equal to the coefficient that controls the growth or decay of the acceleration waves obtained in (B.3.13) and in LSM [1997]:

$$\lim_{\alpha\to 0} k\, \mathcal{I}m\, c(\alpha) = \frac{c_s^e}{L_{\text{dif}}} \lim_{\alpha\to 0} \mathcal{I}m \frac{X(\alpha)}{\alpha} = \frac{c_s^e}{L_{\text{dif}}} \lim_{\alpha\to 0} \frac{d}{d\alpha} \mathcal{I}m\, X(\alpha) = -\frac{\xi}{2\rho^s}\zeta\,. \tag{B.4.18}$$

We can thus state

Proposition B.4.2 Growth and decay of acceleration or harmonic waves, SML [1999].

In the regions of the parameter space where the acceleration waves have real speeds, the corresponding harmonic waves are found to grow or decay at the same rate when their wavelengths are decreased to zero. The transitions from decay to growth are given by (B.3.30).

On the other hand, if the normalized acceleration wave-speed X is not real, then the limit as $\alpha \to 0$ of $\mathcal{I}m\, X(\alpha)$ is equal to $\mathcal{I}m\, X(\alpha=0)>0$ [16] and

$$\lim_{\alpha\to 0} k\, \mathcal{I}m\, c(\alpha) = \frac{c_s^e}{L_{\text{dif}}} \lim_{\alpha\to 0} \mathcal{I}m \frac{X(\alpha)}{\alpha} = \infty\,. \tag{B.4.19}$$

So, complementary to the previous Proposition, we have

Proposition B.4.3 Linear ill-posedness, SML [1999].

Inside the regions of the parameter space where there are some speeds of acceleration waves that are not real, the exponentially growing time behaviour of the harmonic solutions is unboundedly magnified for vanishing small values of the arbitrary wavelengths, and the problem becomes (linearly) ill-posed in the sense of the definitions proposed by Schaeffer [1990], Benallal [1992] and Benallal et al. [1993].

[16] Remember that, if c is an acceleration wave-speed, so is $-c$.

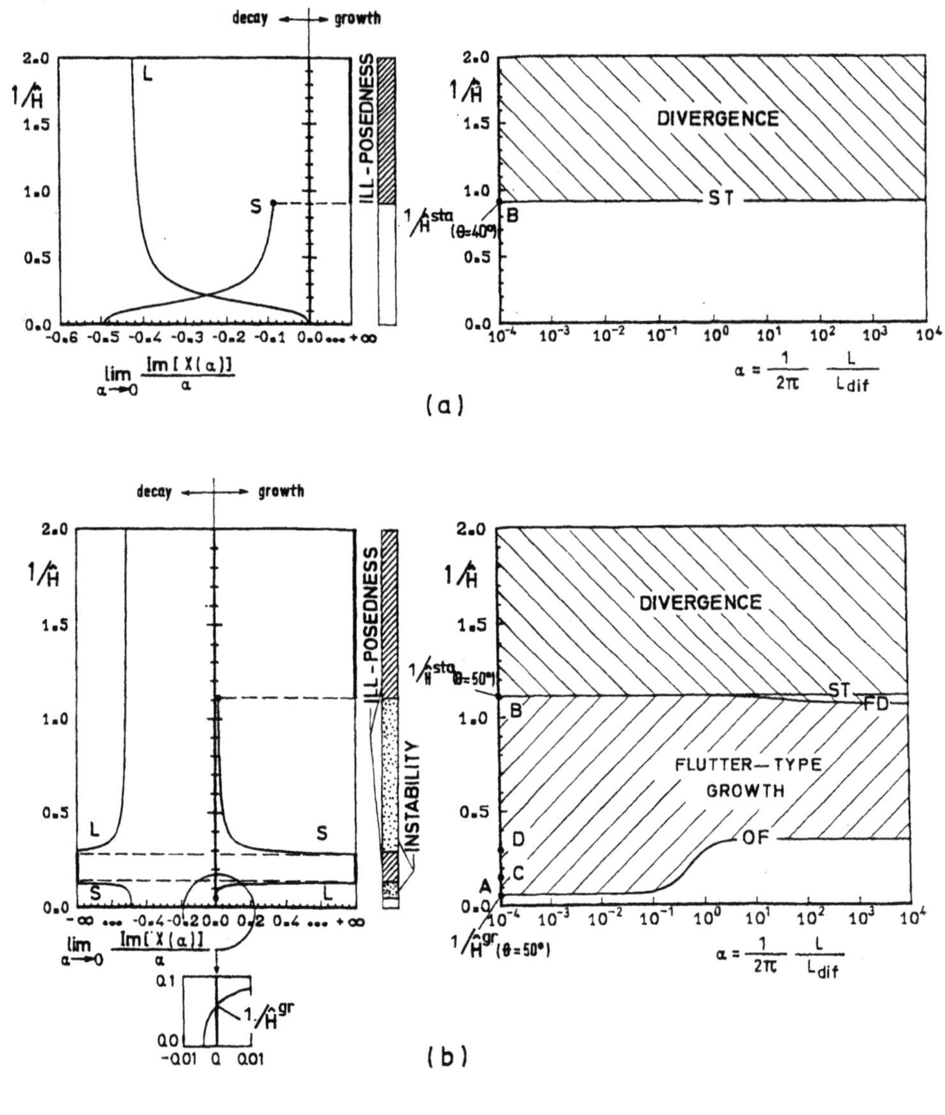
decay
growth
ILL-POSEDNESS
DIVERGENCE
ST
B
FLUTTER—TYPE GROWTH
OF
FD
INSTABILITY
D
C
A
L
S

(a)

(b)

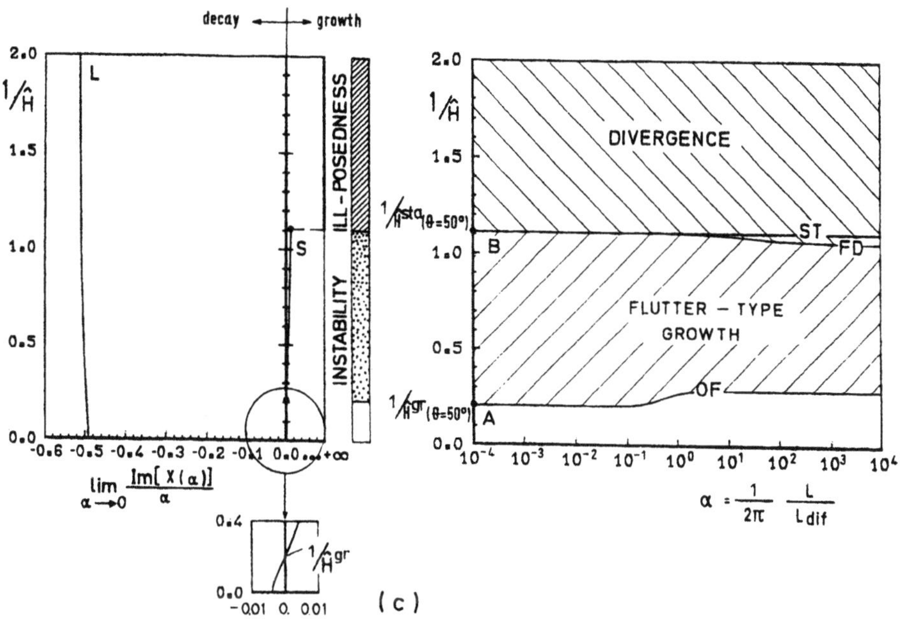

Figure B.4.2 Porous material with incompressible constituents and non-associative plasticity. Regions of divergence and flutter growth in the parameter plane ($1/\hat{H} = 2\mu/H^{\mathrm{DS}}$, α) and limit as $\alpha \to 0$ of the coefficient $\mathcal{I}m\, X(\alpha)/\alpha$ which controls the growth in time of the harmonic waves.

Material parameters: $\chi = 0°$; $\psi = 30°$; $\ell = 20°$ and

(a) $\lambda^{\mathrm{DS}}/\mu = 15$ ($\Rightarrow \tau(r) < 0$), $\theta = 40°$;

(b) $\lambda^{\mathrm{DS}}/\mu = 15$, $\theta = 50°$;

(c) $\lambda^{\mathrm{DS}}/\mu = 25$ ($\Rightarrow \tau(r) > 0$), $\theta = 50°$.

S: Smallest plastic wave-speed.

L: Largest plastic wave-speed, i.e. $|\mathcal{R}e\, X_S| \leq |\mathcal{R}e\, X_L|$.

ST: Stationary discontinuity, $\mathcal{R}e\, X = \mathcal{I}m\, X = 0$.

OF: Onset of flutter, $\mathcal{R}e\, X \neq 0$, $\mathcal{I}m\, X = 0$.

FD: Flutter-divergence transition, $\mathcal{R}e\, X = 0$, $\mathcal{I}m\, X > 0$.

B.4.3 Examples

The theoretical results of the present section are illustrated with the material used in Sect. B.3.3, except that now the elastic moduli are varied in the admissible range $0 \leq \lambda^{DS}/\mu < 30$.

In Fig. B.4.2, we represent the regions of divergence and flutter growth in the parameter plane ($1/\hat{H} = 2\mu/H^{DS}$, α) for various values of λ^{DS}/μ and for different directions of propagation $\mathbf{n}$. The directions of propagation belong to the plane defined by the eigenvectors of $\hat{\mathbf{S}}$ associated to its major and minor eigenvalues; the angle made by the normal $\mathbf{n}$ and the direction of the major eigenvector is denoted by θ. The qualitative evolution with H^{DS} decreasing from ∞ (elasticity) of the corresponding five roots X of (B.4.10) is shown in Figs. B.4.3 and 4. It can be seen that it is not possible to find, for given normalized plastic modulus and material characteristic length, an infimum wavelength which would preclude linear ill-posedness. It can also be seen that the limit as $\alpha \to 0$ of the divergence boundary in the harmonic wave analysis coincides with the onset of stationary waves in the acceleration wave analysis (point B). In contrast, the region of flutter is discontinuous as $\alpha \to 0$: the limit of the flutter region in the harmonic wave analysis (line AB) is larger than the flutter region in the acceleration wave analysis (line CD). This means that in the harmonic wave analysis for the non-associative case, the viscous damping effects due to Darcy's law may anticipate the occurrence of flutter relatively to the acceleration wave analysis (the no-damping case), points A and C in Fig. B.4.2-(b) and Fig. B.4.5 for $\tau(\mathrm{r}) < 0$. In addition, the viscous damping effects may even make flutter possible in situations where it was excluded in the acceleration wave analysis, point A in Figs. B.4.2-(c) and B.4.5 while point C does not exist for $\tau(\mathrm{r}) > 0$. Similarly to what may happen in finite dimensional circulatory systems, Ziegler [1968], Huseyin [1978], the presence of viscous damping has a destabilizing effect on the behaviour of the continuous medium.

In terms of the normalized moduli defining

- for the harmonic wave analysis, the onset of flutter $\hat{H}^{\mathrm{flu}}(\alpha)$, eqn (B.4.15),
- for the acceleration wave analysis, the onset of flutter $\hat{H}^{\mathrm{flu}}(0)$, eqn (B.2.25), and the transition from decay to growth $\hat{H}^{\mathrm{gr}}$, eqn (B.3.30),

the discontinuous behaviour as $\alpha \to 0$ means

$$\lim_{\alpha \to 0} \hat{H}^{\mathrm{flu}}(\alpha) = \hat{H}^{\mathrm{gr}} \neq \hat{H}^{\mathrm{flu}}(0)\,. \tag{B.4.20}$$

On the other hand, when $\alpha \to 0$, the limit values of the harmonic wave-speeds are in agreement with the results of LH [1991] for the corresponding acceleration wave-speeds: for $2\mu/H > 2\mu/H^{\mathrm{sta}}$, the real part of the smallest plastic wave-speed always vanishes as $\alpha \to 0$; for $2\mu/H > 2\mu/H^{\mathrm{flu}}$, the imaginary part of the complex solution X may vanish or not as $\alpha \to 0$, depending on whether flutter is excluded or not for the corresponding acceleration waves.

In Fig. B.4.2, we also represent the limit as $\alpha \to 0$ of the coefficient $\mathcal{I}m\,X(\alpha)/\alpha$ that controls the growing time behaviour of the harmonic waves. That limit coincides with the coefficient that controls the growth or decay of the acceleration waves with real speed of propagation. It can be seen that, although the characteristic length introduced by

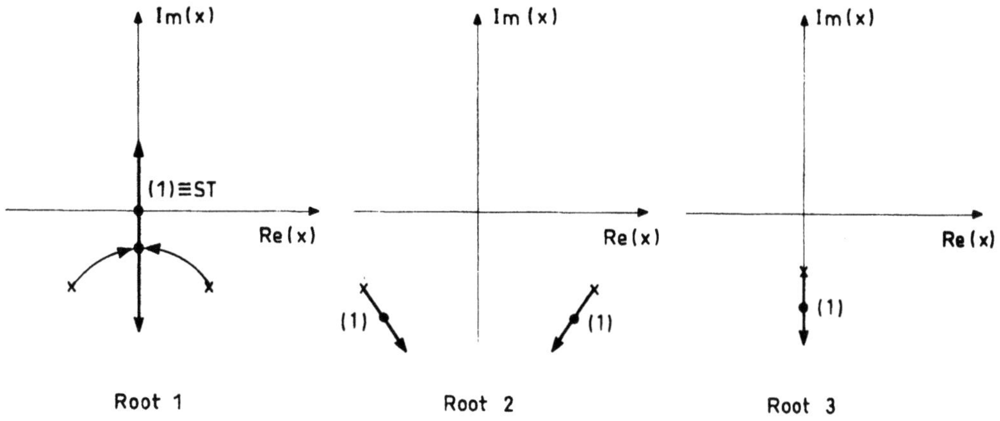

Figure B.4.3 Sketch of the evolution, as the plastic modulus decreases from ∞ (elasticity), of the five roots X of $F(\mathrm{i}\,X)=0$, eqn (B.4.10), in Fig. B.4.2-(a). The two roots '1', initially complex and dissipative, become purely imaginary and then one of them tend 0, corresponding to the onset of a stationary discontinuity, (ST). The two roots '2' are always complex and dissipative, while the root '3' is always imaginary and dissipative.

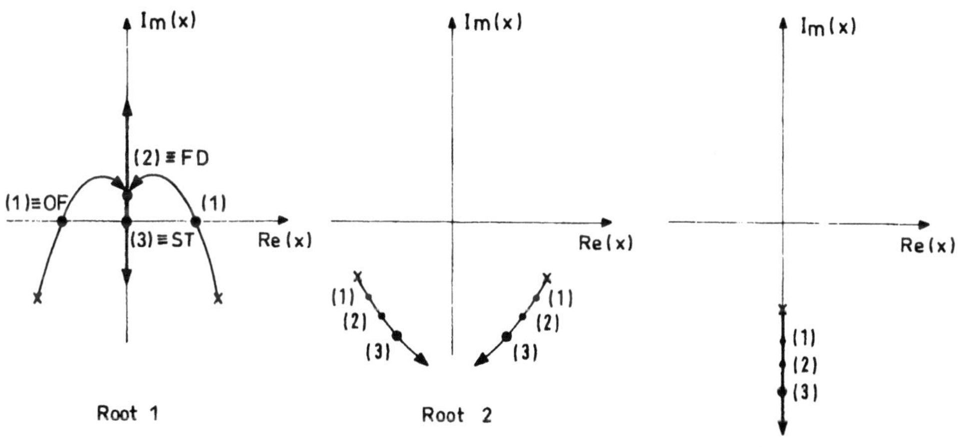

Figure B.4.4 Sketch of the evolution, as the plastic modulus decreases from ∞ (elasticity), of the five roots X of $F(\mathrm{i}\,X)=0$, eqn (B.4.10), in Fig. B.4.2-(b). The two roots '1', initially complex and dissipative, become real ((1)=onset of flutter OF), then remain in the flutter region until they become purely imaginary ((2)= divergence from the flutter region FD), and then one of them tend to vanish ((3)= stationary discontinuity ST) while the imaginary part of the other increases. The two roots '2' are always complex and dissipative, while the root '3' is always imaginary and dissipative.

Darcy's law does not prevent small wavelength unstable modes, the exponentially growing time behaviour of the unstable modes is unboundedly magnified for vanishing small wavelengths only when values of $2\mu/H$ are used such that non-real speeds of propagation exist for the corresponding acceleration waves.

In view of the discussion above and for the linearized problems studied here, in LH [1991] and in LSM [1997], a clear distinction should be made between (linear) ill-posedness and (linear) instability (see also Schaeffer [1990]).

- *Linear instability* is characterized by:
  - the existence of acceleration waves with a real speed of propagation and with an amplitude that grows exponentially in time;
  - the existence of harmonic waves growing exponentially in time at a rate which, in the limit as the wavelengths are decreased to zero, equals the rate of growth of the corresponding acceleration waves.

Regions AC and DB in Fig. B.4.2-(b) are representative of this situation.

- *Linear ill-posedness* is characterized by:
  - the existence of non-real wave-speeds for the acceleration wave analysis;
  - the existence of harmonic wave solutions with an exponential growth in time that is unboundedly magnified as the wavelengths are decreased to zero.

Regions CD and B to ∞ in Fig. B.4.2-(b) are representative of this situation.

B.5 Gradient and Non-Local Plasticity for Fluid-Saturated Porous Media

In the analysis of harmonic waves, we stressed the (linear) ill-posedness character of the equations that govern the behaviour of an elastic-plastic fluid saturated porous medium for some ranges of the material parameters. On the other hand, it has been long recognized that stationarity wave/localization ill-posedness can be fixed by using some enriched constitutive descriptions of the continuum that incorporate internal (or characteristic) lengths of the material. A related study for a case of flutter ill-posedness but in the context of contact problems involving the (non-associative) friction law of Coulomb was recently presented by Simões and Martins [1998]. In the present context of flutter ill-posedness of incompressible fluid-saturated porous media, we discuss now the effect of two classical enrichments of the material behaviour: the gradient plasticity theory, e.g. Mühlhaus and Aifantis [1991], and the non-local plasticity theory, e.g. Bažant and Lin [1988].

B.5.1 Gradient Plasticity

We use here the simple modified form of the plastic behaviour proposed by Mühlhaus and Aifantis [1991] that consists of assuming that the yield function f at a point $\mathbf{x}$ depends not only on the cumulated plastic strain κ,

$$\dot{\kappa} = \| \dot{\epsilon}_p^s \| , \tag{B.5.1}$$

but also on its Laplacian $\Delta\kappa$:

$$f = f(\bar{\sigma}, \kappa, \Delta\kappa) , \tag{B.5.2}$$

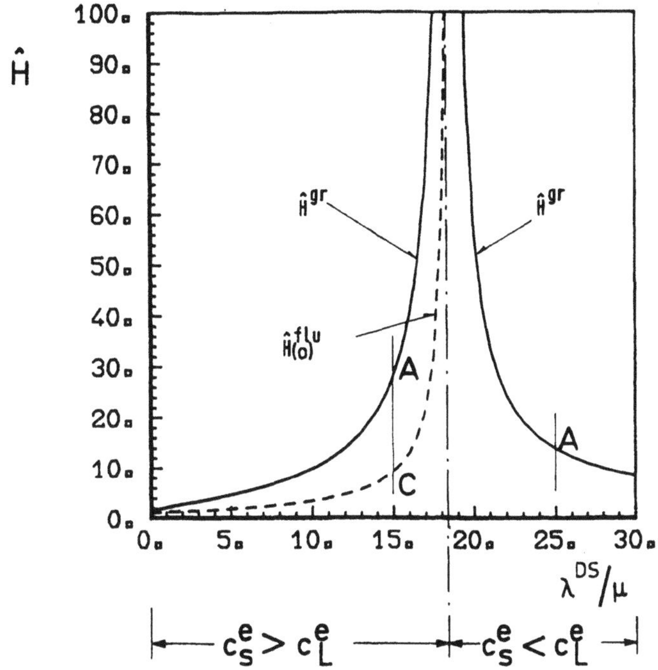

Figure B.4.5 Porous material with incompressible constituents and non-associative plasticity.
—— $\lim_{\alpha\to 0}\hat{H}^{\text{flu}}(\alpha)$: limit as $\alpha \to 0$ of the normalized plastic modulus $\hat{H}^{\text{flu}}(\alpha)$ at the onset of flutter in the harmonic wave analysis as a function of λ^{DS}/μ. This modulus is equal to the normalized plastic modulus $\hat{H}^{\text{gr}}$ at the transition decay/growth in the acceleration wave analysis, as a function of λ^{DS}/μ.
- - - $\hat{H}^{\text{flu}}(0)$: the normalized plastic modulus defining the onset of flutter in the acceleration wave analysis.
Viscous effects due to Darcy's law makes flutter to occur earlier, i.e. $\hat{H}^{\text{flu}}(\alpha)=\hat{H}^{\text{gr}}\geq\hat{H}^{\text{flu}}(0)$, or trigger its existence: recall that Darcy's law does not affect the acceleration wave analysis.
Material parameters: $\psi=30°$, $\chi=0°$, $\ell=20°$.

all three arguments of f being estimated at $\mathbf{x}$. The strain decomposition into elastic and plastic components as well as the flow rule are preserved:

$$\dot{\boldsymbol{\epsilon}}^s = \dot{\boldsymbol{\epsilon}}^s_e + \dot{\boldsymbol{\epsilon}}^s_p\,, \quad \dot{\boldsymbol{\epsilon}}^s_p = \dot{\Lambda}\,\mathbf{P}\,. \tag{B.5.3}$$

Following the usual developments, it is a consequence of the consistency condition

$$\dot{f} = \frac{\partial f}{\partial \boldsymbol{\sigma}} : \dot{\boldsymbol{\sigma}} + \frac{\partial f}{\partial \kappa} : \dot{\kappa} + \frac{\partial f}{\partial \Delta\kappa} : \Delta\dot{\kappa} = 0\,, \tag{B.5.4}$$

that

$$\dot{\Lambda} = \frac{1}{H^{\text{DS}}}(\mathbf{Q} : \mathbb{E}^{\text{DS}} : \dot{\boldsymbol{\epsilon}}^s + b\,\Delta\dot{\kappa})\,, \tag{B.5.5}$$

where

$$\mathbf{Q} = \frac{\partial f}{\partial \sigma}, \; h = -\frac{\partial f}{\partial \kappa}, \; b = \frac{\partial f}{\partial \Delta\kappa} \geq 0, \; H^{\rm DS} = h + \mathbf{Q} : \mathbb{E}^{\rm DS} : \mathbf{P}. \tag{B.5.6}$$

The rate form of the elastic-plastic constitutive law (for the plastic loading regime) reads:

$$\dot{\boldsymbol{\sigma}} = \mathbb{E}^{\rm DS}_{\rm ep} : \dot{\boldsymbol{\epsilon}}^s - \frac{b\,\Delta\dot{\kappa}}{H^{\rm DS}}\,\mathbb{E}^{\rm DS} : \mathbf{P} \tag{B.5.7}$$

with $\mathbb{E}^{\rm DS}_{\rm ep} = \mathbb{E}^{\rm DS} - 1/H^{\rm DS}\,\mathbb{E}^{\rm DS}\!:\!\mathbf{P} \otimes \mathbf{Q}\!:\!\mathbb{E}^{\rm DS}$ exactly as in (B.1.47). These are the equations that we wish to solve, together with the equations (B.1.5) of linear momentum balance and the incompressibility condition (B.1.4). We seek harmonic solutions for them exactly of the form indicated in (B.4.5). Note that the only calculations that are now different from those performed in subsection B.4.3 are those involving the Laplacian of the cumulated plastic strain, present in the term that makes the constitutive equation (B.5.7) different from (B.1.50). For monochromatic velocity $\mathbf{v}_s$ and rate $\dot{\kappa}$,

$$\begin{bmatrix} \mathbf{v}_s \\ \dot{\kappa} \end{bmatrix} = \begin{bmatrix} \hat{\mathbf{v}}_s \\ \hat{\dot{\kappa}} \end{bmatrix} \exp\Big(\mathrm{i}\,k\,(\mathbf{n}\cdot\mathbf{x} - c\,t)\Big), \tag{B.5.8}$$

with wavelength L and wavenumber k linked by (B.4.2), it follows that

$$\dot{\boldsymbol{\epsilon}}^s = \frac{\mathrm{i}\,k}{2}(\mathbf{v}_s \otimes \mathbf{n} + \mathbf{n} \otimes \mathbf{v}_s), \quad \Delta\dot{\kappa} = -k^2\,\dot{\kappa}. \tag{B.5.9}$$

Insertion of these relations in (B.5.5) yields

$$\dot{\kappa} = \frac{\mathrm{i}\,k}{H^{\rm DS}_{\rm grad}}\,\mathbf{q}\cdot\mathbf{v}_s, \quad \Delta\dot{\kappa} = -\frac{\mathrm{i}\,k^3}{H^{\rm DS}_{\rm grad}}\,\mathbf{q}\cdot\mathbf{v}_s, \tag{B.5.10}$$

with $\mathbf{q}$ defined by (B.2.8) and $H^{\rm DS}_{\rm grad}$ is a simple modification of $H^{\rm DS}$:

$$H^{\rm DS}_{\rm grad} = H^{\rm DS} + b\,k^2. \tag{B.5.11}$$

If we now insert the above constitutive equations in the harmonic wave analysis of Sect. B.4.2, we easily observe that only the balance of momentum for the solid phase is modified due to (B.5.7) with $\Delta\dot{\kappa}$ given by (B.5.10). The governing algebraic system of equations (B.2.6) preserves its form, with the only distinction that the acoustic tensor of the drained solid $\mathbf{B}^{\rm DS}_{\rm ep}$, eqn (B.2.7), is replaced by $\mathbf{B}^{\rm DS}_{\rm ep,\,grad}$,

$$\mathbf{B}^{\rm DS}_{\rm ep,\,grad} = \mathbf{B}^{\rm DS} - \frac{1}{H^{\rm DS}_{\rm grad}}\,\frac{\mathbf{p}\otimes\mathbf{q}}{\rho^s}, \tag{B.5.12}$$

with $\mathbf{p}$ and $\mathbf{q}$ defined by (B.2.8). Thus it appears that the sole modification introduced in the plasticity model turns out to be equivalent, for the propagation of harmonic waves, to replace the hardening modulus $H^{\rm DS}$ by the modified value $H^{\rm DS}_{\rm grad}$. For this modified model, it is possible to show that there exist infimum wavelengths for the occurrence of flutter and divergence instability.

The scalar b is assumed positive, eqn (B.5.6), it has dimension stress×length^2 and it introduces an *internal length-scale* L_{grad},

$$L_{\text{grad}} = \sqrt{\frac{b}{\mu}}, \tag{B.5.13}$$

but remember that there is already present another length-scale induced by Darcy's diffusion, namely L_{dif}, eqn (B.4.3).

The dependence of the occurrence of flutter and divergence on the wavelengths is illustrated in Figs. B.5.1 for this gradient plasticity version of the material discussed in subsection B.4.3. The existence of an infimum wavelength for the occurrence of flutter and divergence modes is made clear in those figures, where the critical quantity $1/\hat{H} = 2\,\mu/H^{\text{DS}}$ is plotted as a function of the nondimensional wavelength α, (B.4.11). As the wavelength L, and consequently α, decreases, the modified normalized modulus $\hat{H}_{\text{grad}}$,

$$\hat{H}_{\text{grad}} = \hat{H} + \frac{1}{2\,\alpha^2}\frac{L^2_{\text{grad}}}{L^2_{\text{dif}}}. \tag{B.5.14}$$

becomes larger and larger, which explains the disappearance of flutter and divergence modes. In fact, at a given wavelength, the critical moduli for any of these events for the plasticity model, $\hat{H}^{\text{crit}}(L_{\text{grad}}{=}0)$, and for the gradient plasticity model, $\hat{H}^{\text{crit}}(L_{\text{grad}})$ differ by a simple shift,

$$\hat{H}^{\text{crit}}(L_{\text{grad}}) = \hat{H}^{\text{crit}}(L_{\text{grad}}=0) - \frac{1}{2\,\alpha^2}\frac{L^2_{\text{grad}}}{L^2_{\text{dif}}}. \tag{B.5.15}$$

The assumption of a strictly positive modulus $\hat{H}$ thus implies, unless $\hat{H}^{\text{crit}}(L_{\text{grad}}=0) = +\infty$, the existence of an infimum wavelength $\alpha_{\text{inf}} = (2/\hat{H}^{\text{crit}}(L_{\text{grad}}=0))^{1/2}\,L_{\text{grad}}/L_{\text{dif}}$ for any critical event characterized by $\hat{H}^{\text{crit}}$. Notice that the infimum wavelength $L_{\text{inf}} = 2\,\pi\,(2/\hat{H}^{\text{crit}}(L_{\text{grad}}{=}0))^{1/2}\,L_{\text{grad}}$ is independent of L_{dif}.

B.5.2 Non-Local Plasticity

We provide two non-local modifications, none of which is as satisfactory as the gradient modification. In fact, the point here is to show that the regularizing effects are very sensitive to the quantities that are averaged.

B.5.2.1 **Regularization of the hardening parameter.**

The first modification introduced here in the plastic behaviour consists in assuming that the yield function f at each point of the body depends on a weighted average of the cumulated plastic strains in a neighborhood $V(\mathbf{x})$ of that point:

$$f = f\,(\bar{\sigma}(\mathbf{x}), \tilde{\kappa}(\mathbf{x}))\,, \tag{B.5.16}$$

with

$$\tilde{\kappa}(\mathbf{x}) = \frac{1}{V(\mathbf{x})}\int_{V(\mathbf{x})} w(\mathbf{x}-\mathbf{s})\,\kappa(\mathbf{s})\,dV(\mathbf{s})\,. \tag{B.5.17}$$

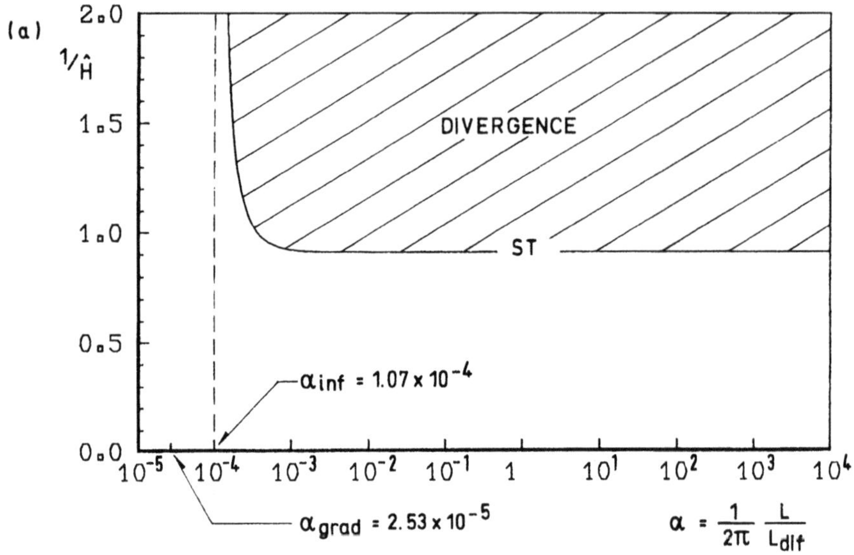

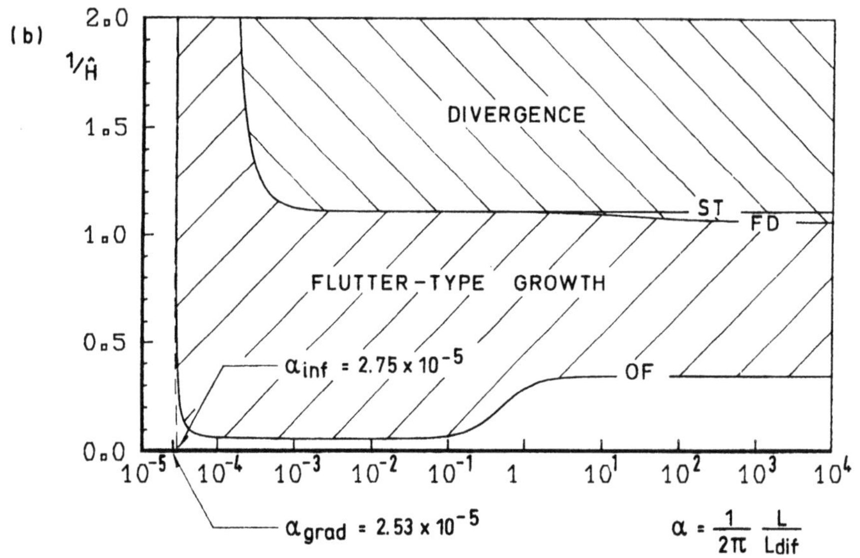

Figure B.5.1 Porous material with incompressible constituents including the effect of a second order gradient. Inverse of critical modulus $1/\hat{H}=2\,\mu/H^{\mathrm{DS}}$ as a function of the nondimensional wavelength α. There exists a minimum wavelength below which (a) neither divergence, (b) nor flutter modes appear. The data are similar to those used for Figs. B.4.2-(a),(b) but with now $L_{\mathrm{grad}}/L_{\mathrm{dif}}=10^{-3}/2\,\pi$ (while the plots in the Figs. B.4.2-(a),(b) correspond to $L_{\mathrm{grad}}/L_{\mathrm{dif}}=0$).

For the present calculations, we shall use as weighting function the normalized Gaussian function

$$w(\mathbf{y}) = \frac{1}{(2\,\pi)^{3/2}} \exp\left(-\frac{\|\,\mathbf{y}\,\|^2}{2\,L_{\rm nl}^2}\right), \tag{B.5.18}$$

where $L_{\rm nl}$ is the characteristic length of this modified plasticity model. To facilitate the comparison of the results of the present model with those of the gradient plasticity, we rewrite the yield function at each point $\mathbf{x}$ of the body as a function of the cumulated plastic strain $\kappa(\mathbf{x})$ at that point and of the difference $\tilde{\kappa}(\mathbf{x}) - \kappa(\mathbf{x})$:

$$f = f\,(\,\bar{\sigma}(\mathbf{x}),\, \kappa(\mathbf{x}),\, \tilde{\kappa}(\mathbf{x}) - \kappa(\mathbf{x})\,)\;. \tag{B.5.19}$$

Again, it is a consequence of the consistency condition

$$\dot{f} = \frac{\partial f}{\partial \sigma} : \dot{\bar{\sigma}} + \frac{\partial f}{\partial \kappa} : \dot{\kappa} + \frac{\partial f}{\partial(\tilde{\kappa} - \kappa)}\,(\,\dot{\tilde{\kappa}} - \dot{\kappa}\,) = 0\,, \tag{B.5.20}$$

that

$$\dot{\Lambda} = \frac{1}{H^{\rm DS}}\Big(\mathbf{Q} : \mathbb{E}^{\rm DS} : \dot{\boldsymbol{\epsilon}}^s + b\,(\dot{\tilde{\kappa}} - \dot{\kappa})\Big)\,, \tag{B.5.21}$$

where $b = \partial f/\partial\tilde{\kappa} \geq 0$. The rate form of the elastic-plastic constitutive law (for the plastic loading regime) reads:

$$\dot{\bar{\sigma}} = \mathbb{E}^{\rm DS}_{\rm ep} : \dot{\boldsymbol{\epsilon}}^s - \frac{b\,(\,\dot{\tilde{\kappa}} - \dot{\kappa}\,)}{H^{\rm DS}}\,\mathbb{E}^{\rm DS} : \mathbf{P}\,. \tag{B.5.22}$$

Using the definition (B.5.17) and the Fourier transform, assuming for simplicity $V(\mathbf{x}) = \mathbb{R}^3$,

$$\frac{1}{V(\mathbf{x})}\int_{V(\mathbf{x})} \frac{1}{(2\,\pi)^{3/2}} \exp\left(-\frac{\|\,\mathbf{s}\,\|^2}{2\,L_{\rm nl}^2} - \mathrm{i}\,k\,\mathbf{n}\cdot\mathbf{s}\right) d\,V(\mathbf{s}) = \exp\left(-\frac{k^2\,L_{\rm nl}^2}{2}\right), \tag{B.5.23}$$

the consideration of harmonic solutions for $\dot{\kappa}$, eqn (B.5.9), leads to

$$\dot{\tilde{\kappa}}(\mathbf{x}) = \dot{\Lambda}(\mathbf{x}) \exp\left(-\frac{k^2\,L_{\rm nl}^2}{2}\right), \tag{B.5.24}$$

and hence,

$$\dot{\tilde{\kappa}}(\mathbf{x}) - \dot{\kappa}(\mathbf{x}) = -\left(1 - \exp\left(-\frac{k^2\,L_{\rm nl}^2}{2}\right)\right)\dot{\kappa}(\mathbf{x})\,. \tag{B.5.25}$$

Then once again, for the harmonic problem, the rate elastic-plastic constitutive equations can be cast in the usual format $\dot{\bar{\sigma}} = \mathbb{E}^{\rm DS}_{\rm ep,nl} : \dot{\boldsymbol{\epsilon}}^s$ with $\mathbb{E}^{\rm DS}_{\rm ep,nl} = \mathbb{E}^{\rm DS} - 1/H^{\rm DS}_{\rm nl,1}\,\mathbb{E}^{\rm DS} : \mathbf{P}\otimes\mathbf{Q} : \mathbb{E}^{\rm DS}$ and the sole modification with respect to the local formulation is due to the shift in modulus,

$$\hat{H}_{\rm nl,1} = \hat{H} + \frac{b}{2\,\mu}\left(1 - \exp\left(-\frac{k^2\,L_{\rm nl}^2}{2}\right)\right). \tag{B.5.26}$$

Notice that an expansion to second order of $H^{\rm DS}_{\rm nl,1}$ yields a modified modulus of the type provided by the second gradient theory, namely,

$$\hat{H}_{\rm nl,1} = \hat{H} + \frac{b}{2\,\mu}\,\frac{L_{\rm nl}^2}{2}\,k^2\,, \tag{B.5.27}$$

that is, for small wavenumber k or large wavelengths L, the non-local and the gradient modifications coincide.

The resulting modification of the governing algebraic system of equations (B.2.6) affects the acoustic tensor of the drained solid only, as in (B.5.12), but with now H^{DS} replaced by $H^{\mathrm{DS}}_{\mathrm{nl},1}$. The normalized moduli defining any critical event are related similarly to (B.5.15),

$$\hat{H}^{\mathrm{crit}}(L_{\mathrm{nl}}) = \hat{H}^{\mathrm{crit}}(L_{\mathrm{nl}}=0) - \frac{b}{2\,\mu}\Big(1 - \exp\Big(-\frac{k^2\,L_{\mathrm{nl}}^2}{2}\Big)\Big). \tag{B.5.28}$$

Unlike for the gradient plasticity model, the modification due to small wavelengths or large wavenumbers is finite and coincides with a shift of plastic modulus equal to b: positive values of b have a stabilizing effect.

However, the non-local problem is not exactly similar to the local one as the shift in apparent modulus increases when the wavelength decreases, so that one can tailor the material parameters b and L_{nl} to imply prescribed critical wavelengths, or conversely. The existence of an infimum wavelength below which neither flutter nor divergence would occur requires $\hat{H} + b/2\,\mu$ to be greater than any of the critical moduli $\hat{H}^{\mathrm{crit}}(L_{\mathrm{nl}}=0)$. The same reasoning applies to individual critical events. For example, in their one-dimensional analysis of strain localization, Vermeer and Brinkgreve [1994] stress that the small wavelength regularization of strain localization requires, in our notation, $h+b$ to be strictly positive. For the material parameters used in Fig. B.4.2, both divergence and flutter are regularized for $b/2\mu$ sufficiently large, no one is for small $b/2\mu$, and for intermediate values only divergence is regularized.

B.5.2.2 **Regularization of the plastic strain.**

In this second non-local modification, the yield function retains its usual local format, namely

$$f = f\,(\,\bar{\sigma}(\mathbf{x}), \kappa(\mathbf{x})\,)\ , \tag{B.5.29}$$

with κ defined by (B.5.1),(B.5.3), but the plastic strain at $\mathbf{x}$ is averaged over a neighborhood $V(\mathbf{x})$ of $\mathbf{x}$ with the weighting function (B.5.18),

$$\dot{\boldsymbol{\epsilon}}^s = \dot{\boldsymbol{\epsilon}}^s_e + \dot{\boldsymbol{\epsilon}}^s_p, \quad \dot{\boldsymbol{\epsilon}}^s_p = \dot{\tilde{\Lambda}}\,\mathbf{P}\,, \tag{B.5.30}$$

with

$$\dot{\tilde{\Lambda}}(\mathbf{x}) = \frac{1}{V(\mathbf{x})}\int_{V(\mathbf{x})} w(\mathbf{x}-\mathbf{s})\,\dot{\Lambda}(\mathbf{s})\;dV(\mathbf{s})\,. \tag{B.5.31}$$

We look now for harmonic solutions of the form (B.5.8),

$$\begin{bmatrix}\mathbf{v}_s\\ \dot{\Lambda}\end{bmatrix} = \begin{bmatrix}\hat{\mathbf{v}}_s\\ \hat{\dot{\Lambda}}\end{bmatrix}\ \exp\Big(\mathrm{i}\,k\,(\mathbf{n}\cdot\mathbf{x} - c\,t)\Big)\,, \tag{B.5.32}$$

so that with (B.5.31),

$$\dot{\tilde{\Lambda}}(\mathbf{x}) = \dot{\Lambda}(\mathbf{x})\ \exp\Big(-\frac{k^2\,L_{\mathrm{nl}}^2}{2}\Big)\,. \tag{B.5.33}$$

Inserting the elastic law $\dot{\sigma}=\mathbb{E}:\dot{\epsilon}^s_e$ in the consistency condition $\dot{f}(\bar{\sigma},\kappa)=0$, eqn (B.5.29), provides the plastic multiplier $\dot{\Lambda}$ from which follows the regularized plastic multiplier $\dot{\tilde{\Lambda}}$. Thus for the harmonic problem, the elastic-plastic rate equations can be cast in the usual format $\dot{\sigma}=\mathbb{E}^{DS}_{ep,nl}:\dot{\epsilon}^s$ with $\mathbb{E}^{DS}_{ep,nl}=\mathbb{E}^{DS}-1/H^{DS}_{nl,2}\,\mathbb{E}^{DS}:\mathbf{P}\otimes\mathbf{Q}:\mathbb{E}^{DS}$ but with $H^{DS}_{nl,2}=2\,\mu\,\hat{H}_{nl,2}$ depending on the wavenumber,

$$\hat{H}_{nl,2}=\hat{H}+\frac{h}{2\,\mu}\left(\exp\left(\frac{k^2\,L^2_{nl}}{2}\right)-1\right). \tag{B.5.34}$$

So now, if the plastic modulus h is strictly positive, similar to the gradient modification of Sect. B.5.1, as the wavelength L and consequently α decrease, the modified normalized modulus $\hat{H}_{nl,2}$,

$$\hat{H}_{nl,2}=\hat{H}+\frac{h}{2\,\mu}\left(\exp\left(\frac{1}{2\alpha^2}\,\frac{L^2_{nl}}{L^2_{dif}}\right)-1\right), \tag{B.5.35}$$

becomes larger and larger, which leads to the disappearance of flutter and divergence modes. On the other hand, this second non-local modification is useless for perfect plasticity, i.e. $h=0$.

References

Achenbach, J.D. (1975). *Wave propagation in elastic solids*, North-Holland, Amsterdam, The Netherlands.

An, L. and Schaeffer, D.G. (1992). The flutter instability in granular flow. *J. Mech. Phys. Solids*, **40**, 683-698.

Asaro, R. and Rice, J.R. (1977). Strain localization in ductile single crystals. *J. Mech. Phys. Solids*, **25**, 309-317.

Atkin, R. (1968). Completeness theorems for linearized theories of interacting continua. *Quart. J. Mech. Appl. Math.*, **21**, 171-193.

Baker, R. and Desai, C.S. (1982). Consequences of deviatoric normality in plasticity with isotropic strain-hardening. *Int. J. Num. Anal. Meth. Geomechanics*, **6**, 383-390.

Bažant, Z.P. and Krizek, R.J. (1975). Saturated sand as an inelastic two-phase medium. *J. of Engng. Mechanics Div.*, Transactions of the ASCE, **101**-EM4, 317-332.

Bažant, Z.P. and Lin, F.B. (1988). Non-local yield limit degradation. *Int. J. Num. Meth. Engineering*, **26**, 1805-1823.

Bažant, Z.P. and Cedolin, L. (1991). *Stability of Structures*, Oxford University Press.

Benallal, A., Billardon, R. and Geymonat, G. (1989). Conditions de bifurcation à l'intérieur et aux frontières pour une classe de matériaux non standards. *Comptes-Rendus Académie des Sciences, Paris*, t. 308, série II, 893-898.

Benallal, A. (1992). Ill-posedness and localization in solid structures. *Computational Plasticity*, D.R.J. Owen et al. eds., Pineridge Press, Swansea, U.K., 581-592.

Benallal, A., Billardon, R. and Geymonat, G. (1993). Bifurcation and localization in rate-independent materials. Some general considerations. *Bifurcation and Stability of Dissipative Systems*, CISM course 327, Q.S. Nguyen ed., Springer-Verlag, Wien-New York, 1-44.

Bigoni, D. and Hueckel, T. (1991). Uniqueness and localization -I. Associative and non-associative elastoplasticity. *Int. J. Solids Structures*, **28**, 197-213.

Bigoni, D. and Willis, J.R. (1994). A dynamical interpretation of flutter instability. *Localisation and Bifurcation Theory for Soils and Rocks*, R. Chambon, J. Desrues and I. Vardoulakis eds., Balkema, Rotterdam, The Netherlands, 51-58.

Bigoni, D. and Loret, B. (1999). Effects of elastic anisotropy on strain localization and flutter instability in plastic solids. *J. Mech. Phys. Solids*, **47**(7), 1409-1436.

Biot, M.A. (1956). Theory of propagation of elastic waves in a fluid saturated porous solid - I. Low-frequency range. *J. of Acoustical Society of America*, **28**(1), 168-179; - II. Higher frequency range. *J. of Acoustical Society of America*, **28**(1), 179-191.

Biot, M.A. and Willis, D.G. (1957). The elastic coefficients of the theory of consolidation. *J. of Applied Mechanics*, Transactions of the ASME, **24**, 594-601.

Biot, M.A. (1973). Nonlinear and semilinear rheology of porous solids. *J. of Geophysical Research*, **78**(23), 4924-4937.

Beskos, D.E. (1989). Dynamics of saturated rocks -I. Equations of motion. *J. of Engng. Mechanics Div.*, Transactions of the ASCE, **115**, 982-995.

Bowen, R.M (1976). Theory of mixtures. *Continuum Physics*, vol. 3, 1-127, A.C. Eringen ed., Academic Press, New York.

Bowen, R.M. and Reinicke, K.M. (1978). Plane progressive waves in a binary mixture of linear elastic materials. *J. of Applied Mechanics*, Transactions of the ASME, **45**, 493-499.

Bowen, R.M. (1982). Compressible porous media models by the use of the theory of mixtures. *Int. J. Eng. Science*, **20**, 697-735.

Bowen, R.M. and Lockett, R.R. (1983). Inertial effects in poroelasticity. *J. of Applied Mechanics*, Transactions of the ASME, **50**, 334-342.

Carroll, M.M. (1979). An effective stress law for anisotropic elastic deformation. *J. of Geophysical Research*, **84**(B13), 7510-7512.

Chen, P.J. (1973). Growth and decay of waves in solids. *Mechanics of Solids*, vol. 3, 303-401, C. Truesdell ed., Springer-Verlag, Berlin-Heidelberg, Germany.

Cho, H. and Barber, J.R. (1999). Stability of the three-dimensional Coulomb friction law. *Proc. R. Soc. Lond.*, A-**455**(1983), 839.

Dowaikh, M.A. and Ogden, R.W. (1990). On surface waves and deformations in a pre-stressed incompressible elastic solid. *IMA J. Appl. Math.*, **44**, 261-284.

Dowaikh, M.A. and Ogden, R.W. (1991). Interfacial waves and deformations in pre-stressed elastic media. *Proc. R. Soc. Lond.*, A-**433**, 313-328.

Eringen, A.C. and Ingram, J.D. (1965). A continuum theory for chemically reacting media -I. *Int. J. Eng. Science*, **3**, 197-212.

Green, A.E. and Naghdi, P.M. (1969). On basic equations for mixtures. *Quart. J. Mech. Appl. Math.*, **22**(4), 427-438.

Hilber, H.M., Hughes, T.J.R. and Taylor, R.M. (1977). Improved numerical dissipation for time integration algorithms in structural dynamics, *Earthquake Engineering and Structural Dynamics*, **5**, 283-292.

Hill, R. (1962). Acceleration waves in solids. *J. Mech. Phys. Solids*, **10**, 1-16.

Holmes, P.J. (1977). Bifurcation to divergence and flutter in flow-induced oscillations: a finite dimensional analysis. *J. of Sound and Vibrations*, Transactions of the ASME, **55**(4), 471-503.

Huseyin, K. (1978). *Vibrations and Stability of Multiple Parameter Systems*, Pitman, London, U.K.

Lade, P.V. and Musante, H.M. (1977). Failure conditions in sand and remolded clay. Proceedings of the 9th International Conference Soil Mechanics and Foundation Engineering, Tokyo, Japan, vol. 1, 181-186.

Loret, B. (1990). Acceleration waves in elastic-plastic porous media: interlacing and separation properties. *Int. J. Eng. Science*, **28**(12), 1315-1320.

Loret, B., Prévost J.H. and Harireche, O. (1990). Loss of hyperbolicity in elastic-plastic solids with deviatoric associativity. *Eur. J. Mechanics-A/Solids*, **9**(3), 225-231.

Loret, B. and Harireche, O. (1991). Acceleration waves, flutter instabilities and stationary discontinuities in inelastic porous media. *J. Mech. Phys. Solids*, **39**(5), 569-606.

Loret, B. and Prévost, J.H. (1991). Dynamic strain localization in fluid-saturated porous media. *J. of Engng. Mechanics Div.*, Transactions of the ASCE, **117**(4), 907-922.

Loret, B. (1992). Does deviation from deviatoric associativity lead to the onset of flutter instability ?. *J. Mech. Phys. Solids*, **40**(6), 1363-1375.

Loret, B., Martins, J.A.C. and Simões, F.M.F. (1995). Surface boundary conditions trigger flutter instability in non-associative elastic-plastic solids, *Int. J. Solids Structures*, **32**(15), 2155-2190.

Loret, B., Simões, F.M.F. and Martins, J.A.C. (1997). Growth and decay of acceleration waves in non-associative elastic-plastic fluid-saturated porous media. *Int. J. Solids Structures*, **34**(13), 1583-1608.

Loret, B. and Rizzi, E. (1997). Qualitative analysis of strain localization. Part II: Transversely isotropic elasticity and plasticity. *Int. J. Plasticity*, **13**(5), 501-519.

Mandel, J. (1962). Ondes plastiques dans un milieu indéfini à trois dimensions. *J. de Mécanique*, **1**, 3-30.

Mandel, J. (1964a). Propagation des surfaces de discontinuité dans un milieu élasto-plastique, IUTAM Symposium *Stress-waves in anelastic solids*, Providence, H. Kolsky and W. Prager eds., Springer-Verlag, Berlin, Germany, 331-340.

Mandel, J. (1964b). Conditions de stabilité et postulat de Drucker, IUTAM Symposium *Rheology and Soil Mechanics*, Grenoble, G. Kravchenko and P. Sirieys eds., Springer-Verlag, Berlin, Germany, 58-68.

Martins, J.A.C. and Faria, L.O. (1991). Friction-induced surface instabilities and oscillations in nonlinear elasticity. *Rencontres Scientifiques du Cinquantenaire*, Rapport interne 124 du Laboratoire de Mécanique et Acoustique, Marseille, France, 31-50.

Martins, J.A.C., Faria, L.O. and Guimarães, J. (1992). Dynamic surface solutions in linear elasticity with frictional boundary conditions. *Friction-Induced Vibration, Chatter, Squeal and Chaos*, DE-vol. 49, R.A. Ibrahim and A. Soom eds., The American Society of Mechanical Engineers, New York, USA, 33-39.

Martins, J.A.C., Faria, L.O. and Guimarães, J. (1995). Dynamic surface solutions in linear elasticity and viscoelasticity with frictional boundary conditions. *J. of Vibration and Acoustics*, Transactions of the ASME, **117**, 445-451.

Martins, J.A.C., Simões, F.M.F, Gastaldi, F. and Monteiro Marques, M. (1995). Dissipative graph solutions for a 2 degree-of-freedom quasistatic frictional contact problem. *Int. J. Eng. Science*, **33**, 1959-1986.

Martins, J.A.C., Barbarin, S., Raous, M. and Pinto da Costa, A. (1999). Dynamic stability of finite dimensional linearly elastic systems with unilateral contact and Coulomb friction. *Computer Methods in Applied Mechanics and Engineering*, **177**(3/4), 289-328.

Moirot, F. (1998). *Étude de la stabilité d'un équilibre en présence de frottement de Coulomb, application au crissement des freins à disque*, Thèse de Doctorat, École Polytechnique, Palaiseau, France.

Molenkamp, F. (1991). Material instability for drained and undrained behaviour -I. Shear-band generation. *Int. J. Num. Anal. Meth. Geomechanics*, **15**, 147-168; -II. Combined uniform deformation and shear-band generation. *Int. J. Num. Anal. Meth. Geomechanics*, **15**, 169-180.

Mühlhaus, H.B. and Aifantis, E. (1991). A variational principle for gradient plasticity. *Int. J. Solids Structures*, **28**, 845-857.

Needleman, A. and Ortiz, M. (1991). Effect of boundaries and interfaces on shear-band localization. *Int. J. Solids Structures*, **28**(7), 859-877.

Neilsen, M.K. and Schreyer, H.L. (1993). Bifurcations in elastic-plastic materials. *Int. J. Solids Structures*, **30**(4), 521-544.

Nguyen, Q.-S. (1994). Bifurcation and stability in dissipative media (plasticity, friction, fracture), Part 1. *Applied Mechanics Review*, Transactions of the ASME, **47**(1), 1-31.

Nur, A. and Byerlee, J.D. (1971). An exact effective stress law for elastic deformation of rock with fluids. *J. of Geophysical Research*, **76**(26), 6414-6419.

O'Malley, R.E., Jr. (1991). *Singular Perturbation Methods for Ordinary Differential Equations.* Applied Mathematical Sciences, 89, Springer-Verlag, New York.

Ottosen, N.S. and Runesson, K. (1991a). Acceleration waves in elastoplasticity. *Int. J. Solids Structures*, **28**, 135-159.

Ottosen, N.S. and Runesson, K. (1991b). Properties of discontinuous bifurcation solutions in elastoplasticity. *Int. J. Solids Structures*, **27**(4), 401-421.

Petryk, H. (1993). General theory of bifurcation and stability in time-independent plasticity. *Bifurcation and Stability of Dissipative Systems*, Q.S. Nguyen, ed., International Centre for Mechanical Sciences, Courses and Lectures, No. 327, Springer-Verlag, Wien, New York.

Rice, J.R. (1975). On the stability of dilatant hardening for saturated rock masses. *J. of Geophysical Research*, **80**, 1531-1536.

Rice, J.R. (1976). The localization of plastic deformation, *Theoretical and Applied Mechanics*, 14th IUTAM Congress, Delft, W.T. Koiter ed., North-Holland, Amsterdam, The Netherlands, 207-220.

Rice, J.R. and Rudnicki, J.W. (1980). A note on some features of the theory of localization of deformation. *Int. J. Solids Structures*, **16**, 597-605.

Rizzi, E. and Loret, B. (1997). Qualitative analysis of strain localization. Part I: Transversely isotropic elasticity and isotropic plasticity. *Int. J. Plasticity*, **13**(5), 461-499.

Rizzi, E. and Loret, B. (1999). Strain localization in fluid-saturated anisotropic elastic-plastic porous media. *Int. J. Eng. Science*, **37**, 235-251.

Rudnicki, J.W. and Rice, J.R. (1975) Conditions for the localization of deformations in pressure-sensitive dilatant materials. *J. Mech. Phys. Solids*, **23**, 371-394.

Schaeffer, D. G. (1990). Instability and ill-posedness in the deformation of granular materials. *Int. J. Num. Anal. Meth. Geomechanics*, **14**, 253-278.

Simo, J.C., and Taylor, R.L. (1985). Consistent tangent operators for rate-independent elastoplasticity. *Computer Methods in Applied Mechanics and Engineering*, **48**, 101-118.

Simo, J.C., and Hughes, T.J.R. (1987). General return mapping algorithms for rate-independent plasticity. *Constitutive Laws for Engineering Materials: Theory and Applications*, C.S. Desai et al. eds., Elsevier Science, Amsterdam, The Netherlands, 221-231.

Simões, F.M.F. (1997). Instabilidades em problemas não associados da Mecânica dos Sólidos. Ph. D. dissertation, 12/12/1997, Universidade Técnica de Lisboa, Portugal.

Simões, F.M.F. and Martins, J.A.C. (1998). Instability and ill-posedness in some friction problems. *Int. J. Eng. Science*, **36**, 1265-1293.

Simões, F.M.F., Martins, J.A.C. and Loret, B. (1999). Instabilities in elastic-plastic fluid-saturated porous media: harmonic wave versus acceleration wave analyses. *Int. J. Solids Structures*, **36**, 1277-1295.

Stoer, J. and Bulirsch, R. (1980). *Introduction to numerical analysis*, Springer-Verlag, New York, USA.

Suo, Z., Ortiz, M. and Needleman, A. (1992). Stability of solids with interfaces. *J. Mech. Phys. Solids*, **40**, 613-640.

Truesdell, C. and Toupin, R. (1960). *The Classical Field Theories*, *Encyclopedia of Physics*, vol. III/1, S. Flügge ed., Springer-Verlag, Berlin, Germany.

Truesdell, C. and Noll, W. (1965). *The Non-Linear Field Theories of Mechanics, Encyclopedia of Physics*, vol. III/3, S. Flügge ed., Springer-Verlag, Berlin, Germany.

Vermeer, P.A. and Brinkgreve, R.B.J. (1994). A new effective non-local strain-measure for softening plasticity. *Localisation and Bifurcation Theory for Soils and Rocks*, R. Chambon, J. Desrues and I. Vardoulakis eds., Balkema, Amsterdam, The Netherlands, 89-100.

Vidyasagar, M. (1993). *Nonlinear Systems Analysis*, Prentice-Hall International Editions, Englewood Cliffs, New Jersey.

Wilkinson, J.H. (1965). *The algebraic eigenvalue problem*, Oxford University Press, Oxford.

Ziegler, H. (1968). *Principles of Structural Stability*, Blaisdell, Waltham, Massachussets, USA.

ELASTIC AND PSEUDO-ELASTIC INSTABILITY AND BIFURCATION

R.W. Ogden
University of Glasgow, Glasgow, UK

Abstract. This article is concerned with the analysis of stability, loss of stability and the associated bifurcation phenomena in elastic and pseudo-elastic solids subject to large deformations. The basic equations of nonlinear elasticity are summarized prior to a discussion of the linearized incremental equations, which are used as a basis for the analysis of (linearized) stability with particular reference to the dead-load boundary-value problem as a key exemplar. Both compressible and incompressible materials are discussed. Global aspects of the general problem are then considered from the viewpoint of inversion of stress-deformation relationships and the theory is applied to two representative problems involving homogeneous deformation. An illustrative example in which bifurcation from a homogeneous state of deformation (namely, simple shear of a plate) is associated with inhomogeneous incremental modes of deformation is then examined in detail. Next, a new theory of *pseudo-elasticity* is discussed. This theory enables changes in the (elastic) constitutive law associated with damage to be invoked. A general form of the theory is described, and aspects of local and global stability associated with the theory are discussed. The theory is then exemplified using a specific model in the context of isotropic material response. The model, which is associated with a discontinuous change in material properties, is applied to the problem of inflation and deflation of a thick-walled spherical shell of incompressible material. Specific results are obtained for the residual deformation and stress distributions remaining after removal of the inflating pressure.

1 The Basic Equations of Nonlinear Elasticity

1.1 Deformation Geometry

In Section 1 and parts of Sections 2 and 3 we follow closely the basic development of the theory as given by Ogden (1991). We consider a continuous body which occupies a connected open subset of a three-dimensional Euclidean point space, and we refer to such a subset as a *configuration* of the body. We identify an arbitrary configuration as a *reference configuration* and denote this by $\mathcal{B}_r$. Let points in $\mathcal{B}_r$ be labelled by their position vectors $\mathbf{X}$ relative to an arbitrarily chosen origin and let $\partial\mathcal{B}_r$ denote the boundary of $\mathcal{B}_r$. Now suppose that the body is deformed quasi-statically from $\mathcal{B}_r$ so that it occupies a new configuration, $\mathcal{B}$ say, with boundary $\partial\mathcal{B}$. We refer to $\mathcal{B}$ as the *current* or *deformed configuration* of the body. The deformation is represented by the mapping $\chi : \mathcal{B}_r \to \mathcal{B}$ which takes points $\mathbf{X}$ in $\mathcal{B}_r$ to points $\mathbf{x}$ in $\mathcal{B}$. Thus,

$$\mathbf{x} = \chi(\mathbf{X}), \qquad \mathbf{X} \in \mathcal{B}_r, \tag{1}$$

where $\mathbf{x}$ is the position vector of the point $\mathbf{X}$ in $\mathcal{B}$. The mapping χ is called the *deformation* from $\mathcal{B}_r$ to $\mathcal{B}$. We require χ to be one-to-one and to satisfy appropriate regularity conditions. Here, it suffices to take χ to be twice continuously differentiable.

For simplicity we consider only Cartesian coordinate systems and let $\mathbf{X}$ and $\mathbf{x}$ respectively have coordinates X_α and x_i, where $\alpha,\ i \in \{1,2,3\}$, so that $x_i = \chi_i(X_\alpha)$. Greek and Roman indices refer respectively to $\mathcal{B}_r$ and $\mathcal{B}$ and the usual summation convention for repeated indices is used.

The *deformation gradient tensor*, denoted $\mathbf{F}$, is given by

$$\mathbf{F} = \text{Grad}\,\mathbf{x} \tag{2}$$

and has Cartesian components $F_{i\alpha} = \partial x_i/\partial X_\alpha$, Grad being the gradient operator in $\mathcal{B}_r$. Local invertibility of $\boldsymbol{\chi}$ requires that $\mathbf{F}$ be non-singular, and we adopt the usual convention that $\det \mathbf{F} > 0$ and the notation

$$J = \det \mathbf{F}. \tag{3}$$

The *polar decompositions*

$$\mathbf{F} = \mathbf{RU} = \mathbf{VR}, \tag{4}$$

where $\mathbf{R}$ is proper orthogonal and $\mathbf{U}$, $\mathbf{V}$ are positive definite and symmetric, are unique. Respectively, $\mathbf{U}$ and $\mathbf{V}$ are called the *right* and *left stretch tensors*.

We also have the *spectral decomposition*

$$\mathbf{U} = \sum_{i=1}^{3} \lambda_i \mathbf{u}^{(i)} \otimes \mathbf{u}^{(i)}, \tag{5}$$

where $\lambda_i > 0,\ i \in \{1,2,3\}$, are the *principal stretches*, $\mathbf{u}^{(i)}$, the (unit) eigenvectors of $\mathbf{U}$, are called the *Lagrangian principal axes* and $\otimes$ denotes the tensor product. Similarly, $\mathbf{V}$ has the spectral decomposition

$$\mathbf{V} = \sum_{i=1}^{3} \lambda_i \mathbf{v}^{(i)} \otimes \mathbf{v}^{(i)}, \tag{6}$$

where $\mathbf{v}^{(i)} = \mathbf{R}\mathbf{u}^{(i)}$, $i \in \{1,2,3\}$. It follows from (3)–(5) that

$$J = \lambda_1\lambda_2\lambda_3. \tag{7}$$

Let $\mathrm{d}\mathbf{A} \equiv \mathbf{N}\mathrm{d}A$ denote a vector surface area element on $\partial\mathcal{B}_r$, where $\mathbf{N}$ is the unit outward normal to the surface, and $\mathrm{d}\mathbf{a} \equiv \mathbf{n}\mathrm{d}a$ the corresponding area element on $\partial\mathcal{B}$. Then, the area elements are connected according to *Nanson's formula*

$$\mathbf{n}\mathrm{d}a = J\mathbf{F}^{-T}\mathbf{N}\mathrm{d}A, \tag{8}$$

where $\mathbf{F}^{-T} = (\mathbf{F}^{-1})^T$ and ^{T} denotes the transpose. If $\mathrm{d}V$ and $\mathrm{d}v$ denote volume elements in $\mathcal{B}_r$ and $\mathcal{B}$ respectively then we also have

$$\mathrm{d}v = J\mathrm{d}V. \tag{9}$$

The identity

$$\mathrm{Div}\,(J\mathbf{F}^{-1}) = \mathbf{0} \tag{10}$$

plays a useful role in transformations between equations associated with the reference and current configurations and can readily be established by integrating (8) over an arbitrary closed surface and applying the divergence theorem. Here, Div is the divergence operator in $\mathcal{B}_r$.

For an *isochoric* (volume preserving) deformation we have

$$J \equiv \det \mathbf{F} \equiv \lambda_1\lambda_2\lambda_3 = 1. \tag{11}$$

The *mass conservation equation* has the form

$$\rho_r = \rho J, \tag{12}$$

where ρ_r and ρ are the *mass densities* in $\mathcal{B}_r$ and $\mathcal{B}$ respectively.

1.2 Stress Tensors and Equilibrium Equations

The surface force per unit area (or *stress vector*) on the vector area element $\mathrm{d}\mathbf{a}$ is denoted by $\mathbf{t}$. It depends on $\mathbf{n}$ according to the formula

$$\mathbf{t} = \boldsymbol{\sigma}^T\mathbf{n}, \tag{13}$$

where $\boldsymbol{\sigma}$, a second-order tensor independent of $\mathbf{n}$, is called the *Cauchy stress tensor*.

By means of (8) the force on $\mathrm{d}\mathbf{a}$ may be written as

$$\mathbf{t}\mathrm{d}a = \mathbf{S}^T\mathbf{N}\mathrm{d}A, \tag{14}$$

where the *nominal stress tensor* $\mathbf{S}$ is related to $\boldsymbol{\sigma}$ by $\mathbf{S} = J\mathbf{F}^{-1}\boldsymbol{\sigma}$. Note that $\mathbf{S}^T$ is often used instead of $\mathbf{S}$ and is called the *first Piola-Kirchhoff stress tensor*.

Let $\mathbf{b}$ denote the body force per unit mass. Then the *equilibrium equation* may be written

$$\mathrm{div}\,\boldsymbol{\sigma} + \rho\mathbf{b} = \mathbf{0}, \tag{15}$$

or, equivalently,

$$\mathrm{Div}\,\mathbf{S} + \rho_r\mathbf{b} = \mathbf{0}, \tag{16}$$

where div and Div denote the divergence operators in $\mathcal{B}$ and $\mathcal{B}_r$ respectively. Note that on use of (10) and (12) equation (16) may be converted immediately to (15). In components, (16) has the form

$$\frac{\partial S_{\alpha i}}{\partial X_\alpha} + \rho_r b_i = 0, \tag{17}$$

and similarly for (15), where $S_{\alpha i}$ are the components of $\mathbf{S}$ and b_i those of $\mathbf{b}$.

Balance of the moments of the forces acting on the body yields simply

$$\boldsymbol{\sigma}^T = \boldsymbol{\sigma}, \tag{18}$$

which may also be expressed as $\mathbf{S}^T\mathbf{F}^T = \mathbf{F}\mathbf{S}$.

For most purposes in this article the Lagrangian formulation based on the use of $\mathbf{S}$ and equation (16), with $\mathbf{X}$ as the independent variable, is preferred to the Eulerian formulation based on use of $\boldsymbol{\sigma}$ and equation (15) with $\mathbf{x}$ as the independent variable. However, the Eulerian formulation is sometimes easier to use when considering incremental deformations relative to a finitely deformed configuration, as we shall see in Section 4.

1.3 Constitutive Law for an Elastic Material

We motivate the definition of elasticity by considering the work done by the surface and body forces in a virtual displacement $\delta\mathbf{x}$ from the current configuration $\mathcal{B}$. By using the divergence theorem and equation (16) we obtain the *virtual work* equation

$$\int_{\mathcal{B}_r} \rho_r \mathbf{b}\cdot\delta\mathbf{x}\,\mathrm{d}V + \int_{\partial\mathcal{B}_r} (\mathbf{S}^T\mathbf{N})\cdot\delta\mathbf{x}\,\mathrm{d}A = \int_{\mathcal{B}_r} \mathrm{tr}\,(\mathbf{S}\delta\mathbf{F})\,\mathrm{d}V, \tag{19}$$

where the left-hand side of (19) represents the virtual work of the body and surface forces, with $\cdot$ signifying the scalar product of two vectors, and in the integrand on the right-hand side tr denotes the trace of a second-order tensor. This work is converted into stored energy if there exists a scalar function, $W = W(\mathbf{F})$ say, defined on the set of deformation gradients, such that $\delta W \approx \mathrm{tr}\,(\mathbf{S}\delta\mathbf{F})$. It follows that

$$\mathbf{S} = \frac{\partial W}{\partial \mathbf{F}}, \tag{20}$$

or, in components, $S_{\alpha i} = \partial W/\partial F_{i\alpha}$. This is the *stress-deformation relation* or *constitutive relation* for an elastic material which possesses a *strain-energy function* W, W being defined per unit volume in $\mathcal{B}_r$.

We take an elastic material to be characterized by the existence of a strain-energy function such that (20) holds. At this point there is no restriction on the form of the function W, but there are several factors limiting the class of functions that are acceptable for the description of the elastic behaviour of materials, as will be discussed in the following sections. First, however, we examine how (20) is modified if the material is subject to the *incompressibility constraint*, so that (11) holds at each point $\mathbf{X} \in \mathcal{B}_r$. By adding to W the constraint term $-p\,(J-1)$ and using the fact that $\partial J/\partial\mathbf{F} = J\mathbf{F}^{-1}$ we see that (20) is modified to give

$$\mathbf{S} = \frac{\partial W}{\partial \mathbf{F}} - p\mathbf{F}^{-1}, \qquad \det\mathbf{F} = 1, \tag{21}$$

where p is a Lagrange multiplier associated with the constraint; p is often referred to as an arbitrary hydrostatic pressure. Equation (21) is the *stress-deformation relation for an incompressible elastic material.*

Alternatively, in terms of the Cauchy stress tensor, (21) becomes

$$\boldsymbol{\sigma} = \mathbf{F}\frac{\partial W}{\partial \mathbf{F}} - p\mathbf{I}, \qquad \det\mathbf{F} = 1, \tag{22}$$

where $\mathbf{I}$ is the identity tensor.

1.4 Objectivity and Material Symmetry

Now consider a rigid deformation superimposed on the deformation (1); this is defined by the transformation $\mathbf{x} \to \mathbf{x}' = \mathbf{Q}\mathbf{x} + \mathbf{c}$, where $\mathbf{Q}$ is a rotation tensor and $\mathbf{c}$ is a translation, both $\mathbf{Q}$ and $\mathbf{c}$ being independent of $\mathbf{x}$. The resulting deformation gradient (relative to $\mathcal{B}_r$) is $\mathbf{F}' = \mathbf{QF}$. We require the elastic stored energy to be independent of such superimposed deformations and it therefore follows that

$$W(\mathbf{QF}) = W(\mathbf{F}) \tag{23}$$

for *all* rotations $\mathbf{Q}$. A strain-energy function satisfying this requirement is said to be *objective* (the term *material frame indifferent* is also used).

Use of the polar decomposition (4) and the choice $\mathbf{Q} = \mathbf{R}^T$ in (23) shows that

$$W(\mathbf{F}) = W(\mathbf{U}). \tag{24}$$

Thus, W depends on $\mathbf{F}$ only through the stretch tensor $\mathbf{U}$ and may therefore be defined on the class of positive definite symmetric tensors. This motivates the introduction of the symmetric stress tensor $\mathbf{T}$ defined by

$$\mathbf{T} = \frac{\partial W}{\partial \mathbf{U}} \tag{25}$$

for an unconstrained material and by

$$\mathbf{T} = \frac{\partial W}{\partial \mathbf{U}} - p\mathbf{U}^{-1}, \qquad \det \mathbf{U} = 1 \tag{26}$$

for an incompressible material. This is called the *Biot stress tensor* (Biot, 1965). In general, $\mathbf{T}$ is related to $\mathbf{S}$ through

$$\mathbf{T} = \frac{1}{2}(\mathbf{SR} + \mathbf{R}^T\mathbf{S}^T), \tag{27}$$

where $\mathbf{R}$ is the rotation tensor arising in the polar decomposition (4).

Further restrictions on the form of W are obtained if the material possesses symmetries in the reference configuration. A material possesses a symmetry if its response is unchanged after a change of reference configuration other than a pure dilatation. Suppose that the reference configuration $\mathcal{B}_r$ is changed to $\mathcal{B}_r'$ so that $\mathbf{X}$ changes to $\mathbf{X}'$. We denote by $\mathbf{P}$ the deformation gradient $\operatorname{Grad}\mathbf{X}'$ of $\mathcal{B}_r'$ relative to $\mathcal{B}_r$ and by $\mathbf{F}'$ the deformation gradient of $\mathcal{B}$ relative to $\mathcal{B}_r'$. Then, we have $\mathbf{F} = \mathbf{F}'\mathbf{P}$. In general, the strain-energy function of the material relative to $\mathcal{B}_r'$, W' say, is different from that relative to $\mathcal{B}_r$. If, however, the change of reference configuration just described does not affect the material response then we write W also for the strain-energy function relative to $\mathcal{B}_r'$, and hence

$$W'(\mathbf{F}) = W'(\mathbf{F}'\mathbf{P}) = W(\mathbf{F}'\mathbf{P}) = W(\mathbf{F}') \tag{28}$$

for *all* deformation gradients $\mathbf{F}'$.

The set of $\mathbf{P}$ for which the right-hand equation in (28) holds forms a group; this is the *symmetry group of the material relative to* $\mathcal{B}_r$.

To be specific we confine attention to *isotropic elastic materials*, for which the symmetry group is the *proper orthogonal group*. Then, we have

$$W(\mathbf{FQ}) = W(\mathbf{F}) \tag{29}$$

for *all* rotations $\mathbf{Q}$. The combination of (23) and (29) then yields

$$W(\mathbf{QUQ}^T) = W(\mathbf{U}) \tag{30}$$

for all rotations $\mathbf{Q}$; note that the rotations in (23) and (29) are independent. Equation (30) states that W is an *isotropic function* of $\mathbf{U}$. It follows from the spectral decomposition (5) that W depends on $\mathbf{U}$ only through the principal stretches $\lambda_1, \lambda_2, \lambda_3$. To avoid introducing additional notation we express this dependence as $W(\lambda_1, \lambda_2, \lambda_3)$; by selecting appropriate values for $\mathbf{Q}$ in (30) we may deduce that W depends symmetrically on $\lambda_1, \lambda_2, \lambda_3$, i.e.

$$W(\lambda_1, \lambda_2, \lambda_3) = W(\lambda_1, \lambda_3, \lambda_2) = W(\lambda_2, \lambda_1, \lambda_3). \tag{31}$$

A consequence of isotropy is that $\mathbf{T}$ is *coaxial* with $\mathbf{U}$ and hence, in parallel with (5), we have

$$\mathbf{T} = \sum_{i=1}^{3} t_i \mathbf{u}^{(i)} \otimes \mathbf{u}^{(i)}, \tag{32}$$

where t_i, $i \in \{1, 2, 3\}$ are the *principal Biot stresses*. For an unconstrained material,

$$t_i = \frac{\partial W}{\partial \lambda_i}, \tag{33}$$

and for an incompressible material this is replaced by

$$t_i = \frac{\partial W}{\partial \lambda_i} - p\lambda_i^{-1}, \qquad \lambda_1\lambda_2\lambda_3 = 1. \tag{34}$$

For later reference, we note here that the principal Cauchy stresses σ_i, $i \in \{1, 2, 3\}$, are given by

$$J\sigma_i = \lambda_i \frac{\partial W}{\partial \lambda_i} \tag{35}$$

and

$$\sigma_i = \lambda_i \frac{\partial W}{\partial \lambda_i} - p, \qquad \lambda_1\lambda_2\lambda_3 = 1 \tag{36}$$

for unconstrained and incompressible materials respectively. Note that in (35) and (36) there is no summation over the repeated index i.

We emphasize that, *for an isotropic elastic material*, $\mathbf{SR}$ *is symmetric* and hence the connection (27) reduces to

$$\mathbf{S} = \mathbf{TR}^T = \sum_{i=1}^{3} t_i \mathbf{u}^{(i)} \otimes \mathbf{v}^{(i)}. \tag{37}$$

The first equation in (37) is a polar decomposition of $\mathbf{S}$ analogous to $(4)_1$ except that it is not unique. This equation has important consequences for considerations of uniqueness and stability and will be discussed in detail in Section 3.

1.5 Boundary-Value Problems

We now take the body force in (19) to be conservative so that we may write

$$\mathbf{b} = -\operatorname{grad}\phi, \tag{38}$$

where ϕ is a scalar field defined on points in $\mathcal{B}$ and grad denotes the gradient operator in $\mathcal{B}$. Since we are considering an elastic material with strain-energy function W, it follows that (19) can be expressed in the form

$$\delta \int_{\mathcal{B}_r} (W + \rho_r\phi)\,\mathrm{d}V - \int_{\partial\mathcal{B}_r} (\mathbf{S}^T\mathbf{N}) \cdot \delta\mathbf{x}\,\mathrm{d}A = 0. \tag{39}$$

At this point we consider the equilibrium equation (16) together with the stress-deformation relation (20) for an unconstrained material, and the deformation gradient (2) coupled with (1):

$$\operatorname{Div}\left(\frac{\partial W}{\partial \mathbf{F}}\right) + \rho_r\mathbf{b} = \mathbf{0}, \quad \mathbf{F} = \operatorname{Grad}\mathbf{x}, \quad \mathbf{x} = \boldsymbol{\chi}(\mathbf{X}), \qquad \mathbf{X} \in \mathcal{B}_r. \tag{40}$$

We supplement (40) with boundary conditions, typical of those arising in problems of nonlinear elasticity, in which $\mathbf{x}$ is specified on part of the boundary, $\partial\mathcal{B}_r^x \subset \partial\mathcal{B}_r$ say, and the stress vector on the remainder, $\partial\mathcal{B}_r^s$, so that $\partial\mathcal{B}_r^x \cup \partial\mathcal{B}_r^s = \partial\mathcal{B}_r$ and $\partial\mathcal{B}_r^x \cap \partial\mathcal{B}_r^s = \emptyset$. We write

$$\mathbf{x} = \boldsymbol{\xi}(\mathbf{X}) \quad \text{on } \partial\mathcal{B}_r^x, \tag{41}$$

$$\mathbf{S}^T\mathbf{N} = \mathbf{s}(\mathbf{X}) \quad \text{on } \partial\mathcal{B}_r^s, \tag{42}$$

where $\boldsymbol{\xi}$ and $\mathbf{s}$ are specified functions. More generally, $\mathbf{s}$ can be allowed to depend on the deformation but, for simplicity, we restrict attention here to the dependence expressed in (42). The surface stress defined by (42) is referred to as a *dead-load traction*. The basic boundary-value problem of nonlinear elasticity is characterized by (40)–(42). In order to analyze this problem additional information about the nature of the function W is needed; this important aspect of the theory is not addressed to any great extent in this article. Reference can be made to, for example, Ciarlet (1988) or Ogden (1997) for discussion of this matter.

In view of the boundary condition (41) we now have $\delta\mathbf{x} = \mathbf{0}$ on $\partial\mathcal{B}_r^x$, and (39) becomes

$$\delta \int_{\mathcal{B}_r} (W + \rho_r\phi)\,\mathrm{d}V - \int_{\partial\mathcal{B}_r^s} \mathbf{s} \cdot \delta\mathbf{x}\,\mathrm{d}A = 0. \tag{43}$$

Since $\mathbf{s}$ is independent of the deformation, we write $\mathbf{s} = \operatorname{grad}(\mathbf{s}\cdot\mathbf{x})$, and (43) then becomes

$$\delta\left\{\int_{\mathcal{B}_r} (W + \rho_r\phi)\,\mathrm{d}V - \int_{\partial\mathcal{B}_r^s} \mathbf{s} \cdot \mathbf{x}\,\mathrm{d}A\right\} = 0. \tag{44}$$

If we interpret $\delta\boldsymbol{\chi}$ as a variation of the function $\boldsymbol{\chi}$, then (44) provides a variational formulation of the boundary-value problem (40)–(42) and can be written $\delta E = 0$, where E is the *functional* defined by

$$E\{\boldsymbol{\chi}\} = \int_{\mathcal{B}_r} \{W(\operatorname{Grad}\boldsymbol{\chi}) + \rho_r\phi(\boldsymbol{\chi})\}\,\mathrm{d}V - \int_{\partial\mathcal{B}_r^s} \mathbf{s} \cdot \boldsymbol{\chi}\,\mathrm{d}A. \tag{45}$$

In (45) χ is taken in some appropriate class of mappings (for example, the twice continuously differentiable mappings we have agreed to consider in this article) and to satisfy the boundary condition (41). Similarly for the admissible variations $\delta\chi$ subject to $\delta\chi = \mathbf{0}$ on $\partial\mathcal{B}_r^x$.

Then, starting from (45) with χ in this class, we obtain, after use of the divergence theorem and the boundary conditions on $\partial\mathcal{B}_r^x$,

$$\delta E = -\int_{\mathcal{B}_r} (\operatorname{Div}\mathbf{S} + \rho_r\mathbf{b}) \cdot \delta\chi \, \mathrm{d}V + \int_{\partial\mathcal{B}_r^s} (\mathbf{S}^T\mathbf{N} - \mathbf{s}) \cdot \delta\chi \, \mathrm{d}A.$$

The variational statement then takes the form $\delta E = 0$ for all admissible $\delta\chi$ if and only if χ is such that, with $\mathbf{F} = \operatorname{Grad}\chi$ and $\mathbf{S}$ given by (20) (for the unconstrained case), the equilibrium equations (16) and the boundary conditions (42) are satisfied. In other words, E is stationary if and only if χ is an actual solution of the boundary-value problem. In general the solution may not be unique.

A similar variational statement can be formulated if the boundary traction in (42) is replaced by a hydrostatic pressure, P say, so that $\boldsymbol{\sigma}\mathbf{n} = -P\mathbf{n}$, in which case $\mathbf{s}$ depends on the deformation in the form

$$\mathbf{s} = -JP\mathbf{F}^{-T}\mathbf{N} \quad \text{on } \partial\mathcal{B}_r^s. \tag{46}$$

As we shall see in Section 2, the energy functional plays an important role in the analysis of stability and bifurcation. Detailed discussion of variational principles in the context of nonlinear elasticity can be found in Ogden (1997); technical mathematical aspects are discussed in Ciarlet (1988), Marsden and Hughes (1983) and the papers by Ball (1977, 1982), for example.

In components, the equation in (40) can be written

$$\mathcal{A}^1_{\alpha i\beta j}\frac{\partial^2 x_j}{\partial X_\alpha \partial X_\beta} + \rho_r b_i = 0, \tag{47}$$

for $i \in \{1, 2, 3\}$, where the coefficients $\mathcal{A}^1_{\alpha i\beta j}$ are defined by

$$\mathcal{A}^1_{\alpha i\beta j} = \frac{\partial^2 W}{\partial F_{i\alpha}\partial F_{j\beta}}. \tag{48}$$

When coupled with suitable boundary conditions, equation (47) forms a system of quasi-linear partial differential equations for $x_i = \chi_i(X_\alpha)$. The coefficients $\mathcal{A}^1_{\alpha i\beta j}$ are, in general, nonlinear functions of the components of the deformation gradient. Only a very limited number of boundary-value problems have been solved for unconstrained materials, and only then for very special choices of the form of W; some of these are discussed in Ogden (1997), for example. For incompressible materials the corresponding equations, obtained by substituting (21) into (16), have yielded more success, and again we refer to Ogden (1997) for relevant citations. We do not discuss such solutions here since our purpose is to examine certain aspects of the structure of the equations.

Equations (47) are said to be *strongly elliptic* if the inequality

$$\mathcal{A}^1_{\alpha i\beta j} m_i m_j N_\alpha N_\beta > 0 \tag{49}$$

holds for all non-zero vectors $\mathbf{m}$ and $\mathbf{N}$.

For an incompressible material, the functional E in (45) is modified by the addition of a constraint term $-p\,[\det(\operatorname{Grad}\chi) - 1]$ to the integrand in the integral over $\mathcal{B}_r$. The counterpart of (47) is then obtained by substitution of (21) into (16). On use of (10) with $J = 1$ this gives

$$\mathcal{A}^1_{\alpha i\beta j}\frac{\partial^2 x_j}{\partial X_\alpha \partial X_\beta} - \frac{\partial p}{\partial x_i} + \rho_r b_i = 0 \tag{50}$$

subject to (11), where the coefficients are again given by (48). The strong ellipticity condition associated with (50) remains in the form (49) but the incompressibility constraint now imposes the restriction

$$\mathbf{m}\cdot\mathbf{n} = 0, \tag{51}$$

on the (non-zero) vectors $\mathbf{m}$ and $\mathbf{n}$, where $\mathbf{n}$ is defined by

$$\mathbf{F}^T\mathbf{n} = \mathbf{N}. \tag{52}$$

Note that $\mathbf{N}$ and $\mathbf{n}$ here are not related to the surface normal vectors defined in Section 1.1.

2 Incremental Deformations and Stability

2.1 Incremental Equations

Let a solution χ to the boundary-value problem (40)–(42) be known and consider the problem of finding solutions near to χ when the boundary conditions are perturbed. Let χ' be a solution for the perturbed problem and write $\mathbf{x}' = \chi'(\mathbf{X})$. Also, we write

$$\dot{\mathbf{x}} = \mathbf{x}' - \mathbf{x} = \chi'(\mathbf{X}) - \chi(\mathbf{X}) \equiv \dot{\chi}(\mathbf{X}) \tag{53}$$

for the difference of the solutions, so that

$$\operatorname{Grad}\dot{\chi} = \operatorname{Grad}\chi' - \operatorname{Grad}\chi = \dot{\mathbf{F}}. \tag{54}$$

In the above and henceforth a superposed dot represents the difference $(\dot{\ }) = (\ )' - (\ )$, while $\delta(\ )$ represents a variation of $(\ )$. Clearly, $\dot{\mathbf{F}}$ is linear in $\dot{\chi}$.

The nominal stress difference is

$$\dot{\mathbf{S}} = \mathbf{S}' - \mathbf{S} = \frac{\partial W}{\partial \mathbf{F}}(\mathbf{F}') - \frac{\partial W}{\partial \mathbf{F}}(\mathbf{F}), \tag{55}$$

which, when $\dot{\mathbf{F}}$ is small (in some appropriate sense), can be approximated as

$$\dot{\mathbf{S}} = \mathcal{A}^1\dot{\mathbf{F}} + \frac{1}{2}\mathcal{A}^2[\dot{\mathbf{F}},\dot{\mathbf{F}}] + \frac{1}{6}\mathcal{A}^3[\dot{\mathbf{F}},\dot{\mathbf{F}},\dot{\mathbf{F}}] + \ldots, \tag{56}$$

where

$$\mathcal{A}^1 = \frac{\partial^2 W}{\partial \mathbf{F}^2}, \quad \mathcal{A}^2 = \frac{\partial^3 W}{\partial \mathbf{F}^3}, \quad \mathcal{A}^3 = \frac{\partial^4 W}{\partial \mathbf{F}^4}. \tag{57}$$

In components, the first of the expressions (57) is given by (48) while the second and third can be written

$$\mathcal{A}^2_{\alpha i\beta j\gamma k} = \frac{\partial^3 W}{\partial F_{i\alpha}\partial F_{j\beta}\partial F_{k\gamma}}, \quad \mathcal{A}^3_{\alpha i\beta j\gamma k\delta l} = \frac{\partial^4 W}{\partial F_{i\alpha}\partial F_{j\beta}\partial F_{k\gamma}\partial F_{l\delta}}, \tag{58}$$

and similarly for higher-order terms if required (assuming that W is sufficiently regular). The component form of (56) is

$$\dot{S}_{\alpha i} = \mathcal{A}^1_{\alpha i\beta j}\dot{F}_{j\beta} + \frac{1}{2}\mathcal{A}^2_{\alpha i\beta j\gamma k}\dot{F}_{j\beta}\dot{F}_{k\gamma} + \frac{1}{6}\mathcal{A}^3_{\alpha i\beta j\gamma k\delta l}\dot{F}_{j\beta}\dot{F}_{k\gamma}\dot{F}_{l\delta} + \dots, \tag{59}$$

and this serves to define the products appearing in (56).

In the linear approximation to (55) only the first term in the expansion (56) is required, so that

$$\dot{\mathbf{S}} = \mathcal{A}^1\dot{\mathbf{F}}. \tag{60}$$

For an incompressible material it follows from (21) that this is modified to

$$\dot{\mathbf{S}} = \mathcal{A}^1\dot{\mathbf{F}} - \dot{p}\mathbf{F}^{-1} + p\mathbf{F}^{-1}\dot{\mathbf{F}}\mathbf{F}^{-1}, \tag{61}$$

whcrc $\dot{p}$ is the linearized incremental form of p, and this is coupled with the linearized incremental incompressibility condition

$$\operatorname{tr}(\dot{\mathbf{F}}\mathbf{F}^{-1}) = 0. \tag{62}$$

From the equilibrium equation (16) and its counterpart

$$\operatorname{Div}\mathbf{S}' + \rho_r\mathbf{b}' = \mathbf{0}$$

for χ', we obtain, by subtraction,

$$\operatorname{Div}\dot{\mathbf{S}} + \rho_r\dot{\mathbf{b}} = \mathbf{0}. \tag{63}$$

This is *exact*, but in the linear approximation $\dot{\mathbf{S}}$ is replaced by either (60), or (61) with (62), as appropriate and with $\dot{\mathbf{b}}$ linearized in $\dot{\chi}$.

Let $\boldsymbol{\xi}'$ and $\mathbf{s}'$ be the prescribed data for χ' so that

$$\mathbf{x}' = \boldsymbol{\xi}' \quad \text{on } \partial\mathcal{B}^x_r, \qquad \mathbf{S}'^T\mathbf{N} = \mathbf{s}' \quad \text{on } \partial\mathcal{B}^s_r, \tag{64}$$

and hence

$$\dot{\mathbf{x}} = \dot{\boldsymbol{\xi}} \quad \text{on } \partial\mathcal{B}^x_r, \qquad \dot{\mathbf{S}}^T\mathbf{N} = \dot{\mathbf{s}} \quad \text{on } \partial\mathcal{B}^s_r. \tag{65}$$

We now consider the linearized problem (63)–(65) with (60) for χ, with $\mathbf{s}$ linearized if it depends on the deformation. For definiteness and simplicity we now assume that there are no body forces and that $\mathbf{s}$ is a dead-load traction, as in (42); thus,

$$\operatorname{Div}\dot{\mathbf{S}} = \mathbf{0}, \quad \dot{\mathbf{S}} = \mathcal{A}^1\dot{\mathbf{F}}, \quad \dot{\mathbf{F}} = \operatorname{Grad}\dot{\chi}. \tag{66}$$

Taken together, equations (65)–(66) constitute the basic boundary-value problem of incremental elasticity for an unconstrained material when the underlying deformation χ is known. We next examine the question of uniqueness of solution of the incremental problem and the associated question of stability of the deformation χ. For definiteness we focus attention on unconstrained materials for purposes of the main theoretical development.

2.2 Uniqueness and Stability

Let $\dot{\chi}$ and $\dot{\chi}^*$ be two possible solutions of the (linearized) incremental problem and let

$$\Delta\dot{\chi} = \dot{\chi}^* - \dot{\chi} \tag{67}$$

denote their difference. More generally, we adopt the notation $\Delta(\cdot)$ to represent the difference $(\cdot)^* - (\cdot)$. Then

$$\Delta\dot{\mathbf{F}} = \operatorname{Grad}\Delta\dot{\chi}, \quad \Delta\dot{\mathbf{S}} = \boldsymbol{\mathcal{A}}^1\Delta\dot{\mathbf{F}}, \tag{68}$$

and hence

$$\operatorname{Div}\Delta\dot{\mathbf{S}} = \mathbf{0} \quad \text{in } \mathcal{B}_r, \tag{69}$$

$$\Delta\dot{\chi} = \mathbf{0} \quad \text{on } \partial\mathcal{B}_r^x, \tag{70}$$

$$\Delta\dot{\mathbf{S}}^T\mathbf{N} = \mathbf{0} \quad \text{on } \partial\mathcal{B}_r^s. \tag{71}$$

On use of the divergence theorem it follows that

$$\int_{\mathcal{B}_r} \operatorname{tr}(\Delta\dot{\mathbf{S}}\Delta\dot{\mathbf{F}})\,\mathrm{d}V = 0. \tag{72}$$

Clearly, a *sufficient condition* for uniqueness of solution of the incremental problem is

$$\int_{\mathcal{B}_r} \operatorname{tr}(\Delta\dot{\mathbf{S}}\Delta\dot{\mathbf{F}})\,\mathrm{d}V > 0 \tag{73}$$

for *all* $\Delta\dot{\chi} \not\equiv \mathbf{0}$ in $\mathcal{B}_r$ and satisfying $\Delta\dot{\chi} = \mathbf{0}$ on $\partial\mathcal{B}_r^x$, or, equivalently,

$$\int_{\mathcal{B}_r} \operatorname{tr}\{(\boldsymbol{\mathcal{A}}^1\Delta\dot{\mathbf{F}})\Delta\dot{\mathbf{F}}\}\,\mathrm{d}V > 0. \tag{74}$$

Note that this sufficient condition applies when the boundary conditions are inhomogeneous (i.e $\dot{\boldsymbol{\xi}} \neq \mathbf{0}, \dot{\mathbf{s}} \neq \mathbf{0}$ in (65)), so in considering incremental uniqueness it suffices henceforth to examine the homogeneous boundary-value problem

$$\operatorname{Div}\dot{\mathbf{S}} = \mathbf{0} \quad \text{in } \mathcal{B}_r, \tag{75}$$

$$\dot{\chi} = \mathbf{0} \quad \text{on } \partial\mathcal{B}_r^x, \tag{76}$$

$$\dot{\mathbf{S}}^T\mathbf{N} = \mathbf{0} \quad \text{on } \partial\mathcal{B}_r^s, \tag{77}$$

with (60) and (66). One solution of this is $\dot{\chi} = \mathbf{0}$. Thus, on setting $\dot{\chi}^* = \mathbf{0}$, the inequality (73) becomes

$$\int_{\mathcal{B}_r} \operatorname{tr}(\dot{\mathbf{S}}\dot{\mathbf{F}})\,\mathrm{d}V > 0, \tag{78}$$

and (74) similarly becomes

$$\int_{\mathcal{B}_r} \operatorname{tr}\{(\boldsymbol{\mathcal{A}}^1\dot{\mathbf{F}})\dot{\mathbf{F}}\}\,\mathrm{d}V > 0 \tag{79}$$

for all $\dot{\chi} \not\equiv \mathbf{0}$ in $\mathcal{B}_r$ satisfying $\dot{\chi} = \mathbf{0}$ on $\partial\mathcal{B}_r^x$.

If (79) holds in the configuration $\mathcal{B}$ then the only solution of the homogeneous incremental problem is the trivial solution $\dot{\chi} \equiv \mathbf{0}$, i.e. non-trivial solutions are excluded. The inequality (79), or equivalently (78), is therefore referred to as the *exclusion condition* (for the dead-load traction boundary condition). Correspondingly, the solution of the inhomogeneous problem is unique (assuming it exists). The exclusion condition requires modification if $\mathbf{s}$ is allowed to depend on the deformation.

Next, we consider the change in the energy functional due to the change in the deformation from χ to χ'. On use of the boundary condition (76) and the divergence theorem this is seen to be

$$E\{\chi'\} - E\{\chi\} = \int_{\mathcal{B}_r} \{W(\mathbf{F}') - W(\mathbf{F}) - \mathrm{tr}\,(\mathbf{S}\dot{\mathbf{F}})\}\,\mathrm{d}V. \tag{80}$$

By application of the Taylor expansion to $W(\mathbf{F}')$ this then becomes

$$E\{\chi'\} - E\{\chi\} = \frac{1}{2}\int_{\mathcal{B}_r} \mathrm{tr}\,\{(\mathcal{A}^1\dot{\mathbf{F}})\dot{\mathbf{F}} + \frac{1}{3}\mathcal{A}^2[\dot{\mathbf{F}},\dot{\mathbf{F}}]\dot{\mathbf{F}} + \frac{1}{12}\mathcal{A}^3[\dot{\mathbf{F}},\dot{\mathbf{F}},\dot{\mathbf{F}}]\dot{\mathbf{F}} + \ldots\}\,\mathrm{d}V, \tag{81}$$

where terms up to fourth order have been included for future reference. Thus, to the second order in $\dot{\mathbf{F}}$, (79) implies that $E\{\chi'\} > E\{\chi\}$ for all admissible $\dot{\chi} \not\equiv \mathbf{0}$ satisfying (76). This inequality states that χ is *locally stable* with respect to perturbations $\dot{\chi}$ from χ, and that χ is a *local minimizer* of the energy functional. This is why we take '> 0' rather than '< 0' in (79). For more discussion of local (*infinitesimal*) stability we refer to Truesdell and Noll (1965).

For the all-round dead-load problem, $\partial\mathcal{B}_r^x = \emptyset, \partial\mathcal{B}_r = \partial\mathcal{B}_r^s$. If the underlying deformation is homogeneous then $\mathbf{F}$ is independent of $\mathbf{X}$ and so, therefore, is $\mathcal{A}^1$. It then follows that (79) is equivalent to the local condition

$$\mathrm{tr}\,\{(\mathcal{A}^1\dot{\mathbf{F}})\dot{\mathbf{F}}\} > 0 \tag{82}$$

for all $\dot{\mathbf{F}} \neq \mathbf{0}$, i.e. $\mathcal{A}^1$ is positive definite at each point $\mathbf{X} \in \mathcal{B}_r$. As we shall make clear in Section 2.3, the inequality (82) *cannot* hold in all configurations.

In order to illustrate loss of uniqueness and stability we now concentrate attention on the dead-load traction boundary-value problem in which the deformation is homogeneous, so that (82) is the relevant exclusion condition.

2.3 Instability and Bifurcation Under Dead Load (Unconstrained Material)

Let us imagine a path of dead loading from the unstressed configuration in $\mathbf{S}$-space resulting in a path of deformation in $\mathbf{F}$-space along which $\mathcal{A}^1$ is positive definite, i.e. along which χ is locally (infinitesimally) stable so that (82) holds. Note, however, as we see shortly, that (82) does not hold at the origin $\mathbf{F} = \mathbf{I}$ in $\mathbf{F}$-space from which the path starts. This point is on the boundary of the region where (82) holds. The inequality first fails, if at all, when a point is reached where

$$\mathrm{tr}\,\{(\mathcal{A}^1\dot{\mathbf{F}})\dot{\mathbf{F}}\} \geq 0 \tag{83}$$

for all $\dot{\mathbf{F}}$, with equality holding for some $\dot{\mathbf{F}} \neq \mathbf{0}$. By the extremal property of quadratic forms it follows that

$$\boldsymbol{\mathcal{A}}^1 \dot{\mathbf{F}}_e = \mathbf{0} \tag{84}$$

for the $\dot{\mathbf{F}}$, $\dot{\mathbf{F}}_e$ say, which minimizes (83), the minimum being zero. Note, however, that in general there may be more than one $\dot{\mathbf{F}}_e$ satisfying (84). Points in $\mathbf{F}$-space where (84) holds are given by

$$\det(\boldsymbol{\mathcal{A}}^1) = 0, \tag{85}$$

where $\boldsymbol{\mathcal{A}}^1$ is regarded as a linear mapping on the vector space of second-order tensors $\dot{\mathbf{F}}$. Equation (85) describes a hypersurface in $\mathbf{F}$-space and is the *stability limit* or *bifurcation surface* since, on this surface,

$$\dot{\mathbf{S}}_e \equiv \boldsymbol{\mathcal{A}}^1 \dot{\mathbf{F}}_e = \mathbf{0}, \tag{86}$$

wherein the notation $\dot{\mathbf{S}}_e$ is defined.

In the terminology of Hill (1957, 1978), $\dot{\mathbf{F}}_e$ is referred to as an *eigenmode* (an eigenvector of $\boldsymbol{\mathcal{A}}^1$ associated with zero eigenvalue) and the surface as an *eigensurface*. Bifurcation occurs since, to the first order in $\dot{\mathbf{F}}_e$, $\mathbf{F}$ and $\mathbf{F}+\dot{\mathbf{F}}_e$ correspond to the same state of stress $\mathbf{S}$. Let $\dot{\mathbf{F}}$ correspond to an increment $\dot{\mathbf{S}}$ at a point $\mathbf{F}$ where (85) holds. Then, by (86), we have $\dot{\mathbf{S}} = \boldsymbol{\mathcal{A}}^1\dot{\mathbf{F}} = \boldsymbol{\mathcal{A}}^1(\dot{\mathbf{F}}+\dot{\mathbf{F}}_e)$. This *bifurcation* of the deformation path as it passes through a point of singularity of $\boldsymbol{\mathcal{A}}^1$ is illustrated in Figure 1, where the (locally) stable region of $\mathbf{F}$-space bounded by surfaces on which (85) holds is depicted. A deformation path emanating from $\mathbf{F} = \mathbf{I}$, traversing this region and then splitting into two branches at a point on the bifurcation surface is shown. Clearly, in

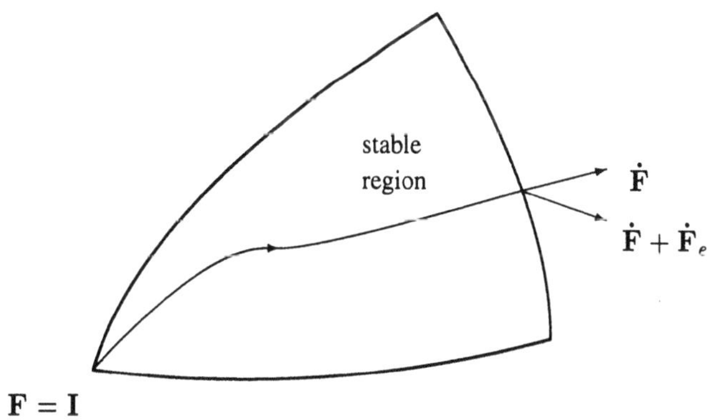

Figure 1. Depiction of the stable region in deformation-gradient space bounded by the singular surfaces (85). A deformation path is shown bifurcating as it meets the boundary.

the dead-load traction problem, *local bifurcation* occurs where $\boldsymbol{\mathcal{A}}^1$ is *singular*. Since, in (83), equality holds for $\dot{\mathbf{F}} = \dot{\mathbf{F}}_e$ the stability criterion (82) is unable to determine the stability status of points on the eigensurface. Referring to (81) we see that stability is determined by the sign of the integrand for $\dot{\mathbf{F}} = \dot{\mathbf{F}}_e$, that is, by the sign of

$$\operatorname{tr}\{\mathcal{A}^2[\dot{\mathbf{F}}_e, \dot{\mathbf{F}}_e]\dot{\mathbf{F}}_e + \frac{1}{4}\mathcal{A}^3[\dot{\mathbf{F}}_e, \dot{\mathbf{F}}_e, \dot{\mathbf{F}}_e]\dot{\mathbf{F}}_e + \ldots\}. \tag{87}$$

If the term in $\mathcal{A}^2$ does not vanish then the considered configuration is *unstable* since, to the third order, its sign is reversed by replacing $\dot{\mathbf{F}}_e$ by $-\dot{\mathbf{F}}_e$. If, on the other hand, it does vanish then stability is determined by the fourth-order term.

Globally, different branches of the solution of the dead-load problem can be found by inverting the stress-deformation relation (20) for an unconstrained material or (21) for an incompressible material. Before examining this inversion, which we do in Section 3, we consider the singularities of $\mathcal{A}^1$ for an isotropic elastic material.

For an isotropic elastic material the non-zero components of $\mathcal{A}^1$ referred to the principal axes $\mathbf{u}^{(i)}$ and $\mathbf{v}^{(i)}$ are given by

$$\mathcal{A}^1_{iijj} = W_{ij}, \tag{88}$$

$$\mathcal{A}^1_{ijij} - \mathcal{A}^1_{ijji} = \frac{W_i + W_j}{\lambda_i + \lambda_j} \quad i \neq j, \tag{89}$$

$$\mathcal{A}^1_{ijij} + \mathcal{A}^1_{ijji} = \frac{W_i - W_j}{\lambda_i - \lambda_j} \quad i \neq j,\ \lambda_i \neq \lambda_j, \tag{90}$$

$$\mathcal{A}^1_{ijij} + \mathcal{A}^1_{ijji} = W_{ii} - W_{ij} \quad i \neq j,\ \lambda_i = \lambda_j, \tag{91}$$

where $W_i = \partial W/\partial\lambda_i$, $W_{ij} = \partial^2 W/\partial\lambda_i\partial\lambda_j$, $i, j \in \{1, 2, 3\}$, and no summation is implied by the repetition of indices. (It is convenient here to drop the convention of using Greek letters for indices relating to Lagrangian components.) For details of the derivation of these components we refer to Ogden (1997), where a method for deriving the corresponding components of $\mathcal{A}^2$ is described. Explicit expressions are given by Fu and Ogden (1999) for the components of $\mathcal{A}^2$ and $\mathcal{A}^3$, but we shall not require them here.

In the classical theory of elasticity, corresponding to the situation in which there is no underlying deformation or stress, the components of $\mathcal{A}^1$ can be written compactly in the form

$$\mathcal{A}^1_{ijkl} = \lambda\delta_{ij}\delta_{kl} + \mu(\delta_{ik}\delta_{jl} + \delta_{il}\delta_{jk}), \tag{92}$$

where λ and μ are the classical Lamé moduli of elasticity and δ_{ij} is the Kronecker delta. Note that λ is not used subsequently in this article so no confusion with the notation used for the stretches should arise. The values of W_{ij} when $\lambda_i = 1$ for $i \in \{1, 2, 3\}$ are simply $W_{ii} = \lambda + 2\mu$, $W_{ij} = \lambda$, $i \neq j$. Also, we take $W_i = 0$ when $\lambda_j = 1$ for $i, j \in \{1, 2, 3\}$ so that the configuration $\mathcal{B}_r$ is stress free; such a configuration is called a *natural configuration*.

We now use the formulas (88)–(90) to express $\operatorname{tr}\{(\mathcal{A}^1\dot{\mathbf{F}})\dot{\mathbf{F}}\}$ explicitly referred to the Lagrangian principal axes. From (4) we calculate the increment $\dot{\mathbf{F}}$ in terms of the increments of $\mathbf{R}$ and $\mathbf{U}$ in the form

$$\mathbf{R}^T\dot{\mathbf{F}} = \mathbf{R}^T\dot{\mathbf{R}}\mathbf{U} + \dot{\mathbf{U}}. \tag{93}$$

Next, we introduce the notations

$$\boldsymbol{\Omega}^R = \mathbf{R}^T\dot{\mathbf{R}}, \quad \boldsymbol{\Omega}^L = \sum_{i=1}^{3} \dot{\mathbf{u}}^{(i)} \otimes \mathbf{u}^{(i)} \tag{94}$$

for the incremental rotations associated with $\mathbf{R}$ and with the Lagrangian principal axes respectively. Then, from (5), we calculate

$$\dot{\mathbf{U}} = \sum_{i=1}^{3} \dot{\lambda}_i \mathbf{u}^{(i)} \otimes \mathbf{u}^{(i)} + \boldsymbol{\Omega}^L\mathbf{U} - \mathbf{U}\boldsymbol{\Omega}^L, \tag{95}$$

and hence the components of $\dot{\mathbf{U}}$ and $\boldsymbol{\Omega}^R$ on the Lagrangian principal axes may be obtained in the form

$$(\dot{\mathbf{U}})_{ii} = \dot{\lambda}_i, \quad (\dot{\mathbf{U}})_{ij} = \Omega^L_{ij}(\lambda_j - \lambda_i) \quad i \neq j, \tag{96}$$

$$(\boldsymbol{\Omega}^R\mathbf{U})_{ij} = \Omega^R_{ij}\lambda_j, \quad (\boldsymbol{\Omega}^R\mathbf{U})_{ji} = -\Omega^R_{ij}\lambda_i \quad i \neq j. \tag{97}$$

Substitution of these expressions into $\mathrm{tr}\,\{(\boldsymbol{\mathcal{A}}^1\dot{\mathbf{F}})\dot{\mathbf{F}}\}$ leads to

$$\mathrm{tr}\,\{(\boldsymbol{\mathcal{A}}^1\dot{\mathbf{F}})\dot{\mathbf{F}}\} = \sum_{i,j=1}^{3} W_{ij}\dot{\lambda}_i\dot{\lambda}_j + \sum_{i\neq j}(W_i - W_j)(\lambda_i - \lambda_j)(\Omega^L_{ij} + \frac{1}{2}\Omega^R_{ij})^2$$
$$+ \frac{1}{4}\sum_{i\neq j}(W_i + W_j)(\lambda_i + \lambda_j)(\Omega^R_{ij})^2. \tag{98}$$

Since $\dot{\lambda}_i$, Ω^L_{ij}, Ω^R_{ij} are independent, necessary and sufficient conditions for (82) to hold are now easily seen to be

$$\text{matrix } (W_{ij}) \text{ is positive definite,} \tag{99}$$

$$W_i + W_j > 0 \qquad i \neq j, \tag{100}$$

$$\frac{W_i - W_j}{\lambda_i - \lambda_j} > 0 \qquad i \neq j \tag{101}$$

jointly for $i, j \in \{1, 2, 3\}$. Note that when $\lambda_i = \lambda_j$, $i \neq j$ (101) reduces to $W_{ii} - W_{ij} > 0$ and that (99) and (101) hold in the natural configuration provided the usual inequalities $\mu > 0$, $3\lambda + 2\mu > 0$ hold. The singularities of $\boldsymbol{\mathcal{A}}^1$ are identified immediately by one or more of the equations

$$(a)\ \det(W_{ij}) = 0, \quad (b)\ W_i + W_j = 0 \quad i \neq j, \quad (c)\ \frac{W_i - W_j}{\lambda_i - \lambda_j} = 0 \quad i \neq j. \tag{102}$$

The form of W is critical in deciding whether (a) or (c) in (102) can hold, but (b) must hold on some surfaces in $\mathbf{F}$-space irrespective of the form of W since the principal Biot stresses $t_i = W_i$ can take positive or negative values (i.e. an elastic material can be subjected to both tensile and compressive stresses). In particular, as indicated above, (b) holds in the natural configuration $\mathbf{F} =$

$\mathbf{I}$, where $W_i = 0$. In passing, we remark that the analogue of (82) for W regarded as a function of $\mathbf{U}$ instead of $\mathbf{F}$ is obtained by setting $\boldsymbol{\Omega}^R = \mathbf{0}$, in which case the relevant inequalities are (99) and (101), i.e. (100) is not required, and (a) and (c) in (102) are the associated singularities.

The eigenmodes $\dot{\mathbf{F}}_e$ corresponding to the above singularities may be calculated by specializing equation (84) as appropriate. The results are as follows. In each case the eigenmodes are determined to within an arbitrary multiplicative factor.

For (a), we have

$$\dot{\mathbf{F}}_e = \sum_{i=1}^{3} \dot{\lambda}_i \mathbf{v}^{(i)} \otimes \mathbf{u}^{(i)} \quad \text{with} \quad \sum_{j=1}^{3} W_{ij} \dot{\lambda}_j = 0. \tag{103}$$

This mode is coaxial with the underlying finite deformation, i.e. there is no rotation associated with the principal axes ($\boldsymbol{\Omega}^L = \boldsymbol{\Omega}^R = \mathbf{0}$).

For (b) there are three possible modes, one for each of the three principal planes, given by

$$\dot{\mathbf{F}}_e = -\Omega_{ij}^L (\lambda_i + \lambda_j)(\mathbf{v}^{(i)} \otimes \mathbf{u}^{(j)} - \mathbf{v}^{(j)} \otimes \mathbf{u}^{(i)}) \quad \text{with} \quad \Omega_{ij}^R = -2\Omega_{ij}^L \quad i \neq j. \tag{104}$$

These correspond to (isochoric) shearing in the principal planes with $\dot{\lambda}_i = 0,\ i \in \{1,2,3\}$, and are accompanied by a change in the relative orientation of the Lagrangian and Eulerian principal axes.

The modes for (c) are similar to those for (b) and are given by

$$\dot{\mathbf{F}}_e = -\Omega_{ij}^L (\lambda_j - \lambda_i)(\mathbf{v}^{(i)} \otimes \mathbf{u}^{(j)} + \mathbf{v}^{(j)} \otimes \mathbf{u}^{(i)}) \quad \text{with} \quad \Omega_{ij}^R = 0 \quad i \neq j. \tag{105}$$

As in (b) these correspond to (isochoric) shearing in the principal planes with $\dot{\lambda}_i = 0,\ i \in \{1,2,3\}$, but there is no accompanying change in the relative orientation of the Lagrangian and Eulerian principal axes.

The special case in which two of the principal stretches are equal is worthy of separate consideration. Suppose $\lambda_i = \lambda_j \neq \lambda_k$, where (i,j,k) is a permutation of $(1,2,3)$. Then, we have the following results. The determinant in (a) decouples into two factors, each of which is associated with a coaxial mode. Thus,

$$(W_{ii} - W_{ij})[(W_{ii} + W_{ij})W_{kk} - 2W_{ik}^2] = 0. \tag{106}$$

The first factor is associated with a pure shear mode in the (i,j)-plane such that

$$\dot{\lambda}_i + \dot{\lambda}_j = 0, \quad \dot{\lambda}_k = 0. \tag{107}$$

This mode destroys the underlying axisymmetry $\lambda_i = \lambda_j$. The second factor corresponds to an equibiaxial (symmetry-preserving) mode and is given by

$$\dot{\lambda}_i = \dot{\lambda}_j, \quad \dot{\lambda}_k = -2W_{ik}\dot{\lambda}_i / W_{kk}. \tag{108}$$

The specialization of (b) corresponds to uniaxial stress ($W_i = W_j = 0$) with $\dot{\lambda}_i = 0,\ i \in \{1,2,3\}$, and Ω_{ij}^L and Ω_{ij}^R arbitrary. In the limit $\lambda_i \to \lambda_j$ (c) is associated with vanishing of the first factor in (106) but the mode is different from that in (a) and is given by

$$\dot{\mathbf{F}}_e = \Sigma_{ij} \lambda_i (\mathbf{v}^{(i)} \otimes \mathbf{u}^{(j)} + \mathbf{v}^{(j)} \otimes \mathbf{u}^{(i)}), \tag{109}$$

with $\dot{\lambda}_i = 0$, $i \in \{1,2,3\}$, and $\Omega^R_{ij} = 0$, where Σ_{ij} are the components of the tensor $\boldsymbol{\Sigma}$ on the Lagrangian principal axes, $\boldsymbol{\Sigma}$ being defined as the symmetric part of $\dot{\mathbf{F}}\mathbf{F}^{-1}$, and we have used the connection $2\lambda_i\lambda_j\Sigma_{ij} = (\lambda_j^2 - \lambda_i^2)\Omega^L_{ij}$ in its limiting form. Thus, in this specialization, two distinct eigenmodes are associated with the singularity $W_{ii} - W_{ij} = 0$. We do not discuss here the degenerate case in which all three stretches coincide.

2.4 Results for Incompressible Materials

For an incompressible material, bearing in mind the difference between (60) and (61), the components of $\mathcal{A}^1$ have the same form as in (88)–(91), but are subject to (11). In this case, however, when $\lambda_i = 1$ for $i \in \{1,2,3\}$ the components are given by

$$\mathcal{A}^1_{iiii} = \mathcal{A}^1_{ijij} = \mu, \quad \mathcal{A}^1_{iijj} = \mathcal{A}^1_{ijji} = 0 \quad i \neq j, \tag{110}$$

instead of by (92), and $W_{ii} = W_i = \mu$, $W_{ij} = 0$, where μ is the *shear modulus* in $\mathcal{B}_r$. These expressions are not unique because they depend on the point at which $\lambda_1\lambda_2\lambda_3$ is set to unity in the differentiation. The differences between (110) and any alternative expressions are embraced by the incremental Lagrange multiplier $\dot{p}$ in (61).

On use of (61) and (62) we may deduce that the analogue of the stability inequality (82) for an incompressible material is

$$\operatorname{tr}\{(\mathcal{A}^1\dot{\mathbf{F}})\dot{\mathbf{F}}\} + p\operatorname{tr}(\mathbf{F}^{-1}\dot{\mathbf{F}}\mathbf{F}^{-1}\dot{\mathbf{F}}) > 0, \tag{111}$$

subject to (62). Note that this does not depend on $\dot{p}$.

By defining the modified strain-energy function $\hat{W}(\lambda_1, \lambda_2)$ by

$$\hat{W}(\lambda_1, \lambda_2) = W(\lambda_1, \lambda_2, \lambda_1^{-1}\lambda_2^{-1}), \tag{112}$$

making use of the expressions (88)–(90), (62) and (93)–(97), with p eliminated using (34), the inequality (111) becomes explicitly

$$(\lambda_1^2\hat{W}_{11} - 2\sigma_3)\left(\frac{\dot{\lambda}_1}{\lambda_1}\right)^2 + (\lambda_2^2\hat{W}_{22} - 2\sigma_3)\left(\frac{\dot{\lambda}_2}{\lambda_2}\right)^2 + 2(\lambda_1\lambda_2\hat{W}_{12} - \sigma_3)\frac{\dot{\lambda}_1\dot{\lambda}_2}{\lambda_1\lambda_2}$$
$$+ \sum_{i\neq j}(t_i - t_j)(\lambda_i - \lambda_j)(\Omega^L_{ij} + \frac{1}{2}\Omega^R_{ij})^2 + \frac{1}{4}\sum_{i\neq j}(t_i + t_j)(\lambda_i + \lambda_j)(\Omega^R_{ij})^2 > 0, \tag{113}$$

where the subscripts on $\hat{W}$ denote partial derivatives with respect to λ_1 and λ_2. Implicitly we have used the connections

$$\sigma_1 - \sigma_3 = \lambda_1\hat{W}_1, \quad \sigma_2 - \sigma_3 = \lambda_2\hat{W}_2 \tag{114}$$

obtained from (36) and (112).

The analogues of (99)–(101) in this case are read off as

$$\text{matrix} \begin{bmatrix} \lambda_1^2\hat{W}_{11} - 2\sigma_3 & \lambda_1\lambda_2\hat{W}_{12} - \sigma_3 \\ \lambda_1\lambda_2\hat{W}_{12} - \sigma_3 & \lambda_2^2\hat{W}_{22} - 2\sigma_3 \end{bmatrix} \text{ is positive definite,} \tag{115}$$

$$t_i + t_j > 0 \quad i \neq j, \tag{116}$$

$$\frac{t_i - t_j}{\lambda_i - \lambda_j} > 0 \quad i \neq j. \tag{117}$$

The eigenmodes associated with the corresponding singularities are just as for the compressible case except that the incremental incompressibility condition

$$\frac{\dot{\lambda}_1}{\lambda_1} + \frac{\dot{\lambda}_2}{\lambda_2} + \frac{\dot{\lambda}_3}{\lambda_3} = 0 \tag{118}$$

must be satisfied. Note also that p occurs in (115)–(117) through σ_3 and t_i, $i \in \{1,2,3\}$.

3 Global Considerations

3.1 Global Non-Uniqueness

The singularities of $\mathcal{A}^1$ considered in Section 2 are local manifestations of non-uniqueness in the relationship between $\mathbf{S}$ and $\mathbf{F}$ expressed through the constitutive equation

$$\mathbf{S} = \frac{\partial W}{\partial \mathbf{F}}(\mathbf{F}) \tag{119}$$

(for an unconstrained material) or (21) in the case of an incompressible material. It is very instructive to examine the global inversion of (119), which we now do for an isotropic material.

Recall from $(37)_1$ the polar decomposition

$$\mathbf{S} = \mathbf{T}\mathbf{R}^T. \tag{120}$$

It follows that

$$\mathbf{T}^2 = \mathbf{S}\mathbf{S}^T = \sum_{i=1}^{3} t_i^2 \mathbf{u}^{(i)} \otimes \mathbf{u}^{(i)} \tag{121}$$

and

$$\det \mathbf{S} = \det \mathbf{T} = t_1 t_2 t_3. \tag{122}$$

Unlike (4), however, the polar decomposition (120) is not in general unique since $\mathbf{T}$ is not sign definite. For a given $\mathbf{S}$ there are four distinct polar decompositions of the form (120) if $t_i^2 \neq t_j^2$, $i,j \in \{1,2,3\}$, and infinitely many when $t_i^2 = t_j^2$, $i \neq j$, but at most one of these satisfies the stability inequalities (100). For each such polar decomposition each $\mathbf{R}$ and $\mathbf{T}$ pair is determined uniquely if $t_i^2 \neq t_j^2$, $i,j \in \{1,2,3\}$ and to within an arbitrary rotation about the $\mathbf{u}^{(k)}$ principal axis if $t_i^2 = t_j^2$, $i \neq j$, where (i,j,k) is a permutation of $(1,2,3)$. Then, for each $\mathbf{T}$, $\mathbf{U}$ can in principle be found by inverting (25), or (26) for an incompressible material. Since we have

$$\mathbf{T} = \sum_{i=1}^{3} t_i \mathbf{u}^{(i)} \otimes \mathbf{u}^{(i)}, \tag{123}$$

and $\mathbf{T}$ is coaxial with $\mathbf{U}$, this inversion is equivalent to inverting the scalar equations (33) or (34) for λ_i, $i \in \{1,2,3\}$, when t_i, $i \in \{1,2,3\}$, are given. For convenience, we repeat these equations here. For an unconstrained material, we have

$$t_i = \frac{\partial W}{\partial \lambda_i} \quad i \in \{1,2,3\}, \tag{124}$$

and for an incompressible material

$$t_i = \frac{\partial W}{\partial \lambda_i} - p\lambda_i^{-1} \quad i \in \{1,2,3\}, \quad \lambda_1\lambda_2\lambda_3 = 1. \tag{125}$$

The resulting deformation gradients are then calculated from $\mathbf{F} = \mathbf{RU}$ since $\mathbf{u}^{(i)}$, $i \in \{1,2,3\}$, are determined from $\mathbf{T}$. It is tempting to assume that the inversions of (124) and (125) are unique, but, as we show in Section 3.2, this is definitely not a valid assumption, at least for incompressible materials.

We now justify the statements made in the above paragraph.

Case A: $t_i^2 \neq t_j^2$, $i,j \in \{1,2,3\}$. For a given $\mathbf{S}$, the tensors $\mathbf{u}^{(i)} \otimes \mathbf{u}^{(i)}$, $i \in \{1,2,3\}$, are determined uniquely by (121) (and hence each $\mathbf{u}^{(i)}$ is determined to within sign), and t_i^2, $i \in \{1,2,3\}$, are also determined uniquely. Thus, we may write $\mathbf{T}$ in the form (123) with the sign of each t_i uncertain. However, from (122), the sign of the product $t_1t_2t_3$ is known. The following possibilities for the signs of the t_i therefore arise.

If $t_1t_2t_3 \equiv \det \mathbf{S} \geq 0$ then there are four possibilities: *either* (*a*) $t_i > t_j > t_k \geq 0$ for some permutation (i,j,k) of $(1,2,3)$ *or* (*b*) $t_i > 0 \geq t_j > t_k$ with $i = 1, 2$ or 3.

If $t_1t_2t_3 \equiv \det \mathbf{S} < 0$ there are again four possibilities: *either* (c) $0 > t_i > t_j > t_k$ for some permutation (i,j,k) of $(1,2,3)$ *or* (*d*) $t_i > t_j > 0 > t_k$ for $k = 1, 2$ or 3.

Thus, each case is associated with four possible tensors $\mathbf{T}$ of the form (123) and each $\mathbf{T}$ has associated with it a unique $\mathbf{R}$. When $\det \mathbf{S} \geq 0$, only (*a*) is consistent with the stability inequalities (100) or, for an incompressible material, (116). When $\det \mathbf{S} < 0$ only (*d*), and for only one value of k, is so consistent. Thus, *when the stability inequalities are imposed* the distribution of signs of the t_i is *uniquely* determined; hence $\mathbf{T}$, and therefore $\mathbf{R}$, is uniquely determined in (120).

Case B: $t_i^2 = t_j^2 \neq t_k^2$ for some permutation (i,j,k) of $(1,2,3)$. In this case $\mathbf{u}^{(k)} \otimes \mathbf{u}^{(k)}$ is determined uniquely and we may have either $t_i = t_j > 0$, $t_i = t_j < 0$ or $t_i = -t_j$ with the sign of t_k dependent on that of $\det \mathbf{S}$ in each case. There is no essential distinction between the possibilities $t_i = -t_j > 0$ and $t_i = -t_j < 0$. Thus, for a given value of $\det \mathbf{S}$ there are essentially three forms of the polar decomposition, but the orientation of the (orthogonal) vectors $\mathbf{u}^{(i)}$ and $\mathbf{u}^{(j)}$ is arbitrary in the plane normal to $\mathbf{u}^{(k)}$ and $\mathbf{R}$ is determined only to within an arbitrary rotation about $\mathbf{u}^{(k)}$ in each case. If $\det \mathbf{S} \geq 0$ then only one of the possibilities, namely $t_i = t_j > 0, t_k \geq 0$ satisfies (100), or (116) in the case of an incompressible material, as appropriate. If $\det \mathbf{S} < 0$ the only possibility corresponds to $t_i = t_j > 0, t_k < 0$, which requires $t_i + t_k > 0$.

If $t_i = t_j = 0 \neq t_k$ then t_k may be either positive of negative, but the stability inequalities fail since $t_i + t_j = 0$.

Case C: $t_i^2 = t_j^2 = t_k^2$. For completeness we note that in this degenerate case the orientation of the Lagrangian principal axes is arbitrary for each possible set of sign combinations for (t_i, t_j, t_k), and $\mathbf{R}$ is determined to within an arbitrary rotation in the reference configuration. The stability inequalities are violated for all these possibilities except for $t_i = t_j = t_k > 0$. These remarks also apply in the trivial case $t_i = t_j = t_k = 0$ except that $\mathbf{R}$ is a completely arbitrary rotation.

Bearing in mind the above discussion we now focus on incompressible materials and consider the inversion of equations (125) in order to illustrate the non-uniqueness in the relationship between $\mathbf{T}$ and $\mathbf{U}$, which is a separate issue from non-uniqueness of the polar decomposition (120). More particularly, we consider in Section 3.2 the pure strain $\mathbf{F} = \mathbf{U}$ ($\mathbf{R} = \mathbf{I}$) in which $\mathbf{u}^{(i)}$ are fixed in direction as the load increases. To this end we eliminate p from (125) and obtain

$$\lambda_1 t_1 - \lambda_1 \frac{\partial W}{\partial \lambda_1} = \lambda_2 t_2 - \lambda_2 \frac{\partial W}{\partial \lambda_2} = \lambda_3 t_3 - \lambda_3 \frac{\partial W}{\partial \lambda_3}, \qquad \lambda_1 \lambda_2 \lambda_3 = 1. \tag{126}$$

This provides three equations from which $\lambda_1, \lambda_2, \lambda_3$ can, in principle, be determined when t_1, t_2, t_3 are prescribed.

For $t_1 = t_2 = t_3$, equations (126) govern the cube problem of Rivlin (1974), that is the problem in which a cube of elastic material is subjected to equal normal forces per unit reference area on its three pairs of faces.

3.2 A Two-Dimensional (Plane-Strain) Example

For simplicity we now restrict attention to the two-dimensional (plane strain) specialization of equations (126) with $\lambda_3 = 1$ and t_1 and t_2 prescribed. It then remains to determine λ_1 and λ_2 from the two equations

$$\lambda_1 t_1 - \lambda_1 \frac{\partial W}{\partial \lambda_1} = \lambda_2 t_2 - \lambda_2 \frac{\partial W}{\partial \lambda_2}, \qquad \lambda_1 \lambda_2 = 1, \tag{127}$$

with $\lambda_3 = 1$ (t_3 can be found from the third equation in (126) if required once (127) have been solved for λ_1 and λ_2).

At this point we return to the stability inequality (113) and obtain the appropriate specialization for plane strain increments, so that (118) holds with $\dot{\lambda}_3 = 0$. This leads to

$$\lambda_1^2 \hat{W}_{11} + \lambda_2^2 \hat{W}_{22} - 2\lambda_1 \lambda_2 \hat{W}_{12} - 2t_3 > 0 \tag{128}$$

(noting that, since $\lambda_3 = 1$, $t_3 = \sigma_3$), together with

$$t_1 + t_2 > 0, \qquad \frac{t_1 - t_2}{\lambda_1 - \lambda_2} > 0. \tag{129}$$

On introducing the change of variables $\lambda_1 = \lambda, \lambda_2 = \lambda^{-1}$ and the notation

$$\tilde{W}(\lambda) = \hat{W}(\lambda, \lambda^{-1}), \tag{130}$$

so that, from (114),

$$\sigma_1 - \sigma_2 = \lambda \tilde{W}'(\lambda), \tag{131}$$

we may recast (128) and (129) together as

$$-\frac{\lambda\tilde{W}'(\lambda)}{\lambda+\lambda^{-1}} < t_2 < \frac{\lambda\tilde{W}'(\lambda)}{\lambda-\lambda^{-1}}, \quad 2t_2 < \lambda^3\tilde{W}''(\lambda), \tag{132}$$

where a prime signifies differentiation with respect to λ. Note that $\sigma_1 = \lambda t_1$, $t_2 = \lambda\sigma_2$ and that (131) may be used to express (132) as inequalities on t_1 instead of t_2. We shall make use of these inequalities shortly.

In terms of (130) equations (127) may be reduced to the single equation

$$\lambda t_1 - \lambda^{-1}t_2 = \lambda\tilde{W}'(\lambda), \tag{133}$$

subject to

$$\tilde{W}'(1) = 0. \tag{134}$$

For prescribed t_1 and t_2 this is an implicit equation for λ. The further specialization $t_1 = t_2$ (= t say) reduces (133) to

$$(\lambda - \lambda^{-1})t = \lambda\tilde{W}'(\lambda). \tag{135}$$

Since the problem is two dimensional we may confine consideration to a material having a square cross-section in the $(1,2)$-plane. The objective now is to determine the solutions λ of (135) when the dead-load traction t is prescribed.

Noting (134), we see that, independently of t, $\lambda = 1$ is always one solution of (135); this corresponds to the situation in which the square cross-section remains unchanged as t increases from zero. We wish, however, to consider the possible emergence of solutions of (135) for which $\lambda \neq 1$. To this end we rewrite (135) in the form

$$t = \frac{\lambda^2\tilde{W}'(\lambda)}{\lambda^2-1}. \tag{136}$$

In the limit $\lambda \to 1$ this becomes

$$t_c = \frac{1}{2}\tilde{W}''(1) = 2\mu, \tag{137}$$

where μ is the (positive) shear modulus of the material. By specializing (132) for $t_2 = t$ and $\lambda = 1$ we obtain simply $0 < t < 2\mu$ so that t_c is a *critical value* of t in the sense that it is the value at which the right-hand inequality fails and bifurcation from the symmetrical solution $\lambda = 1$ can occur. This can be seen by expanding (136) as a Taylor series near $\lambda = 1$ to give

$$t = t_c + \eta(\lambda-1)^2 + \ldots, \tag{138}$$

where

$$\eta = \frac{1}{12}\tilde{W}''''(1) - \tilde{W}''(1). \tag{139}$$

Note that $\mathrm{d}t/\mathrm{d}\lambda = 0$ at $\lambda = 1$. In (139) the result $\tilde{W}'''(1) = -3\tilde{W}''(1)$ has been used.

Equation (138) shows that a symmetry-breaking bifurcation in which the square cross-section becomes rectangular is possible, with homogeneity of the deformation maintained. Note that at the same critical value shearing modes corresponding to failure of the second inequality in (132) are also possible but we have excluded these from consideration since they are associated with rotation of the principal axes and the inclusion of $\mathbf{R}$ in the equations. Inclusion of rotations would impose the additional stability inequality $(t_1-t_2)/(\lambda-\lambda^{-1}) > 0$, which, of course, is violated on the asymmetrical path since $t_1 = t_2 = t$. (A study of the stability of asymmetric paths allowing for rotations requires consideration of higher-order terms such as (87) appropriately modified for incompressible materials and specialized for the considered problem.) For $\lambda \neq 1$ use of (135) with $t_2 = t$ enables the remaining stability inequalities in (132) to be reduced to

$$\lambda^2\tilde{W}''(\lambda) > \frac{2\lambda\tilde{W}'(\lambda)}{\lambda^2-1} > 0. \tag{140}$$

We now illustrate the above results by choosing a strain-energy function of the form

$$W(\lambda_1,\lambda_2,\lambda_3) = 2\mu(\lambda_1^m+\lambda_2^m+\lambda_3^m-3)/m^2, \tag{141}$$

where m is a constant and μ is again the shear modulus. Then, we have

$$\tilde{W}(\lambda) = 2\mu(\lambda^m+\lambda^{-m}-2)/m^2. \tag{142}$$

From (139) and (142) we obtain

$$\eta = \frac{1}{3}\mu(m^2-1) \tag{143}$$

and the inequalities (140) show that the solution branches corresponding to $\lambda \neq 1$ are stable for $m^2 > 1$ and unstable for $m^2 < 1$. Moreover, the symmetrical solution $\lambda = 1$ is unstable for $t > t_c$. We emphasize that bifurcation is predicted for *every* form of W consistent with the requirement $\tilde{W}''(1) = 4\mu$ of the classical theory. Thus, inversion of the principal stress-stretch relations (125) is not in general unique. We illustrate these results in Figure 2, where $\lambda\,(\geq 1)$ is plotted against t/μ.

3.3 Energy Considerations

For the dead-load traction boundary-value problem the energy functional (45) can be written, on use of the divergence theorem, as

$$E = \int_{\mathcal{B}_r} \{W - \mathrm{tr}\,(\mathbf{SF})\}\mathrm{d}V, \tag{144}$$

arguments being omitted. We now specialize this to the considered plane-strain problem so that $\mathcal{B}_r$ becomes the relevant plane cross-section of the body. Since the deformation is homogeneous, it suffices to consider the integrand of (144), which we write as

$$\bar{E} = W - \mathrm{tr}\,(\mathbf{SF}) = W - (t_1\lambda_1 + t_2\lambda_2) = \tilde{W}(\lambda) - t(\lambda+\lambda^{-1}). \tag{145}$$

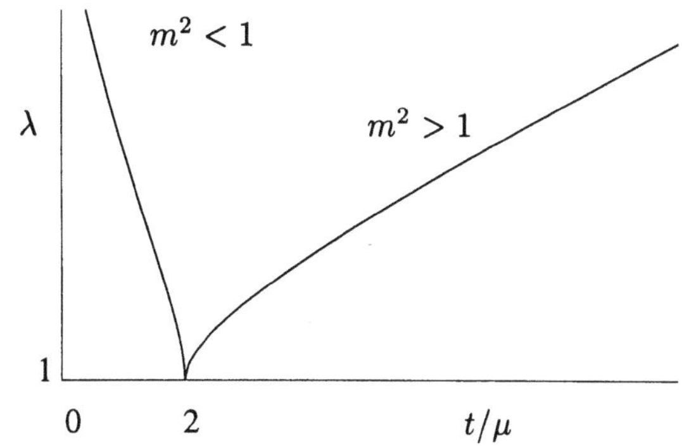

Figure 2. Bifurcation curves branching from the symmetric path at $t/\mu = 2$ for the plane-strain example with λ plotted against t/μ for $m^2 < 1$ (unstable) and $m^2 > 1$ (stable).

This is the energy per unit area of $\mathcal{B}_r$ and per unit length normal to the plane.

Now let E_{as} and E_s respectively denote the forms of $\bar{E}$ on the asymmetric and symmetric solution paths discussed above. Then,

$$E_{as} = \tilde{W}(\lambda) - t(\lambda + \lambda^{-1}), \quad E_s = -2t,$$

and hence

$$E_{as} - E_s = \tilde{W}(\lambda) - t(\lambda + \lambda^{-1} - 2). \tag{146}$$

On the asymmetric path we can regard λ as a function of t through (136); note that $E_{as} = E_s$ when $t = t_c$. On use of (136) we then obtain

$$\frac{\mathrm{d}}{\mathrm{d}t}(E_{as} - E_s) = -(\lambda^{1/2} - \lambda^{-1/2})^2 \le 0. \tag{147}$$

This shows that on a stable (asymmetric) path of deformation with t increasing from t_c the energy is less than on the corresponding symmetric path. Thus, the asymmetric path provides the (global) energy-minimizing solution. In particular, near $t = t_c$, we have

$$E_{as} - E_s = -\frac{1}{\eta}(t - t_c)^2 + \ldots, \tag{148}$$

where η is given by (139). If $\eta > 0$ the asymmetric solution is stable and branches to the right, while if $\eta < 0$ it is unstable and branches to the left, as shown in Figure 2 in respect of the specialization (143). In the latter case the symmetric path provides the energy minimizer, but t cannot exceed t_c.

As an explicit example, we take $m = 2$ in (142). Then $\eta > 0$,

$$t = \mu(\lambda + \lambda^{-1}), \tag{149}$$

which applies for $t \ge 2\mu$, and (146) reduces to

$$E_{as} - E_s = -\frac{1}{2\mu}(2\mu - t)^2, \tag{150}$$

which is exact and should be compared with the approximation (148). For $m = 2$ the strain energy (142) is called the *neo-Hookean* strain-energy function; for $m = 1$ we have the *Varga* strain-energy function; both these will be used in Section 4.

3.4 A Plane Stress Example

It is interesting to compare the results for plane strain with those for plane stress where, instead of fixing λ_3 at unity, we set $\sigma_3 = 0$ so that λ_3 varies in accordance with the incompressibility condition $\lambda_1\lambda_2\lambda_3 = 1$. On use of the strain-energy function $\hat{W}(\lambda_1, \lambda_2)$ defined by (112), equations (126) then reduce to

$$t_1 = \hat{W}_1, \quad t_2 = \hat{W}_2, \tag{151}$$

and the stability inequalities (115)–(117) to

$$\hat{W}_{11} > 0, \quad \hat{W}_{11}\hat{W}_{22} - \hat{W}_{12}^2 > 0, \tag{152}$$

$$\hat{W}_1 > 0, \quad \hat{W}_2 > 0, \tag{153}$$

$$\lambda_1^2\lambda_2 > 1, \quad \lambda_1\lambda_2^2 > 1, \quad \frac{\hat{W}_1 - \hat{W}_2}{\lambda_1 - \lambda_2} > 0. \tag{154}$$

For the strain-energy function (141) we obtain

$$\lambda_1\hat{W}_1 = \frac{2\mu}{m}(\lambda_1^m - \lambda_3^m), \quad \lambda_2\hat{W}_2 = \frac{2\mu}{m}(\lambda_2^m - \lambda_3^m), \tag{155}$$

$$\frac{\hat{W}_1 - \hat{W}_2}{\lambda_1 - \lambda_2} = \frac{2\mu}{m}\left[\frac{\lambda_1^{m-1} - \lambda_2^{m-1}}{\lambda_1 - \lambda_2} + \lambda_3^{m+1}\right], \tag{156}$$

$$\lambda_1^2\hat{W}_{11} = \frac{2\mu}{m}[(m-1)\lambda_1^m + (m+1)\lambda_3^m], \tag{157}$$

$$\begin{aligned}\lambda_1^2\lambda_2^2(\hat{W}_{11}\hat{W}_{22} - \hat{W}_{12}^2) = \left(\frac{2\mu}{m}\right)^2 [(m-1)^2\lambda_1^m\lambda_2^m \\ + (m^2-1)(\lambda_1^m + \lambda_2^m)\lambda_3^m + (2m+1)\lambda_3^{2m}],\end{aligned} \tag{158}$$

where $\lambda_3 = \lambda_1^{-1}\lambda_2^{-1}$.

Clearly, the inequalities (153) follow from the first two inequalities in (154) for any positive or negative m, the expression in (157) is positive for all positive λ_1 and λ_2 if $|m| \geq 1$ and (156) and (158) are positive for all positive λ_1 and λ_2 if $m \geq 1$. Thus, if $m \geq 1$ all the stability inequalities are satisfied for all λ_1 and λ_2 in the convex region of (λ_1, λ_2)-space defined by $\lambda_1^2\lambda_2 > 1$, $\lambda_1\lambda_2^2 > 1$. If $m < 1$ then one or more of the inequalities may fail within this region. This is exemplified by $m = -1$, in which case $\hat{W}_{11} > 0$, $\hat{W}_{22} > 0$, but (156) and (158) are positive if and only if

$$\lambda_1^2\lambda_2^2 < \lambda_1 + \lambda_2 \quad \text{and} \quad \lambda_1^3\lambda_2^3 < 4, \tag{159}$$

respectively.

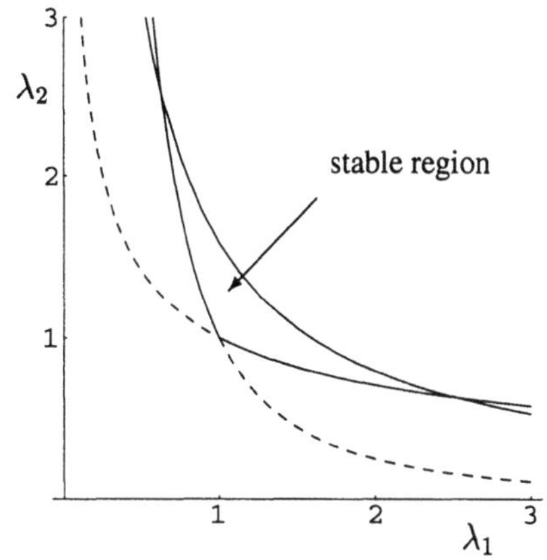

Figure 3. Plot of the stable region, bounded by the continuous curves, in (λ_1, λ_2)-space.

In Figure 3 we plot in (λ_1, λ_2)-space the boundaries of the region of stability defined by

$$\lambda_1^2\lambda_2 = 1, \quad \lambda_1\lambda_2^2 = 1, \quad \lambda_1\lambda_2 = 2^{2/3},$$

and we note, in particular, that the stable region is not convex and that within this region the first inequality in (159) is satisfied.

For discussion of related problems we refer to, for example, Ball and Schaeffer (1983), Ogden (1984, 1985, 1997), Rivlin (1974), MacSithigh (1986) and MacSithigh and Chen (1992*a*, *b*).

The non-uniqueness in the relation between $\mathbf{S}$ and $\mathbf{F}$, which arises from both the multiple polar decompositions of $\mathbf{S}$ and the lack of uniqueness in the relation between t_i and λ_i has implications for the solution of more general boundary-value problems than the homogeneous dead-load problems discussed here. In Section 4 we consider a further example of the onset of bifurcation from a homogeneously-deformed configuration, but in this case the resulting mode of bifurcation is not homogeneous.

3.5 Strong Ellipticity in Two Dimensions

In general the stability inequality (82) is stronger than the strong ellipticity inequality (49). This can be seen by making the specialization $\dot{\mathbf{F}} = \mathbf{m} \otimes \mathbf{N}$ in (82), which then reduces to (49). The same is true for incompressible materials since use of (51) and (52) removes the term in p from (111). We now obtain an explicit form for the strong ellipticity inequality in two dimensions for incompressible materials for later reference. Using (52) we rewrite (49) as

$$\mathcal{A}^1_{0ijkl} n_i n_k m_j m_l > 0, \tag{160}$$

where the notation $\mathcal{A}^1_{0ijkl}$ is defined by

$$\mathcal{A}^1_{0ijkl} = F_{i\alpha} F_{k\beta} \mathcal{A}^1_{\alpha j \beta l}. \tag{161}$$

For an incompressible material we then use (51) in two dimensions to write $m_1 = n_2, m_2 = -n_1$, so that (49) reduces to

$$\mathcal{A}^1_{01212}n_1^4 + (\mathcal{A}^1_{01111} + \mathcal{A}^1_{02222} - 2\mathcal{A}^1_{01122} - 2\mathcal{A}^1_{02112})n_1^2 n_2^2 + \mathcal{A}^1_{02121}n_2^4 > 0, \tag{162}$$

where, on use of (88)–(90) and the specialization of (161) to principal axes, we obtain

$$\mathcal{A}^1_{01111} = \lambda_1^2 W_{11}, \quad \mathcal{A}^1_{01122} = \lambda_1\lambda_2 W_{12}, \quad \mathcal{A}^1_{02222} = \lambda_2^2 W_{22}, \tag{163}$$

$$\lambda_1^{-2}\mathcal{A}^1_{01212} = \lambda_2^{-2}\mathcal{A}^1_{02121} = \frac{\lambda_1 W_1 - \lambda_2 W_2}{\lambda_1^2 - \lambda_2^2}, \quad \mathcal{A}^1_{02112} = \frac{\lambda_2 W_1 - \lambda_1 W_2}{\lambda_1^2 - \lambda_2^2}\lambda_1\lambda_2. \tag{164}$$

Necessary and sufficient conditions for (162) to hold are then easily seen to be

$$\mathcal{A}^1_{01212} > 0, \quad \mathcal{A}^1_{01111} + \mathcal{A}^1_{02222} - 2\mathcal{A}^1_{01122} - 2\mathcal{A}^1_{02112} > -2\sqrt{\mathcal{A}^1_{01212}\mathcal{A}^1_{02121}}, \tag{165}$$

and we note that these are independent of p.

For plane strain with $\lambda_3 = 1$ and the notation defined in (130), the inequalities simplify to

$$\lambda^2 \tilde{W}''(\lambda) + \frac{2\tilde{W}'(\lambda)}{\lambda + \lambda^{-1}} > 0, \quad \frac{\tilde{W}'(\lambda)}{\lambda - \lambda^{-1}} > 0. \tag{166}$$

It can be seen immediately that these follow from the stability inequalities (132).

4 Stability of an Elastic Plate

4.1 Homogeneous Simple Shear of a Plate

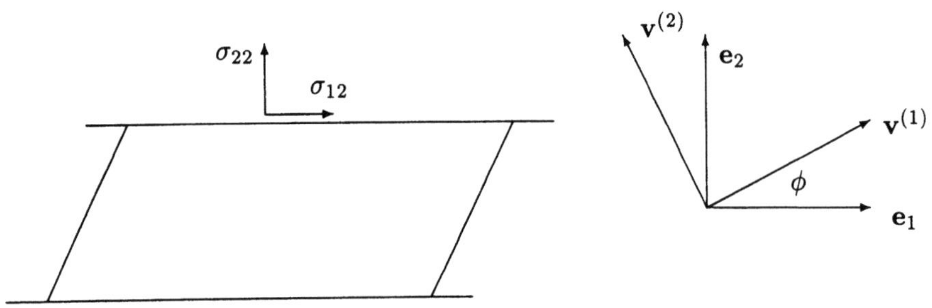

Figure 4. Simple shear of a horizontal plate showing the normal and shear components of the Cauchy stress and the orientation of the Eulerian principal axes.

We consider an elastic plate which, in its undeformed stress-free configuration, is defined by

$$-\infty < X_1 < \infty, \quad -h \le X_2 \le h, \quad -\infty < X_3 < \infty,$$

in Cartesian coordinates. In the (X_1, X_2)-plane we apply to the plate a simple shear deformation defined by

$$x_1 = X_1 + \epsilon X_2, \quad x_2 = X_2, \quad x_3 = X_3, \tag{167}$$

where ϵ is the *amount of shear*. We write the principal stretches for this plane deformation as

$$\lambda_1 = \lambda, \quad \lambda_2 = \lambda^{-1}, \quad \lambda_3 = 1, \tag{168}$$

and note that λ is related to ϵ through

$$\epsilon = \lambda - \lambda^{-1}, \tag{169}$$

where, without loss of generality, we have taken $\epsilon \geq 0,\ \lambda \geq 1$. The associated left-stretch tensor $\mathbf{V}$ may be written

$$\mathbf{V} = \lambda \mathbf{v}^{(1)} \otimes \mathbf{v}^{(1)} + \lambda^{-1}\mathbf{v}^{(2)} \otimes \mathbf{v}^{(2)} + \mathbf{v}^{(3)} \otimes \mathbf{v}^{(3)}, \tag{170}$$

where $\mathbf{v}^{(i)},\ i \in \{1,2,3\}$, are the principal directions of $\mathbf{V}$ identified in (6). Here $\mathbf{v}^{(3)}$ is the Cartesian axis normal to the considered plane. In terms of the angle ϕ defined by

$$\tan\phi = 2/\epsilon, \qquad 0 < \phi \leq \pi/4, \tag{171}$$

the axes $\mathbf{v}^{(1)}$ and $\mathbf{v}^{(2)}$ are given by

$$\mathbf{v}^{(1)} = \cos\phi\,\mathbf{e}_1 + \sin\phi\,\mathbf{e}_2, \quad \mathbf{v}^{(2)} = -\sin\phi\,\mathbf{e}_1 + \cos\phi\,\mathbf{e}_2, \tag{172}$$

where $\mathbf{e}_1$ and $\mathbf{e}_2$ are the Cartesian axes corresponding to x_1 and x_2 respectively.

For an isotropic elastic material the Cauchy stress tensor is coaxial with $\mathbf{V}$ and has the spectral decomposition

$$\boldsymbol{\sigma} = \sigma_1 \mathbf{v}^{(1)} \otimes \mathbf{v}^{(1)} + \sigma_2 \mathbf{v}^{(2)} \otimes \mathbf{v}^{(2)} + \sigma_3 \mathbf{v}^{(3)} \otimes \mathbf{v}^{(3)}, \tag{173}$$

where

$$\sigma_1 = \lambda\frac{\partial W}{\partial\lambda_1} - p, \quad \sigma_2 = \lambda^{-1}\frac{\partial W}{\partial\lambda_2} - p, \quad \sigma_3 = \frac{\partial W}{\partial\lambda_3} - p, \tag{174}$$

are the principal Cauchy stresses evaluated for the stretches given by (168). The Cartesian components of $\boldsymbol{\sigma}$ are then given by

$$\sigma_{11} = \sigma_1\cos^2\phi + \sigma_2\sin^2\phi, \quad \sigma_{22} = \sigma_1\sin^2\phi + \sigma_2\cos^2\phi, \tag{175}$$

$$\sigma_{12} = (\sigma_1 - \sigma_2)\sin\phi\cos\phi, \quad \sigma_{33} = \sigma_3, \quad \sigma_{13} = \sigma_{23} = 0, \tag{176}$$

but we note that σ_3 plays no role in the subsequent analysis. Expressions for $\cos\phi$ and $\sin\phi$ may be obtained from (171) with (169) as

$$\cos\phi = \lambda/\sqrt{\lambda^2+1}, \quad \sin\phi = 1/\sqrt{\lambda^2+1}. \tag{177}$$

For the considered deformation (168) we define the strain energy in terms of λ through

$$\tilde{W}(\lambda) \equiv W(\lambda, \lambda^{-1}, 1), \tag{178}$$

which, on recalling the definition (112), is seen to be identical to (130). We then have

$$\sigma_1 - \sigma_2 = \lambda \tilde{W}'(\lambda), \tag{179}$$

as in (131), and it follows from (175) that

$$\sigma_{22} = \sigma_2 + \tilde{W}'(\lambda)/(\lambda + \lambda^{-1}). \tag{180}$$

Since the material is incompressible, superposition of an arbitrary hydrostatic pressure does not affect the deformation. Accordingly, σ_2 may be set arbitrarily, or, equivalently, the normal stress σ_{22} on the boundary $x_2 = h$ may be imposed independently of ϵ.

In the next section we examine the stability of the simply-sheared configuration of the plate with respect to independent changes in the parameters ϵ and σ_{22}. For this purpose it is useful to define Cartesian coordinates (x_1', x_2') associated with the Eulerian principal axes $\mathbf{v}^{(1)}$ and $\mathbf{v}^{(2)}$ and related to (x_1, x_2) by

$$x_1' = x_1 \cos\phi + x_2 \sin\phi, \quad x_2' = -x_1 \sin\phi + x_2 \cos\phi. \tag{181}$$

4.2 Incremental Equations

We now consider, in the notation of Section 2, a small increment $\dot{\mathbf{x}}$ superimposed on the finite deformation considered in Section 4.1. We represent $\dot{\mathbf{x}}$ as a function of the position vector $\mathbf{x}$ in the deformed configuration by the displacement vector $\mathbf{u}$. We again restrict attention to the $(1, 2)$-plane and write

$$u_1 = u_1(x_1, x_2), \quad u_2 = u_2(x_1, x_2), \quad u_3 = 0. \tag{182}$$

Since $\dot{\mathbf{F}} = \operatorname{Grad} \dot{\mathbf{x}} = (\operatorname{grad} \mathbf{u})\mathbf{F}$, where grad is the gradient operator with respect to $\mathbf{x}$, the linearized incompressibility condition (62) takes the form $\operatorname{div} \mathbf{u} = 0$, and therefore, on specializing to (182), we may write u_1 and u_2 in terms of a scalar 'potential' function $\psi = \psi(x_1, x_2)$, so that

$$u_1 = \frac{\partial \psi}{\partial x_2} \equiv \psi_{,2}, \quad u_2 = -\frac{\partial \psi}{\partial x_1} \equiv -\psi_{,1}. \tag{183}$$

Since the material is incompressible, the incremental nominal stress tensor is given by (61). We now update the reference configuration to the finitely deformed configuration so that the independent variable $\mathbf{X}$ is changed to $\mathbf{x}$. Let $\dot{\mathbf{F}}_0$, $\dot{\mathbf{S}}_0$, $\mathcal{A}_0^1$ be the resulting values of $\dot{\mathbf{F}}$, $\dot{\mathbf{S}}$, $\mathcal{A}^1$. These are given respectively by

$$\dot{\mathbf{F}}_0 = \dot{\mathbf{F}}\mathbf{F}^{-1}, \quad \dot{\mathbf{S}}_0 = \mathbf{F}\dot{\mathbf{S}}, \tag{184}$$

and, in component form, (161), and we note that

$$\dot{\mathbf{F}}_0 = \operatorname{grad} \mathbf{u}. \tag{185}$$

The resulting incremental constitutive law has the form

$$\dot{\mathbf{S}}_0 = \mathcal{A}_0^1\dot{\mathbf{F}}_0 + p\dot{\mathbf{F}}_0 - \dot{p}\mathbf{I}. \tag{186}$$

The equilibrium equation in the absence of body forces is then written

$$\operatorname{div}\dot{\mathbf{S}}_0 = \mathbf{0}.$$

Since the principal axes are associated with the Cartesian coordinates (x_1', x_2') it is convenient to regard ψ alternatively as a function of (x_1', x_2') and we therefore write

$$\psi'(x_1', x_2') \equiv \psi(x_1, x_2)$$

to express this dependence. Let $u_1' = \psi'_{,2}$, $u_2' = -\psi'_{,1}$ be the components of $\mathbf{u}$ referred to the same axes, where the derivatives of ψ' are with respect to x_1' and x_2'. Let $\mathcal{A}_{0ijkl}^{1'}$ be the corresponding components of $\mathcal{A}_0^1$. Expressions for these components are given by the formulas (163) and (164). The equilibrium equations may now be written

$$\mathcal{A}_{0piqj}^{1'}u'_{j,pq} - \dot{p}'_{,i} = 0 \quad i \in \{1,2\}, \tag{187}$$

where again the differentiation is with respect to x_1' and x_2'. Elimination of $\dot{p}'$ by cross differentiation and subtraction enables the equations to be combined as a single equation for ψ', namely

$$\alpha\psi'_{,1111} + 2\beta\psi'_{,1122} + \gamma\psi'_{,2222} = 0, \tag{188}$$

where the material constants α, β, γ are defined by

$$\alpha = \mathcal{A}_{01212}^{1'}, \quad \gamma = \mathcal{A}_{02121}^{1'}, \quad 2\beta = \mathcal{A}_{01111}^{1'} + \mathcal{A}_{02222}^{1'} - 2\mathcal{A}_{01122}^{1'} - 2\mathcal{A}_{01221}^{1'},$$

or, via (163) and (164), in terms of the strain energy specialized for the considered deformation, as

$$\alpha = \frac{\lambda^5\tilde{W}'(\lambda)}{\lambda^4 - 1} = \lambda^4\gamma, \quad 2\beta + 2\gamma = \lambda^2\tilde{W}''(\lambda). \tag{189}$$

We emphasize that the derivatives in (188) are with respect to x_1' and x_2'.

In terms of the notation (189) the strong ellipticity conditions (165) become simply

$$\alpha > 0, \quad \beta > -\sqrt{\alpha\gamma}. \tag{190}$$

These inequalities arise naturally in Section 4.3 during the derivation of the bifurcation condition.

A number of possible sets of boundary conditions on $x_2 = \pm h$ may be considered but for definiteness we take the incremental tractions to vanish. If $\mathbf{n}$ denotes the unit normal vector in the deformed configuration, as discussed in Section 1.1, then the incremental traction vector is $\dot{\mathbf{S}}_0^T\mathbf{n}$ per unit deformed surface area. When this is specialized to the geometry shown in Section 4.1, the boundary conditions become

$$\dot{S}_{021} = \dot{S}_{022} = 0 \quad \text{on } x_2 = \pm h. \tag{191}$$

By referring these to the principal axes through use of the components $\dot{S}'_{0ij}$ and the angle ϕ we obtain

$$\begin{aligned}\dot{S}_{021} &= 2(\beta+\gamma-\sigma_2)\psi'_{,12}\sin\phi\cos\phi \\ &+ [\gamma\psi'_{,22}-(\gamma-\sigma_2)\psi'_{,11}]\cos^2\phi \\ &+ [\alpha\psi'_{,11}-(\gamma-\sigma_2)\psi'_{,22}]\sin^2\phi = 0 \quad \text{on } x_2 = \pm h. \end{aligned} \tag{192}$$

In order to express the second boundary condition in terms of ψ' we first differentiate with respect to x_1, then express, through (181), this derivative in terms of derivatives with respect to x'_1 and x'_2 and finally use the equilibrium equation (187) to eliminate the derivatives of $\dot{p}'$ with respect to x'_1 and x'_2. Thus, we replace the second equation in (191) by

$$\begin{aligned}\frac{\partial \dot{S}_{022}}{\partial x_1} &= [(\alpha+2\gamma-2\sigma_2)\psi'_{,112}-(3\gamma-\sigma_2)\psi'_{,222}]\sin^2\phi\cos\phi \\ &+ [(3\gamma-2\sigma_2)\psi'_{,122}-(2\alpha+\gamma-\sigma_2)\psi'_{,111}]\sin\phi\cos^2\phi \\ &- [(2\beta+\gamma-\sigma_2)\psi'_{,112}+\gamma\psi'_{,222}]\cos^3\phi \\ &- [(2\beta+\gamma-\sigma_2)\psi'_{,122}+\alpha\psi'_{,111}]\sin^3\phi = 0 \quad \text{on } x_2 = 0. \end{aligned} \tag{193}$$

Again we note that the derivatives in these expressions are with respect to x'_1 and x'_2. There are various different ways of writing these boundary conditions, either in terms of x'_1 and x'_2 or in terms of x_1 and x_2, but all are equally complicated. The ones given here are different from, but equivalent to, those used in Connor and Ogden (1996), for example. The coefficients in the boundary conditions depend, as do those in (188), on α, β and γ, but, additionally, they depend on σ_2, or equivalently σ_{22}.

4.3 Bifurcation Equations

We now seek Fourier-type solutions of equation (188) of the form

$$\psi = A\exp(iksx_2 - ikx_1), \tag{194}$$

where k is the 'wave' number and s is to be determined. Substitution of (194) into the equation of equilibrium (188) leads to the quartic equation

$$s^4 - 2\epsilon s^3 + [4\delta + 2 + \epsilon^2]s^2 - 2(1+2\delta)\epsilon s + 1 + (1+\delta)\epsilon^2 = 0 \tag{195}$$

for s, where δ is a material constant defined by

$$\delta = \frac{\alpha+\gamma-2\beta}{2(\beta+\sqrt{\alpha\gamma})}. \tag{196}$$

Thus, the material constants appear only through the combination δ in the expressions for s. Note that it follows from the strong ellipticity conditions (190) that $\delta > -1$.

Let s_1, s_2 and their complex conjugates $\bar{s}_1, \bar{s}_2$ be the solutions for s. Then, it can be shown that these are given explicitly as

$$s_1 = \frac{1}{2}\epsilon - \frac{1}{2}\sqrt{\epsilon^2 - 4\delta} - \mathrm{i}\sqrt{1+\delta}, \quad \bar{s}_1 = \epsilon - s_2,$$
$$s_2 = \frac{1}{2}\epsilon + \frac{1}{2}\sqrt{\epsilon^2 - 4\delta} - \mathrm{i}\sqrt{1+\delta}, \quad \bar{s}_2 = \epsilon - s_1 \tag{197}$$

for the case in which $\epsilon^2 \geq 4\delta$, and similarly for $\epsilon^2 < 4\delta$.

We now write the general solution in the form

$$\psi = (A_1 \mathrm{e}^{\mathrm{i}ks_1x_2} + A_2 \mathrm{e}^{\mathrm{i}ks_2x_2} + A_3 \mathrm{e}^{\mathrm{i}k\bar{s}_1x_2} + A_4 \mathrm{e}^{\mathrm{i}k\bar{s}_2x_2})\mathrm{e}^{-\mathrm{i}kx_1}, \tag{198}$$

where $\mathbf{A} = (A_1, A_2, A_3, A_4)$ is a set of constants. Substitution of (198) into the boundary conditions (192) and (193) leads, after some manipulation, to an equation of the form

$$\mathcal{M}\mathbf{A} = \mathbf{0}, \tag{199}$$

where $\mathcal{M}$ is a 4×4 matrix having the structure

$$\mathcal{M} = \begin{bmatrix} f_1 & f_2 & \bar{f}_1 & \bar{f}_2 \\ \bar{f}_2 & \bar{f}_1 & f_2 & f_1 \\ f_1 a_1 & f_2 a_2 & \bar{f}_1/\bar{a}_1 & \bar{f}_2/\bar{a}_2 \\ \bar{f}_2 a_1 & \bar{f}_1 a_2 & f_2/\bar{a}_1 & f_1/\bar{a}_2 \end{bmatrix}. \tag{200}$$

The terms in $\mathcal{M}$ are defined by

$$f_i = f(s_i) = (s_i - \lambda)(s_i + \lambda^{-1})^2/(1+\delta) + 3s_i - \lambda$$
$$+ 2\lambda^{-1} + \sigma(s_i - \lambda), \quad i \in \{1, 2\}, \tag{201}$$

with

$$\sigma = 1 - \sigma_{22}/\sqrt{\alpha\gamma}, \tag{202}$$

and

$$a_i = \mathrm{e}^{-\mathrm{i}khs_i}, \quad i \in \{1, 2\}, \tag{203}$$

and the overbar signifies complex conjugate.

For (199) to yield a non-trivial solution for $\mathbf{A}$, that is for bifurcation into an inhomogeneous incremental mode of deformation, we must have

$$\det \mathcal{M} = 0. \tag{204}$$

This equation, after substitution of the values of s_1 and s_2 from (197), yields a *bifurcation equation*, i.e. a relationship between the parameters λ, σ, δ and kh. For parameter values satisfying this equation bifurcation can occur. After some manipulation this equation can be made explicit and, in fact, decouples into the two equations

$$(1 + 2\bar{\beta} + 2\sigma - \sigma^2)\sinh\kappa = \pm(\sigma^2 + 2\sigma\bar{\beta} + 1)\frac{\sin\left[\kappa\sqrt{(\bar{\beta}-1)/2}\right]}{\sqrt{(\bar{\beta}-1)/2}}, \tag{205}$$

in which we have introduced the notations

$$\bar{\beta} = \beta/\sqrt{\alpha\gamma}, \quad \kappa = kh\sqrt{1+\delta}. \tag{206}$$

Note that it follows from the strong ellipticity conditions (190) that $\bar{\beta} > -1$. Equation (205) applies when $\bar{\beta} > 1$. For $\bar{\beta} = 1$ it is replaced by

$$(\sigma+1)(3-\sigma)\sinh\kappa = \pm(\sigma+1)^2\kappa, \tag{207}$$

and for $\bar{\beta} < 1$ by

$$(1+2\bar{\beta}+2\sigma-\sigma^2)\sinh\kappa = \pm(\sigma^2+2\sigma\bar{\beta}+1)\frac{\sinh\left[\kappa\sqrt{(1-\bar{\beta})/2}\right]}{\sqrt{(1-\bar{\beta})/2}}. \tag{208}$$

Collectively, equations (205), (207) and (208) are valid for a general form of strain-energy function. They provide relationships between σ, $\bar{\beta}$ and κ, and, by comparison with (204), the number of parameters has therefore effectively been reduced by one. For a given value of $\bar{\beta}$, for example, the equations determine the values of σ for which bifurcation can occur at any given value of κ. In view of the connection

$$(\epsilon^2-4\delta)/(1+\delta) = 2(\bar{\beta}-1), \tag{209}$$

we see that δ then depends on ϵ, and neither variable appears explicitly in the equations. The dependence on δ and hence on ϵ is reflected in the value of κ as defined by $(206)_2$. Likewise, the variable σ reflects changes in material properties through the scaling $\sqrt{\alpha\gamma}$ in (202).

4.4 Examples

In the unstrained configuration $\epsilon = 0$ we have $\bar{\beta} = 1, \delta = 0$ and equation (207) is applicable. It provides a relationship between $\sigma = 1 - \sigma_{22}/\mu$ and $\kappa = kh$. This result is shown in Figure 5.

The + and − solutions in (207) in this case correspond to *flexural* (or antisymmetric) and *barrelling* (or symmetric) modes respectively. Note that in the natural configuration, where $\sigma = 1$, (207) reduces to $\sinh\kappa = \pm\kappa$, which cannot hold except in the trivial situation $\kappa = 0$. Moreover, for $\kappa > 0$, $\sinh\kappa > \pm\kappa$. For the considered problem we may therefore give an *exclusion condition* which requires, for $\epsilon = 0$, that the left-hand side of (207) is greater than the right-hand side. The resulting inequalities may be rearranged as

$$(3-\sigma)\frac{\sinh\kappa}{\kappa} > \sigma + 1 > 0. \tag{210}$$

The upper − curve in the figure lies outside the region of stability.

For $\epsilon = 0$, we have $\kappa = kh$, but (207) and hence (210) is also valid for $\bar{\beta} = 1$ if $\epsilon \neq 0$, in which case $4\delta = \epsilon^2$ and κ is given by $(206)_2$. The bifurcation curves are the same as in Figure 5 but both the horizontal and vertical axes depend on ϵ. In fact, it can be shown, using (189), that when $\bar{\beta} = 1$ holds identically the strain-energy function must be of the form

$$\tilde{W}(\lambda) = 2\mu(\lambda + \lambda^{-1} - 2)$$

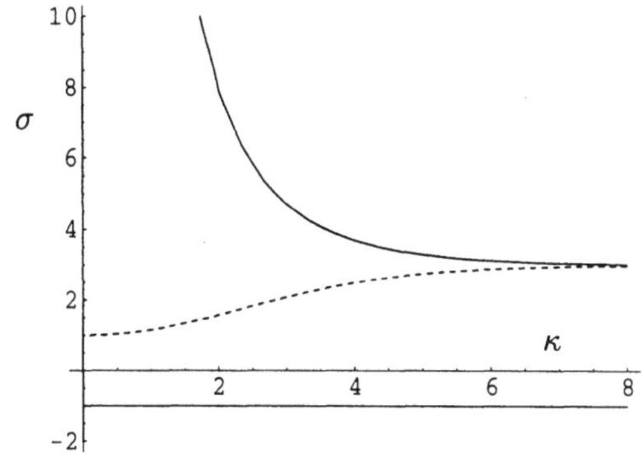

Figure 5. Plot of σ against κ for the bifurcation curves associated with the flexural modes (dashed curves) and barrelling modes (continuous curves): note that the two modes coincide for $\sigma = -1$ and that these are independent of κ.

for any plane deformation (with $\lambda_3 = 1$). This is the specialization to this case of the Varga material, which corresponds to $m = 1$ in (141), as mentioned in Section 3.3. For this material we have $\sqrt{\alpha\gamma} = 2\mu/\sqrt{4+\epsilon^2}$ and the dependence of σ on ϵ is then given by (202).

Another special case corresponds to $\delta = 0$, for which the material is necessarily neo-Hookean with strain energy for the present deformation given by

$$\tilde{W}(\lambda) = \frac{1}{2}\mu(\lambda^2 + \lambda^{-2} - 2).$$

Then we have $\bar{\beta} = 1 + \epsilon^2/2$ and $\sqrt{\alpha\gamma} = \mu$.

More generally than for these special cases the exclusion condition associated with (205) is

$$(1 + 2\bar{\beta} + 2\sigma - \sigma^2)\sinh\kappa > \pm(\sigma^2 + 2\sigma\bar{\beta} + 1)\frac{\sin\left[\kappa\sqrt{(\bar{\beta}-1)/2}\right]}{\sqrt{(\bar{\beta}-1)/2}}, \tag{211}$$

and similarly for (208).

In Figure 6 we plot σ against κ for each of the fixed values $0, 3, 6, 9$ of $\bar{\beta}$ in $(a), (b), (c), (d)$ respectively on the basis of equations (205) and (208). In each case the $+$ and $-$ mode curves are shown together for comparison. In general, for each value of $\bar{\beta}$ results for different values of ϵ (or, equivalently, δ) may be read off by adjusting the horizontal scale and, for different values of $\sqrt{\alpha\gamma}$, by adjusting the vertical scale. For the neo-Hookean material fixing $\bar{\beta}$ is equivalent to fixing ϵ. The results in (a) are similar to those for $\bar{\beta} = 1$, but in (b)–(d) they are oscillatory and reflect the presence of the trigonometric term in (205). Thus, whether a flexural or barrelling type of mode of bifurcation has priority depends on the value of κ. Actually, when $\epsilon \neq 0$ the symmetry is destroyed and the $+$ and $-$ modes can no longer be interpreted strictly as flexural and barrelling modes.

The analysis in this section is based on that given by Ogden and Connor (1995) and Connor and Ogden (1996). Some references to other papers dealing with the linear theory for incompressible elastic bodies are Sawyers and Rivlin (1974), Dowaikh and Ogden (1990) and Ogden

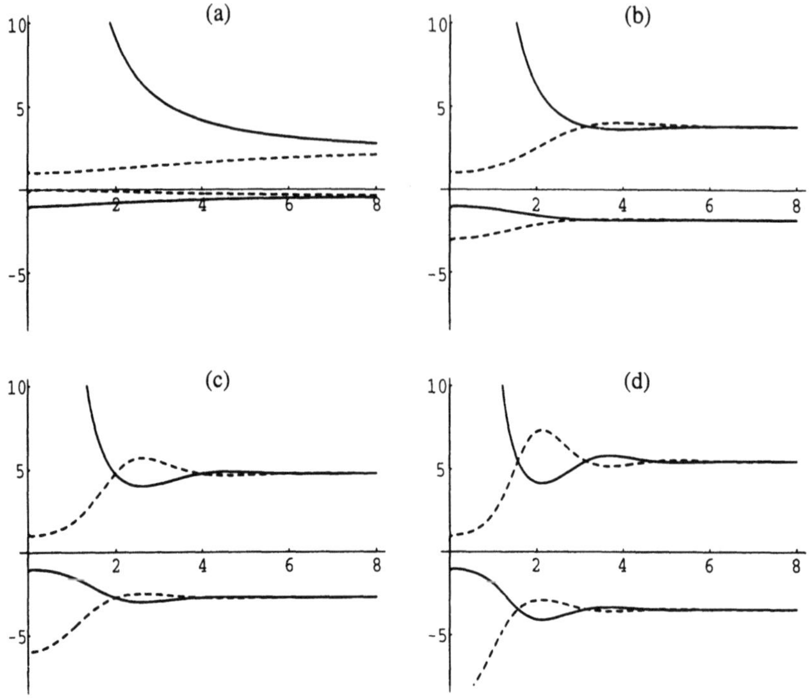

Figure 6. Plot of σ (vertical scale) against κ (horizontal scale) for the bifurcation curves associated with the flexural modes (dashed curves) and barrelling modes (continuous curves) with the following values of $\bar{\beta}$: (a) 0, (b) 3, (c) 6, (d) 9.

and Roxburgh (1993). Nonlinear stability analysis for such problems has been reviewed in the recent paper by Fu and Ogden (1999) in which detailed references to the literature for both the linear and nonlinear theory are given.

5 Pseudo-Elasticity and Global Stability

5.1 The Equations of Pseudo-Elasticity

The term *pseudo-elasticity* is used in different contexts for somewhat different concepts. For example, Fung (1981) used the term to describe the hysteretic response of soft biological tissues in which the loading path was modelled using an elastic stress-strain law and the unloading path by a different elastic stress-strain law. It is also used in connection with phase transitions where there is a discontinuity in the deformation gradient across the phase boundary. This is often modelled using a non-convex form of the (elastic) strain-energy function (Ericksen, 1991). Here, we use the term for situations where there is a change in the form of strain-energy function (i.e. a change in the material properties) occasioned by, for example, deformation-induced damage. This change may occur continuously, as in the progressive stress softening arising in the so-called

Mullins effect in rubberlike solids and associated with breakage of molecular crosslinks, detachment of attachments between long-chain molecules and filler particles or breakage of molecules at their fully-extended length. This effect is illustrated for simple tension in Figure 7(a). The unloading curve shown falls below the loading curve and, for any given value of the stretch, the stress on unloading is therefore less than that on loading. This is referred to as *stress softening*. Alternatively, the change may occur discontinuously, where sudden partial failure of the material is accompanied by a drop in stress when a critical value of the deformation is reached. This is illustrated in Figure 7(b). Note that the (strain) energy (the area under the curve in this example) is unchanged at the point of discontinuity. In each case there is generally a residual strain remaining when the stress is removed.

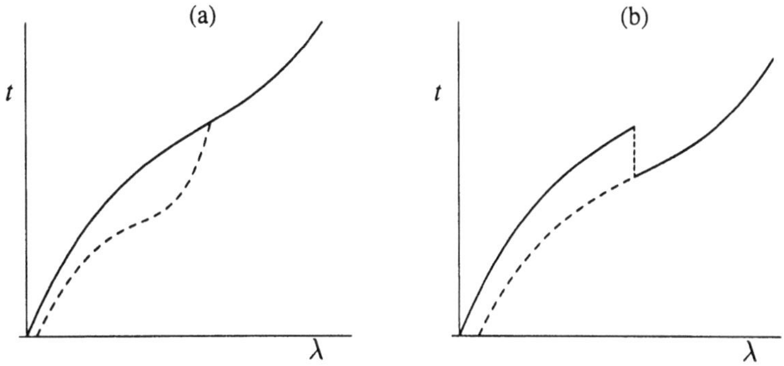

Figure 7. Sketch of the nominal stress t against stretch λ for loading (continuous curves) and unloading (dashed curves) in simple tension: (a) continuous stress, (b) discontinuous stress.

For a detailed discussion of the Mullins effect we refer to the recent paper by Ogden and Roxburgh (1999a). Here we focus attention on the case in which the change of material properties is discontinuous. In order to describe this behaviour we use a model introduced by Lazopoulos and Ogden (1998, 1999) and adapted for the continuous case by Ogden and Roxburgh (1999a, b). In this we modify the strain-energy function W, a function of the deformation gradient $\mathbf{F}$, by incorporating an additional variable η into the function. This is referred to as a *damage* or *softening variable*. Thus, we write

$$W = W(\mathbf{F}, \eta), \tag{212}$$

which we call a *pseudo-energy function*.

The variable η may be either active or inactive and may be switched on or off according to some appropriate criterion, which will be discussed below. In the case of the Mullins effect it is switched on at the start of unloading. When η is inactive, it is set at a constant value, which, without loss of generality, we take as unity. Then, (212) yields the strain-energy function W_0 defined by

$$W_0(\mathbf{F}) = W(\mathbf{F}, 1). \tag{213}$$

When η is active it is taken to depend on the deformation $\mathbf{F}$, $\eta = \eta(\mathbf{F})$ say. In this case, (212) becomes another strain-energy function, which we write as

$$w(\mathbf{F}) = W(\mathbf{F}, \eta(\mathbf{F})), \tag{214}$$

and the associated nominal stress tensor is

$$\mathbf{S} = \frac{\partial w}{\partial \mathbf{F}} - p\mathbf{F}^{-1} = \frac{\partial W}{\partial \mathbf{F}}(\mathbf{F}, \eta) + \frac{\partial W}{\partial \eta}(\mathbf{F}, \eta)\frac{\partial \eta}{\partial \mathbf{F}}(\mathbf{F}) - p\mathbf{F}^{-1}, \quad \det \mathbf{F} = 1, \tag{215}$$

for an incompressible material. The corresponding expression for an unconstrained material is obtained from (215) by dropping the term in p and the constraint. Clearly, for w to be an objective function of $\mathbf{F}$, we require that $\eta(\mathbf{F})$ be objective.

As yet we have not specified the form of $\eta(\mathbf{F})$, nor have we defined the criterion for switching η on or off when it is discontinuous. Before discussing this we mention a more general formulation that admits the possibility of the relationship between $\mathbf{F}$ and η being non-unique. We write this relationship in the general implicit form

$$C(\mathbf{F}, \eta) = 0, \tag{216}$$

where C is some function which provides a constraint between the variables $\mathbf{F}$ and η. Associated with this constraint we introduce a Lagrange multiplier, q say, so that, instead of (215), we may write

$$\mathbf{S} = \frac{\partial W}{\partial \mathbf{F}}(\mathbf{F}, \eta) + q\frac{\partial C}{\partial \mathbf{F}}(\mathbf{F}, \eta) - p\mathbf{F}^{-1}, \quad \det \mathbf{F} = 1. \tag{217}$$

Comparison of (215) and (217) after use of the connection

$$\frac{\partial \eta}{\partial \mathbf{F}} = -\frac{\partial C}{\partial \mathbf{F}}\frac{1}{C_\eta},$$

obtained from (216), shows that

$$q = -W_\eta / C_\eta, \tag{218}$$

where the subscript indicates partial differentiation with respect to η.

Since the function C in (216) is quite arbitrary (subject only to the objectivity requirement), i.e. it is not a specific constraint, and, moreover, the dependence of W on η is similarly at our disposal, no generality is lost by setting $C = W_\eta$, so that the expression (215), or equivalently (217), reduces to

$$\mathbf{S} = \frac{\partial W}{\partial \mathbf{F}}(\mathbf{F}, \eta) - p\mathbf{F}^{-1}. \tag{219}$$

Thus, we adopt the implicit relationship

$$\frac{\partial W}{\partial \eta}(\mathbf{F}, \eta) = 0 \tag{220}$$

between $\mathbf{F}$ and η. Then, the nominal stress is given by (219) whether η is active or inactive.

The theory can be generalized to allow for more than one η-type variable, as in the formulation described by Lazopoulos and Ogden (1998), but this is not necessary here. In the formulation based on (220), $\eta(\mathbf{F})$ can be selected so that the material symmetry of $w(\mathbf{F})$ in (214) either inherits the material symmetry of $W_0(\mathbf{F})$ or has material symmetry independent of that of $W_0(\mathbf{F})$. This is discussed in Ogden (2000). Hence switching on or off of η can change the material symmetry. Here, however, we use (220) and we restrict attention to isotropic material response.

When η is active the exclusion condition (82) for an unconstrained material is modified, on use of (220) and the chain rule for partial derivatives, to give

$$\operatorname{tr}\{(\boldsymbol{\mathcal{A}}^1\dot{\mathbf{F}})\dot{\mathbf{F}}\} - [\operatorname{tr}(\mathbf{S}_\eta\dot{\mathbf{F}})]^2/W_{\eta\eta} > 0, \tag{221}$$

which may be modified for the incompressible case by addition of the term in p appearing in (111). Clearly, the stability is strengthened or weakened according as the sign of $W_{\eta\eta}$ is negative or positive.

Henceforth in this article we restrict attention to incompressible materials. In the absence of body forces the equilibrium equation has the form

$$\operatorname{Div}\mathbf{S} = \mathbf{0}, \tag{222}$$

with $\mathbf{S}$ given by (219) whether or not η is active. If η is continuous as a function of position then (220) and (222) are the relevant field equations. If the relationship (220) between η and $\mathbf{F}$ is not in general one-to-one, as we are assuming, then there is the possibility that a given deformation $\mathbf{F}$ may correspond to different values of η. In particular, as $\mathbf{F}$ varies there may be generated in the reference configuration $\mathcal{B}_r$ a surface across which η may jump but across which $\mathbf{F}$ is continuous, that is a surface which acts as a switch for η. Let such a surface be denoted by $\mathcal{S}$. This surface then separates the parts of $\mathcal{B}_r$ corresponding to different forms of the strain-energy function. During the deformation process $\mathcal{S}$ will appear at some critical value of the deformation (if such a critical value is reached) and then move in $\mathcal{B}_r$ as the deformation proceeds. Let η^+ and η^- be the values of η on the two sides of $\mathcal{S}$.

In addition to the field equations (220) and (222) we now require jump conditions across $\mathcal{S}$. These were derived by Lazopoulos and Ogden (1998). Together, we have four equations, namely

$$\operatorname{Div}\mathbf{S} = \mathbf{0} \quad \text{in } \mathcal{B}_r - \mathcal{S}, \tag{223}$$

$$\frac{\partial W}{\partial \eta} = 0 \quad \text{in } \mathcal{B}_r - \mathcal{S}, \tag{224}$$

$$[W]_-^+ = 0 \quad \text{on } \mathcal{S}, \tag{225}$$

$$[\mathbf{S}^T\mathbf{N}]_-^+ = \mathbf{0} \quad \text{on } \mathcal{S}, \tag{226}$$

where $\mathbf{N}$ is the unit normal to $\mathcal{S}$ (in either sense). In (225) and (226) $[\cdot]_-^+$ denotes the difference in the enclosed quantity on the two sides of $\mathcal{S}$. For example,

$$[W]_-^+ = W(\mathbf{F},\eta^+) - W(\mathbf{F},\eta^-),$$

evaluated on $\mathcal{S}$.

Equation (226) is simply a statement that the traction is continuous across $\mathcal{S}$, while (225), which states that there is no jump in the strain energy across $\mathcal{S}$, was foreshadowed in the simple tension example at the beginning of this section. It provides the criterion for a switch in the material properties, i.e. for a jump in η. In general, the stress will be discontinuous across $\mathcal{S}$ even though the traction is continuous. Appropriate boundary conditions are appended to the equations as required, but they are not discussed at this point.

5.2 Isotropy

For an isotropic pseudo-elastic material we write the energy in the form

$$W(\lambda_1, \lambda_2, \lambda_3, \eta),$$

subject to the constraint $\lambda_1\lambda_2\lambda_3 = 1$. The principal Cauchy stresses are then given by

$$\sigma_i = \lambda_i \frac{\partial W}{\partial \lambda_i} - p \quad i \in \{1, 2, 3\},$$

as in the case when η is absent, and equation (224) becomes

$$\frac{\partial W}{\partial \eta}(\lambda_1, \lambda_2, \lambda_3, \eta) = 0, \tag{227}$$

so that η depends implicitly on the stretches. Since the material is incompressible we define the modified pseudo-energy function $\hat{W}(\lambda_1, \lambda_2, \eta)$ by

$$\hat{W}(\lambda_1, \lambda_2, \eta) = W(\lambda_1, \lambda_2, \lambda_1^{-1}\lambda_2^{-1}, \eta) \tag{228}$$

analogously to (112), so that

$$\sigma_1 - \sigma_3 = \lambda_1 \hat{W}_1, \quad \sigma_2 - \sigma_3 = \lambda_2 \hat{W}_2, \tag{229}$$

where the subscripts 1 and 2 attached to $\hat{W}$ denote partial derivatives with respect to λ_1 and λ_2 respectively. Equation (227) is modified to

$$\frac{\partial \hat{W}}{\partial \eta}(\lambda_1, \lambda_2, \eta) = 0. \tag{230}$$

We now specialize to the case of equibiaxial deformations and we set $\lambda_2 = \lambda_3 = \lambda, \lambda_1 = \lambda^{-2}$ with $\sigma_2 = \sigma_3$ and define the function $\bar{W}$ by

$$\bar{W}(\lambda, \eta) = \hat{W}(\lambda^{-2}, \lambda, \eta) = W(\lambda^{-2}, \lambda, \lambda, \eta). \tag{231}$$

This should be distinguished from its counterpart for plane strain defined by (130) in the absence of η. It follows from (229) that

$$2(\sigma_2 - \sigma_1) = \lambda \bar{W}_\lambda(\lambda, \eta) \tag{232}$$

and $\hat{W}_2 = 0$, while (230) becomes simply

$$\bar{W}_\eta(\lambda, \eta) = 0 \tag{233}$$

where the subscripts λ and η denote partial derivatives.

If we set $\sigma_1 = 0$ in (232) and write $t = \sigma/\lambda$ for the principal Biot stress associated with the equibiaxial deformation, where $\sigma = \sigma_2 = \sigma_3$, then, using (233), we obtain

$$2\frac{\mathrm{d}t}{\mathrm{d}\lambda} = \bar{W}_{\lambda\lambda} - \bar{W}_{\lambda\eta}^2/\bar{W}_{\eta\eta}. \tag{234}$$

This is a one-dimensional specialization of the coefficient of the quadratic term in $\dot{\mathbf{F}}$ in the expression on the left-hand side of (221). Again, the term $W_{\eta\eta}$ modifies the stability requirement $(\mathrm{d}t/\mathrm{d}\lambda > 0)$ as compared with the situation in which η is inactive.

5.3 An Illustrative Model

To illustrate the theory we consider a pseudo-energy function of the form

$$W(\lambda_1, \lambda_2, \lambda_3, \eta) = W_0(\lambda_1, \lambda_2, \lambda_3) + (\eta - 1)N(\lambda_1, \lambda_2, \lambda_3) + \phi(\eta), \tag{235}$$

where W_0 and N are independent of η. We set

$$\phi(1) = 0, \tag{236}$$

so that when $\eta = 1$ (235) is just the strain energy W_0 of an isotropic elastic material with η inactive. For equibiaxial deformations we write

$$\bar{W}(\lambda, \eta) = \bar{W}_0(\lambda) + (\eta - 1)\bar{N}(\lambda) + \phi(\eta) \tag{237}$$

in accordance with (231), and equation (233) then gives

$$\phi'(\eta) = -\bar{N}(\lambda), \tag{238}$$

where the prime indicates differentiation with respect to η.

For an elastic material with $\eta \equiv 1$ we have, when $\sigma_1 = 0$, $2t = \bar{W}_0'(\lambda)$ and it is physically sensible to take $t > 0$ to correspond to $\lambda > 1$. Thus, we assume that

$$\bar{W}_0(1) = 0, \quad \bar{W}_0'(1) = 0, \tag{239}$$

$$\bar{W}_0'(\lambda) > 0\,(< 0) \quad \text{for} \quad \lambda > 1\,(< 1). \tag{240}$$

We assume further that $\bar{N}(\lambda)$ also satisfies the conditions (239) and (240).

When η is active we have

$$\bar{W}_\lambda(\lambda, \eta) = \bar{W}_0'(\lambda) + (\eta - 1)\bar{N}'(\lambda). \tag{241}$$

We now consider $-\phi'(\eta)$ to have the non-convex form shown in Figure 8 so that for certain values of $-\phi'(\eta)$ there correspond two or three values of η.

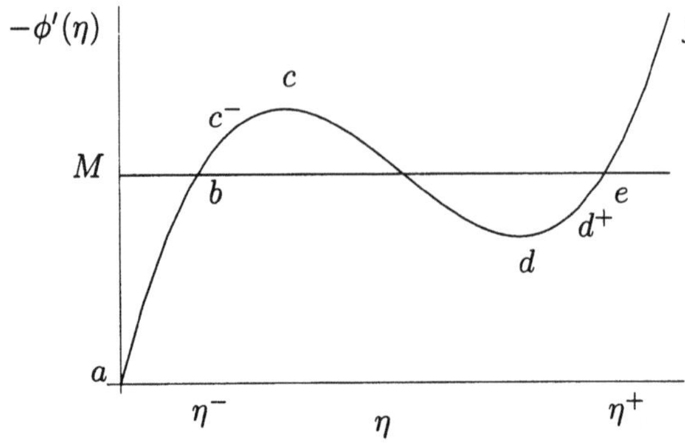

Figure 8. Model form of the function $-\phi'(\eta)$ showing the Maxwell value M.

For the model (237) the jump condition (225) becomes

$$[\bar{W}]_-^+ \equiv [\eta\bar{N}(\lambda) + \phi(\eta)]_-^+ = M[\eta]_-^+ + [\phi(\eta)]_-^+ = 0. \tag{242}$$

For a given function ϕ this defines the *Maxwell value*, M say, of $\bar{N}(\lambda)$ for which the energies associated with the values η^- and η^+ shown in Figure 8 are equal. It also defines the values of η^- and η^+ uniquely. For the model (237) it should be noted that it is necessary to take η to be decreasing from 1 to the right along the horizontal axis, so that $\eta^- > \eta^+$. Otherwise, there is no restriction on the value of η and, in particular, it may become negative. The areas cut off the curve by the *Maxwell line* (the horizontal line with ordinate M) are equal since $M(\eta^- - \eta^+) = [\phi(\eta)]_-^+$. The left-hand side of this expression is the area of the rectangle of height M and length $\eta^- - \eta^+$ while the right-hand side is the (positive) area under the curve $-\phi'(\eta)$ between the values η^- and η^+. Let λ_m denote the value of λ corresponding to $\bar{N}(\lambda) = M$. Then, since it is assumed to satisfy (240), $\bar{N}(\lambda)$ is monotonic increasing in λ and it follows from equation (238) that λ is uniquely determined by $-\phi'(\eta)$.

To examine the global stability of the different parts of the curve we consider the energy function (237). As λ increases from 1 it is associated with one, two or three values of η corresponding to different parts of the curve. Let η_1 and η_2 be two distinct values of η satisfying (238) and such that $\eta_1 > \eta_2$ (bearing in mind that η decreases to the right along the horizontal axis). Then,

$$\bar{W}(\lambda, \eta_1) - \bar{W}(\lambda, \eta_2) = (\eta_1 - \eta_2)\bar{N}(\lambda) + \phi(\eta_1) - \phi(\eta_2). \tag{243}$$

It is easy to see that $(\eta_1 - \eta_2)\bar{N}(\lambda)$, which is positive, is the area of the rectangle of height $\bar{N}(\lambda)$ between η_1 and η_2, while $\phi(\eta_1) - \phi(\eta_2)$ is *minus* the area under the curve between the same values. Thus, if η_1 corresponds to the segment *ac* and η_2 to *cd* the expression in (243) is negative. Similarly, if η_1 corresponds to *cd* and η_2 to *df* it is positive. If η_1 corresponds to *ab* and η_2 to *de* (with $\lambda < \lambda_m$) it is negative, while for η_1 on *bc* and η_2 on *ef* (with $\lambda > \lambda_m$) it is positive. This means that *ab* is globally stable for $\lambda < \lambda_m$ and *ef* is globally stable for $\lambda > \lambda_m$ since these segments are associated with least energy.

To examine local stability we substitute (237) into (234) to obtain

$$2\frac{\mathrm{d}t}{\mathrm{d}\lambda} = \bar{W}_0''(\lambda) + (\eta - 1)\bar{N}''(\lambda) - \frac{\bar{N}'(\lambda)^2}{\phi''(\eta)} \tag{244}$$

and note that $\phi''(\eta)$ is positive on ac and df, negative on cd and zero at c and d. Hence, (244) is positive on ac up to some point, c^- say, just to the left of c and on df to the right of some point, d^+ say, near d, at each of which $\mathrm{d}t/\mathrm{d}\lambda = 0$. We therefore have local stability on ac^- and d^+f. As λ increases, stability is lost at c^-, which is after the point b where global stability is lost. The segment d^+f is also locally stable but c^-c and dd^+ are locally unstable. Moreover, the segment cd is locally stable but not globally stable.

Figure 9(a) shows a schematic plot of $2(\sigma_2 - \sigma_1)/\lambda \equiv \bar{W}_\lambda$ against λ corresponding to the considered form of ϕ, but with the part of the curve between c^- (local maximum) and d^+ (local minimum) omitted. The precise form of the part bc^-cdd^+ need not be specified since there is a jump to e when the point b is reached on loading. Figure 9(b) shows an example in which the axis $\bar{W}_\lambda = 0$ is reached on unloading prior to the point d^+ (which is not shown in this case) and a residual value, λ_r say, of the stretch is associated with $\bar{W}_\lambda = 0$.

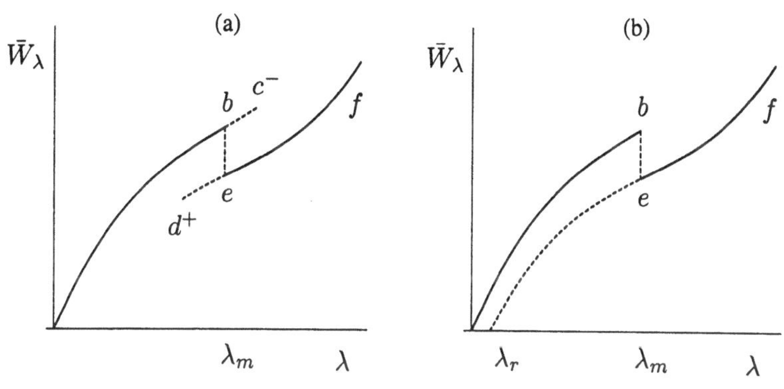

Figure 9. Schematic plot of $\bar{W}_\lambda$ against λ: (a) corresponding to the model shown in Figure 8, (b) model with unloading curve meeting the axis $\bar{W}_\lambda = 0$.

As λ increases from 1 the branch ab is traversed since it is both locally and globally stable. At the value λ_m the points b and e are energetically equivalent and are both equally possible states of globally stable equilibrium, but as λ increases beyond λ_m the branch ef is the only globally and locally stable segment. Thus, η must jump from η^- to η^+ as λ passes through the value λ_m (a negative jump). This is analogous to the procedure described by Ericksen (1975) in respect of coexistent phases associated with extension of an elastic bar with a nonconvex strain-energy function. In Ericksen's theory λ is discontinuous, the phase boundary separates regions with different deformation gradients and the stress is continuous. Here, the deformation gradient is continuous but there is discontinuity in the stress described by the (negative) jump in $\bar{W}_\lambda$. As this jump takes place the dependence of η on λ changes and hence the form of the constitutive law changes.

We may interpret this switch as being associated with a discontinuous change in material properties induced by damage arising at the critical value λ_m of the stretch. With this interpretation, the segment bc is not accessible. On the other hand, d^+e is accessible on unloading since once the material properties have changed they will not in general change back because the damage is not in general recoverable. The unloading path therefore corresponds to fd^+. If d^+ is reached on unloading then a jump back to ab is possible and this may be associated with partial recovery of damage. With a different choice of the function ϕ the form of the dependence of $\bar{W}_\lambda$ on λ can be arranged so that there is no jump reversal, as exemplified in Figure 9(b). In the problem discussed in Section 6, a model of this kind will be used.

6 Inflation-Deflation of a Pseudo-Elastic Spherical Shell

6.1 Geometry and Equilibrium Equations

We consider a spherical shell of incompressible isotropic pseudo-elastic material with initial geometry defined by

$$A \leq R \leq B, \quad 0 \leq \Theta \leq \pi, \quad 0 \leq \Phi \leq 2\pi, \tag{245}$$

where (R, Θ, Φ) are spherical polar coordinates.

The spherically-symmetric deformation is defined by

$$r = Rf(R), \quad \theta = \Theta, \quad \varphi = \Phi, \tag{246}$$

where (r, θ, φ) are spherical polar coordinates in the current configuration, the function f is defined by

$$f(R) = \left(1 + \frac{a^3 - A^3}{R^3}\right)^{1/3} \equiv \lambda, \tag{247}$$

and the notation $\lambda = \lambda_2 = \lambda_3$ is introduced for the equibiaxial stretch associated with the principal directions corresponding to the θ and φ coordinates. From the incompressibility condition (11) we have $\lambda_1 = \lambda^{-2}$.

We use the notations

$$\lambda_a = a/A, \quad \lambda_b = b/B \tag{248}$$

respectively for the values of λ at the inner and outer boundaries of the spherical shell. Then, using (247), we obtain

$$\lambda_a^3 - 1 = \left(\frac{R}{A}\right)^3 (\lambda^3 - 1) = \left(\frac{B}{A}\right)^3 (\lambda_b^3 - 1), \tag{249}$$

as in Ogden (1997). Since we are considering inflation, we have

$$\lambda_a \geq \lambda \geq \lambda_b \geq 1, \tag{250}$$

with equality holding in (250) if and only if $\lambda \equiv 1$, i.e. the shell is undeformed.

For the considered spherically-symmetric deformation, the only equilibrium equation not satisfied trivially is the radial equation

$$\frac{\mathrm{d}\sigma_1}{\mathrm{d}r}+\frac{2}{r}(\sigma_1-\sigma_2)=0, \tag{251}$$

where σ_1, σ_2 are the principal Cauchy stresses associated with the directions r and θ (and φ by symmetry). Using (246) and (247) we change variables from r to λ by means of the formula

$$r\frac{\mathrm{d}\lambda}{\mathrm{d}r}=-\lambda(\lambda^3-1) \tag{252}$$

and replace (251) by

$$\lambda\frac{\mathrm{d}\sigma_1}{\mathrm{d}\lambda}=2\frac{\sigma_1-\sigma_2}{\lambda^3-1}. \tag{253}$$

The boundary conditions are taken in the form

$$\sigma_1=\begin{cases}-P & \text{on } r=a\\ T & \text{on } r=b\end{cases} \tag{254}$$

corresponding to pressure loading on the inside of the sphere and hydrostatic tension on the outside.

On integration of (253) and use of (232) and the boundary conditions (254) we obtain

$$P+T=\int_{\lambda_b}^{\lambda_a}\frac{\bar{W}_\lambda(\lambda,\eta)}{\lambda^3-1}\mathrm{d}\lambda. \tag{255}$$

For an elastic material this formula can be found in Ogden (1997), but here it includes the additional variable η, which depends on λ when η is active. Note that from (255) we may deduce that the effect of a pressure on the inside of the sphere is equivalent to a tension on the outside.

6.2 A Simple Model

For simplicity of application we consider here a pseudo-energy function which, for equibiaxial deformations, has the form

$$\bar{W}(\lambda,\eta)=\eta\bar{N}(\lambda)+\phi(\eta). \tag{256}$$

This is obtained by setting $\bar{W}_0=\bar{N}$ in (237) and we note that the connection between η and λ is still given by (238) when η is active. We now simplify the form of ϕ by taking η to be inactive for λ below the Maxwell value λ_m and we take ϕ to be such that $\phi'(\eta)$ is linear in η when the latter is active. The form of the function ϕ is fixed by choosing the gradient of the straight line and the value η_0 for which $\phi'(\eta_0)=0$. Since η decreases to the right along the horizontal axis, as in Section 5, the gradient is positive and we denote it by m. The Maxwell value M of $\bar{N}(\lambda)$ is then obtained from the jump condition (242), with $\eta^-=1$ and η^+ now denoted by η_m. This gives

$$(1-\eta_m)M=\phi(\eta_m). \tag{257}$$

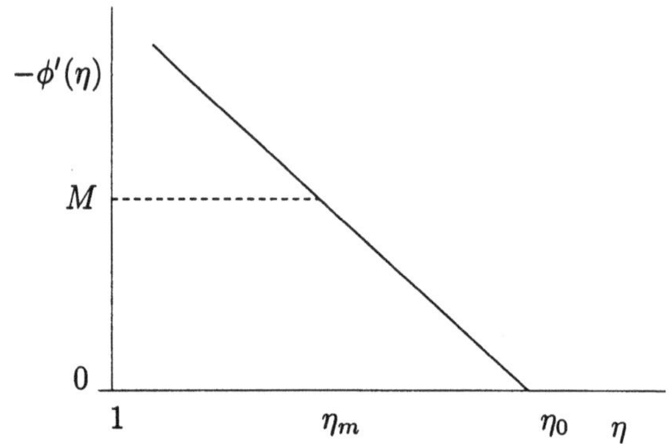

Figure 10. Linear model for $-\phi'(\eta)$.

The function $-\phi'(\eta)$ now has the form shown in Figure 10 and we note that $\phi''(\lambda) < 0$. The choice of model here removes the need to allow the inverse of ϕ' to be multivalued.

The argument concerning global stability is similar to that in Section 5 but less involved. As λ increases from 1, η is fixed at 1 until λ reaches λ_m. Then, at $\lambda = \lambda_m$, η jumps to the lower value η_m (> 0). As λ increases from λ_m, $-\phi'(\eta)$ follows the linear path shown with η increasing. Subsequently, when λ is decreased it passes through λ_m and η decreases continuously, following the linear path. It does not jump back to 1 when $\lambda = \lambda_m$ for the reasons discussed in Section 5.

The difference $\bar{N}(\lambda) - [\eta\bar{N}(\lambda) + \phi(\eta)]$ of the energies on the deformation path with $\eta \equiv 1$ and the linear path with η active is then calculated using the expression for $\phi(\eta)$ given by

$$\phi(\eta) = (1-\eta)M - \frac{1}{2}m(\eta - \eta_m)^2. \tag{258}$$

Note that, unlike for the model used in Section 5, we do *not* set $\phi(1) = 0$ since ϕ does not become operative until η jumps from 1 to η_m. It is possible, however, that η could reach the value 1 on the linear path at a suitably large value of λ.

From (258) and (238) we obtain η explicitly in terms of λ in the form

$$\eta - \eta_m = [\bar{N}(\lambda) - M]/m. \tag{259}$$

The energy difference is then obtained as

$$\frac{1}{2}m(\eta - \eta_m)(2 - \eta - \eta_m). \tag{260}$$

This is clearly positive for $\lambda > \lambda_m$, i.e. for $\eta > \eta_m$, provided $\eta < 2 - \eta_m$, which is greater than 1. (This restriction is a reflection of the very simple model adopted and can be removed by using a slightly less simple model.) The energy difference is negative for $\lambda < \lambda_m$, as in the corresponding situation discussed in Section 5.

For the model (256) we have

$$\bar{W}_\lambda = \eta\bar{N}'(\lambda). \tag{261}$$

According to the inequalities (240), with $\bar{W}_0 = \bar{N}$, we have $\bar{N}'(\lambda) > 0$ for $\lambda > 1$ and the right-hand side of (261) then has the sign of η. We now impose the condition $\eta_0 < 0$ so that as η decreases it vanishes at some $\lambda > 1$ and becomes negative as λ is decreased further. This is important in consideration of the equation (255) since as the load $P+T$ is removed the integrand can become negative in part of the interval of integration. It is then possible that the integral will vanish before the strain vanishes (i.e. before λ_a reduces to 1). In this case there will be a residual strain distribution throughout the sphere, and vanishing of (261) is then associated with a (local) residual value of λ, λ_r say, given (uniquely) by $\bar{N}(\lambda_r) = -\phi'(0)$, which is obtained from (238) by setting $\eta = 0$. For this situation Figure 9(b) illustrates the behaviour of $\bar{W}_\lambda$ as a function of λ. The (local) residual value of λ, denoted λ_r, corresponds to $\bar{W}_\lambda = 0$.

It is important to point out that the strong ellipticity inequality $2(\sigma_2 - \sigma_1)/(\lambda_2 - \lambda_1) \equiv \lambda\bar{W}_\lambda(\lambda,\eta)/(\lambda - \lambda^{-2}) > 0$ fails at $\eta = 0$. Moreover, it follows that local stability must therefore fail on the linear path at some value of λ between λ_m and λ_r. Because the deformation in the sphere is inhomogeneous this does not necessarily mean that overall stability of the spherical shell is lost. Investigation of such stability is quite complicated and will not be considered here.

6.3 Loading-Unloading Response

Since, by (249), λ_b is given in terms of λ_a, we regard the stretch λ_a at the inner boundary as a prescribed parameter. As described in Lazopoulos and Ogden (1999) for the same problem but for a different model, we take the loading path to start from $\lambda_a = 1$, at which point $\lambda \equiv 1$ throughout the sphere, and to increase from 1, so that $\lambda_b < \lambda_a$ is then given by (249). For $\lambda_a < \lambda_m$ the material properties throughout the sphere are described by the strain-energy function $\bar{N}(\lambda)$. When λ_a reaches and passes through λ_m the value of η jumps from 1 to η_m and the material in the inner part of the sphere for which $\lambda > \lambda_m$ is governed by the constitutive law (256), i.e. by the strain-energy function

$$\bar{w}(\lambda) = \frac{1}{2m}[\bar{N}(\lambda) - M]^2 + \eta_m[\bar{N}(\lambda) - M] + M, \tag{262}$$

which is obtained from (256) by use of (258) and (259). The material in the outer part of the sphere, where $\lambda_b < \lambda < \lambda_m$ is still governed by $\bar{N}(\lambda)$. When λ_b reaches and exceeds the value λ_m the whole sphere has material properties corresponding to (262).

Thus, during the loading process three cases need to be considered separately.

Case L(i): $\lambda_a \le \lambda_m$. In this case equation (255) is simply

$$P + T = \int_{\lambda_b}^{\lambda_a} \frac{\bar{N}'(\lambda)}{\lambda^3 - 1}\mathrm{d}\lambda, \tag{263}$$

and the whole sphere is governed by the strain energy $\bar{N}(\lambda)$, i.e. no change in material properties has occurred.

Case L(ii): $\lambda_b < \lambda_m < \lambda_a$. In this case we have

$$P + T = \int_{\lambda_b}^{\lambda_m} \frac{\bar{N}'(\lambda)}{\lambda^3 - 1}\mathrm{d}\lambda + \int_{\lambda_m}^{\lambda_a} \frac{\eta\bar{N}'(\lambda)}{\lambda^3 - 1}\mathrm{d}\lambda. \tag{264}$$

The Maxwell value λ_m of λ determines the value R_m of the radius R in the reference configuration that separates the (damaged and undamaged) parts of the material with different properties through the equation

$$R_m^3 = A^3(\lambda_a^3 - 1)/(\lambda_m^3 - 1), \tag{265}$$

which is obtained from (249) by setting $R = R_m$ and $\lambda = \lambda_m$. Clearly, as λ_a increases R_m increases, subject to the restriction in this case that $R_m \le B$. Thus, for $R_m < R \le B$ the material properties are described by $\bar{N}(\lambda)$ and for $A \le R < R_m$ by $\bar{w}(\lambda)$.

Case L(iii): $\lambda_b \ge \lambda_m$. For this case we have

$$P + T = \int_{\lambda_b}^{\lambda_a} \frac{\eta \bar{N}'(\lambda)}{\lambda^3 - 1} \mathrm{d}\lambda, \tag{266}$$

and the properties of the whole sphere are governed by $\bar{w}(\lambda)$.

We now consider unloading associated with each of the three loading cases above, with $P+T$ reducing to zero as λ_a decreases from its maximum value. Let λ_{al} and λ_{bl} respectively be the maximum values of λ_a and λ_b achieved during the loading process. Then, as for loading, we consider three separate cases.

Case UL(i): $\lambda_{al} \le \lambda_m$. During unloading $P + T$ is again given by (263) and the undeformed configuration is recovered when $P+T$ reaches zero, i.e. the behaviour is that of an (undamaged) elastic material.

Case UL(ii): $\lambda_{bl} < \lambda_m < \lambda_{al}$. Let λ^* be the value of λ at the radius corresponding to the boundary between the damaged and undamaged material. Then, from (249), if λ_a is the value during unloading from λ_{al} we have, using (265),

$$\lambda^{*3} - 1 = \frac{A^3}{R_m^3}(\lambda_a^3 - 1) = \frac{\lambda_m^3 - 1}{\lambda_{al}^3 - 1}(\lambda_a^3 - 1). \tag{267}$$

Since λ_m and λ_{al} are known this determines λ^* as λ_a decreases. Then, for the material in $R_m \le R \le B$, λ lies in the range $\lambda_b \le \lambda \le \lambda^*$, while for $A \le R \le R_m$ we have $\lambda^* \le \lambda \le \lambda_a$. The value of $P + T$ is then given by

$$P + T = \int_{\lambda_b}^{\lambda^*} \frac{\bar{N}'(\lambda)}{\lambda^3 - 1} \mathrm{d}\lambda + \int_{\lambda^*}^{\lambda_a} \frac{\eta \bar{N}'(\lambda)}{\lambda^3 - 1} \mathrm{d}\lambda, \tag{268}$$

so that λ_m in (264) is replaced by λ^*.

Case UL(iii): $\lambda_m \le \lambda_{bl}$. During unloading $P+T$ is again given by the formula (266), but it now applies, unlike in the loading case, when λ_b (and also λ_a if it reaches this value on unloading) drops below λ_m.

In either Case UL(ii) or UL(iii), $P + T$ may reduce to zero before λ_a reaches the value 1. This will definitely happen in Case UL(iii) since, as already noted, $\bar{N}'(\lambda)/(\lambda^3 - 1) > 0$ (which, incidentally, is a consequence of strong ellipticity for an elastic material), and hence the integrand in (266), while positive for $\lambda \ge \lambda_m$, must become zero at some $\lambda > 1$ such that $\lambda < \lambda_m$ because it is negative for $\lambda = 1$. When this happens the sphere retains a residual strain distribution in the absence of the boundary loading, and this is accompanied by a residual stress distribution.

The residual value, λ_{ar} say, of λ_a is found by setting $P + T$ to zero in either (268) or (266) as appropriate, performing the integration and then solving the resulting algebraic equation for λ_{ar}, making use of the connection

$$\lambda_{ar}^3 - 1 = \frac{B^3}{A^3}(\lambda_{br}^3 - 1) \tag{269}$$

between λ_{ar} and the corresponding residual value λ_{br} of λ_b.

For Case UL(ii) the residual radial stress is then calculated as

$$\sigma_1 = \int_\lambda^{\lambda_r^*} \frac{\bar{N}'(\lambda)}{\lambda^3 - 1}\mathrm{d}\lambda + \int_{\lambda_r^*}^{\lambda_{ar}} \frac{\eta\bar{N}'(\lambda)}{\lambda^3 - 1}\mathrm{d}\lambda \tag{270}$$

for $\lambda < \lambda_r^*$ and as

$$\sigma_1 = \int_\lambda^{\lambda_{ar}} \frac{\eta\bar{N}'(\lambda)}{\lambda^3 - 1}\mathrm{d}\lambda \tag{271}$$

for $\lambda > \lambda_r^*$, where λ_r^* is the residual value of λ^* obtained from

$$\lambda_r^{*3} - 1 = \frac{\lambda_m^3 - 1}{\lambda_{al}^3 - 1}(\lambda_{ar}^3 - 1). \tag{272}$$

For Case UL(iii) the residual radial stress is given by the formula (271) for all λ in the interval $\lambda_{br} \leq \lambda \leq \lambda_{ar}$.

Note that σ_1 is continuous across the boundary $R = R_m$. Thus, the traction is continuous across the boundary in accordance with the requirement (226). Locally, the stress difference is given by

$$2(\sigma_2 - \sigma_1) = \eta\lambda\bar{N}'(\lambda), \tag{273}$$

which is obtained from (232) with (256). Since σ_1 is continuous the hoop stress σ_2 suffers a discontinuity across the boundary R_m given by

$$2[\sigma_2(1) - \sigma_2(\eta_m)] = (1 - \eta_m)\lambda\bar{N}'(\lambda), \tag{274}$$

where $\sigma_2(1)$ and $\sigma_2(\eta_m)$ are the values of σ_2 evaluated at $R = R_m$ for the indicated values of η. Since this is positive the hoop stress suffers a reduction as the material properties change, i.e. as damage occurs.

6.4 A Numerical Example

To be specific we consider the strain-energy function given by (141) for $m = -3$, so that

$$\bar{N}(\lambda) = \frac{2\mu}{9}(\lambda^6 + 2\lambda^{-3} - 3)$$

and

$$\bar{N}'(\lambda)/(\lambda^3-1)=\frac{4\mu}{3}(\lambda^2+\lambda^{-1}+\lambda^{-4}).$$

For convenience, we now define the notations

$$f(\lambda)=\lambda^3-\lambda^{-3}+3\log\lambda, \tag{275}$$

$$g(\lambda)=\frac{1}{9}\lambda^9+\frac{1}{6}\lambda^6+\frac{1}{3}\lambda^3+2\log\lambda-\frac{2}{3}\lambda^{-3}-\frac{1}{3}\lambda^{-6} \tag{276}$$

to enable compact expressions for the integrals in $P+T$ to be given for the different cases, as follows.

Case L(i):

$$9(P+T)/4\mu=f(\lambda_a)-f(\lambda_b). \tag{277}$$

Case L(ii):

$$9(P+T)/4\mu=f(\lambda_m)-f(\lambda_b)+(\eta_0-\alpha_0)[f(\lambda_a)-f(\lambda_m)]+\alpha_0[g(\lambda_a)-g(\lambda_m)]. \tag{278}$$

Case L(iii):

$$9(P+T)/4\mu=(\eta_0-\alpha_0)[f(\lambda_a)-f(\lambda_b)]+\alpha_0[g(\lambda_a)-g(\lambda_b)]. \tag{279}$$

In (277)–(279) we have introduced the dimensionless constant α_0 defined by $\alpha_0=2\mu/3m$.

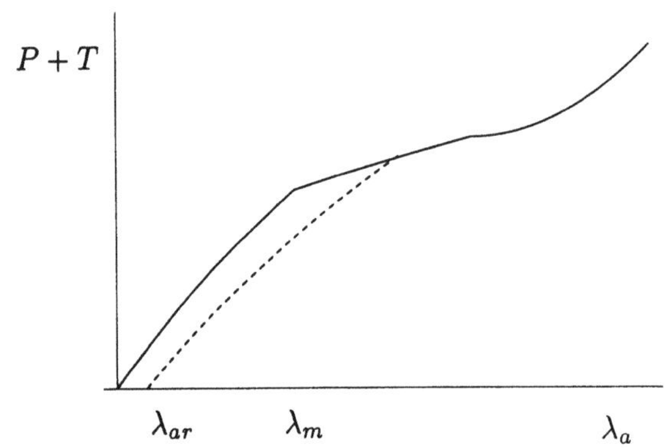

Figure 11. Representative plot of $P+T$ against λ_a for loading (continuous curve) and unloading (dashed curve): note the discontinuities in the tangent to the loading curve corresponding to $\lambda_a=\lambda_m, \lambda_b=\lambda_m$.

For the unloading Case UL(i), equation (277) again applies. For Case UL(ii), the result is obtained from (278) by replacing λ_m by λ^*, which is given by (267). For Case UL(iii) the formula

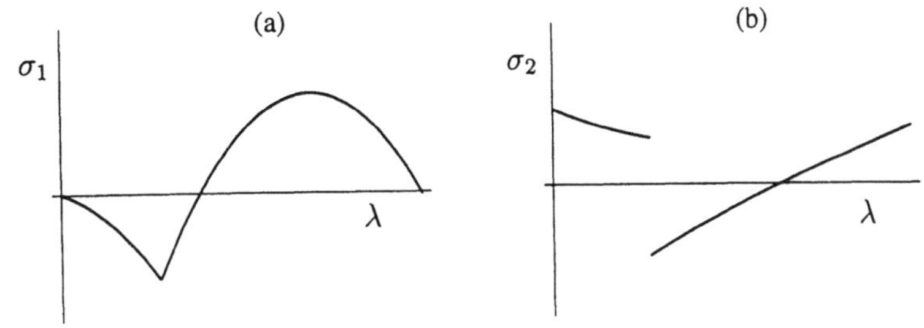

Figure 12. Typical dependence of residual stress on λ, with $\lambda_{br} \leq \lambda \leq \lambda_{ar}$, for Case UL(ii): (a) radial stress σ_1, (b) hoop stress σ_2.

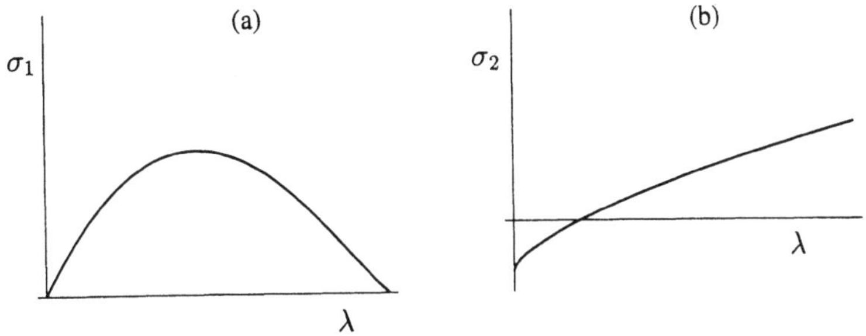

Figure 13. Typical dependence of residual stress on λ, with $\lambda_{br} \leq \lambda \leq \lambda_{ar}$, for Case UL(iii): (a) radial stress σ_1, (b) hoop stress σ_2.

(279) applies, as for Case L(iii), but with the proviso noted earlier that it holds for values of λ_b and λ_a associated with η through (259).

Figure 11 shows the plot of $9(P + T)/4\mu$ against λ_a for representative values of the parameters for both loading and unloading. The loading curve has discontinuities in the tangent where $\lambda_a = \lambda_m$ and $\lambda_b = \lambda_m$. The residual value λ_{ar} of λ_a is shown. In Figure 12 the residual radial stress distribution σ_1 and the associated distribution of the hoop stress σ_2 are shown as functions of λ for Case UL(ii) on the basis of equations (270)–(272), while Figure 13 shows the corresponding results for Case UL(iii). In each figure the range of λ values is from λ_{br} to λ_{ar}. Note that in Case UL(ii) the tangent to the radial stress curve has a point of discontinuity and that the hoop stress is discontinuous at the same point (where $\lambda = \lambda_r^*$). In Case UL(iii), on the other hand, there are no such discontinuities. The local maximum in the σ_1 curves is where $\eta = 0$.

References

Ball, J.M. (1977). Convexity conditions and existence theorems in non-linear elasticity. *Arch. Ration. Mech. Anal.* 63:337–403.

Ball, J.M. (1982). Discontinuous equilibrium solutions and cavitation in non-linear elasticity. *Phil. Trans. R. Soc. Lond.* A 306:557–611.

Ball, J.M. and Schaeffer, D.G. (1983). Bifurcation and stability of homogeneous equilibrium configurations of an elastic body under dead-load tractions. *Math. Proc. Camb. Phil. Soc.* 94:315–339.

Biot, M.A. (1965). *Mechanics of Incremental Deformations*. New York: Wiley.

Ciarlet, P.G. (1988). *Mathematical Elasticity Vol. 1: Three-dimensional Elasticity*. Amsterdam: North Holland.

Connor, P. and Ogden, R.W. (1996). The influence of shear strain and hydrostatic stress on stability and waves in a layer. *Int. J. Engng Sci.* 34:375–397.

Dowaikh, M.A. and Ogden, R.W. (1990). On surface waves and deformations in a pre-stressed incompressible elastic solid. *IMA J. Appl. Math.* 44:261–284.

Ericksen, J.L. (1975). Equilibrium of bars. *J. Elasticity* 5:191–201.

Ericksen, J.L. (1991). *Introduction to the Thermodynamics of Solids*. London: Chapman and Hall.

Fu, Y.B. and Ogden, R.W. (1999). Nonlinear stability analysis of pre-stressed elastic bodies. *Cont. Mech. Thermodyn.* 11:141–172.

Fung, Y.C. (1981). *Biomechanics: Mechanical Properties of Living Tissue*. New York: Springer.

Hill, R. (1957). On uniqueness and stability in the theory of finite elastic strain. *J. Mech. Phys. Solids* 5:229–241.

Hill, R. (1978). Aspects of invariance in solid mechanics. *Adv. Appl. Mech.* 18:1–75.

Lazopoulos, K.A. and Ogden, R.W. (1998). Nonlinear elasticity theory with discontinuous internal variables. *Math. Mech. Solids* 3:29–51.

Lazopoulos, K.A. and Ogden, R.W. (1999). Spherically-symmetric solutions for a spherical shell in finite pseudo-elasticity. *European J. Mech. A/Solids* 18:617–632.

MacSithigh, G.P. (1986). Energy-minimal finite deformations of a symmetrically loaded elastic sheet. *Q. J. Mech. Appl. Math.* 39:111–123.

MacSithigh G.P. and Chen, Y.C. (1992a). Bifurcation and stability of an incompressible elastic body under homogeneous dead loads with symmetry. 1. General isotropic materials. *Q. J. Mech. Appl. Math.* 45:277–291.

MacSithigh G.P. and Chen, Y.C. (1992b). Bifurcation and stability of an incompressible elastic body under homogeneous dead loads with symmetry. 2. Mooney-Rivlin materials. *Q. J. Mech. Appl. Math.* 45:293–313.

Marsden, J.E. and Hughes, T.J.R. (1983). *Mathematical Foundations of Elasticity*. Englewood Cliffs, NJ: Prentice Hall. .

Ogden, R.W. (1984). On non-uniqueness in the traction boundary-value problem for a compressible elastic solid. *Quart. Appl. Math.* 42:337–344.

Ogden, R.W. (1985). Local and global bifurcation phenomena in plane strain finite elasticity. *Int. J. Solids Structures* 21:121–132.

Ogden, R.W. (1991). Nonlinear elasticity: incremental equations and bifurcation phenomena. In Ames, W.F. and Rogers, C., eds., *Nonlinear Equations in the Applied Sciences*. New York: Academic Press. 437–468.

Ogden, R.W. (1997). *Non-linear Elastic Deformations*. New York: Dover Publications.

Ogden, R.W. (2000). On a theory of pseudo-elasticity. Manuscript in preparation.

Ogden, R.W. and Connor, P. (1995). On the stability of shear bands. In Parker, D.F. and England, A.H., eds.,*Proceedings of the IUTAM & ISIMM Symposium on Anisotropy, Inhomogeneity and Nonlinearity in Solid Mechanics*. Dordrecht: Kluwer. 217–222.

Ogden, R.W. and Roxburgh, D.G. (1993). The effect of pre-stress on the vibration and stability of elastic plates. *Int. J. Engng Sci.* 30:1611–1639.

Ogden, R.W. and Roxburgh, D.G. (1999a). A pseudo-elastic model for the Mullins effect in filled rubber. *Proc. R. Soc. Lond. A* 455:2861–2877.

Ogden, R.W. and Roxburgh, D.G. (1999b). An energy-based model of the Mullins effect. In Dorfmann, A. and Muhr, A., eds., *Proceedings of the First European Conference on Constitutive Models for Rubber*. Rotterdam: Balkema. 23–28.

Rivlin, R.S. (1974). Stability of pure homogeneous deformations of an elastic cube under dead loading. *Quart. Appl. Math.* 32:265–271.

Sawyers, K.N. and Rivlin, R.S. (1982). Stability of a thick elastic plate under thrust. *J. Elasticity* 12:101–125.

Truesdell, C.A. and Noll, W. (1965). In Flügge, S., ed., *The Nonlinear Field Theories of Mechanics: Handbuch der Physik Vol. III/3*. Berlin: Springer.

THEORY OF MATERIAL INSTABILITY IN INCREMENTALLY NONLINEAR PLASTICITY

H. Petryk
Polish Academy of Sciences, Warsaw, Poland

Abstract. Material instability in time-independent elastic-plastic solids is studied as a phenomenon strictly related to qualitative properties of an incremental constitutive law. A broad class of incrementally nonlinear material models is considered which encompasses classical elastoplasticity with a single smooth yield surface, as well as multi-mode plasticity with many internal mechanisms of inelastic deformation which give the yield-surface vertex effect. Three types of instability are investigated: with respect to internal microstructural rearrangements, for deviations from uniform deformation under boundary displacement control, and under flexible constraints corresponding to deformation-sensitive loading. It is shown that the respective instability criteria are different and, moreover, dependent on whether the instability concerns a single equilibrium state or a process of quasi-static deformation. Instability of equilibrium is of dynamic type, while instability of a process going through stable equilibrium states is related to a continuous spectrum of quasi-static bifurcation points along the deformation path. Basic concepts are outlined by the example of a one-dimensional discretized tensile bar and extended to incipient localization of deformation in a three-dimensional continuum. Under symmetry restrictions imposed on an incrementally nonlinear constitutive law, a unified approach to material instabilities of various types is presented which is based on the single energy criterion. By specifying the incremental energy consumption to second-order terms and determining the circumstances in which it fails to be minimized along a fundamental deformation path, the onset of material instability of a selected type can be estimated.

1 Introduction

1.1 Preliminary Remarks

An observation based on long-standing experience is that an elastic-plastic body deformed under conditions which could result in a quasi-static process of macroscopically uniform deformation, in certain circumstances starts to deform in a significantly different manner. The term 'instability' is commonly used in such cases, although frequently in a more intuitive than precise sense. Typical examples are provided by buckling or necking phenomena understood as a 'geometric' instability, snap-through phenomena as a 'dynamic' instability, or localization of deformation into shear bands as a 'material' instability.

It is well known that the concept of instability of *equilibrium* is too narrow to account for such a variety of instability phenomena in elastic-plastic solids. A suitable framework becomes available if the instability concept is extended to plastic deformation *processes*. For this purpose, a fundamental process is considered which represents a model response of an elastic-plastic body to a given program of slowly varying loading. In reality, there are always certain disturbing influences which are neglected in a theoretical description of the fundamental process; such influences may be represented by additional perturbing forces. If vanishingly small perturbing forces can influence the body response and induce, within a small loading increment, a finite distance from the fundamental process then the latter may be regarded as unstable and, consequently, practically unrealizable in a physical system. This introductory remark indicates how the term instability will be understood below; a more detailed discussion will follow. In plasticity it is frequently more convenient (and physically not less natural) to work with 'persistent disturbances' represented by the perturbing forces mentioned above rather than with 'initial disturbances' common in mathematics.

This article is concerned with *material instability* of various types which may appear independently of the body shape. The emphasis is placed on the influence of the inherent *incremental nonlinearity* of elastic-plastic materials on the onset of material instability and incipient post-critical behaviour. The principal aim is to present in a systematic way the basic theoretical aspects of material instability in slowly loaded incrementally nonlinear solids; thermal and viscous effects are not examined here. No attempt is made to discuss the effect of an internal length scale in the material. This current topic is being examined extensively in the literature (cf. Fleck and Hutchinson, 1997, de Borst and van der Giessen, 1998), and at present it is not easy to summarize the obtained results in the context of material instability.

The considerations below are limited to continuum mechanics. Instabilities occurring in a more refined scale where discrete nature of a solid becomes relevant (e.g., jumps in a position of a single dislocation line in a metal crystal) are treated here as physical sources of macroscopic plastic deformation and not as instability in the elastic-plastic material.

This paper is to much extent based on the present author's own work in the last years. Among a number of important contributions to the literature which cannot all be mentioned here, the concepts developed by Biot (1965), Drucker (1950, 1964), Hadamard (1903), Hill (1958, 1959, 1962, 1967b, 1968, 1978, 1983), Maier (1969), Mandel (1966), Nguyen (1993) and Rice (1971, 1977) (in alphabetic order) were particularly helpful in this work. More references can be found in the survey papers (Needleman and Tvergaard, 1992, Tomita, 1994, Petryk, 1997b) and in the previous CISM Lecture Notes by the author (Petryk, 1993a, 1993b). Although concerned with related topics, the author's present and previous Lecture Notes were intended to be more complementary to than overlapping each other.

1.2 On the Concept of Material Instability

Consider a solid body of an elastic-plastic material subjected to a given loading program parameterized by a scalar parameter λ which varies in time t in a prescribed manner; the distinction between λ and t will be useful below. Suppose that in the macroscopic scale of observation the material may be regarded as homogeneous and the loading conditions admit a fundamental process of macroscopically uniform deformation. In reality, there are disturbing influences which are neglected when defining the fundamental behaviour of the material under a loading program. Possible instability against physically unavoidable disturbances is to be examined. One may try to give a characterization of material instability as follows: *Material instability is met when finite deviations from a fundamental behaviour of the material in a small interval of λ can be induced*

by imposing arbitrarily small disturbances, [1] *irrespectively of the body shape.* This formulation is not free of ambiguities and has to be made precise by defining (cf. Petryk, 1993b)
(i) the fundamental behaviour whose stability is examined,
(ii) the class of disturbances considered,
(iii) the measure of a disturbance 'strength',
(iv) the measure of a 'distance' from the fundamental behaviour.

When making the specifications (i)-(iv), some arbitrariness cannot be avoided. We come thus to the conclusion that the concept of material instability is not an absolute one and always involves certain conventions. Moreover, depending on the conventions adopted, the *criteria* of material instability may differ substantially from each other.

At (i), the basic decision to be made is whether we wish to investigate
• instability of *equilibrium* at a fixed loading parameter λ, or
• instability of a *deformation process* at varying λ.

In incrementally nonlinear materials, the corresponding points of stability loss may be quite distinct. It is worth mentioning that instability concerns a *theoretical* behaviour; processes that are unstable in a physically meaningful sense should not be observable in reality. The proposed use of a small interval of λ is motivated by the wish to avoid the possibility that a predicted instability would concern only processes of unrealizable duration.

At (ii), there are various possibilities in perturbing only initial conditions (which does not seem realistic) or also boundary conditions or field equations by applying perturbing forces that may vary in time. In this article, the following three classes of disturbances are distinguished :
• perturbing forces at a micro-level, corresponding to disturbances of the internal state of a material point in the macroscopic description,
• perturbing body forces at a macro-level, corresponding to local disturbances of strain uniformity inside a homogenized material element,
• perturbing forces at a material element boundary, corresponding to disturbances of overall stress and strain under assumed flexible constraints.

Again, the respective stability conditions can take substantially different forms. Since each class of disturbances can be used to examine stability of an equilibrium state or of a deformation process, the combinations generate already six possible types of material instability.

At (iii), decision is to be made which disturbances are unavoidable from a physical point of view by attributing to them a vanishingly small measure of their 'strength'. For instance, the magnitude of perturbing forces or the energy supplied by a disturbance may be used to generate such measures. As a special case, the possibility of deviations from the fundamental behaviour of the material in the absence of *any* disturbances can be examined; this characterizes the *bifurcation* approach.

At (iv), the choice may depend on the examined type of instability. There is the possibility of neglecting deviations which are trivially possible and unessential for material stability, e.g., rigid-body translations or rotations about a symmetry axis. A careful choice of the measures (iii) and (iv) may be of fundamental importance when precise mathematical statements about stability in a continuum are aimed at.

[1] Disturbances should not be confused with *imperfections* which are not considered here. The distinction is that in the stability analysis a perturbed motion of a given material system is investigated while the initial imperfections change the system itself.

It must be stressed that the above classification does not cover all aspects of material instability, for instance dynamic propagation of weak or strong disturbances.

It can be concluded that in research on material instability, the adopted meaning of instability should always be clearly specified to avoid misunderstandings. On the other hand, it would be desirable to develop a physically sound and sufficiently wide concept of material instability to provide a common basis for differently oriented investigations of particular aspects.

1.3 Intrinsic Instability

The conditions for instability in a finite solid body can be influenced by the body shape or depend on the assumed boundary conditions. In contrast, material instability should rather be viewed as an intrinsic property of the deforming material. Some arbitrariness is again unavoidable here; for instance, a surface instability that leads to waviness of an initially smooth or planar free surface might be regarded as a material instability. That kind of instability is *not* examined in this article.

The fundamental behaviour of the material (cf. point 1.2 (i)) may be defined as a *process of macroscopically uniform straining*[2] of a representative volume element *embedded in a continuum* formed by other elements of the material. Uniform perturbations of the deformation gradient within a *bounded* element are generally incompatible with an unperturbed strain in the surrounding material, except in one-dimensional case. Therefore, the loss of material stability in the sense of Drucker (1950, 1964) need not have consequences in a physical system under sufficiently 'stiff' boundary constraints.

An intrinsic instability in the material is associated below with the possibility of deviation from the fundamental uniform process *within some interior domain* (a localized band in particular) without affecting the behaviour of the surrounding material (cf. Hill, 1962, Rice, 1977, Petryk, 1992a).

According to the discussion on 1.2 (ii) above, we may distinguish three types of so-understood intrinsic instability in the material, either deforming or being at rest.

First, instability against *internal structural rearrangements* in the material can be examined, while the macroscopic deformation gradient is not perturbed at all (cf. Sect. 4.2, 5.2). The internal rearrangements can be described by changes in *internal state variables*, or by a *history* of activity of internal mechanisms of inelastic deformations. The perturbing forces are then imagined to act at a micro-level, and in the macroscopic description they are represented by perturbations of thermodynamic driving forces. Instability for such perturbations may physically correspond to spontaneous movement, generation or annihilation of dislocations in metal crystals, to rearrangements of the packing of grains in granular materials, to martensitic phase transformations in austenitic steels or shape memory alloys, to uncontrolled propagation of microcracks in a variety of materials, etc.

Second, instability of *uniform straining* within a finite domain can be investigated, while the displacements over the domain boundary remain unaffected by disturbances (cf. Sect. 4.3, 5.3). In particular, strain localization in a disk-like subdomain may be examined. If a material element is treated as an infinite homogeneous continuum then localization of deformation within a planar band or bands corresponds to that type of intrinsic instability. A special case arises when

[2] In particular, instability of a single equilibrium state may be regarded as instability of a degenerate deformation process in which nothing changes.

the uniform straining is of zero rate, i.e. instability concerns a given equilibrium state and is associated with the possibility of dynamic localization of deformation.

Third, the material in a perturbed process may still deform uniformly in an interior *part* of the domain mentioned above, while the remaining outer part (possibly extended up to a specimen surface) compensates that deformation and constitutes flexible constraints for the interior part (cf. Sect. 4.4, 5.4). This is associated with the concept of stability of the deformation process or equilibrium within a material element under *deformation-sensitive* loading exerted by an outer part. The problem of determining a constitutive law for such loading is not addressed here; rather, we will discuss stability conditions for a whole class of *linearized* constitutive equations for deformation-sensitive loading. Consequently, the associated instability in general need not be 'intrinsic' in the sense adopted above.

There are other types of material instability in elastic-plastic solids which are not discussed here. This concerns, for instance, flutter instability (e.g., Loret, 1992, Bigoni and Loret, 1999) and shock waves (e.g., Drugan and Shen, 1990, Drugan, 1998) examined in other Chapters of this volume.

In material instability studies it is of importance to predict not only the onset of instability but also the post-critical behaviour of the material. Attention is restricted below to the *incipient* behaviour after stability loss, whose character is determined by the encountered type of instability. Analysis of post-critical deformation pattern significantly beyond the first instant of instability is presented by K. Thermann in another Chapter of this volume.

1.4 Stability or Instability in the First Approximation

To simplify the task of establishing conditions or criteria of instability, the classical approach is to use the linear stability analysis. However, in elastic-plastic materials obeying always an *incrementally nonlinear* constitutive law due to the existence of at least two (loading and unloading) constitutive branches, the concept of fully linear (in)stability is too simplified.

The linear (in)stability concept can be generalized to the (in)stability *in the first approximation* by linearizing the material response only along *direct* paths of deformation starting from a given state. More specifically, in the first approximation the increments in forces are assumed proportional to the associated increments in displacements if the latter follow a *straight* segment in the displacement space. In a three-dimensional continuum, this corresponds to the replacement of *rates* of the nominal stress and deformation gradient in the incremental constitutive law by small but finite *increments*, provided the increments are reached on a direct path of constant direction of the velocity gradient. In many cases only such segments need be considered to prove instability. In doing that, the material nonlinearity in transition from one such segment to another differently oriented, as well as the important effect of *incremental* changes in geometry, are still taken into account.

A more detailed characterization of the stability in the first approximation will be given later. In particular, a stability property will be distinguished, called 'directional' stability, restricted to possible departures from the examined equilibrium state or fundamental process along direct paths only. Direct paths may be defined as paths of bounded curvature; if the path length tends to zero then the influence of the curvature on certain quantities essential in stability considerations such as the second-order work becomes negligible. An extension of stability conditions to arbitrarily circuitous paths will also be discussed.

It must be expected that certain instability phenomena (e.g. involving shock waves or cavitation) will not be covered by the concept of instability in the first approximation.

2 Illustrative Example: Outline of Basic Concepts

Before proceeding to a three dimensional elastic-plastic continuum, it seems preferable to discuss first the basic concepts by referring to a possibly simple example. The tensile bar with two degrees of freedom, analysed in this section, can hardly be further simplified. Nevertheless, many (perhaps unexpectedly many) aspects and different approaches to material instability can be illustrated by this elementary example. It may be remarked that most of the equations apply, by changing only the meaning of the quantities involved, to the problem of simple shear of a layer in a three-dimensional elastic-plastic continuum discretized (drastically, in fact) across the layer thickness. Of course, the one-dimensional example, in addition to restrictions due to the discretization, has its own peculiarities which do not all transmit to a genuine three-dimensional case.

2.1 Example

Bar with two degrees of freedom. As a substitute of an elastic-plastic material element, let us consider a one-dimensional *representative bar* of current length $\bar{l}$ under a tensile load P. In an *arbitrarily chosen* but fixed reference configuration, the bar length is denoted by l. According to the remarks in Sect. 1.3, suppose that the relative displacement $\bar{l}(t) - l = \hat{u}_o(\lambda(t)) \equiv u_o(t)$ of the end points of the bar is a given function of time t (Fig. 1A); analysis of the case of a given or deformation-sensitive load P is deferred until Sect. 2.4. To simplify the analysis as much as possible, we shall assume that the bar consists of two parts (a) and (b) of given reference lengths l_a and l_b, respectively, $l_a + l_b = l$, $l_b \leq l_a$, such that each part is enforced to deform uniformly. The part (b) will represent the strain localization zone. For the part (b) of a current length $\bar{l}_b$, the relative displacement of its end points is denoted by $u_b = \bar{l}_b(t) - l_b$, the engineering longitudinal strain by $\epsilon_b = u_b/l_b$, and the volume fraction occupied in the reference configuration by $\eta_b = l_b/l$. The notation for the part (a) is analogous, with an index 'b' replaced by 'a'.

The bar has two degrees of freedom, ϵ_b and ϵ_a, say. They are constrained by the kinematic compatibility condition

$$\eta_b\epsilon_b + \eta_a\epsilon_a = \epsilon_o \,, \tag{2.1}$$

where $\epsilon_o = u_o/l$ is the overall engineering strain of the representative bar, being a given function of time. The remaining degree of freedom is to be determined from the equation of either dynamic or quasi-static motion.

One-dimensional constitutive equations. The incremental constitutive framework to be used involves infinitesimal increments of basic variables. An infinitesimal increment will be replaced for convenience by the *forward* rate, i.e. the one-sided *right-hand* derivative with respect to time t, denoted by a dot superimposed over the symbol of a differentiated quantity.

The material of the bar may deform elastically or plastically. Attention is restricted to extensional plastic straining. In the simplest case discussed, the rate of plastic straining is measured

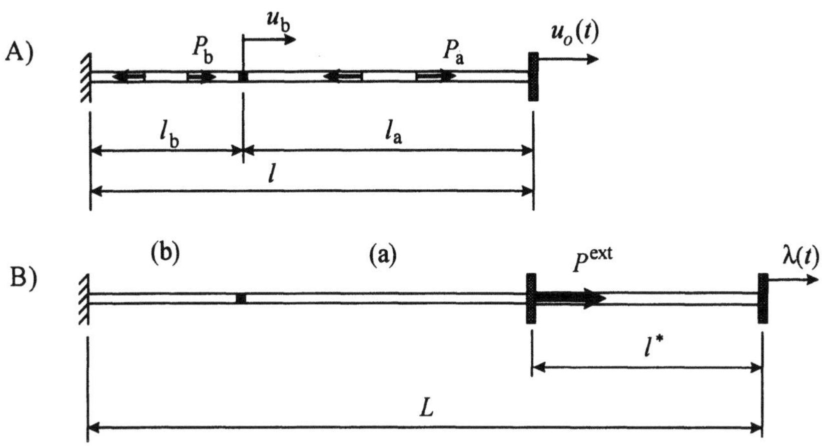

Figure 1. An illustratory example of a representative bar composed of two uniform-strain segments: (A) under end-displacement control, (B) as a part of a longer bar. All dimensions are in the reference configuration.

by a scalar plastic multiplier $\dot{\gamma} \geq 0$. In the time-independent framework, $\dot{\gamma}$ can be non-zero only if a loading variable f has reached from below a threshold value, without loss of generality set equal to zero. The basic rule for plastic activity reads

$$\dot{\gamma} \geq 0, \qquad f \leq 0, \qquad f\dot{\gamma} = 0. \tag{2.2}$$

As the simplest identification for the elastic-plastic bar, we take

$$f = P - P^{\mathrm{y}}, \qquad \dot{\gamma} = \dot{\epsilon}^{\mathrm{p}}, \tag{2.3}$$

where $P^{\mathrm{y}} = P^{\mathrm{y}}(\gamma)$ is the actual yield limit, and $\dot{\epsilon}^{\mathrm{p}}$ is the plastic part of strain-rate $\dot{\epsilon}$, which in general is *different* from the rate of plastic strain. The actual value of $\dot{\gamma}$ at $P = P^{\mathrm{y}}$ is to be found from the consistency condition

$$\dot{f} \leq 0, \quad \dot{f}\dot{\gamma} = 0 \qquad \text{if } f = 0 \tag{2.4}$$

which follows from (2.2) by differentiation. The constitutive rate equations read

$$\dot{f} = \dot{P} - h\dot{\gamma}, \qquad \dot{\epsilon} = (C^{\mathrm{e}})^{-1}\dot{P} + \dot{\gamma}, \tag{2.5}$$

where $C^{\mathrm{e}} = \partial P(\epsilon, \gamma)/\partial \epsilon > 0$ is the elastic modulus, and $h = dP^{\mathrm{y}}/d\gamma$, $-C^{\mathrm{e}} < h < C^{\mathrm{e}}$, is the hardening modulus. On eliminating $\dot{\gamma}$ between the equations in (2.5) with the help of (2.4), the rate (or incremental) constitutive relationship between a tensile load $P \geq 0$ and strain ϵ for the elastic-plastic bar takes the form

$$\dot{P} = \begin{cases} C^{\mathrm{p}}\,\dot{\epsilon} & \text{if } P = P^{\mathrm{y}} \text{ and } \dot{\epsilon} > 0, \\ C^{\mathrm{e}}\,\dot{\epsilon} & \text{otherwise,} \end{cases} \tag{2.6}$$

where $C^p = hC^e/(h + C^e) < C^e$ is the tangent (elastoplastic) modulus. In general, all quantities in the above constitutive equations may be different in the parts (b) and (a) of the bar, and will be distinguished by the respective indices. However, the bar is assumed to be initially homogeneous, so that the load-strain relationship that results from (2.6) for a given history $\epsilon(t)$ is the same for the parts (b) and (a).

Fundamental deformation process. Suppose now that the controlled relative end displacement u_o, which may be identified with λ, varies linearly in time with a constant speed $v_o \geq 0$, $u_o = v_o t$, $t \geq 0$, and that the external end point of the part (b) is fixed. Then the standard discretization procedure yields the discretized equation of motion of the bar in the form

$$\begin{aligned} P_a(t) - P_b(t) &= \int_0^l \rho_l \ddot{u}(x,t)\,\varphi(x)\,dx \\ &= \frac{1}{3} m l_b \ddot{\epsilon}_b(t)\,, \end{aligned} \tag{2.7}$$

where P_a (or P_b) is a uniform load carried by the part (a) (or (b)) of the bar, ρ_l is the mass density per unit length of the bar in the reference configuration, and φ is a triangular 'shape function' which takes the value 1 at $x = l_b$ and vanishes at the bar ends. The total mass of the representative bar is denoted by m, and $\frac{1}{3}m$ may be regarded as a reduced mass placed between the parts (b) and (a).

Assuming that the bar is in a uniform state at $t = 0$, the fundamental solution to the equation of motion is given by

$$\epsilon_b(t) = \epsilon_a(t) = v_o t/l\,, \qquad t \geq 0\,, \tag{2.8}$$

irrespectively whether the motion is regarded as quasi-static or dynamic; note that (2.8) implies $P_a = P_b$ and $\ddot{u}_b = \frac{d}{dt}(\eta_b v_o) = 0$. In a dynamic analysis it is essential that the velocity in the fundamental motion is constant in time at any material point.

We are primarily interested in instability of the process of slow uniform deformation of the bar. Consequently, v_o in (2.8) is understood as a small parameter, and the increments in time t and in strain ϵ_o are of different order of magnitude in the limit as $v_o \to 0$.

The first approximation. We will examine only such perturbations of the fundamental process that satisfy $\Delta_{max} > \Delta\epsilon_b(t) \geq \Delta\epsilon_o \geq 0$ in the examined interval $\mathcal{T}$ of time $t \geq t_0$, where

$$\Delta\psi(t) \equiv \psi(t) - \psi(t_0) \tag{2.9}$$

denotes an increment of a quantity ψ in time, and t_0 defines the deformation stage at which stability is examined. Moreover, we will restrict ourselves to examine stability or instability in the first approximation at t_0. That is, the number Δ_{max} is fixed and taken sufficiently small to allow any changes in C^e, h and C^p to be neglected within the examined time interval $\mathcal{T}$ starting from a fixed t_0. However, C^e, h and C^p may depend on t_0.

Time integration of $(2.5)_2$ now yields

$$\Delta f = \Delta P - h(t_0)\Delta\gamma\,, \qquad \Delta P = C^e(t_0)(\Delta\epsilon - \Delta\gamma) \tag{2.10}$$

(generally, to first order in $\Delta_{\max}$) no matter how the strain at a point of the bar varies in the time interval $\mathcal{T}$. Equations (2.10) provide a definition of the first approximation in the analyzed example. Unless stated otherwise, the first approximation will be tacitly assumed in the following. In particular, $\Delta\epsilon_{\rm o}(t) < \Delta_{\max}$ for $t \in \mathcal{T}$, which means that path stability will be examined within a small interval of the loading parameter λ.

In two special cases of the deviation on a *direct* path from the fundamental process, the following *partial* linearization follows from (2.10):

$$\Delta P_{\rm a} - \Delta P_{\rm b} = \begin{cases} C^{\rm p}\,(\Delta\epsilon_{\rm a} - \Delta\epsilon_{\rm b}) & \text{if } \dot\gamma_{\rm a} > 0 \text{ and } \dot\gamma_{\rm b} > 0 \text{ at every } t \geq t_0\,, \\ C^{\rm e}\Delta\epsilon_{\rm a} - C^{\rm p}\Delta\epsilon_{\rm b} & \text{if } \dot\gamma_{\rm a} = 0 \text{ and } \dot\gamma_{\rm b} > 0 \text{ at every } t \geq t_0\,. \end{cases} \tag{2.11}$$

2.2 Dynamic Instability

Linear stability analysis. The classical approach to examining stability of motion is the *linear stability analysis*. For that purpose, the governing equations of motion are fully linearized about the fundamental solution, and a class of perturbed solutions close to the fundamental solution is investigated for those linear equations. The conclusion about (Lyapunov) stability or instability is conventionally drawn according to whether the deviation of perturbed solutions from the fundamental one over the *entire* time domain (usually infinite) tends or does not tend, respectively, to zero with the vanishing strength of an *initial* disturbance.

Let us apply that approach to the fundamental process (2.8). By neglecting the possibility of unloading, the linearized version of the equation of motion (2.7), with the strain difference[3] $\delta\epsilon_{\rm b}(t) = \epsilon_{\rm b}(t) - \epsilon_{\rm o}(t)$ with respect to the fundamental solution taken as the unknown, on substituting $(2.11)_1$ and (2.1) is obtained in the form

$$\frac{1}{3} m l_{\rm b} \delta\ddot\epsilon_{\rm b} + \frac{1}{\eta_{\rm a}} C^{\rm p}\delta\epsilon_{\rm b} = 0\,, \tag{2.12}$$

which has an elementary general solution

$$\delta\epsilon_{\rm b}(t) = \begin{cases} c_1 \sin ct + c_2 \cos ct & \text{if } C^{\rm p} > 0 \\ c_1 t + c_2 & \text{if } C^{\rm p} = 0 \\ c_1 e^{ct} + c_2 e^{-ct} & \text{if } C^{\rm p} < 0 \end{cases} \qquad c = \sqrt{\frac{3|C^{\rm p}|}{m l_{\rm b}\eta_{\rm a}}} \tag{2.13}$$

with integration constants c_1, c_2 adjusted to a small initial perturbation. As an initial disturbance consistent with the assumed absence of unloading, we may take $\delta\dot\epsilon_{\rm b}(t_0) \neq 0$ and $\delta\epsilon_{\rm b}(t_0) = 0$ at an initial instant t_0 of a perturbed motion. According to (2.13), the fundamental process of uniform deformation of the bar might appear stable if $C^{\rm p} > 0$ and unstable if $C^{\rm p} \leq 0$.

That conclusion is, however, not so obvious. It is true that for $C^{\rm p} > 0$ a perturbed solution following a small initial disturbance will oscillate in time about the fundamental solution, with the amplitude of oscillations as small as we please if the initial disturbance measured by $|\delta\dot\epsilon_{\rm b}(t_0)|$ tends to zero. The loading condition is clearly satisfied everywhere for disturbances sufficiently

[3] Note the essential difference between the meaning of the prefixes Δ and δ adopted here: $\Delta\psi$ denotes an increment in ψ between two time instants, and $\delta\psi$ denotes a variation of ψ at a single instant of time.

small. Moreover, if viscous effects are included then the oscillations are expected to diminish asymptotically in time. Nevertheless, stability for perturbations that involve temporary unloading remains yet to be proved.

Further, the solution for $C^{\mathrm{p}} = 0$ may change qualitatively if higher-order effects are included; this is why the usual term 'neutral stability' is more appropriate in this case. To show that the conclusion about instability need not be valid, one may consider a special case when C^{p} is positive along the fundamental straining path, with the exception for $C^{\mathrm{p}}(t_0) = 0$.

Next, the perturbed solutions at fixed c depend continuously on the initial data over any *bounded* interval of time. It follows that the conclusion about instability for $C^{\mathrm{p}} < 0$ requires formally an unbounded interval of time. This can be made consistent with the assumption of small strain increments bounded by $\Delta_{\max}$ (and with the linearization assumption) if and only if $\dot{\epsilon}_0$ is vanishingly small. Moreover, $\delta\epsilon_{\mathrm{b}}(t_0) = 0$ in the last solution in (2.13) implies $c_1 e^{ct_0} = -c_2 e^{-ct_0}$ so that

$$\begin{aligned} |\delta\dot{\epsilon}_{\mathrm{b}}(t)| &= c\,|(c_1 e^{ct} - c_2 e^{-ct})| \\ &= c\,|c_1\,(e^{ct} + e^{c(2t_0 - t)})| \\ &> c\,|c_1\,(e^{ct} - e^{c(2t_0 - t)})| \\ &= c\,|\delta\epsilon_{\mathrm{b}}(t)|\,. \end{aligned} \tag{2.14}$$

From (2.14) and (2.1) it follows that either $\delta\dot{\epsilon}_{\mathrm{b}}(t) < -c\,|\delta\epsilon_{\mathrm{b}}(t)|$ or $\delta\dot{\epsilon}_{\mathrm{a}}(t) < -c\frac{l_{\mathrm{b}}}{l_{\mathrm{a}}}|\delta\epsilon_{\mathrm{b}}(t)|$. Consequently, the exponential growth of $|\delta\epsilon_{\mathrm{b}}(t)|$ at $\dot{\epsilon}_0 \to 0$ enforces the solution to cross the limit of validity of the no-unloading assumption, before a finite distance in strain can be reached. The conclusion about instability for $C^{\mathrm{p}} < 0$ drawn on the basis of (2.13) is thus not fully convincing.

If c is allowed to be unbounded by taking $\eta_{\mathrm{b}} \to 0$ (rather than being constant as above) then the use of an infinite time interval and infinitesimal $\dot{\epsilon}_0$ is not required in the instability analysis. However, the estimate (2.14) implies again, now on account of $c \to \infty$, that the limit of validity of the no-unloading assumption must be crossed before a finite distance in strain can be reached following a vanishingly small disturbance.

One may conclude that the nonlinearity of the incremental elastic-plastic constitutive law must be taken into account to provide a more satisfactory picture of stability or instability of the fundamental process.

Instability of equilibrium. The necessity of taking into account the incremental nonlinearity of the material is evident when examining instability of a degenerate fundamental process with vanishing velocities, i.e. of an equilibrium state at constant ϵ_0. For, if the ends of the representative elastic-plastic bar are fixed then further plastic straining in one part of the bar necessarily requires elastic unloading in the other part.

Consider a motion at constant ϵ_0 (i.e. at fixed λ) that follows an initial disturbance in velocities superimposed on a uniform equilibrium state at $P = P^{\mathrm{y}}$. Attention is restricted to a *direct* path of departure from the equilibrium state, such that in each part of the bar only one constitutive branch is activated, i.e. here $\dot{\epsilon}_{\mathrm{b}}(t) > 0$ and $\dot{\epsilon}_{\mathrm{a}}(t) < 0$ at every $t \geq t_0$. Accordingly, on applying $(2.11)_2$ and (2.6), the equation of motion (2.7) reduces to another linear differential equation

$$\frac{1}{3} m l_{\mathrm{b}} \Delta\ddot{\epsilon}_{\mathrm{b}} + C^{\mathrm{r}} \Delta\epsilon_{\mathrm{b}} = 0\,, \qquad C^{\mathrm{r}} \equiv C^{\mathrm{p}} + \frac{l_{\mathrm{b}}}{l_{\mathrm{a}}} C^{\mathrm{e}}\,, \tag{2.15}$$

which has an elementary general solution analogous to (2.13). The oscillatory solution for $C^{\mathrm{r}} > 0$ is now inconsistent with the non-unloading assumption $\Delta\dot{\epsilon}_{\mathrm{b}} = \dot{\epsilon}_{\mathrm{b}} > 0$, so that stability of equilibrium cannot be inferred; we will see later that the energy argument can be used for that purpose. The transitory case $C^{\mathrm{r}} = 0$ is also not conclusive and corresponds to the 'neutral' stability (or instability) of equilibrium, cf. the similar remark above. However, the condition

$$C^{\mathrm{r}} < 0 \tag{2.16}$$

is sufficient for dynamic instability of equilibrium (in the first approximation) since an arbitrarily small initial perturbation can grow exponentially in time in a free inertial motion within the assumed constitutive branches. As a possible perturbed solution one can take

$$\Delta\epsilon_{\mathrm{b}} = \bar{\delta}\,\sinh\left(\sqrt{\frac{-3C^{\mathrm{r}}}{ml_{\mathrm{b}}}}\,\Delta t\right) \tag{2.17}$$

with an arbitrarily small constant $\bar{\delta} > 0$. The requirement of an infinite time interval creates no problem here: the second time scale, related previously to given variations of λ and ϵ_{o} in time, does not appear at $\epsilon_{\mathrm{o}} = \mathrm{const}$.

The linearized analysis of dynamic instability in this simple example, although not complete as discussed above, does indicate that the conditions for stability of the fundamental deformation process and of an equilibrium state must be expected to be different in general. The distinction between the two concepts of stability will be demonstrated below by a nonlinear analysis of stability of a *quasi-static* fundamental process.

Exercise: Extend the instability condition (2.16) to the case of an arbitrary sign of $\dot{\epsilon}_{\mathrm{b}}$.

2.3 Instability of a Quasi-Static Deformation Process

Persistent disturbances. The disturbances examined above influenced the initial conditions while the subsequent motion was assumed free of further disturbances. That assumption, although made frequently, is not realistic. In every mechanical model, there are always certain influences neglected in comparison with the behaviour of the modeled physical system, which are perpetual *during* the examined deformation process. Such influences are called *persistent disturbances* and may be represented by additional, unknown terms in the governing equations.

In the illustratory example, persistent disturbances may be represented by a small additional load $P^{\mathrm{dist}}(t)$ varying in time, applied by an external agency at the point between the parts (a) and (b) of the bar and taken positive when pointed towards the part (a). Let us examine stability of the uniform deformation process (2.8) with respect to such a disturbance.

Suppose that the overall strain ϵ_{o} varies slowly with respect to a natural time. The inertia term on the right-hand side of (2.7) is regarded as a small part of the perturbing load P^{dist}. The perturbed equation of a quasi-static motion reads

$$P_{\mathrm{a}}(t) - P_{\mathrm{b}}(t) + P^{\mathrm{dist}}(t) = 0\,, \tag{2.18}$$

while all kinematic constraints remain unperturbed.

A special case is now examined when P^{dist} varies in time proportionally to Δt, and thus to $\Delta\epsilon_{\mathrm{o}}$, starting from zero value at an instant t_0. A general rate solution to (2.18) satisfies

$$(C_{\mathrm{b}} + \frac{l_{\mathrm{b}}}{l_{\mathrm{a}}}C_{\mathrm{a}})\delta\dot{\epsilon}_{\mathrm{b}} = \dot{P}^{\mathrm{dist}} + (C_{\mathrm{a}} - C_{\mathrm{b}})\dot{\epsilon}_0\,, \qquad \delta\dot{\epsilon}_{\mathrm{a}} = -\frac{l_{\mathrm{b}}}{l_{\mathrm{a}}}\delta\dot{\epsilon}_{\mathrm{b}}\,. \tag{2.19}$$

The incremental stiffness moduli C_a and C_b in the part (a) and (b) of the bar, respectively, must be consistent with (2.6). Among possible combinations, consider the following two solutions corresponding to *direct* paths:

(i) (fully active loading) $C_\mathrm{b} = C_\mathrm{a} = C^\mathrm{p}$, $\dot\epsilon_\mathrm{b} > \dot\epsilon_\mathrm{o} > 0$, $\dot\epsilon_\mathrm{a} > 0$, so that, in the first approximation,

$$C^\mathrm{p}(\Delta\epsilon_\mathrm{b} - \Delta\epsilon_0) = \eta_\mathrm{a} P^\mathrm{dist}\,, \qquad C^\mathrm{p}(\Delta\epsilon_\mathrm{a} - \Delta\epsilon_0) = -\eta_\mathrm{b} P^\mathrm{dist}\,. \tag{2.20}$$

(ii) (elastic unloading in the part (a)) $C_\mathrm{b} = C^\mathrm{p}$, $C_\mathrm{a} = C^\mathrm{e}$, $C^\mathrm{r} > 0$, $\dot\epsilon_\mathrm{b} > \dot\epsilon_\mathrm{o} > 0$, $\dot\epsilon_\mathrm{a} < 0$, so that, in the first approximation,

$$\Delta\epsilon_\mathrm{b} - \Delta\epsilon_\mathrm{o} = (C^\mathrm{r})^{-1}(P^\mathrm{dist} + (C^\mathrm{e} - C^\mathrm{p})\Delta\epsilon_0)\,, \qquad \dot\epsilon_\mathrm{a} = \frac{l_\mathrm{b}}{l_\mathrm{a}}(C^\mathrm{r})^{-1}(-\dot P^\mathrm{dist} + \frac{l}{l_\mathrm{b}} C^\mathrm{p}\dot\epsilon_0)\,. \tag{2.21}$$

In the case (i), if $C^\mathrm{p} \neq 0$ then the distance between the perturbed and fundamental solutions for a given $\Delta\epsilon_\mathrm{o}$ decreases to zero as the final perturbing force ΔP^dist tends to zero. This is so no matter whether $C^\mathrm{p} > 0$ or $C^\mathrm{p} < 0$, so that the expected instability range $C^\mathrm{p} < 0$ is not found in that way. We can only conclude that the rate solution becomes indefinitely sensitive to $\dot P^\mathrm{dist}$ as $C^\mathrm{p} \to 0$, and is indeterminate when $C^\mathrm{p} = 0$. The latter case is called 'bifurcation in velocities' since more than one velocity solution exists. This corresponds to bifurcation of the quasi-static deformation process (path) in the first approximation when C^p is assumed constant. However, if C^p varies along the path then the bifurcation in velocities at $C^\mathrm{p} = 0$ usually does *not* imply a genuine path branching *without unloading* in the absence of any perturbing force. One can check that the fundamental solution (2.8) remains a unique unperturbed solution *without unloading*, satisfying (2.18) for $P^\mathrm{dist} \equiv 0$, if the slope of the load-strain curve changes sign and vanishes at a *single* value of strain. The conclusion about path bifurcation can then only be obtained by appealing to the incremental nonlinearity of the material, as discussed below.

In the case (ii) with partial unloading, definite conclusions can be drawn. The consistency with the assumed unloading ($\dot\epsilon_\mathrm{a} < 0$) in the part (a) at $\dot P^\mathrm{dist} \to 0$ is ensured if $C^\mathrm{p} < 0$, by the second equation in (2.21). In that case the left-hand difference between the perturbed and fundamental solutions in the first equation in (2.21) does not tend to zero as $|P^\mathrm{dist}| \to 0$ but remains finite for any finite overall strain increment $\Delta\epsilon_\mathrm{o}$. Also without appealing to the first-order approximation, if $C^\mathrm{p} < 0$ at t_0 then the solution path obtained by time integration of (2.19) under the assumptions of (ii) does not tend to the fundamental solution (2.8) as $\dot P^\mathrm{dist} \to 0$ but deviates from it with a finite rate, which in the limit reads

$$\dot\epsilon_\mathrm{b} - \dot\epsilon_\mathrm{o} = \frac{C^\mathrm{e} - C^\mathrm{p}}{C^\mathrm{r}}\dot\epsilon_\mathrm{o} > 0\,, \qquad \dot P_\mathrm{b} - \dot P_\mathrm{o} = \frac{C^\mathrm{e} - C^\mathrm{p}}{C^\mathrm{r}}\dot P_\mathrm{o} < 0\,, \qquad \dot P_\mathrm{o} = C^\mathrm{p}\dot\epsilon_\mathrm{o}\,. \tag{2.22}$$

We thus conclude that the fundamental process of slow uniform straining of the representative bar in the range

$$C^\mathrm{p} < 0 \qquad \text{and} \qquad C^\mathrm{r} > 0 \tag{2.23}$$

is unstable for persistent disturbances of infinitesimal strength defined by $|P^\mathrm{dist}|$. The use of the incremental nonlinearity of the elastic-plastic constitutive law has been essential in arriving at that conclusion.

Exercise: Show that the above conclusion remains valid if $P^\mathrm{dist} \neq 0$ only within a vanishingly short time interval.

Quasi-static bifurcations. From (2.22) it follows that the existence of *exact bifurcations* of a quasi-static deformation path can be established by taking the incremental nonlinearity of the material into account. *At every point* on a solution path with C^{p} varying continuously within the range defined by (2.23), there exists a non-trivial, secondary rate solution given by (2.22) to the quasi-static *unperturbed* rate-problem. That solution initiates a quasi-static branching path with elastic unloading in the part (a), which can be constructed by time integration of (2.22) also if C^{p} is varying in time. Consequently, we have a *continuous spectrum* of genuine bifurcation points along the fundamental solution path within the whole range (2.23). If C^{p} is continuously decreasing at the critical instant when $C^{\mathrm{p}} = 0$ then, by the same argument as above, at this instant the primary bifurcation takes place with neutral loading, followed by *elastic unloading just beyond the critical instant.* The secondary branch emanating at that point is the only post-critical quasi-static deformation path that can be stable with respect to small persistent disturbances.

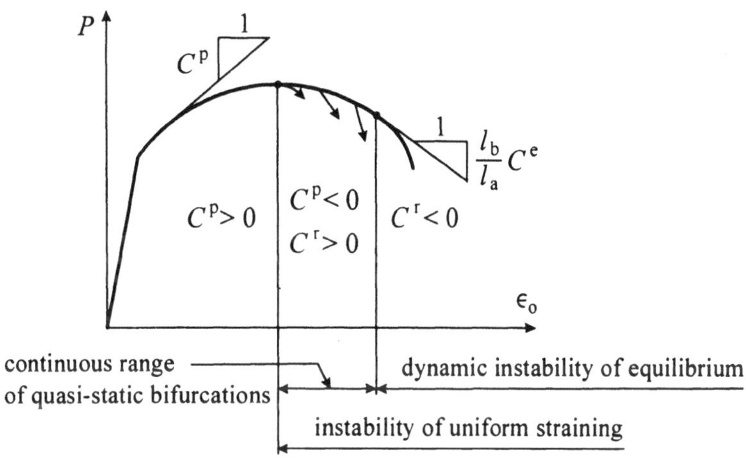

Figure 2. Instability ranges for the discretized bar in quasi-static tension under end displacement control.

The concept of instability of a quasi-static deformation process for persistent disturbances is thus closely related to the existence of a continuous range of bifurcation points on the deformation path. The instability concept is, however, more general than the bifurcation concept. If the fundamental process is governed by the dynamic equation of motions as in Sect. 2.2, and not by its approximate quasi-static version as above, then there is no exact bifurcation. Any deviation from the fundamental solution (2.8) requires some disturbance, irrespectively of the sign of C^{p} and no matter how slow the fundamental motion is. However, if $C^{\mathrm{p}} < 0$ then a finite deviation from the fundamental motion (2.8) can be induced within a given increment $\Delta\epsilon_{\mathrm{o}}$ by a perturbing force $(P^{\mathrm{dist}}(t) - \frac{1}{3}ml_{\mathrm{b}}\ddot{\epsilon}_{\mathrm{b}})$ whose magnitude tends to zero as $\dot{\epsilon}_{\mathrm{o}} \to 0$. This can be shown by 'smoothing out' in time the initial discontinuous change in velocities in the quasi-static bifurcation solution; details are omitted here.

2.4 The Energy Criterion

Analysis of particular perturbations may not always be effective, especially for more complex systems. Moreover, to obtain a condition sufficient for stability, all possible disturbances from a specified class should be taken into account. For a class of materials, encompassing that characterized in Sect. 2.1 which will be generalized later, the approach based on energy considerations can be effectively applied.

The second-order work. Consider a uniform equilibrium state $\mathcal{G}_0$ of the bar at an instant t_0 at which

$$P_a(t_0) = P_b(t_0) = P^y(\gamma(t_0)) . \tag{2.24}$$

On any path starting from $\mathcal{G}_0$, the work of deformation in the representative bar in a time interval $[t_0, t]$ reads

$$\begin{aligned} \Delta W(t) &\equiv l_b \int_{t_0}^{t} P_b(\theta)\dot{\epsilon}_b(\theta)\, d\theta + l_a \int_{t_0}^{t} P_a(\theta)\dot{\epsilon}_a(\theta)\, d\theta \\ &= \Delta_1 W(t) + l_b \int_{t_0}^{t} \Delta P_b(\theta)\dot{\epsilon}_b(\theta)\, d\theta + l_a \int_{t_0}^{t} \Delta P_a(\theta)\dot{\epsilon}_a(\theta)\, d\theta . \end{aligned} \tag{2.25}$$

where

$$\begin{aligned} \Delta_1 W &= l_b P_b(t_0)\Delta\epsilon_b(t) + l_a P_a(t_0)\Delta\epsilon_a(t) \\ &= l P^y(t_0)\Delta\epsilon_o \end{aligned} \tag{2.26}$$

is the *first-order* work, with the later expression obtained from (2.1) and (2.24). Further,

$$\int_{t_0}^{t} \Delta P(\theta)\dot{\epsilon}(\theta)\, d\theta = \Delta_2 w(t) + o((\Delta_{\max})^2) , \tag{2.27}$$

where $\Delta_2 w$ is the *second-order* work density per unit length of the bar in the reference configuration, and o is the classical order symbol used to indicate higher-order terms; recall that, by definition, $o(r^2)/r^2 \to 0$ as $r \to 0$. We thus have

$$\Delta W = \Delta_1 W + \Delta_2 W \qquad \text{(to second order)} , \tag{2.28}$$

where the second-order work

$$\Delta_2 W = l_b\, \Delta_2 w_b + l_a\, \Delta_2 w_a \tag{2.29}$$

is generally *dependent on the strain path.*[4] By using the formulae (2.10) of the first approximation and rearranging with the help of $(2.2)_3$, (2.5) and (2.24), one can show (Petryk, 1991a) that

$$\Delta_2 w = \tfrac{1}{2}\Delta P \Delta\epsilon - \tfrac{1}{2}\Delta f \Delta\gamma \tag{2.30}$$

for an *arbitrary* straining path (with or without unloading) starting from $\mathcal{G}_0$. Along *direct* paths to be examined below, we have $\dot{\epsilon}_b(\theta) \geq 0$ and either $\dot{\epsilon}_a(\theta) \geq 0$ or $\dot{\epsilon}_a(\theta) < 0$ at all instants. Along

[4] In this one-dimensional example, strain paths may only differ by involving or not segments of compression. In a three-dimensional solid, paths may be more diverse.

a direct path the last term in (2.30) vanishes, and the associated second-order work density, distinguished by a superscript 'd', reduces to

$$\Delta_2 w^{\mathrm{d}} = \begin{cases} \frac{1}{2} C^{\mathrm{p}} (\Delta\epsilon)^2 & \text{if } \dot{\epsilon} > 0\,, \\ \frac{1}{2} C^{\mathrm{e}} (\Delta\epsilon)^2 & \text{if } \dot{\epsilon} < 0\,. \end{cases} \tag{2.31}$$

Accordingly, for $C_{\mathrm{b}} = C^{\mathrm{p}}$ and either $C_{\mathrm{a}} = C^{\mathrm{p}}$ or $C_{\mathrm{a}} = C^{\mathrm{e}}$, the second-order work for the bar deformed along a direct path reads

$$\Delta_2 W^{\mathrm{d}} = \frac{1}{2} l_{\mathrm{b}} C^{\mathrm{p}} (\Delta\epsilon_{\mathrm{b}})^2 + \frac{1}{2} l_{\mathrm{a}} C_{\mathrm{a}} (\Delta\epsilon_{\mathrm{a}})^2 \tag{2.32}$$

which is dependent only on the final strain increments.

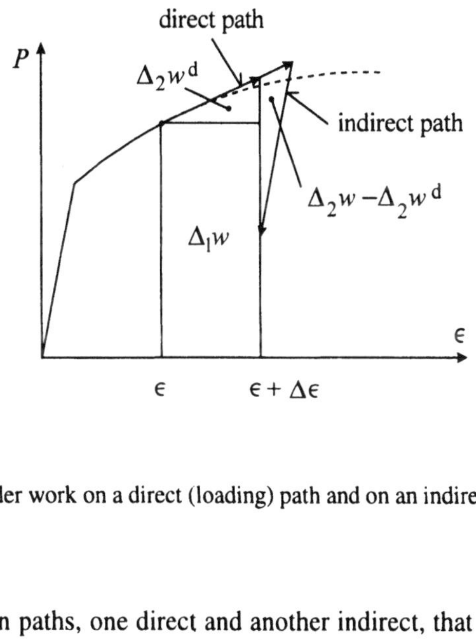

Figure 3. The second-order work on a direct (loading) path and on an indirect path leading to the same final strain increment $\Delta\epsilon$.

Consider two strain paths, one direct and another indirect, that lead to the same final strain increment. From (2.2) and (2.24) we obtain $\Delta f \Delta\gamma \leq 0$. One can show that for the assumed constitutive law, and within the class of all (arbitrarily short) paths leading to the same final strain increment $\Delta\epsilon$, the second-order work density $\Delta_2 w$ attains a minimum $\Delta_2 w^{\mathrm{d}}$ on a direct path (Petryk, 1991a):

$$\Delta_2 w \geq \Delta_2 w^{\mathrm{d}}\,. \tag{2.33}$$

This obviously implies that $\Delta_2 W \geq \Delta_2 W^{\mathrm{d}}$.

Exercise: Prove validity of (2.30) and (2.33) for the assumed one-dimensional constitutive law.

Stability of equilibrium. Without specifying the type of disturbance, the disturbance strength may be defined by determining the amount of energy ΔW^* supplied to the system due to the disturbance. Consider the homogeneous representative bar from Fig.1A with fixed ends, ϵ_o =const, initially at equilibrium under load $P = P^y$ at $\epsilon_a = \epsilon_b$, set in motion at $t = t_0$ by a disturbance. At fixed boundary displacements, the mechanical energy balance implies

$$\Delta W + \Delta K = \Delta W^* \qquad \text{at fixed } \lambda\,, \tag{2.34}$$

where K is the kinetic energy. On account of (2.1), the second-order work (2.32) evaluated on a *direct* path now reads

$$\Delta_2 W^{\mathrm{d}} = \frac{1}{2} l_{\mathrm{b}} (\Delta\epsilon_{\mathrm{b}})^2\, C^{\mathrm{r}}\,. \tag{2.35}$$

According to (2.28), (2.26) and (2.33), we have $\Delta W = \Delta_2 W \geq \Delta_2 W^{\mathrm{d}}$. Moreover, $\Delta K \geq 0$ since the kinetic energy is non-negative and vanishes at equilibrium. It follows that

$$\Delta W^* \geq \Delta_2 W^{\mathrm{d}}\,. \tag{2.36}$$

Consequently, if

$$\Delta_2 W^{\mathrm{d}} > 0 \qquad \text{at fixed } \lambda\,, \tag{2.37}$$

or equivalently, if

$$C^{\mathrm{r}} > 0 \tag{2.38}$$

then any departure from equilibrium at fixed ϵ_0 requires a positive amount of energy ΔW^* to be supplied by a disturbing agency. If ΔW^* tends to zero then the maximum possible distance from the equilibrium configuration, measured by $|\Delta\epsilon_{\mathrm{b}}(t)|$, tends also to zero by (2.36) and (2.35). This means that (2.37) is *sufficient* for stability of equilibrium of the bar with fixed ends, for any disturbances whose strength is defined by the energy measure.

Exercise: Derive a condition sufficient for stability of equilibrium of a bar divided into an arbitrary finite number of uniformly deforming segments.

Stability of uniform straining. The object of stability analysis is now the uniform, *quasi-static* plastic straining of the bar. To extend the energy criterion of stability of equilibrium to the uniform straining, imagine our representative bar as a part of a longer bar (Fig.1B) viewed as a substitute of a continuum surrounding a material element. Attention is restricted first to local disturbances (cf. Sect. 1.3) that leave the motion outside the representative bar unperturbed, so that $\bar{l}(t) = \bar{l}^0(t)$, where $\bar{l}^0$ is the current length of the representative bar in the fundamental process. Consider a perturbed deformation process in the bar that deviates from the fundamental process of uniform deformation, starting from a common configuration at t_0. The mechanical energy balance implies

$$\Delta W + \Delta K - \Delta W^0 = \Delta W^* \qquad \text{at varying } \lambda\,. \tag{2.39}$$

In comparison with (2.34), the energy balance includes now additionally the work done by the force P^{ext} applied *from the outside* to the movable end point of the representative bar, which under the above assumptions is equal to the deformation work ΔW^0 within the representative bar in the quasi-static fundamental process.

The work difference in the perturbed and fundamental processes involves only the second-order works,

$$\Delta W - \Delta W^0 = \Delta_2 W - \Delta_2 W^0\,, \tag{2.40}$$

since the first-order terms cancel each other and higher order terms have been neglected. Further, $\Delta_2 W$ is minimized on direct paths by (2.33), and $\Delta K \geq 0$ since the kinetic energy is nonnegative and vanishes in the quasi-static motion. We thus obtain

$$\begin{aligned} \Delta_2 W^* &\geq \Delta_2 W^{\mathrm{d}} - \Delta_2 W^0 \\ &= \tfrac{1}{2} l_{\mathrm{b}} C^{\mathrm{p}} (\Delta\epsilon_{\mathrm{b}})^2 + \tfrac{1}{2} l_{\mathrm{a}} C_{\mathrm{a}} (\Delta\epsilon_{\mathrm{a}})^2 - \tfrac{1}{2} l C^{\mathrm{p}} (\Delta\epsilon_{\mathrm{o}})^2 \\ &\geq \tfrac{1}{2} C^{\mathrm{p}} (l_{\mathrm{b}} (\Delta\epsilon_{\mathrm{b}})^2 + l_{\mathrm{a}} (\Delta\epsilon_{\mathrm{a}})^2 - l (\Delta\epsilon_{\mathrm{o}})^2) \\ &= \tfrac{1}{2} C^{\mathrm{p}} \left(l_{\mathrm{b}} (\delta\epsilon_{\mathrm{b}})^2 + l_{\mathrm{a}} (\delta\epsilon_{\mathrm{a}})^2 \right). \end{aligned} \tag{2.41}$$

The latter inequality follows from $C_{\mathrm{a}} \geq C^{\mathrm{p}}$, and the final equality from $l_{\mathrm{b}}(\Delta\epsilon_{\mathrm{b}} - \Delta\epsilon_{\mathrm{o}}) + l_{\mathrm{a}}(\Delta\epsilon_{\mathrm{a}} - \Delta\epsilon_{\mathrm{o}}) = l_{\mathrm{b}}\delta\epsilon_{\mathrm{b}} + l_{\mathrm{a}}\delta\epsilon_{\mathrm{a}} = 0$, cf. (2.1). It follows that the fulfillment of the condition

$$\Delta_2 W^{\mathrm{d}} > \Delta_2 W^0 \qquad \text{at varying } \lambda \tag{2.42}$$

for $\Delta\epsilon_{\mathrm{o}} \neq \Delta\epsilon_{\mathrm{b}} \geq 0$, which is equivalent to

$$C^{\mathrm{p}} > 0\,, \tag{2.43}$$

ensures positiveness of the expression (2.39) for all examined deviations from the fundamental process within the representative bar. Hence, if (2.42) holds (or if $C^{\mathrm{p}} > 0$) then any deviation from the homogeneous plastic flow within the representative bar, embedded in an otherwise unperturbed longer bar, requires a positive amount of energy ΔW^* to be supplied by a disturbing agency. If ΔW^* tends to zero then the maximum possible distance from the fundamental process, measured by $|\delta\epsilon_{\mathrm{b}}(t)|$, tends also to zero by (2.41). This means that (2.42) or (2.43) is *sufficient* for stability of the quasi-static uniform plastic straining of a bar for perturbations restricted to its *interior* representative part. It is clear that the dimensions of the parts (a) and (b) of the representative bar are irrelevant for this conclusion.

Stability under deformation-sensitive loading. Consider again the representative bar as a part of a longer bar of reference length $L = l + l^*$, say (cf. Fig. 1B), but now with the flexibility of the outer part of the length l^* taken into account. This allows the current length $\bar{l}$ of the representative bar, previously fixed at λ=const, to vary arbitrarily in a perturbed process. The force P^{ext} applied to the representative bar from the side of the outer part depends on $\bar{l}$, constituting thus *deformation-sensitive loading.*

Consider stability of equilibrium under the assumption that there are no perturbations outside the *longer* bar of length L. Assume for simplicity that the part of the length l^* deforms uniformly and only elastically with the elastic modulus C^{e}, and denote by ϵ^* an increment of the respective engineering strain. The previous energy balance (2.34) must now be completed by changes in the potential energy $\Delta\Omega$ in that part, viz.

$$\Delta W + \Delta K + \Delta\Omega = \Delta W^* \qquad \text{at fixed } \lambda\,, \tag{2.44}$$

where

$$\Delta\Omega = l^* P^{\text{ext}}(t_0)\Delta\epsilon^* + \Delta_2\Omega\,, \qquad \Delta_2\Omega = \frac{1}{2} l^* \Delta P^{\text{ext}} \Delta\epsilon^*\,, \qquad \Delta P^{\text{ext}} = C^{\text{e}}\Delta\epsilon^*\,. \quad (2.45)$$

Displacement continuity at λ=const requires $l^*\Delta\epsilon^* = -l\Delta\epsilon_{\text{o}}$, which means that in an equilibrium state at t_0 the first-order terms of ΔW and $\Delta\Omega$ cancel each other. By proceeding similarly as previously, in place of (2.37) as the condition sufficient for stability of equilibrium of the longer bar we obtain

$$\Delta_2 W^{\text{d}} + \Delta_2\Omega > 0 \qquad \text{at fixed } \lambda\,, \quad (2.46)$$

equivalent to

$$C^{\text{p}} + \frac{l_{\text{b}}}{l_{\text{a}} + l^*} C^{\text{e}} > 0 \quad (2.47)$$

or to

$$C^{\text{p}} + \frac{l}{l^*} C^{\text{e}} > 0 \qquad \text{under restriction } \epsilon_{\text{a}} \equiv \epsilon_{\text{b}}\,. \quad (2.48)$$

The case of a prescribed load P^{ext} =const is met in the limit as $l/l^* \to 0$, $\Delta_2\Omega \to 0$.

Similarly, in placc of (2.42) wc now obtain

$$\Delta_2 W^{\text{d}} + \Delta_2\Omega > \Delta_2 W^0 + \Delta_2\Omega^0 \qquad \text{at varying } \lambda \quad (2.49)$$

as a condition of stability of the quasi-static fundamental *process* in the longer bar. It can be checked that (2.49) is still equivalent to (2.43). However, under restriction $\epsilon_{\text{a}} \equiv \epsilon_{\text{b}}$, (2.49) becomes equivalent to (2.48).

Exercise: Extend the analysis from Sect. 2.3 to the present case of deformation-sensitive loading.

Stability against internal rearrangements. Examine now disturbances at a micro-level in the representative bar, associated with internal rearrangements *not* satisfying the consistency condition (2.4). Suppose that the actual loading level P within the bar at equilibrium is locally increased by $\hat{P}^{\text{dist}}$ by superimposing a *self-equilibriated* system of internal perturbing forces. The rule for plastic activity $(P - P^{\text{y}})\dot{\gamma} = 0$ (cf. $(2.2)_3$) is accordingly extended to

$$(P + \hat{P}^{\text{dist}} - P^{\text{y}})\dot{\gamma} = 0\,. \quad (2.50)$$

To simplify the presentation, assume that C^{e} is constant; the corresponding elastic strain energy density ϕ^{e}, per unit reference length of the bar, satisfies

$$\phi^{\text{e}} = \frac{1}{2}(C^{\text{e}})^{-1}P^2\,, \qquad \dot{\phi}^{\text{e}} = (C^{\text{e}})^{-1}P\dot{P} = P\dot{\epsilon} - P\dot{\gamma}\,, \quad (2.51)$$

the last equality being obtained from $(2.5)_2$.

The rate of work to be supplied to the deforming bar, denoted by $\dot{w}'$ and defined per unit reference length, can be expressed as the sum of rates of the elastic and plastic work,

$$\dot{w}' = \dot{\phi}^{\text{e}} + \dot{w}^{\text{p}}\,, \qquad \dot{w}^{\text{p}} = P^{\text{y}}\dot{\gamma}\,. \quad (2.52)$$

Due to the additional work w^{dist} done by $\hat{P}^{\text{dist}}$, the supplied work-rate $\dot{w}'$ is no longer equal to $\dot{w} = P\dot{\epsilon}$ but on substituting (2.50), (2.51) and $(2.5)_1$ reads

$$\dot{w}' = \dot{w} + \dot{w}^{\text{dist}}\,, \qquad \dot{w} = P\dot{\epsilon}\,, \qquad \dot{w}^{\text{dist}} = \hat{P}^{\text{dist}}\dot{\gamma} = -f\dot{\gamma}\,. \tag{2.53}$$

Stability of equilibrium with respect to internal rearrangements within the representative bar is now examined under the restriction to deformations which are *uniform* both plastically and elastically. To limit attention to infinitesimal perturbing forces, we shall assume that in the current equilibrium state of the bar in which stability is investigated, $P = P^{\text{y}}$ and $\hat{P}^{\text{dist}} = 0$ at t_0. The first-order work expression remains thus the same as previously. The second-order work is split, according to (2.52), into two parts: $\Delta_2\phi^{\text{e}}$ dependent on ΔP, and $\Delta_2 w^{\text{p}}$ dependent on $\Delta\gamma$. Its final value (in this one-dimensional example) is thus independent of the actual path in (P,γ)-plane leading to the final state. On using (2.53) and (2.5) or (2.10) along a direct path in (P,γ)-plane corresponding to a proportional variation of $\hat{P}^{\text{dist}}$ in time, the second-order work expression reads

$$\begin{aligned} \Delta_2 W &= \tfrac{1}{2} l\Delta_2 w' \qquad \text{(at uniform strain)} \\ &= \tfrac{1}{2} l(\Delta P\Delta\epsilon - \Delta f\Delta\gamma) \\ &= \tfrac{1}{2} l(\Delta P(\Delta\epsilon - \Delta\gamma) + h(\Delta\gamma)^2) \\ &= \tfrac{1}{2} l(C^{\text{e}}\Delta\epsilon(\Delta\epsilon - 2\Delta\gamma) + (h + C^{\text{e}})(\Delta\gamma)^2)\,. \end{aligned} \tag{2.54}$$

Note that an analogous expression for the second-order work was obtained previously in (2.30) for arbitrary strain paths in the absence of internal perturbing forces $\hat{P}^{\text{dist}}$.

Under a fixed load P or strain ϵ, the condition sufficient for stability of equilibrium against internal rearrangements is $\Delta_2 W > 0$, from the energy balance as done previously. On using appropriate expressions in (2.54), this reduces to

$$\begin{aligned} h + C^{\text{e}} &> 0 \qquad \text{at fixed } \epsilon_{\text{o}}\,, \\ h &> 0 \qquad \text{at fixed } P\,. \end{aligned} \tag{2.55}$$

Note that $C^{\text{p}} \to -\infty$ as $h + C^{\text{e}} \to 0$. The analysis of stability of a *process* of internal rearrangements is deferred until Sect. 5.2.

Unified approach based on the energy criterion: a summary. The results obtained for different types of instability in the illustrative example can be put in a unified form as follows. Detailed assumptions given previously are not repeated.

Denote by

$$\Delta E = \Delta W + \Delta\Omega \qquad \text{(to second order)} \tag{2.56}$$

the increment of energy to be supplied to the representative bar *and* its outer (elastic) part along a kinematically admissible deformation path. If stability with respect to internal rearrangements is studied then the deformation work ΔW should include the work done by internal perturbing forces, e.g. by integrating (2.52) or (2.53). The part $\Delta\Omega$ expresses a change in the potential energy of the load P^{ext} applied to the representative bar from the outside; if λ is varying in time then $\Delta\Omega$ *cannot* be expressed by the work done by P^{ext} alone.

The energy condition sufficient for stability of *equilibrium* of the bar can be written in the form

$$\Delta E > 0 \qquad \text{at fixed } \lambda \tag{2.57}$$

to hold along any path of departure from equilibrium. Along a *direct* path of departure we may write

$$\Delta E^{\mathrm{d}} = \dot{E}\Delta t + \frac{1}{2}\ddot{E}(\Delta t)^2 \qquad \text{(to second order)}, \tag{2.58}$$

where the (right-hand) time derivatives are calculated in the examined equilibrium state. Since $\dot{E}$ vanishes in equilibrium under zero perturbing forces, the condition sufficient for directional stability of equilibrium takes the form

$$\ddot{E} > 0 \qquad \text{for all admissible directions at fixed } \lambda\,. \tag{2.59}$$

On account of (2.33), arbitrarily circuitous paths require not less energy supply (to second order) than direct paths, and therefore (2.59) is sufficient for stability of equilibrium, and not only for directional stability. This condition reduces to (2.37), (2.46) or (2.55) under the assumptions given in the respective sections. 'Directions' mean here non-zero vectors of four components $(\dot{\epsilon}_{\mathrm{a}}, \dot{\epsilon}_{\mathrm{b}}, \dot{\gamma}_{\mathrm{a}}, \dot{\gamma}_{\mathrm{b}})$, always restricted by $\dot{\gamma} \geq 0$ and with further restrictions dependent on the examined type of stability.

On the basis of the result of Sect. 2.2 and in view of the equality (2.35), under assumptions adopted therein the condition

$$\ddot{E} \geq 0 \qquad \text{for all admissible directions at fixed } \lambda \tag{2.60}$$

is *necessary* for stability of equilibrium. However, in more general circumstances this assertion requires a separate proof.

As shown in Sect. 2.4, stability (in the first approximation) of the quasi-static *process* of uniform straining of the *interior* representative bar is ensured for perturbations vanishing outside the bar if the second-order work inequality (2.42) holds. In particular, this excludes quasi-static bifurcations of the deformation path. The extended inequality (2.49), rewritten as,

$$\ddot{E} > \ddot{E}^0 \qquad \text{for all (except fundamental) admissible directions at varying } \lambda\,, \tag{2.61}$$

may be regarded as a condition sufficient for directional stability of the quasi-static fundamental process under deformation-sensitive loading. The superscript '0' refers to the fundamental direction of straining.

In turn, the results of Sect. 2.3 along with the equivalence (2.42) $\Leftrightarrow$ (2.43) indicate that the condition

$$\ddot{E} \geq \ddot{E}^0 \qquad \text{for all admissible directions at varying } \lambda \tag{2.62}$$

is *necessary* for directional stability of the quasi-static fundamental process in the representative bar under the end displacement control. Interpretation of the path instability predicted by failure of (2.62) will be discussed in Sect. 5 in more general circumstances.

Four versions of the second-order energy criterion of material stability have been thus distinguished. However, it can be shown (cf. Petryk, 1985a) that (2.62) implies (2.60), which indicates that the energy conditions for stability of a quasi-static process are generally more restrictive than those for stability of equilibrium. In other words, if the former are satisfied then the latter are not needed. On the other hand, if the former fail then the latter can provide the information whether the resulting instability is quasi-static in character, as in Sect. 2.3, or is of dynamic type.

Moreover, the quasi-static process can be parameterized by a time-like parameter t in a special way such that an equilibrium state is kept fixed at constant λ in some interval of $t \geq t_0$, so that $\ddot{E}^0 = 0$. The conditions for stability of a quasi-static process within such a particular but arbitrarily long period *reduce* to those for stability of equilibrium. Consequently, the conditions (2.59) and (2.60) may be regarded as particular cases of (2.61) and (2.62), respectively. Finally, since the critical deformation stage associated with failure of (2.61) or (2.62) along a deformation path in usual circumstances will be the same, we arrive in effect at *a single energy criterion of material instability*, associated with the inequality (2.62).

All this concerned the simple example of the bar as a (one-dimensional and discretized) substitute of a material element. Generalization of the energy criterion to material instability in a three-dimensional case is not immediate, and will be discussed in Sections 4 and 5, following earlier papers by the author (Petryk, 1982, 1985a, 1991b, 1992a, 1999a). The above considerations limited to a highly idealized example may, however, be helpful in conceptually similar but mathematically more complex investigations of three-dimensional problems.

2.5 Second-Order Geometric Effect

The critical condition

$$C^{\mathrm{p}} = 0 \tag{2.63}$$

which, according to (2.23), marks the onset of instability of uniform straining of the bar, coincides with the classical Considère criterion. It states that necking in a slender tensile bar begins when the *load* (and not the current stress) attains a maximum. At a first glance it could seem that the difference between the maxima of the load and stress may be neglected in the range of sufficiently small strains. That conjecture is incorrect. Denote by σ the true tensile stress, by $\bar{\rho}$ the mass density per unit current volume, and by $\bar{A}$ the current cross-section area of the bar. During uniform straining of the bar, we have

$$\begin{aligned} P &= \sigma \bar{A} \\ &= \frac{\tau}{1+\epsilon} A \, , \end{aligned} \tag{2.64}$$

where A is the (fixed) cross-section area of the bar in the reference configuration, and $\tau = \sigma\rho/\bar{\rho}$ is the uniaxial Kirchhoff stress with ρ as the mass density per unit reference volume. Consequently, for the plastic loading, we have

$$C^{\mathrm{p}} = \frac{dP}{d\epsilon} = \left(\frac{d\tau}{d\epsilon} - \frac{\tau}{1+\epsilon}\right)\frac{A}{1+\epsilon} \, . \tag{2.65}$$

The critical condition takes thus the form

$$C^{\mathrm{p}} = 0 \qquad \Leftrightarrow \qquad \frac{d\tau}{d\varepsilon} = \tau \, , \qquad \varepsilon = \ln(1+\epsilon) \, , \tag{2.66}$$

where ε is the true (logarithmic) strain. If volume changes are neglected then the Kirchhoff stress τ reduces to the true (Cauchy) stress σ, and the condition in (2.66) reduces to $d\sigma/d\varepsilon =$

σ, the classical condition due to Considère. It is evident from (2.66) that even for vanishingly small strain ϵ (which can always be met by choosing the reference configuration as coinciding momentarily with the current configuration of the bar), instability of the uniform deformation process in the bar occurs when the (Kirchhoff) stress-strain curve is still increasing, with the slope equal to the current stress.

One might argue that the geometric effect discussed above is strictly connected with the necking instability. This is true; however, the effect of changes in geometry is also important for intrinsic instability in the material. The difference between current values of the Cauchy and nominal stresses (per unit current and reference area, respectively) may by attributed to the first-order geometric effect. If the strain and rotation are is sufficiently small then the difference can safely be neglected in calculations of the *first-order* work. However, in calculations of the *second-order* work, which is essential in stability analysis, the *second-order* geometric effect is not negligible in general. It is related to the difference between the *rates* (or increments) of the Cauchy and nominal stresses which persists irrespectively whether the current strain is small or not. For instance, if the second-order geometric effect were not taken into account, the right-hand term τ in (2.66) would be erroneously set equal to zero. Similar errors can appear in a three-dimensional analysis. The recommended procedure of calculating the rates or differentials of loads and stresses is thus to perform the needed time differentiation within a geometrically exact framework *before* introducing any small-strain simplifications, and not conversely. As in the simple calculations above, in more general studies of material instabilities this helps in avoiding the over-simplification caused by neglecting the important second-order geometric effect.

Exercise: Carry out a detailed proof of (2.64) to arrive at the equivalence in (2.66).

3 Constitutive Framework in Incrementally Nonlinear Plasticity

Notation. From now on we extend investigations to three-dimensional solids. In the tensor notation employed, bold-face letters denote vectors (in $\mathcal{R}^3$) or second-order tensors, and 'sans serif' bold capitals (like A) stand for fourth-order tensors. Direct juxtaposition of two vectors or tensors denotes simple contraction, separation by a dot - double contraction, and the symbol $\otimes$ - the tensor product. In components on a fixed orthonormal basis, which are denoted by lower case Latin subscripts with the summation convention for repeated indices, it means that $s_j n_j = \mathbf{sn}$, $F_{ij}n_j = (\mathbf{Fn})_i$, $\sigma_{ij}W_{jk} = (\boldsymbol{\sigma}\mathbf{W})_{ik}$, $m_{ij}W_{ij} = \mathbf{m}\cdot\mathbf{W}$, $C_{ijkl}F_{kl} = (\mathsf{C}\cdot\mathbf{F})_{ij}$, $C_{ijkl}P_{klpq} = (\mathsf{C}\cdot\mathsf{P})_{ijpq}$, $s_i n_j = (\mathbf{s}\otimes\mathbf{n})_{ij}$, $\lambda_{ij}\mu_{kl} = (\boldsymbol{\lambda}\otimes\boldsymbol{\mu})_{ijkl}$. $\mathbf{I}$ stands for the second-order unit tensor, an upper index T denotes a transpose, $(\mathbf{F}^{\mathrm{T}})_{ij} = F_{ji}$ or $(\mathsf{A}^{\mathrm{T}})_{ijkl} = A_{klij}$, $\operatorname{tr}\mathbf{a} = a_{qq}$ denotes the trace, and subscripts 's' and 'a' denote the symmetric and antisymmetric part, $\frac{1}{2}(\mathbf{a}+\mathbf{a}^{\mathrm{T}})$ and $\frac{1}{2}(\mathbf{a}-\mathbf{a}^{\mathrm{T}})$, respectively, of a second-order tensor $\mathbf{a}$. A superimposed dot over a symbol denotes the *forward* rate, i.e. the one-sided time derivative from the right with respect to a time-like variable t at a fixed material point.

3.1 Rates of Strain and Stress

Constitutive rate equations in plasticity involve forward rates of strain and stress as basic quantities. Any objective measures of strain-rate and stress-rate which do not depend on the material spin may be employed. As a basic pair of spin-insensitive, symmetric measures of strain-rate and stress-rate, it is convenient to take $\mathbf{D}$, the Eulerian strain-rate, and $\overset{\nabla}{\boldsymbol{\tau}}$, the Zaremba-Jaumann flux of the Kirchhoff stress based on the *current* configuration. Recall that the Kirchhoff stress $\boldsymbol{\tau}$ is related to the true or Cauchy stress $\boldsymbol{\sigma} = \boldsymbol{\sigma}^{\mathrm{T}}$ by $\boldsymbol{\tau} = J\boldsymbol{\sigma}$, where J is the ratio of reference-to-current material density, and that the Zaremba-Jaumann flux is corotational with the material spin $\mathbf{W}$. If $\boldsymbol{\tau}$ is based on the current configuration then $\boldsymbol{\tau}$ and $\boldsymbol{\sigma}$ momentarily coincide, but their rates are still different if $J\,\mathrm{tr}\,\mathbf{D} = \dot{J} \neq 0$. The relevant standard formulae are

$$\mathbf{D} = (\bar{\nabla}\mathbf{v})_{\mathrm{s}}\,, \qquad \mathbf{W} = (\bar{\nabla}\mathbf{v})_{\mathrm{a}}\,, \qquad \overset{\nabla}{\boldsymbol{\tau}} = \overset{\nabla}{\boldsymbol{\sigma}} + \boldsymbol{\sigma}\mathrm{tr}\,\mathbf{D}\,, \qquad \overset{\nabla}{\boldsymbol{\sigma}} = \dot{\boldsymbol{\sigma}} + \boldsymbol{\sigma}\mathbf{W} - \mathbf{W}\boldsymbol{\sigma}\,, \tag{3.1}$$

where $\bar{\nabla}\mathbf{v}$ is the spatial gradient of a material velocity $\mathbf{v}$ in the current configuration.

Consider a class of of finite strain and stress measures provided by symmetric second-order tensors $\mathbf{e}$ and $\mathbf{t}$, respectively, which are objective in the Lagrangean sense (Hill, 1978, Ogden, 1984a). Denote by $\dot{w}$ the rate of virtual work by current tractions on the material element per unit *reference* volume, assuming that the tractions are in equilibrium with the internal state of stress. The identity

$$\dot{w} \equiv J\boldsymbol{\sigma} \cdot \mathbf{D} = \mathbf{t} \cdot \dot{\mathbf{e}} \tag{3.2}$$

is used to define the stress $\mathbf{t}$ *work-conjugate to* $\mathbf{e}$ (Hill, 1968, 1978). The material (right-hand) time derivatives $(\dot{\mathbf{e}}, \dot{\mathbf{t}})$ provide a class of objective measures of strain-rate and stress-rate, respectively. The basic pair $(\mathbf{D}, \overset{\nabla}{\boldsymbol{\tau}})$ belongs to this class, which motivated the use of $\overset{\nabla}{\boldsymbol{\tau}}$ rather than $\overset{\nabla}{\boldsymbol{\sigma}}$ as the basic stress-rate measure. If the reference configuration is chosen to coincide momentarily with the current configuration then the rate of *any* strain measure $\bar{\mathbf{e}}$ based on that reference configuration is assumed equal to $\mathbf{D}$, while $\bar{\mathbf{e}} = \mathbf{0}$ instantaneously. In turn, $\overset{\nabla}{\boldsymbol{\tau}}$ is then equal to the rate of the *particular* stress measure $\bar{\mathbf{t}}^{(0)}$ that is work-conjugate to the logarithmic strain $\bar{\mathbf{e}}^{(0)}$ based on the current configuration, although every stress $\bar{\mathbf{t}}$ work-conjugate to $\bar{\mathbf{e}}$ coincides momentarily with $\boldsymbol{\sigma}$ (Hill, 1968).

Transformation formulae under the transition from one work-conjugate pair to another were given by Hill (1978, 1983), cf. also (Ogden, 1984b). In particular, it has been shown that (3.2) implies[5]

$$\mathbf{t} = J\bar{\mathbf{t}}^{(0)} \cdot \frac{\partial \bar{\mathbf{e}}^{(0)}}{\partial \mathbf{e}}\,, \qquad \dot{\mathbf{t}} = J\overset{\nabla}{\boldsymbol{\tau}} \cdot \frac{\partial \bar{\mathbf{e}}^{(0)}}{\partial \mathbf{e}} + J\bar{\mathbf{t}}^{(0)} \cdot \frac{\partial^2 \bar{\mathbf{e}}^{(0)}}{\partial \mathbf{e} \partial \mathbf{e}} \cdot \dot{\mathbf{e}}\,. \tag{3.3}$$

Hence,

$$\dot{\mathbf{e}} = \mathsf{A} \cdot \mathbf{D} \quad \text{and} \quad \dot{\mathbf{t}} = \mathsf{B}_{\mathrm{w}} \cdot \overset{\nabla}{\boldsymbol{\tau}} + \mathsf{Z}_{\mathrm{w}} \cdot \mathbf{D}\,, \tag{3.4}$$

with the following identification

$$\mathsf{A} = \frac{\partial \mathbf{e}}{\partial \bar{\mathbf{e}}^{(0)}}\,, \qquad \mathsf{B}_{\mathrm{w}}^{\mathrm{T}} = J\mathsf{A}^{-1}\,, \qquad \mathsf{Z}_{\mathrm{w}} = J\mathsf{Y} \cdot \mathsf{A}\,, \qquad \mathsf{Y} = \mathsf{Y}^{\mathrm{T}} = \boldsymbol{\sigma} \cdot \frac{\partial^2 \bar{\mathbf{e}}^{(0)}}{\partial \mathbf{e} \partial \mathbf{e}}\,. \tag{3.5}$$

A, B_{w}, Z_{w} are fourth-order tensors with minor symmetries, independent of $\mathbf{D}$ and $\overset{\nabla}{\boldsymbol{\tau}}$ and dependent in general on the material state; the index 'w' is used to indicate that we are dealing with

[5] The dot on the left (right) side of a partial derivative denotes the contraction over two first (last) indices of the numerator (denominator), respectively.

work-conjugate stress measures. A and B_w are assumed invertible when treated as linear operators acting in the space of symmetric second-order tensors. That is, $\mathsf{A}\cdot\mathsf{A}^{-1} = \mathsf{A}^{-1}\cdot\mathsf{A} = \mathsf{I}_s$, where I_s is the unit tensor with minor symmetries, of Cartesian components $(\mathsf{I}_s)_{ijkl} = \frac{1}{2}(\delta_{ik}\delta_{jl}+\delta_{il}\delta_{jk})$ on a fixed orthonormal basis, and δ_{ij} denotes the Kronecker delta. If $\mathbf{F} = \mathbf{I}$ momentarily (one of the assumptions of the 'small strain' theory) then $\mathsf{A} = \mathsf{B}_w = \mathsf{I}_s$, but Z_w does not vanish and is generally of order of the stress $\boldsymbol{\sigma}$. It can be seen that contrary to another assumption of the 'small strain' theory, the resulting differences between various stress-rate measures are generally *not* negligible in plasticity even at small strains.

While the above rate-variables are useful in defining objective incremental constitutive laws, in the analysis of bifurcation and instability problems it is convenient to use the non-symmetric rate-variables $(\dot{\mathbf{F}}, \dot{\mathbf{S}})$ that include the effect of a material spin. They are generated by the material time derivatives of the deformation gradient $\mathbf{F}$ (restricted by $J \equiv \det \mathbf{F} > 0$) and nominal stress $\mathbf{N}$ or its transpose $\mathbf{S} = \mathbf{N}^{\mathrm{T}}$ that satisfy

$$\mathbf{F} = \frac{\partial \mathbf{x}}{\partial \boldsymbol{\xi}}, \qquad \mathbf{S}\mathbf{F}^{\mathrm{T}} = J\boldsymbol{\sigma} = \mathbf{F}\mathbf{N}, \tag{3.6}$$

where $\mathbf{x}$ and $\boldsymbol{\xi}$ denote the position vector of a material point in the current and reference configurations, respectively. Below the tensor $\mathbf{S}$ will be used and for brevity referred to as the Piola stress (rather than the first Piola-Kirchhoff stress tensor). Time differentiation of (3.6) after some rearrangements shows that the rate-variables $(\dot{\mathbf{S}}, \dot{\mathbf{F}})$ are related to these in (3.4) through

$$\dot{\mathbf{F}} = (\mathbf{D} + \mathbf{W})\mathbf{F}, \qquad \dot{\mathbf{S}} = J(\overset{\nabla}{\boldsymbol{\tau}} + \mathbf{W}\boldsymbol{\sigma} - \boldsymbol{\sigma}\mathbf{D})\mathbf{F}^{-1^{\mathrm{T}}}. \tag{3.7}$$

3.2 The Second-Order Work

The concept of the second-order work plays an important role in material stability investigations. If all changes in geometry of a material element are disregarded then an infinitesimal increment of work per unit volume is usually written as

$$\sigma \cdot d\epsilon + \frac{1}{2} d\sigma \cdot d\epsilon + \ldots, \tag{3.8}$$

where ϵ stands for the small-strain tensor and a prefix d denotes a differential. The former term which is a linear form with respect to an infinitesimal strain increment $d\epsilon$ is called the first-order work. The second term with $d\sigma$ related to $d\epsilon$ is called the second-order work. Positiveness of the second-order work:

$$d\sigma \cdot d\epsilon > 0 \qquad \text{or} \qquad \dot{\sigma} \cdot \dot{\epsilon} > 0 \qquad \text{for all } d\epsilon = \dot{\epsilon}\, dt \tag{3.9}$$

was introduced by Drucker (1950) in his postulate of stability of material in relation to the definition of work-hardening.

As shown by Hill (1968), the above postulate is not sufficiently precise unless the measure of stress or stress-rate is specified to define the symbols $d\sigma$ or $\dot{\sigma}$, respectively. The correct formula for an increment Δw of the deformation work per unit reference volume evaluated to second order along a *direct* path[6] in strain space is the 'trapezoid rule of quadrature' (Hill, 1978)

$$\Delta w^{\mathrm{d}} = \Delta_1 w + \Delta_2 w^{\mathrm{d}} + o(|\Delta\mathbf{e}|^2)\,, \qquad \Delta_1 w = \mathbf{t}_0 \cdot \Delta\mathbf{e}\,, \qquad \Delta_2 w^{\mathrm{d}} = \frac{1}{2}\Delta\mathbf{t}\cdot\Delta\mathbf{e}\,, \tag{3.10}$$

where $(\mathbf{e}, \mathbf{t})$ stands for any *work-conjugate* pair of strain and stress measures, $\Delta\mathbf{t}$ is the stress increment corresponding to a strain increment $\Delta\mathbf{e}$ along the path, and $\mathbf{t}_0$ is the initial stress. Under transition to another work-conjugate pair $(\mathbf{e}, \mathbf{t})$ the expression for Δw^{d} in (3.10) remains invariant to second order. However, the decomposition (3.10) itself is not invariant since *both* terms $\Delta_1 w$ and $\Delta_2 w$ change by a second-order quantity with opposite signs. As explained by Hill, the lack of invariance is related to the 'curvature' term in the expression for $\Delta\mathbf{e}$ correct to *second* order (e.g. $\Delta\mathbf{e} = \dot{\mathbf{e}}\Delta t + \frac{1}{2}\ddot{\mathbf{e}}(\Delta t)^2 + o(|\Delta\mathbf{e}|^2)$) which should be substituted into $\Delta_1 w$ when used in (3.10). Accordingly, positiveness of the second-order work written as

$$d\mathbf{t}\cdot d\mathbf{e} > 0 \qquad \text{or} \qquad \dot{\mathbf{t}}\cdot\dot{\mathbf{e}} > 0 \qquad \text{for all } d\mathbf{e} = \dot{\mathbf{e}}\,dt \tag{3.11}$$

is *not* a measure-invariant concept (Hill, 1968).

The distinction between (3.9) and (3.11) is generally not negligible even if the current configuration is taken as reference, i.e. when the current strain is exactly zero and $\dot{\mathbf{e}} = \mathbf{D}$. For instance, if $\dot{\boldsymbol{\sigma}}$ in (3.9) is literally understood as the material time derivative of the Cauchy stress then

$$\overset{\nabla}{\boldsymbol{\tau}}\cdot\mathbf{D} - \dot{\boldsymbol{\sigma}}\cdot\dot{\boldsymbol{\epsilon}} = \boldsymbol{\sigma}\cdot(\mathbf{D}\,\mathrm{tr}\,\mathbf{D} + 2\mathbf{W}\mathbf{D})\,. \tag{3.12}$$

More generally, from (3.4) with (3.5) it follows that

$$\frac{1}{J}\dot{\mathbf{t}}\cdot\dot{\mathbf{e}} - \overset{\nabla}{\boldsymbol{\tau}}\cdot\mathbf{D} = \dot{\mathbf{e}}\cdot\mathbf{Y}\cdot\dot{\mathbf{e}}\,. \tag{3.13}$$

In turn, for the work-conjugate pair of non-symmetric measures $(\mathbf{F}, \mathbf{S})$ useful in boundary-value problems, we have

$$\frac{1}{J}\dot{\mathbf{S}}\cdot\dot{\mathbf{F}} - \overset{\nabla}{\boldsymbol{\tau}}\cdot\mathbf{D} = -\boldsymbol{\sigma}\cdot(\mathbf{D}\mathbf{D} + \mathbf{W}\mathbf{W} + 2\mathbf{W}\mathbf{D})\,. \tag{3.14}$$

The differences make explicit the second-order geometric effect. It is generally not negligible for a non-zero current stress being of the order of a tangent stiffness modulus, no matter whether the current strain is 'small' or not.

3.3 Constitutive Rate Equations in Plasticity

Incremental stiffness moduli. In a given state of the material, denoted here by a symbol $\mathcal{G}$, a stress-rate is assumed to be a continuous function of a strain-rate[7],

$$\dot{\mathbf{t}} = \dot{\mathbf{t}}(\dot{\mathbf{e}}, \mathcal{G}) \qquad \text{such that} \qquad \dot{\mathbf{t}}(r\dot{\mathbf{e}}, \mathcal{G}) = r\dot{\mathbf{t}}(\dot{\mathbf{e}}, \mathcal{G}) \quad \text{for all } r > 0\,. \tag{3.15}$$

[6] A strain path is understood here as a directed, continuous and piecewise smooth line in a strain space. A direct path in strain-space may be identified with a class of paths of uniformly bounded curvature, connecting given initial and final strains arbitrarily close to each other. The concept of *path-dependent* second-order work applicable to indirect, arbitrarily circuitous paths, which will be needed later, was developed in (Petryk, 1991a).

[7] Note that the function itself might be dependent on strain *gradients* or on the size of a material element. This way of introducing an internal length scale in the constitutive theory deserves separate analysis, not attempted here.

The above homogeneity property results from the assumption that the mechanical properties of the material do not depend in any way on a natural time (*time-independent material*). For the sake of simplicity of the notation, the same symbol is used for a function and its value. By Euler's theorem on homogeneous functions applied at every point of differentiability of $\dot{\mathbf{t}}$ it follows that

$$\dot{\mathbf{t}} = \mathsf{L}(\dot{\mathbf{e}}, \mathcal{G}) \cdot \dot{\mathbf{e}}\,, \qquad \mathsf{L} \equiv \frac{\partial \dot{\mathbf{t}}}{\partial \dot{\mathbf{e}}}\,, \qquad \mathsf{L}(r\dot{\mathbf{e}}, \mathcal{G}) = \mathsf{L}(\dot{\mathbf{e}}, \mathcal{G}) \quad \text{for all } r > 0\,. \tag{3.16}$$

Components of the fourth-order tensor L are called *incremental stiffness moduli.* The tensor L depends in general on the *direction* (but not magnitude) of strain-rate $\dot{\mathbf{e}}$; the dependence may be discontinuous and nonlinear. A domain (necessarily cone) in the strain-rate space within which L does not depend on $\dot{\mathbf{e}}$ is called *a constitutive cone* or a constitutive branch. On the interface between two constitutive cones, L is not uniquely defined. Henceforth, it will be tacitly assumed that the incremental stiffness moduli tensor, whenever appears in the formulae, is well defined.

Irrespectively of the variables in which it is originally formulated to ensure objectivity, the constitutive rate-equation (3.16) can be expressed as

$$\dot{\mathbf{S}} = \dot{\mathbf{S}}(\dot{\mathbf{F}}, \mathcal{G}) = \mathsf{C}(\dot{\mathbf{F}}, \mathcal{G}) \cdot \dot{\mathbf{F}}, \qquad \mathsf{C} = \frac{\partial \dot{\mathbf{S}}}{\partial \dot{\mathbf{F}}}, \tag{3.17}$$

e.g., by using the formulae given in Sect. 3.1. The moduli tensor C will play an essential role later on.

The simplest example of a non-linear constitutive rate equation (3.16) or (3.17) is provided by the classical elastoplasticity with two constitutive branches, where

$$\overset{\nabla}{\tau} = \bar{\mathsf{L}} \cdot \mathbf{D}\,, \qquad \bar{\mathsf{L}} = \bar{\mathsf{L}}^{\mathrm{e}} - \frac{\kappa}{g} \bar{\boldsymbol{\lambda}}^{\mathrm{p}} \otimes \bar{\boldsymbol{\lambda}}\,, \qquad \kappa = \begin{cases} 1 \text{ if } \bar{\boldsymbol{\lambda}} \cdot \mathbf{D} > 0 & \text{(loading)}, \\ 0 \text{ if } \bar{\boldsymbol{\lambda}} \cdot \mathbf{D} < 0 & \text{(unloading)}. \end{cases} \tag{3.18}$$

The notation will be explained in more detail below when discussing the extension to multi-mode plasticity. At the level of generality adopted here, there is no need to appeal to the usual multiplicative decomposition of $\mathbf{F}$. The present incremental description follows Hill and is more general since the plastic deformation gradient need not even be defined.

To make a model complete, the constitutive rate equations must be complemented with evolution rules for the variables dependent on a material state. This means that the dependence of $\mathsf{L}(\dot{\mathbf{e}})$ on $\mathcal{G}$ must be specified. However, in the analysis of material instability in elastic-plastic solids, as already illustrated in Sect. 2, of primary importance are the incremental stiffness moduli, and definite conclusions can be drawn without specifying how L as a *function* of $\dot{\mathbf{e}}$ varies along a deformation path. To simplify the notation, the dependence of L on $\mathcal{G}$ will not be indicated explicitly in the following.

Multi-mode plasticity and the vertex effect. A basic feature of an incremental constitutive law of the classical bi-linear form (3.18) is that the *direction* of the plastic part of strain-rate, $\mathbf{D}^{\mathrm{p}} \equiv \mathbf{D} - (\bar{\mathsf{L}}^{\mathrm{e}})^{-1}\overset{\nabla}{\tau}$, in a given state is prescribed (for the *fixed* rate-measures) and hence independent of the stress-rate. According to experimental and micromechanical studies, not reviewed here, that assumption may be an oversimplification for real materials, for instance metal crystals or polycrystals as well as geomaterials. A more accurate approximation of the incremental elastoplastic behaviour can be obtained by considering activity of *multiple internal mechanisms of plastic deformation*. The constitutive relations of classical elastoplasticity are included as the special case $N = 1$.

Let the rate of activity of K-th mechanism, $K = 1, ..., N$, be measured by a scalar rate-variable $\overset{\circ}{\gamma}_K$, taken nonnegative by convention. It can be nonzero only if the respective loading variable f_K has reached from below a threshold value, without loss of generality set equal to zero. Thus

$$\overset{\circ}{\gamma}_K \geq 0, \qquad f_K \leq 0, \qquad f_K \overset{\circ}{\gamma}_K = 0. \tag{3.19}$$

No implicit summation convention is adopted for the capital Latin indices of the mechanisms: the summation over an index K will always be indicated explicitly. If all the $\overset{\circ}{\gamma}_K$'s are zero then the material undergoes purely elastic deformation. A circle rather than a dot over γ_K is used to emphasize that $\overset{\circ}{\gamma}_K$'s need not represent rates of any internal state variables (cf. Rice, 1971).

Current values of the strain, stress and f_K's are obtained by the integration of their rates along a straining path with given initial conditions. The basic constitutive expressions for the rate-variables $\dot{\mathbf{t}}$ and $\dot{f}_K$ in a given state of the material are assumed as linear forms in $(\dot{\mathbf{e}}, \overset{\circ}{\gamma}_K)$:

$$\dot{\mathbf{t}} = \mathbf{L}^{\mathrm{e}} \cdot \dot{\mathbf{e}} - \sum_K \boldsymbol{\lambda}_K^{\mathrm{p}} \overset{\circ}{\gamma}_K, \qquad \dot{f}_K = \boldsymbol{\lambda}_K \cdot \dot{\mathbf{e}} - \sum_L g_{KL} \overset{\circ}{\gamma}_L. \tag{3.20}$$

A dual form, fully equivalent to (3.20) under the sole additional assumption that $\mathbf{L}^{\mathrm{e}}$ represents an invertible linear operator acting in the space of symmetric second-order tensors, reads

$$\dot{\mathbf{e}} = \mathbf{M}^{\mathrm{e}} \cdot \dot{\mathbf{t}} + \sum_K \boldsymbol{\mu}_K^{\mathrm{p}} \overset{\circ}{\gamma}_K, \qquad \dot{f}_K = \boldsymbol{\mu}_K \cdot \dot{\mathbf{t}} - \sum_L h_{KL} \overset{\circ}{\gamma}_L. \tag{3.21}$$

The elastic moduli tensor $\mathbf{L}^{\mathrm{e}}$, direction tensors $\boldsymbol{\lambda}_K^{\mathrm{p}}$ of inelastic stress-rate, normals $\boldsymbol{\lambda}_K$ to loading hyperplanes in $\dot{\mathbf{e}}$-space, and interaction moduli g_{KL}, depend on the material state and not on values of any rate-variables. In the dual formulation, the elastic compliance tensor $\mathbf{M}^{\mathrm{e}}$, direction tensors $\boldsymbol{\mu}_K^{\mathrm{p}}$ of inelastic strain-rate, normals $\boldsymbol{\mu}_K$ to loading hyperplanes in $\dot{\mathbf{t}}$-space, and *effective* hardening (or softening) moduli h_{KL}, are likewise state-dependent, with the mutual relationships:

$$\mathbf{M}^{\mathrm{e}} = (\mathbf{L}^{\mathrm{e}})^{-1}, \qquad \boldsymbol{\mu}_K^{\mathrm{p}} = \mathbf{M}^{\mathrm{e}} \cdot \boldsymbol{\lambda}_K^{\mathrm{p}}, \qquad \boldsymbol{\mu}_K = \boldsymbol{\lambda}_K \cdot \mathbf{M}^{\mathrm{e}}, \qquad h_{KL} = g_{KL} - \boldsymbol{\lambda}_K \cdot \mathbf{M}^{\mathrm{e}} \cdot \boldsymbol{\lambda}_L^{\mathrm{p}}. \tag{3.22}$$

The existence of an elastic strain energy function is assumed, which jointly with the work-conjugacy of $(\mathbf{e}, \mathbf{t})$ implies that $\mathbf{L}^{\mathrm{e}} = \mathbf{L}^{\mathrm{e}\mathrm{T}}$, $\mathbf{M}^{\mathrm{e}} = \mathbf{M}^{\mathrm{e}\mathrm{T}}$.

In the finite-strain theory of single crystals deformed plastically by multislip, $\overset{\circ}{\gamma}_K$ is identified as the *invariant* rate of shearing on K-th slip system, and $f_K = \tau_K - \tau_K^{\mathrm{c}}$, where τ_K is the *generalized* resolved shear stress on that system and τ_K^{c} is its critical value; cf. Hill and Rice (1972), Asaro (1983), Havner (1992).

The nonnegative $\overset{\circ}{\gamma}_K$'s for potentially active mechanisms are to be determined from the consistency conditions that follow from (3.19) by differentiation, viz.

$$\dot{f}_K \leq 0, \quad \dot{f}_K \overset{\circ}{\gamma}_K = 0 \qquad \text{for } K \in \mathcal{P} \equiv \{K \,:\, f_K = 0\}, \tag{3.23}$$

while $\overset{\circ}{\gamma}_K = 0$ for $K \notin \mathcal{P}$. A vector of $\overset{\circ}{\gamma}_K$'s with $K \in \mathcal{P}$ will be denoted by $[\overset{\circ}{\gamma}]$.

If $\dot{\mathbf{e}}$ is prescribed then (3.20) and (3.23) lead to the following rate problem to be solved:

Problem (P1). *In a given material state and for given* $\dot{\mathbf{e}}$, *find* $\overset{\circ}{\gamma}_K$*'s such that*

$$\dot{f}_K = \boldsymbol{\lambda}_K \cdot \dot{\mathbf{e}} - \sum_{L\in\mathcal{P}} g_{KL}\overset{\circ}{\gamma}_L\,, \quad \overset{\circ}{\gamma}_K \geq 0\,, \quad \dot{f}_K \leq 0\,, \quad \dot{f}_K\overset{\circ}{\gamma}_K = 0 \qquad \text{for } K \in \mathcal{P}\,. \tag{3.24}$$

If (P1) is solved then the stress-rate is determined from $(3.20)_1$.

Eqs. (3.21) and (3.23) lead to the dual problem $(\mathbf{P1})^{\text{dual}}$ formulated analogously to (P1) if $\dot{\mathbf{t}}$ is prescribed, with obvious replacements.

Suppose that Problem (P1) has a unique solution $\overset{\circ}{\gamma}_K(\dot{\mathbf{e}})$; a condition sufficient for that will be given in Sect. 6.1. Denote by $\mathcal{A} \subseteq \mathcal{P}$ the set of indices of the active mechanisms ($\overset{\circ}{\gamma}_K > 0$ if and only if $K \in \mathcal{A}$) in the solution for a given $\dot{\mathbf{e}}$, and by $(g_{\mathcal{A}})$ the submatrix g_{KL} with the indices $K, L \in \mathcal{A}$. It is clear that the non-zero $\overset{\circ}{\gamma}_K$'s must satisfy the set of linear equations

$$\dot{f}_K = \boldsymbol{\lambda}_K \cdot \dot{\mathbf{e}} - \sum_{K,L\in\mathcal{A}} (g_{\mathcal{A}})_{KL}\overset{\circ}{\gamma}_L = 0 \qquad \text{for } K \in \mathcal{A}\,. \tag{3.25}$$

On solving this under the assumption $\det(g_{\mathcal{A}}) \neq 0$ and substituting into $(3.20)_1$, we obtain

$$\dot{\mathbf{t}} = \mathbf{L}(\dot{\mathbf{e}}) \cdot \dot{\mathbf{e}}\,, \qquad \mathbf{L} = \mathbf{L}^{\mathrm{e}} - \sum_{L\in\mathcal{A}} (g_{\mathcal{A}})^{-1}_{KL}\boldsymbol{\lambda}^{\mathrm{p}}_K \otimes \boldsymbol{\lambda}_L\,, \tag{3.26}$$

where the dependence of $\mathbf{L}$ on $\dot{\mathbf{e}}$ is implicit and solely due to the dependence of $\mathcal{A}$ on $\dot{\mathbf{e}}$.

Analogous conclusions can formally be drawn from the dual formulation (3.21) by regarding $\dot{\mathbf{t}}$ as given. However, in this case there are problems, to be clarified later, connected with the fact that a given $\dot{\mathbf{t}}$ for one choice of the stress-rate measure results in a *deformation-sensitive* $\dot{\mathbf{t}}$ for the same loading condition expressed through another stress-rate measure.

The problem (P1) admits a purely elastic solution with all $\overset{\circ}{\gamma}_K = 0$ if and only if

$$\boldsymbol{\lambda}_K \cdot \dot{\mathbf{e}} \leq 0 \qquad \text{for all } K \in \mathcal{P}\,. \tag{3.27}$$

Thus, (3.27) with the strict inequality for all K is the condition for elastic unloading of all potentially active mechanisms. The set of all $\dot{\mathbf{e}}$ that satisfy the strict inequalities in (3.27), if non-empty, constitutes the *elastic unloading cone* (Fig. 4), which can be identified with a *corner* on the theoretical yield surface representing the boundary of an elastic domain in strain-space. The actual strain represents a *vertex* of this domain. In the dual formulation, the associated corner on the yield surface in stress-space is obtained.

Another constitutive cone in $\dot{\mathbf{e}}$-space in a given state under the assumption $\det(g_{\mathcal{P}}) \neq 0$ is defined by

$$\boldsymbol{\lambda}^{\mathrm{a}}_K \cdot \dot{\mathbf{e}} > 0 \qquad \text{for all } K \in \mathcal{P}\,, \qquad \boldsymbol{\lambda}^{\mathrm{a}}_K = \sum_{L\in\mathcal{P}} (g_{\mathcal{P}})^{-1}_{KL}\boldsymbol{\lambda}_L\,. \tag{3.28}$$

Every $\dot{\mathbf{e}}$ from this cone corresponds to a solution with $\overset{\circ}{\gamma}_K > 0$ for all $K \in \mathcal{P}$, which can be verified by substituting $\mathcal{A} = \mathcal{P}$ into (3.25) and (3.26). The set of all $\dot{\mathbf{e}}$ that satisfy (3.28) constitutes the *cone of fully active loading* in strain-rate space; cf. (Hill, 1967a, Petryk, 1989) for a discussion of this concept.

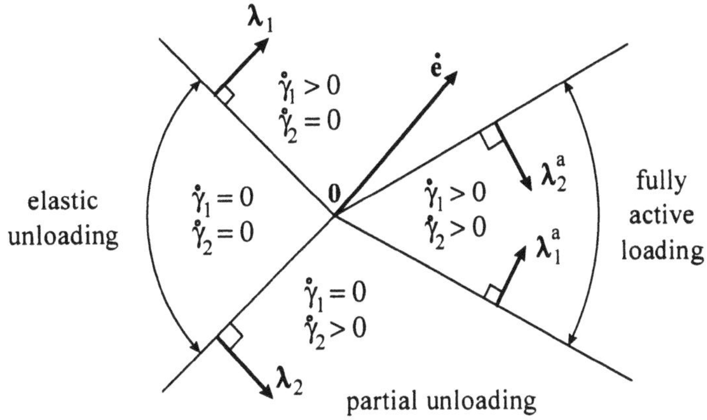

Figure 4. Decomposition of strain-rate space in double-mode plasticity.

It can be checked that if more than one mechanisms are simultaneously active then the direction of the plastic part of strain-rate varies generally with the stress-rate direction. This effect is known as the *yield-surface vertex effect*, of fundamental importance for material stability since it can significantly decrease the material stiffness against an abrupt change of the direction of straining (cf. Hutchinson, 1970).

Transformation rules. An important question, investigated thoroughly by Hill (1967c, 1978, 1983) and more recently in (Petryk, 2000) is the following: how the quantities in constitutive relations for a given material change under transition to another pair of strain-rate and stress-rate measures.

It is assumed here [8] that

$$J^{-1} f_K \text{ and } \mathring{\gamma}_K \text{ are measure-invariant.} \tag{3.29}$$

Let the quantities appearing in the constitutive rate equations above be distinguished by a superimposed bar when associated with the pair $(\mathbf{D}, \overset{\triangledown}{\tau})$. Then, in order for the material properties to be invariant under the transition to any other pair of strain-rate and stress-rate measures from the class (3.4), it can be shown along the lines in (Hill, 1978) and (Petryk, 2000) that the following transformation rules must hold:

For the incremental stiffness moduli :

$$\mathbf{L} = (\mathbf{B}_w \cdot \bar{\mathbf{L}} + \mathbf{Z}_w) \cdot \mathbf{A}^{-1} \qquad \text{if } \dot{\mathbf{t}} = \mathbf{L} \cdot \dot{\mathbf{e}}, \;\; \overset{\triangledown}{\tau} = \bar{\mathbf{L}} \cdot \mathbf{D}\,. \tag{3.30}$$

[8] In comparison with (Petryk, 2000), the factor J^{-1} is introduced here in order to ensure identical transformation rules for $\boldsymbol{\lambda}_K$ and $\boldsymbol{\lambda}_K^{\mathrm{p}}$ (cf. $(3.5)_2$), at the cost of loosing full invariance of g_{KL}.

For the quantities in (3.20):

$$\mathbf{L}^{\mathrm{e}} = (\mathbf{B}_{\mathrm{w}} \cdot \bar{\mathbf{L}}^{\mathrm{e}} + \mathbf{Z}_{\mathrm{w}}) \cdot \mathbf{A}^{-1}\,, \qquad \boldsymbol{\lambda}^{\mathrm{p}}_{K} = \mathbf{B}_{\mathrm{w}} \cdot \bar{\boldsymbol{\lambda}}^{\mathrm{p}}_{K}\,, \qquad \boldsymbol{\lambda}_{K} = J\bar{\boldsymbol{\lambda}}_{K} \cdot \mathbf{A}^{-1}\,, \qquad g_{KL} = J\bar{g}_{KL}\,. \tag{3.31}$$

For the hardening (softening) moduli in (3.21):

$$h_{KL} = J(\bar{h}_{KL} + \boldsymbol{\mu}_{K} \cdot \mathbf{Z}_{\mathrm{w}} \cdot \bar{\boldsymbol{\mu}}^{\mathrm{p}}_{L})\,. \tag{3.32}$$

If the non-symmetric measures (3.7) are used then:
Equations (3.20) transform to

$$\dot{\mathbf{S}} = \mathbf{C}^{\mathrm{e}} \cdot \dot{\mathbf{F}} - \sum_{K} \boldsymbol{\Lambda}^{\mathrm{p}}_{K} \overset{\circ}{\gamma}_{K}\,, \qquad \dot{f}_{K} = \boldsymbol{\Lambda}_{K} \cdot \dot{\mathbf{F}} - \sum_{L} g_{KL} \overset{\circ}{\gamma}_{L}\,, \tag{3.33}$$

with the relationships

$$\boldsymbol{\Lambda}^{\mathrm{p}}_{K} = J\bar{\boldsymbol{\lambda}}^{\mathrm{p}}_{K} \mathbf{F}^{-1^{\mathrm{T}}}\,, \qquad \boldsymbol{\Lambda}_{K} = \bar{\boldsymbol{\lambda}}_{K} \mathbf{F}^{-1^{\mathrm{T}}}\,. \tag{3.34}$$

Equations (3.26) transforms to

$$\dot{\mathbf{S}} = \mathbf{C}(\dot{\mathbf{F}}) \cdot \dot{\mathbf{F}}, \qquad \mathbf{C} = \mathbf{C}^{\mathrm{e}} - \sum_{K,L \in \mathcal{A}} (g_{\mathcal{A}})^{-1}_{KL} \boldsymbol{\Lambda}^{\mathrm{p}}_{K} \otimes \boldsymbol{\Lambda}_{L}\,. \tag{3.35}$$

Incremental stiffness moduli (elastic or elastoplastic) transform as follows:

$$J^{-1} F_{ip} F_{kq} C_{jplq} = \bar{L}_{ijkl} - \frac{1}{2}(\sigma_{jl}\delta_{ik} + \sigma_{jk}\delta_{il} + \sigma_{il}\delta_{jk} - \sigma_{ik}\delta_{jl}) \qquad \text{if } \dot{\mathbf{S}} = \mathbf{C} \cdot \dot{\mathbf{F}}\,,\ \overset{\nabla}{\boldsymbol{\tau}} = \bar{\mathbf{L}} \cdot \mathbf{D}\,. \tag{3.36}$$

Exercise: Show that substitution of the above transformation rules leads to the constitutive rate equations (3.20) and (3.21) expressed in terms of $(\mathbf{D}, \overset{\nabla}{\boldsymbol{\tau}})$ and the respective barred quantities.

3.4 Thermodynamic Framework

The material element treated at the macroscopic level as a material point in a continuum, in a more refined scale exhibits an internal structure which may be very complex and strongly heterogeneous. Macroscopic plastic deformation is considered as a consequence of microstructural rearrangements (Rice, 1971). From a physical point of view this is an inherently nonuniform process, with local instabilities such as jumps of dislocation loops at a micro-level. Consequently, the concept of a macroscopic strain-rate or stress-rate involves averaging over a *representative volume* of the material, as well as over a *representative time increment.*

In a thermodynamic description, changes in the internal structure of the material are frequently described with the help of *internal state variables*. In this article, internal state variables play only an auxiliary role. Their number and character (scalar or tensorial) are left unspecified; they are denoted collectively by $\boldsymbol{\alpha}$ and considered in the *reference* configuration. As in the classical theory of irreversible processes (cf. de Groot and Mazur, 1962), a material element is assumed to be in local (constrained) thermodynamic equilibrium when its absolute temperature (T), macroscopic strain ($\mathbf{e}$) and internal microstructure ($\boldsymbol{\alpha}$) are given. The triple $(T, \mathbf{e}, \boldsymbol{\alpha})$ may define the material state $\mathcal{G}$.

In the following the temperature T is assumed constant and removed from the list of variables.

To 'smooth out' the influence of possible discontinuous jumps in the material state at a micro-level and to admit the rate-description, the internal variables $\boldsymbol{\alpha}$ are defined as *averaging* variables over *both* a representative volume and a representative time increment (Petryk, 1995). The rate of the isothermal Helmholtz free energy density ϕ per unit reference volume may be written in the usual form as

$$\phi = \phi(\mathbf{e}, \boldsymbol{\alpha}), \qquad \dot{\phi} = \mathbf{t} \cdot \dot{\mathbf{e}} - \boldsymbol{A} \cdot \dot{\boldsymbol{\alpha}}, \qquad \mathbf{t} = \frac{\partial \phi}{\partial \mathbf{e}}, \qquad \boldsymbol{A} = -\frac{\partial \phi}{\partial \boldsymbol{\alpha}}. \tag{3.37}$$

The symbol $\boldsymbol{A}$ (not to be confused with the fourth-order tensor $\mathbf{A}$ used above) denotes collectively the set of thermodynamic forces conjugate to $\boldsymbol{\alpha}$, and the central dot between $\boldsymbol{A}$ and $\dot{\boldsymbol{\alpha}}$ denotes full contraction over all components of internal variables.

Identification of appropriate internal variables $\boldsymbol{\alpha}$ may be not easy. Parameters γ_K obtained by formal time integration of $\overset{\circ}{\gamma}_K$ in the multi-mode plasticity theory are generally not internal state variables since the actual pattern of internal rearrangement may also depend on the prior history of the $\overset{\circ}{\gamma}_K$'s. Generally, the internal microstructure may be regarded as a *functional* of the prior history $\mathcal{H}$ of non-elastic deformation. Then the material state $\mathcal{G}$ may be identified with the triple $(T, \mathbf{e}, \mathcal{H})$. If the relation between $\mathbf{t}$ and $\mathbf{e}$ at fixed $\mathcal{H}$ is one-to-one then $\mathcal{G}$ may also be identified with $(T, \mathbf{t}, \mathcal{H})$. In the isothermal treatment below, any dependence on T is omitted.

Following Rice (1971) and Hill and Rice (1973), ϕ in the multi-plasticity framework is treated below as a known function of $\mathbf{e}$ at each fixed $\mathcal{H}$, while the dependence of ϕ on $\mathcal{H}$ is prescribed only incrementally, viz.

$$\phi = \phi(\mathbf{e}, \mathcal{H}), \qquad \dot{\phi} = \mathbf{t} \cdot \dot{\mathbf{e}} - \sum_K \pi_K \overset{\circ}{\gamma}_K, \qquad \mathbf{t} = \frac{\partial \phi}{\partial \mathbf{e}}, \qquad \pi_K = \pi_K(\mathcal{G}), \tag{3.38}$$

instead of (3.37). Here, a material state is defined by $\mathcal{G} = (\mathbf{e}, \mathcal{H})$, and π_K are state-dependent scalar thermodynamic forces conjugate to $\overset{\circ}{\gamma}_K$. When writing (3.38), no explicit reference to internal *state* variables is needed. Links between (3.38) and (3.37) are obtained if every $\overset{\circ}{\gamma}_K$ is regarded as a state-dependent, positively homogeneous of degree one function of $\dot{\boldsymbol{\alpha}}$ (or conversely). We assume that that function is independent of elastic strain or stress, that is

$$\overset{\circ}{\gamma}_K = \overset{\circ}{\gamma}_K(\boldsymbol{\alpha}, \dot{\boldsymbol{\alpha}}), \qquad \overset{\circ}{\gamma}_K(\boldsymbol{\alpha}, r\dot{\boldsymbol{\alpha}}) = r\overset{\circ}{\gamma}_K(\boldsymbol{\alpha}, \dot{\boldsymbol{\alpha}}) \qquad \text{for all } r > 0. \tag{3.39}$$

The loading variables f_K are given the thermodynamic interpretation:

$$f_K = \pi_K - \pi_K^{c}, \qquad \pi_K^{c} \geq 0, \qquad \pi_K^{c} = \pi_K^{c}(\mathcal{G}), \tag{3.40}$$

where a threshold value π_K^{c} may depend here not only on $\mathcal{H}$ (and T) as in Rice's (1971) theory, but also on *elastic* changes in strain or stress. In that way non-associative plasticity is included here. Current values of π_K and π_K^{c} may be determined by the integration of their rates, e.g., assumed in the form analogous to that for $\dot{f}_K$ in (3.20), along a straining path with given initial conditions.

As in the case of ϕ, any other state-dependent variable is regarded as an explicit function of $\mathbf{e}$ (or $\mathbf{t}$, alternatively) while its dependence in $\mathcal{H}$ may be prescribed only incrementally. The *plastic* part of the stress-rate satisfies (Rice, 1971, Hill and Rice, 1973)

$$\dot{\mathbf{t}}^{\mathrm{p}} \equiv \dot{\mathbf{t}} - \mathbf{L}^{\mathrm{e}} \cdot \dot{\mathbf{e}} = \frac{\partial}{\partial \mathbf{e}}(\dot{\phi} - \mathbf{t} \cdot \dot{\mathbf{e}}) \,. \tag{3.41}$$

By combining this result with (3.20), (3.38) and using (3.40), and proceeding analogously in the dual formulation, one can show (Petryk, 1999b) that

$$\begin{aligned} &\lambda_K = f_{K,\mathbf{e}} \qquad \lambda^{\mathrm{p}}_K = \pi_{K,\mathbf{e}} \qquad \lambda^{\mathrm{p}}_K - \lambda_K = \pi^{\mathrm{c}}_{K,\mathbf{e}} \qquad \text{if } \mathcal{G} = (\mathbf{e}, \mathcal{H}) \,, \\ &\mu_K = f_{K,\mathbf{t}} \qquad \mu^{\mathrm{p}}_K = \pi_{K,\mathbf{t}} \qquad \mu^{\mathrm{p}}_K - \mu_K = \pi^{\mathrm{c}}_{K,\mathbf{t}} \qquad \text{if } \mathcal{G} = (\mathbf{t}, \mathcal{H}) \,. \end{aligned} \tag{3.42}$$

Here, the usual short-hand notation is employed, where a comma as a lower index denotes partial differentiation with respect to the arguments that follow the comma. Also, a single symbol is used for the same physical *quantity* dependent on either **e** or **t**, remembering that the respective *functions* are mathematically different. The formulae (3.42) provide thermodynamic interpretation to the quantities involved in (3.20) and (3.21).

In the spirit of the theory of 'standard' materials based on two thermodynamic potentials (cf. Nguyen, 1993) but without the need of using internal *state* variables, another modification to Rice's theory is introduced by defining the *dissipation function* D:

$$D = D(\mathcal{G}, \mathring{\gamma}_K) \equiv \sum_K \pi^{\mathrm{c}}_K(\mathcal{G})\, \mathring{\gamma}_K \,. \tag{3.43}$$

The dissipation function D divided by T defines the rate of entropy production due to *virtual* $\mathring{\gamma}_K$'s.

The first condition in (3.19) now ensures that any $\mathring{\gamma}_K$ satisfying it corresponds to a non-negative intrinsic dissipation rate in view of the assumed non-negativeness of the π^{c}_K's, the second condition defines an admissible (elastic) domain in the space of thermodynamic forces π_K where $D(\mathcal{G}, \mathring{\gamma}_K) \geq \sum_K \pi_K(\mathcal{G})\, \mathring{\gamma}_K$, and the third condition ensures that the actual plastic dissipation rate in an unperturbed process is equal to D, also when each mechanism is considered separately.

To clarify the meaning of the virtual dissipation rate D, consider (possibly infinitesimal) perturbations of thermodynamic forces π_K by some additional disturbing forces π^{dist}_K. A perturbed loading variable f'_K should now be determined by using a perturbed value of π_K, that is $f'_K = \pi_K + \pi^{\mathrm{dist}}_K - \pi^{\mathrm{c}}_K$. The perturbed activity condition in place of $(3.19)_3$ reads

$$f'_K \mathring{\gamma}_K = (\pi_K + \pi^{\mathrm{dist}}_K - \pi^{\mathrm{c}}_K)\mathring{\gamma}_K = 0 \,. \tag{3.44}$$

The total rate of work to be supplied to a deforming material element, per unit reference volume, includes the rate of work done by the π^{dist}_K's, viz.

$$\dot{w}' = \dot{w} + \dot{w}^{\mathrm{dist}} , \qquad \dot{w} = \mathbf{t} \cdot \dot{\mathbf{e}} \,, \qquad \dot{w}^{\mathrm{dist}} = \sum_K \pi^{\mathrm{dist}}_K \mathring{\gamma}_K \,. \tag{3.45}$$

From (3.45), (3.44) and (3.40) it follows that

$$\dot{w}' = \mathbf{t} \cdot \dot{\mathbf{e}} - \sum_K f_K \mathring{\gamma}_K \,. \tag{3.46}$$

In turn, by combining (3.43), (3.44), (3.45) and (3.38), we obtain

$$D = \dot{w}' - \dot{\phi} \tag{3.47}$$

in agreement with the standard definition of isothermal dissipation.

3.5 Normality and Symmetry

It will be shown in later sections that much more can be said about material stability or instability if a general constitutive law is restricted by the normality and symmetry assumptions. This motivated a more detailed study of these assumptions presented in this section.

Basic assumptions. The normality flow rule in classical elastoplasticity correctly extended to finite strain can be written down as follows (cf. Hill, 1978)

$$\dot{\mathbf{e}}^{\mathrm{p}} = \overset{\circ}{\gamma} \frac{\partial \hat{f}(\mathbf{t}, \mathcal{H})}{\partial \mathbf{t}} \quad \Leftrightarrow \quad \dot{\mathbf{t}}^{\mathrm{p}} = -\overset{\circ}{\gamma} \frac{\partial f(\mathbf{e}, \mathcal{H})}{\partial \mathbf{e}}, \qquad \text{where } \dot{\mathbf{e}}^{\mathrm{p}} \equiv \dot{\mathbf{e}} - \mathbf{M}^{\mathrm{e}} \cdot \dot{\mathbf{t}}, \ \ \dot{\mathbf{t}}^{\mathrm{p}} \equiv \dot{\mathbf{t}} - \mathbf{L}^{\mathrm{e}} \cdot \dot{\mathbf{e}} \tag{3.48}$$

and $f(\mathbf{e}, \mathcal{H}) = \hat{f}(\mathbf{t}, \mathcal{H})$. It is essential that the stress and strain measures are work-conjugate and that the plastic parts $\dot{\mathbf{e}}^{\mathrm{p}}$ of strain-rate and $\dot{\mathbf{t}}^{\mathrm{p}}$ of stress-rate are defined in the way shown above. The equivalence in (3.48) follows from the chain rule of differentiation of f at fixed $\mathcal{H}$ along with the assumed invertibility and diagonal symmetry of $\mathbf{L}^{\mathrm{e}}$ and $\mathbf{M}^{\mathrm{e}}$. The normality rule (3.48) can be deduced from the well-known Ilyushin postulate applied at finite strain (Hill, 1968).

In multi-mode plasticity, the normality rule can be extended to each mechanism separately by adding an index K at both f and $\overset{\circ}{\gamma}$ in (3.48). Comparison with (3.21) and (3.20) shows that the extended normality rule is equivalent to

$$\boldsymbol{\mu}_K^{\mathrm{p}} = \boldsymbol{\mu}_K \qquad \Longleftrightarrow \qquad \boldsymbol{\lambda}_K^{\mathrm{p}} = \boldsymbol{\lambda}_K \, . \tag{3.49}$$

In the thermodynamic description, this means that π_K^{c} and the dissipation function D are independent of elastic changes in strain or stress, cf. (3.42). In the case of a single mechanism of plastic deformation as in (3.18), the normality rule (3.48) is equivalent to imposing on (3.17) the symmetry restriction

$$\mathbf{C} = \mathbf{C}^{\mathrm{T}} \qquad \Longleftrightarrow \qquad \mathbf{L} = \mathbf{L}^{\mathrm{T}} \, . \tag{3.50}$$

In turn, (3.50) means that the relationship (3.17) admits a potential, $U(\dot{\mathbf{F}}, \mathcal{G})$ say, such that

$$\dot{\mathbf{S}} = \frac{\partial U}{\partial \dot{\mathbf{F}}}, \qquad \mathbf{C} = \frac{\partial^2 U}{\partial \dot{\mathbf{F}} \partial \dot{\mathbf{F}}}, \qquad U = \frac{1}{2} \dot{\mathbf{S}} \cdot \dot{\mathbf{F}} \, . \tag{3.51}$$

More generally, one may postulate the symmetry property (3.50) for an arbitrary (also infinite) number of internal mechanisms of plastic deformation. This leads to (3.51) where U may be any homogeneous of degree two, continuously differentiable and piecewise continuously twice differentiable function of $\dot{\mathbf{F}}$ (Hill, 1959, 1978).

In multi-mode plasticity, normality does not imply the symmetry property (3.50). From (3.26) it follows that (3.50) is ensured if besides normality it is additionally required that

$$g_{KL} = g_{LK} \qquad \Longleftrightarrow \qquad h_{KL} = h_{LK} \qquad \text{for } K, L \in \mathcal{A} \, , \tag{3.52}$$

i.e. for K, L from the set of indices of *active* moduli.

It can be shown, e.g., by applying the transformation rules given in Sect. 3.3, that the above normality and symmetry properties are invariant under transition to another pair of *work-conjugate* stress and strain measures (Hill, 1978).

Minimum of path-dependent dissipation. It will be shown below that the above normality and symmetry assumptions can be deduced from the following single postulate (Petryk, 1999b).

Postulate of minimum incremental dissipation. *The dissipation evaluated to the* second-order *terms along a path connecting any two states arbitrarily close to each other is minimized on a* straight *(or direct) path* •

A straight path for a chosen set of state variables, $(\mathbf{e}, \boldsymbol{\alpha})$ say, generally becomes a curved but *direct* path after transition to other variables. A finite curvature will not influence the value of dissipation up to the second-order with respect to the path length. The dissipation is understood here as *virtual*, i.e. obtained by time integration of the dissipation function D along a path.

If the postulate is not satisfied then the existence of alternative, more complex but energetically more preferable paths *might* lead to an instability of the fundamental smooth path. This is indicated only as a possibility, and the above postulate, even if it seems to be natural, is not obligatory.

The above postulate is analogous to the 'radial path lemma' formulated by Nguyen (1993) for generalizing the stability condition to arbitrarily complex paths. Validity of the lemma was stated for a state-independent convex dissipation function (and obviously also for path independent dissipation). Petryk (1993b) extended validity of the lemma to a convex dissipation function dependent linearly on internal variables under the symmetry assumption (3.61). These previous results are completed in the Appendix, following (Petryk, 1999b), by establishing necessary and sufficient conditions for the above postulate of minimum incremental dissipation.

The preliminary condition for the fulfillment of the postulate is, of course, that the virtual dissipation evaluated to the *first* order be minimized on a direct path. In the formalism of multi-mode plasticity, the first-order dissipation on a *direct* path starting from $\mathcal{G} = (\mathbf{e}, \boldsymbol{\alpha})$ reads

$$\Delta_1 \mathcal{D}^{\mathrm{d}} \equiv \sum_K \pi_K^{\mathrm{c}}(\mathcal{G}) \Delta\gamma_K = D(\mathcal{G}, \Delta\gamma_K), \tag{3.53}$$

where $\Delta\gamma_K = \overset{\circ}{\gamma}_K \, \Delta t$. An arbitrary small increment $\Delta\gamma_K$ can also be imagined to be produced by imposing first a small elastic increment $\delta\mathbf{e}$ in strain (of order $\Delta\gamma_K$), then applying the increment $\Delta\gamma_K$ and next removing $\delta\mathbf{e}$ elastically. The final state will be the same, to first order, on account of (3.39). If some π_K^{c} depends on the elastic strain then the resulting value of the first-order dissipation

$$\Delta_1 \mathcal{D} = D(\mathcal{G}, \Delta\gamma_K) + \delta\mathbf{e} \cdot D_{,\mathbf{e}}(\mathbf{e}, \boldsymbol{\alpha}, \Delta\gamma_K) \tag{3.54}$$

can be made less than (3.53) by selecting $\Delta\gamma_K$ and $\delta\mathbf{e}$ such that the term involving $\delta\mathbf{e}$ be negative. The postulate excludes this, so that the π_K^{c}'s in (3.42) must be independent of elastic changes in strain or stress. This implies the normality rule (3.49) as a consequence of the postulate.

Alternatively, from (3.47), (3.46) and (3.37) it follows that the dissipation function expressed in terms of the rates of internal variables takes the form

$$D(\mathcal{G},\dot{\boldsymbol{\alpha}}) = \boldsymbol{A}(\mathcal{G}) \cdot \dot{\boldsymbol{\alpha}} - \sum_K f_K(\mathcal{G})\overset{\circ}{\gamma}_K(\boldsymbol{\alpha},\dot{\boldsymbol{\alpha}})\,. \tag{3.55}$$

Partial differentiation of this expression with respect to $\mathbf{e}$, on using (3.37), (3.41) and (3.20), yields

$$D_{,\mathbf{e}} = \boldsymbol{A}_{,\mathbf{e}} \cdot \dot{\boldsymbol{\alpha}} - \sum_K f_{K,\mathbf{e}}\overset{\circ}{\gamma}_K = -\dot{\mathbf{t}}^{\mathrm{p}} - \sum_K \boldsymbol{\lambda}_K\overset{\circ}{\gamma}_K = \sum_K(\boldsymbol{\lambda}^{\mathrm{p}}_K - \boldsymbol{\lambda}_K)\overset{\circ}{\gamma}_K\,, \tag{3.56}$$

so that the normality rule in the form (3.49) follows again as a consequence of $D_{,\mathbf{e}} = 0$ if $\overset{\circ}{\gamma}_K \geq 0$ is regarded as arbitrary.

The condition $D_{,\mathbf{e}} = \mathbf{0}$ leads to

$$D = D(\boldsymbol{\alpha},\dot{\boldsymbol{\alpha}})\,, \qquad D(\boldsymbol{\alpha},r\dot{\boldsymbol{\alpha}}) = rD(\boldsymbol{\alpha},\dot{\boldsymbol{\alpha}}) \quad \text{for all } r > 0\,, \tag{3.57}$$

as it is assumed in the theory of time-independent 'standard' materials (cf. Nguyen, 1993). The following lemma holds (cf. Appendix).

Lemma A1. *Under the assumptions (3.57), the first-order dissipation is always minimized on a straight path if and only if*

$$D(\boldsymbol{\alpha},\dot{\boldsymbol{\alpha}}) \textit{ is convex in } \dot{\boldsymbol{\alpha}}\,. \tag{3.58}$$

All $\dot{\boldsymbol{\alpha}}$ compatible with the current thermodynamic forces $\boldsymbol{A}(\mathcal{G})$ constitute a cone $\mathcal{L}(\mathcal{G})$ in $\dot{\boldsymbol{\alpha}}$-space within which the dissipation function D yields the actual value of the dissipation rate, that is

$$D(\boldsymbol{\alpha},\dot{\boldsymbol{\alpha}}) = \boldsymbol{A}(\mathcal{G}) \cdot \dot{\boldsymbol{\alpha}} \qquad \textit{for all } \dot{\boldsymbol{\alpha}} \in \mathcal{L}(\mathcal{G})\,. \tag{3.59}$$

Thus, D is linear in $\dot{\boldsymbol{\alpha}}$ within that cone.

Suppose that at least one $\dot{\boldsymbol{\alpha}} = \dot{\boldsymbol{\alpha}}^{\mathrm{a}}$ in (3.59) is associated with $\dot{\mathbf{e}}^{\mathrm{a}}$ from the cone of *fully active loading* in $\dot{\mathbf{e}}$-space. For such a pair $(\dot{\mathbf{e}}^{\mathrm{a}},\dot{\boldsymbol{\alpha}}^{\mathrm{a}})$ the cone $\mathcal{L}(\mathcal{G})$ is assumed to vary right-hand continuously in time, so that as a consequence of (3.59) we obtain

$$\dot{\boldsymbol{\alpha}} \cdot D_{,\boldsymbol{\alpha}}(\boldsymbol{\alpha},\hat{\boldsymbol{\alpha}}) = \dot{\boldsymbol{A}} \cdot \hat{\boldsymbol{\alpha}} \qquad \text{for all } \hat{\boldsymbol{\alpha}} \in \mathcal{L}(\mathcal{G}) \quad \text{if } (\dot{\mathbf{e}},\dot{\boldsymbol{\alpha}}) = (\dot{\mathbf{e}}^{\mathrm{a}},\dot{\boldsymbol{\alpha}}^{\mathrm{a}})\,. \tag{3.60}$$

It is assumed here that the dissipation function $D(\boldsymbol{\alpha},\dot{\boldsymbol{\alpha}})$ is continuously (Fréchet) differentiable with respect to $\boldsymbol{\alpha}$ at each $\dot{\boldsymbol{\alpha}}$, and the derivative is denoted by $D_{,\boldsymbol{\alpha}}$.

In the Appendix the following result is proven.

Lemma A2. *Under the assumptions (3.57), (3.58), (3.59) and (3.60), the dissipation evaluated to second-order is always minimized on a straight path if and only if*

$$\hat{\boldsymbol{\alpha}} \cdot D_{,\boldsymbol{\alpha}}(\boldsymbol{\alpha},\dot{\boldsymbol{\alpha}}) = \dot{\boldsymbol{\alpha}} \cdot D_{,\boldsymbol{\alpha}}(\boldsymbol{\alpha},\hat{\boldsymbol{\alpha}}) \quad \textit{for all } \dot{\boldsymbol{\alpha}},\hat{\boldsymbol{\alpha}} \in \mathcal{L}(\mathcal{G})\,. \tag{3.61}$$

The symmetry property (3.61) is thus obtained as a consequence of the postulate of minimum incremental dissipation. The symmetry (3.50) of the incremental stiffness moduli can be deduced from (3.61) by appealing to an integral constitutive inequality (Petryk, 1989, Nguyen and Petryk, 1991).

The symmetry (3.52) of the moduli matrix in the multi-mode plasticity can be deduced more directly as follows. A 'plastic' variation of (3.55) in a direction $\hat{\boldsymbol{\alpha}}$ at fixed $\dot{\boldsymbol{\alpha}}$ gives

$$\hat{\boldsymbol{\alpha}} \cdot D_{,\boldsymbol{\alpha}}(\mathcal{G}, \dot{\boldsymbol{\alpha}}) = \hat{\boldsymbol{\alpha}} \cdot \boldsymbol{A}_{,\boldsymbol{\alpha}}(\mathcal{G}) \cdot \dot{\boldsymbol{\alpha}} - \hat{\boldsymbol{\alpha}} \cdot \sum_{K} (f_{K,\boldsymbol{\alpha}}(\mathcal{G}) \mathring{\gamma}_K(\boldsymbol{\alpha}, \dot{\boldsymbol{\alpha}}) + f_K(\mathcal{G}) \mathring{\gamma}_{K,\boldsymbol{\alpha}}(\boldsymbol{\alpha}, \dot{\boldsymbol{\alpha}})) \quad (3.62)$$

with $\mathcal{G} = (\mathbf{e}, \boldsymbol{\alpha})$. It is assumed here that the functions $f_K(\mathbf{e}, \boldsymbol{\alpha})$ and $\mathring{\gamma}_K(\boldsymbol{\alpha}, \dot{\boldsymbol{\alpha}})$ are continuous and (Gateaux) differentiable with respect to $\boldsymbol{\alpha}$.

Consider admissible directions of inelastic straining in the multi-mode plasticity framework, subject to (3.19) at every instant. Then $f_K(\mathcal{G}) < 0$ implies $\mathring{\gamma}_K = 0$ in some neighborhood of $\mathcal{G}$, by $(3.19)_3$ and continuity of f_K. Consequently, the last term in (3.62) vanishes. From $(3.20)_2$ the 'plastic' variation of f_K reads

$$\hat{\boldsymbol{\alpha}} \cdot f_{K,\boldsymbol{\alpha}}(\mathbf{e}, \boldsymbol{\alpha}) = -\sum_{KL} g_{KL}(\mathbf{e}, \boldsymbol{\alpha}) \mathring{\gamma}_L(\boldsymbol{\alpha}, \hat{\boldsymbol{\alpha}}) . \quad (3.63)$$

On account of symmetry of $\phi_{,\boldsymbol{\alpha}\boldsymbol{\alpha}} = -\boldsymbol{A}_{,\boldsymbol{\alpha}}$, from (3.62) we thus obtain

$$\hat{\boldsymbol{\alpha}} \cdot D_{,\boldsymbol{\alpha}}(\mathcal{G}, \dot{\boldsymbol{\alpha}}) - \dot{\boldsymbol{\alpha}} \cdot D_{,\boldsymbol{\alpha}}(\mathcal{G}, \hat{\boldsymbol{\alpha}}) = \sum_{KL} (g_{KL}(\mathcal{G}) - g_{LK}(\mathcal{G})) \mathring{\gamma}_K(\boldsymbol{\alpha}, \dot{\boldsymbol{\alpha}}) \mathring{\gamma}_L(\boldsymbol{\alpha}, \hat{\boldsymbol{\alpha}}) \quad (3.64)$$

for any two directions $\dot{\boldsymbol{\alpha}}$, $\hat{\boldsymbol{\alpha}}$ admissible in $\mathcal{G}$. On applying $(3.57)_1$ and (3.61), as a consequence of the postulate of minimum incremental dissipation we conclude that the expression (3.64) vanishes. The symmetry property (3.52) follows if $\dot{\boldsymbol{\alpha}}$, $\hat{\boldsymbol{\alpha}}$ vary within the constitutive cone where $\mathcal{A}$ is the set of indices of active mechanisms.

Linear comparison solid. A widely used approach to deriving conditions for stability or uniqueness is to linearize the incremental problem formulated in a given state. The question is whether such linearized conditions pertain to the actual elastic-plastic material which is incrementally nonlinear. An affirmative answer can be given for a class of materials for which suitable constitutive inequalities are established.

In the analysis of bifurcation in materials obeying the constitutive law (3.51) with a potential U, the 'relative convexity property' introduced by Hill (1958, 1959) is very useful. It states that the actual potential U for an elastic-plastic material is 'more convex' than the quadratic potential U^{L} for an incrementally linear *comparison solid*, in the sense that $(U - U^{\mathrm{L}})$ in a given state is a convex function of $\dot{\mathbf{F}}$. This can be written as

$$(\dot{\mathbf{S}} - \dot{\mathbf{S}}^*) \cdot (\dot{\mathbf{F}} - \dot{\mathbf{F}}^*) \geq (\dot{\mathbf{F}} - \dot{\mathbf{F}}^*) \cdot \mathbf{C}^{\mathrm{L}} \cdot (\dot{\mathbf{F}} - \dot{\mathbf{F}}^*) \qquad \text{for every } \dot{\mathbf{F}}, \dot{\mathbf{F}}^* , \quad (3.65)$$

where the quantities in each pair $(\dot{\mathbf{F}}, \dot{\mathbf{S}})$ and $(\dot{\mathbf{F}}^*, \dot{\mathbf{S}}^*)$ are mutually related by (3.17). The moduli tensor $\mathbf{C}^{\mathrm{L}}$ of the comparison solid does not depend on the actual strain-rate and must be suitably defined for a given material model. From (3.7) and (3.4) it follows that (3.65) is equivalent to

$$(\dot{\mathbf{t}} - \dot{\mathbf{t}}^*) \cdot (\dot{\mathbf{e}} - \dot{\mathbf{e}}^*) \geq (\dot{\mathbf{e}} - \dot{\mathbf{e}}^*) \cdot \mathbf{L}^{\mathrm{L}} \cdot (\dot{\mathbf{e}} - \dot{\mathbf{e}}^*) \qquad \text{for every } \dot{\mathbf{e}}, \dot{\mathbf{e}}^* \quad (3.66)$$

for *all* work-conjugate measures $(\mathbf{e}, \mathbf{t})$ (Hill, 1978), with the analogous meaning of the quantities involved. Hill (1958) showed that (3.65) holds for the classical plasticity (3.18) with normality (3.48) when L^{L} is taken as the moduli tensor from the loading branch. For the classical elasto-plasticity *without* normality, Raniecki (1979) (cf. also Raniecki and Bruhns, 1981) constructed a comparison solid that verifies (3.65) and whose moduli do not coincide with the actual stiffness moduli from the loading branch.

Validity of (3.65) was extended to multi-mode plasticity by Sewell (1972) under the normality and symmetry assumptions given above and under the additional assumption that

$$\text{the matrix } (g_{KL}) \text{ with } K, L \in \mathcal{P} \text{ is positive definite.} \tag{3.67}$$

In that case, the comparison solid is defined by the moduli from the branch of fully active loading, that is, by (3.26) with $\mathcal{A} = \mathcal{P}$.

If a fundamental solution is known, which is usually the case, then the *tangent moduli tensor* $\mathsf{L}^0 \equiv \mathsf{L}(\dot{\mathbf{e}}^0)$ corresponding to the fundamental strain-rate $\dot{\mathbf{e}}^0$ defines the *tangent comparison solid.* Instead of (3.66), for the analogous purpose the following constitutive inequality has been proposed (Petryk, 1989)

$$(\dot{\mathbf{t}} - \dot{\mathbf{t}}^0) \cdot (\dot{\mathbf{e}} - \dot{\mathbf{e}}^0) \geq (\dot{\mathbf{e}} - \dot{\mathbf{e}}^0) \cdot \mathsf{L}^0 \cdot (\dot{\mathbf{e}} - \dot{\mathbf{e}}^0) \qquad \text{for every } \dot{\mathbf{e}}\,, \tag{3.68}$$

equivalent to

$$(\dot{\mathbf{S}} - \dot{\mathbf{S}}^0) \cdot (\dot{\mathbf{F}} - \dot{\mathbf{F}}^0) \geq (\dot{\mathbf{F}} - \dot{\mathbf{F}}^0) \cdot \mathbf{C}^0 \cdot (\dot{\mathbf{F}} - \dot{\mathbf{F}}^0) \qquad \text{for every } \dot{\mathbf{F}}\,, \tag{3.69}$$

where the fundamental pair $(\dot{\mathbf{e}}^0, \dot{\mathbf{t}}^0)$ or $(\dot{\mathbf{F}}^0, \dot{\mathbf{S}}^0)$ is known and fixed. This is a less restrictive requirement than (3.65) with $\mathbf{C}^{\mathrm{L}} = \mathbf{C}^0$ since now only one pair $(\dot{\mathbf{t}}, \dot{\mathbf{e}})$ is variable. Validity of (3.69) has been extended beyond the classical and multi-mode plasticity to an unspecified constitutive law (3.51) under the assumption that the framework of multi-mode plasticity with the normality and symmetry conditions (3.49) and (3.52) (but without assuming (3.67)) applies *at a micro-level.* This was done by appealing to the inequality

$$\dot{\mathbf{t}}^0 \cdot \dot{\mathbf{e}} - \dot{\mathbf{t}} \cdot \dot{\mathbf{e}}^0 \geq 0 \qquad \text{for every } \dot{\mathbf{e}} \tag{3.70}$$

transmissible from the micro- to macro-level. The inequality (3.68) applies at a *regular* point of a straining path where L^0 does not correspond to any abrupt unloading at the micro-level; for more details, cf. (Petryk, 1989, 1993a).

Path-dependent second-order work. In stability investigations based on the energy criterion, an estimate is needed for the work done on arbitrary circuitous paths of departure from a given state (Hill, 1958). Generally, the work must be calculated with accuracy to second-order terms. It might seem at a first glance that piecewise-regular strain paths become 'direct' if their length is sufficiently small, so that there are no problems in using the expression (3.10). This is really so when any path is regarded as *fixed* as its length tends to zero. However, when proving stability we should also consider a *sequence* of paths which may substantially differ in shape from a smooth arc of bounded curvature, no matter how short they may be. More specifically, a given path in strain space may be scaled down such that it length $\Delta\theta$ tends to zero while its complexity is preserved. Then, in place of $(3.10)_1$ we can write

$$\Delta w = \Delta_1 w + \Delta_2 w + o(|\Delta\theta|^2)\,, \qquad \Delta\theta = \int (\mathrm{d}\mathbf{e}\cdot\mathrm{d}\mathbf{e})^{1/2}\,, \tag{3.71}$$

where $\Delta_2 w$ is the *path-dependent* second-order work. The concept is discussed in more detail in (Petryk, 1991a) where the following result, useful in stability investigations for justifying the convenient restriction to direct strain paths, has been proven.

Theorem on minimum second-order work. *In the framework of multi-mode plasticity from Sect. 3.3 under the normality (3.49) and symmetry (3.52) assumptions, the positive-definiteness property (3.67) implies*

$$\Delta_2 w \geq \Delta_2 w^{\mathrm{d}} \tag{3.72}$$

within the class of all paths leading to the same final strain increment $\Delta\mathbf{e}$, to second-order with respect to $\Delta\theta$ • [9]

Although the value of the second-order work on each side in (3.72) depends on the chosen pair of work-conjugate measures of strain and stress, the inequality itself is measure-invariant.

3.6 Constitutive Equations for External Loading

A material element surrounded by similar elements is generally subjected to tractions dependent on the element configuration, also if a larger sample of the material is tested under the control of external surface tractions or displacements. The actual interaction rule between the configuration of a material element and its loading environment may be difficult to determine, especially if the surrounding material undergoes nonuniform, somewhere plastic and elsewhere elastic deformation. In general investigations of the influence of such flexible internal constraints on material instability, it seems natural to restrict attention in the first instance to a *linearized* interaction rule. The surrounding material is accordingly treated as nonuniformly deformed but incrementally *linear* (not necessarily elastic), while the interior material element is incrementally nonlinear but assumed to deform *uniformly*. A special case of interest arises when the interior element is imagined to be of ellipsoidal shape and embedded in an infinite homogeneous linear continuum, which is consistent with uniform straining of the ellipsoid (Eshelby, 1957). In that case the external stiffness moduli (distinguished below by a star) can be effectively determined in a manner which is not discussed here.

Following Petryk (2000), the rate of work done by tractions applied to the element boundary from the outside is assumed as a linear form in $\dot{\mathbf{F}}$,

$$\dot{w}^{\mathrm{ext}} = \mathbf{S}^{\mathrm{ext}}\cdot\dot{\mathbf{F}}\,, \tag{3.73}$$

per unit reference volume of the material. In analogy to Hill's concept of a *symmetric* work-conjugate stress (cf. Sect. 3.1), the rate of work is understood here as a basic quantity, and (3.73)

[9] In contrast to (A.9), the final *state* (including the final stress) need not be the same for the paths involved in (3.72). In obtaining (3.72), the conditions (2.4) were assumed to hold along every strain path under consideration, and no reference to thermodynamics was needed.

serves as a definition of the *external* Piola stress $\mathbf{S}^{\text{ext}}$. The product in (3.73) can be decomposed as

$$\mathbf{S}^{\text{ext}} \cdot \dot{\mathbf{F}} = J(\boldsymbol{\sigma}^{\text{ext}} \cdot \mathbf{D} + \mathbf{m}^{\text{ext}} \cdot \mathbf{W}),$$
$$\boldsymbol{\sigma}^{\text{ext}} = \boldsymbol{\sigma}^{\text{ext}\,\mathrm{T}} = J^{-1}(\mathbf{S}^{\text{ext}}\mathbf{F}^{\mathrm{T}})_{\mathrm{s}}, \qquad \mathbf{m}^{\text{ext}} = -\mathbf{m}^{\text{ext}\,\mathrm{T}} = J^{-1}(\mathbf{S}^{\text{ext}}\mathbf{F}^{\mathrm{T}})_{\mathrm{a}}. \tag{3.74}$$

The need for distinguishing between the external ($\mathbf{S}^{\text{ext}}$) and internal ($\mathbf{S}$) Piola stress can be seen from the absence of a counterpart to $\mathbf{m}^{\text{ext}}$ in the constitutive framework for the material without couple stresses. The requirement $\mathbf{S}\mathbf{F}^{\mathrm{T}} = \mathbf{F}\mathbf{S}^{\mathrm{T}}$ (equivalent to $\boldsymbol{\sigma} = \boldsymbol{\sigma}^{\mathrm{T}}$) of rotational equilibrium need not apply *in advance* to $\mathbf{S}^{\text{ext}}$, so that $\mathbf{m}^{\text{ext}} \neq \mathbf{0}$ in general. A simple illustration is provided by 'dead' loading with prescribed $\mathbf{S}^{\text{ext}}$, when $\mathbf{S}^{\text{ext}}\mathbf{F}^{\mathrm{T}}$ need not be equal to $\mathbf{F}\mathbf{S}^{\text{ext}\,\mathrm{T}}$ for an arbitrary $\mathbf{F}$.

The actual history of the deformation gradient $\mathbf{F}$ in a quasi-static motion must be calculated from the equilibrium condition. Suppose that $\mathbf{F}$ is kinematically unconstrained. Then in a state of mechanical equilibrium we must have

$$\mathbf{S} = \mathbf{S}^{\text{ext}} \qquad \Leftrightarrow \qquad (\,\boldsymbol{\sigma} = \boldsymbol{\sigma}^{\text{ext}} \quad \text{and} \quad \mathbf{m}^{\text{ext}} = \mathbf{0}\,), \tag{3.75}$$

by the principle of virtual power. In view of the incremental character of the constitutive law for the material, the basic condition of continuing mechanical equilibrium of a material element is formulated for the stress rates:

$$\dot{\mathbf{S}} = \dot{\mathbf{S}}^{\text{ext}}. \tag{3.76}$$

To specify the loading conditions, we need a constitutive expression for $\dot{\mathbf{S}}^{\text{ext}}$ (which is different from that for $\dot{\mathbf{S}}$). In a given equilibrium state, we shall consider a class of deformation-sensitive incremental loadings of a material element, defined by

$$\dot{\mathbf{S}}^{\text{ext}} = \dot{\mathbf{S}}^{*} - \mathsf{C}^{*} \cdot \dot{\mathbf{F}}. \tag{3.77}$$

Here, a fourth-order tensor C^{*} of external stiffness moduli and an applied stress-rate $\dot{\mathbf{S}}^{*}$ are given quantities in a given state of the material element. $\dot{\mathbf{S}}^{*}$ is a controllable part and $(-\mathsf{C}^{*} \cdot \dot{\mathbf{F}})$ is a deformation-sensitive part of $\dot{\mathbf{S}}^{\text{ext}}$. For example, for the loading by all-around hydrostatic pressure ($\boldsymbol{\sigma}^{\text{ext}} = -p\mathbf{I}$) and in the current configuration taken as reference ($\mathbf{F} = \mathbf{I}$ instantaneously), we have $\dot{\mathbf{S}}^{*} = -\dot{p}\mathbf{I}$ and $C^{*}_{ijkl} = (\delta_{il}\delta_{jk} - \delta_{ij}\delta_{kl})p$.

Eq. (3.76) provides nine equations for nine unknowns: six components of the strain-rate and three components of the material spin. Suppose that the components of the material spin are either known or expressible uniquely in terms of the strain-rate components (cf. Hill, 1967b, Petryk, 2000). Then the equation (3.77) can be reduced to the following form not involving material spin

$$\dot{\mathbf{t}}^{\text{ext}} = \dot{\mathbf{t}}^{*} - \mathsf{L}^{*} \cdot \dot{\mathbf{e}}, \tag{3.78}$$

where the external stress-rate measure $\dot{\mathbf{t}}^{\text{ext}}$ is defined by $\dot{\mathbf{t}}^{\text{ext}} = \mathsf{B}_{\mathrm{w}} \cdot \overset{\triangledown}{\boldsymbol{\tau}}{}^{\text{ext}} + \mathsf{Z}_{\mathrm{w}} \cdot \mathbf{D}$ with the same B_{w} and Z_{w} as in (3.4). Here, $\dot{\mathbf{t}}^{*}$ and L^{*} are regarded as known quantities. The following transformation rules analogous to those in (3.31) apply:

$$\dot{\mathbf{t}}^{*} = \mathsf{B}_{\mathrm{w}} \cdot \overset{\triangledown}{\boldsymbol{\tau}}{}^{*}, \qquad \mathsf{L}^{*} = (\mathsf{B}_{\mathrm{w}} \cdot \bar{\mathsf{L}}^{*} - \mathsf{Z}_{\mathrm{w}}) \cdot \mathsf{A}^{-1} \qquad \text{if } \overset{\triangledown}{\boldsymbol{\tau}}{}^{\text{ext}} = \overset{\triangledown}{\boldsymbol{\tau}}{}^{*} - \bar{\mathsf{L}}^{*} \cdot \mathbf{D}. \tag{3.79}$$

Note that if $\mathsf{L}^{*} = \mathbf{0}$ for one choice of a pair $(\dot{\mathbf{e}}, \dot{\mathbf{t}}^{\text{ext}})$ then generally $\mathsf{L}^{*} \neq \mathbf{0}$ for another choice. This means that if one stress-rate measure is fully controlled then another stress-rate measure is generally dependent on the strain-rate, cf. Hill (1967c, 1978).

4 Stability of Equilibrium

4.1 The Energy Condition for Stability of Equilibrium

The concepts of stability of equilibrium in the energy sense, presented in Sect. 2 with reference to the one-dimensional example, are now generalized to a three-dimensional continuum.

There are different ways to derive a criterion of stability of equilibrium based on the energy considerations. Without using thermodynamics, the basic way (cf. Hill, 1958) is to begin with the mechanical energy balance for a solid body taken away from equilibrium, viz. (cf. (2.44))

$$\Delta W - \Delta W^{\text{ext}} + K = \Delta W^* \,, \tag{4.1}$$

where ΔW is the work to be supplied to the deformed body along a path of departure from the examined equilibrium state, ΔW^{ext} is the work done by external loads along that path, and K is the (non-negative) kinetic energy at the path end. The unbalanced amount of energy ΔW^* must be supplied to the deformed body by a disturbing agency. ΔW is to be determined by integration of the work-rate density (3.45) over the reference volume V of a material element and over a time interval, viz.

$$\Delta W = \int \dot{W}(t)\,dt\,, \qquad \dot{W} = \int_V (\dot{w} + \dot{w}^{\text{dist}})\,dV\,. \tag{4.2}$$

Equivalently, in the thermodynamic framework with the dissipation function D, we can use the following decomposition (cf. Nguyen, 1993)

$$\Delta W = \int_V (\Delta\phi + \Delta\mathcal{D})\,dV\,, \qquad \Delta\mathcal{D} = \int D(t)\,dt\,. \tag{4.3}$$

The time integrals above are generally path-dependent. Further specifications of the quantities involved in (4.1) may depend on the kind of material stability as discussed below.

If

$$\Delta W - \Delta W^{\text{ext}} > 0 \tag{4.4}$$

along every kinematically admissible deformation path that starts from the examined equilibrium state then any departure from that state requires a positive amount of energy ΔW^* to be supplied by a disturbing agency. The condition (4.4) may be only imposed in a certain neighborhood of the equilibrium state and for a specified class of paths of departure from equilibrium. This is usually regarded as a condition sufficient for dynamic stability of equilibrium (for the specified class of paths). It is justified provided the (positive) left-hand quantity in (4.4) is accepted as the measure of the disturbance strength as well as of the distance from the equilibrium state. If another distance measure $d \geq 0$ is used (e.g., geometric or in state variables) then (4.4) should be strengthened to

$$\Delta W - \Delta W^{\text{ext}} \geq a(d) \qquad \text{if } d \leq R, \tag{4.5}$$

where $a(\cdot)$ is an arbitrary continuous and monotonically increasing function with $a(0) = 0$, and $R = \text{const} > 0$. In turn, for a quasi-static path of departure from equilibrium in a time interval $[t_0, t]$, the quantity

$$d_0(t) = \sup_{t_0 \leq \theta \leq t} (\Delta W(\theta) - \Delta W^{\text{ext}}(\theta)) \tag{4.6}$$

may serve as a natural measure for a disturbance strength. Clearly, for a dynamic path the disturbance strength is not less on account of a non-negative kinetic energy.

The stability of equilibrium with respect to two measures (metrics) d_0, d in the rigorous sense of the Lyapunov-Movchan theory follows from (4.5), as discussed in more detail in other CISM Lecture Notes (Petryk, 1993b); cf. also Nguyen and Radenkovic (1975). An extension to thermodynamic stability of equilibrium in plasticity under possible perturbations of the temperature field was given in (Petryk, 1993b, 1995).

A natural first step in verifying (4.4) is to consider *direct* paths of departure from equilibrium by developing the work expressions into a Taylor series with respect to 'time' t along a path,

$$\Delta W = \dot{W}\,\Delta t + \frac{1}{2}\ddot{W}\,(\Delta t)^2 + \dots, \qquad \Delta W^{\text{ext}} = \dot{W}^{\text{ext}}\,\Delta t + \frac{1}{2}\ddot{W}^{\text{ext}}\,(\Delta t)^2 + \dots, \tag{4.7}$$

where the superimposed dots denote the right-hand rates evaluated in the equilibrium state. If (4.4) holds after substituting the series (4.7) truncated to the needed (usually the second) order terms then the resulting stability will be called 'directional'. The second-order condition for directional stability reads

$$\ddot{W} - \ddot{W}^{\text{ext}} > 0\,, \tag{4.8}$$

to hold for all admissible directions of departure from equilibrium. A strengthened version of (4.8) obtained from (4.5) can also be examined.

It must be noted that, in some analogy to structural stability under follower loads, directional stability does not necessarily imply dynamic stability for arbitrarily circuitous paths.

4.2 Stability for Internal Rearrangements

In this section, the material stability is examined at uniform macroscopic strain against possible spontaneous internal structural rearrangements at a micro-level, cf. Sect. 1.3. We will study parallely two cases when either the strain $\mathbf{e}$ is fixed (a material element with a clamped boundary), or a stress measure $\mathbf{t} = \mathbf{t}^{\text{ext}}$ is fixed (irrespectively of unrestricted changes in the uniform material strain). Internal rearrangements are described macroscopically in the framework of multi-mode plasticity, cf. Sect. 3.3 and in the thermodynamic framework, cf. Sect. 3.4.

Consider a departure from a given equilibrium state, induced by some disturbances at a micro-level which correspond to perturbations of thermodynamic forces in the macroscopic description. Since $\mathbf{t}\cdot\dot{\mathbf{e}} = \mathbf{t}^{\text{ext}}\cdot\dot{\mathbf{e}}$ by assumption, the rate of extra work $(W - W^{\text{ext}})$ to be supplied to the element of a reference volume V reduces to

$$\frac{1}{V}(\dot{W} - \dot{W}^{\text{ext}}) = \dot{w}' - \dot{w} = -\sum_K f_K \overset{\circ}{\gamma}_K \tag{4.9}$$

according to (3.46), or to

$$\frac{1}{V}(\dot{W} - \dot{W}^{\text{ext}}) = \dot{\phi} + D - \dot{w} = D - \boldsymbol{A}\cdot\dot{\boldsymbol{\alpha}} \tag{4.10}$$

from (3.47) and (3.37).

A prerequisite for (4.4) is that in the examined equilibrium state

$$\dot{W} - \dot{W}^{\text{ext}} \geq 0 \tag{4.11}$$

for all possible rearrangements. Since all $\overset{\circ}{\gamma}_K \geq 0$ by assumption, in this way from (4.11) and (4.9) we recover the condition $f_K \leq 0$ which has been assumed in advance in (3.19) as the restriction on admissible states in the multi-mode plasticity framework. Similarly, on using (4.10) we obtain the inequality

$$D(\boldsymbol{\alpha}, \dot{\boldsymbol{\alpha}}) - \boldsymbol{A}(\mathcal{G}) \cdot \dot{\boldsymbol{\alpha}} \geq 0 \quad \text{for all } \dot{\boldsymbol{\alpha}} \tag{4.12}$$

as the restriction on admissible states $\mathcal{G}$ in the framework with internal state variables $\boldsymbol{\alpha}$.

However, this is not a 'proof' of these restrictions on admissible states since the status of (4.11) as a condition necessary for material stability is not fully clear.

In an admissible ($f_K \leq 0$) equilibrium state, on account of (4.11) we may restrict attention to the directions of straining that make the expression (4.9) equal to zero. That is, all the conditions in the basic rule (3.19) are now assumed to hold *in the examined equilibrium state*, but the consistency conditions (3.23) may be violated. This means that any vector $[\overset{\circ}{\gamma}]$ of non-negative components represents an admissible perturbation (recall that $[\overset{\circ}{\gamma}]$ denotes a vector $(\overset{\circ}{\gamma}_K)$ with $K \in \mathcal{P}$), while $\overset{\circ}{\gamma}_K$ is assumed to vanish if $f_K < 0$. According to the discussion in Sect. 3.4, we consider in this way directional stability of equilibrium for *arbitrarily small* perturbing forces π_K^{dist} (if $K \in \mathcal{P}$) that have zero values at equilibrium and vary proportionally in time (to first order) with non-zero *rates*.

Since (4.11) holds universally in all admissible equilibrium states, the condition sufficient for directional stability of equilibrium takes the form (4.8) with

$$\frac{1}{V}(\ddot{W} - \ddot{W}^{\text{ext}}) = \ddot{w}^{\text{dist}} = -\sum_K \dot{f}_K \overset{\circ}{\gamma}_K - \sum_K f_K \frac{d}{dt}\overset{\circ}{\gamma}_K . \tag{4.13}$$

At equilibrium we have $f_K \frac{d}{dt}\overset{\circ}{\gamma}_K \leq 0$ since (3.19) holds initially and $\overset{\circ}{\gamma}_K \geq 0$ at subsequent instants. The value of $\dot{f}_K$ is determined from $(3.20)_2$ or $(3.21)_2$ for $\dot{\mathbf{e}} = \mathbf{0}$ or $\dot{\mathbf{t}} = \mathbf{0}$, respectively. In that way from (4.8) we arrive at the following condition sufficient for directional stability of equilibrium with respect to spontaneous internal rearrangements :

$$\sum_{K,L \in \mathcal{P}} g_{KL} \overset{\circ}{\gamma}_K \overset{\circ}{\gamma}_L > 0 \quad \text{for every } [\overset{\circ}{\gamma}] \geq [0],\ [\overset{\circ}{\gamma}] \neq [0] \quad \text{at fixed } \mathbf{e}, \tag{4.14}$$

or

$$\sum_{K,L \in \mathcal{P}} h_{KL} \overset{\circ}{\gamma}_K \overset{\circ}{\gamma}_L > 0 \quad \text{for every } [\overset{\circ}{\gamma}] \geq [0],\ [\overset{\circ}{\gamma}] \neq [0] \quad \text{at fixed } \mathbf{t}. \tag{4.15}$$

The inequalities between vectors should be read componentwise (for each $K \in \mathcal{P}$). Under the assumptions introduced, the condition (4.8) is equivalent to (4.14) or (4.15).

The inequalities (4.14) and (4.15) as conditions for material stability at finite strain against spontaneous internal rearrangements under fixed strain or stress were derived in (Petryk, 1993b) and (Petryk, 1999a), respectively. A precursor to (4.14) in the formalism with internal state variables reads (cf. Nguyen, 1993)

$$\dot{\boldsymbol{\alpha}} \cdot D_{,\boldsymbol{\alpha}}(\boldsymbol{\alpha}, \dot{\boldsymbol{\alpha}}) + \dot{\boldsymbol{\alpha}} \cdot \phi_{,\boldsymbol{\alpha}\boldsymbol{\alpha}} \cdot \dot{\boldsymbol{\alpha}} > 0 \qquad \text{for all } \dot{\boldsymbol{\alpha}} \in \mathcal{L}(\mathcal{G}), \quad \dot{\boldsymbol{\alpha}} \neq \mathbf{0}. \tag{4.16}$$

It is pointed out that in distinction to (4.16), the stability conditions (4.14) and (4.15) can be used within the purely mechanical, phenomenological framework of multi-mode plasticity presented in Sect. 3.3, and their derivation does not require specific reference to internal state variables.

The inequalities in (4.14) and (4.15) appeared earlier in the literature on crystal plasticity in a different context, cf. Havner (1992). It should be noted that while the inequality in (4.14) holds either for all or for none of work-conjugate pairs $(\mathbf{t}, \mathbf{e})$ of stress and strain measures, this is not so for the inequality in (4.15) involving h_{KL}. This can be seen from the transformation formulae (3.31) and (3.32) given in Sect. 3.3. The lack of invariance of (4.15) is related to the dependence of stability on *which* stress measure is kept constant. In spite of this, for a reasonable choice of the measures, the condition (4.15) is acceptable for effectively hardening solids, for instance, it holds automatically if all h_{KL} are positive. The stronger condition of positive definiteness of the matrix (h_{KL}) fails in crystals when effective latent hardening of the slip systems is greater than their self-hardening. The invariant condition imposed on g_{KL} in (4.14) is usually still less restrictive than (4.15) and refers to the most rigid boundary constraints. The inequality (4.14) may thus be regarded as the condition for intrinsic (thermodynamic) stability of the material with respect to internal microstructural rearrangements.

It should be added that (4.14) is sufficient for the existence of a solution to the constitutive rate-problem (P1) of multi-mode plasticity, which follows from the mathematical theory of the linear complementarity problem (cf. Cottle et al., 1992). Similarly, (4.15) is sufficient for the existence of a solution to the dual rate-problem $(\mathrm{P1})^{\mathrm{dual}}$ at given $\dot{\mathbf{t}}$.

4.3 Dynamic Localization of Deformation

Intrinsic stability of equilibrium in a homogeneous material element in a uniform state of deformation and stress is examined below under fixed *overall* deformation gradient. In particular, attention is primarily restricted to the case of fixed displacements over the element boundary. The restriction to strain uniformity within the element, adopted in Sect. 4.2, is now removed, and the possibility of dynamic instability of equilibrium due to spontaneous localization of deformation is investigated.

Suppose that instability of equilibrium with respect to internal structural rearrangements has been excluded. We shall assume that constitutive equations for the material have been transformed to the form (3.17). This could have been done by transforming an objective constitutive law (3.15) or (3.16) with the help of the transformation formulae from Sect. 3.1 and 3.3. Starting from multi-mode plasticity or from a thermodynamic formalism, that transformation could be preceded by elimination of the inelastic multipliers or rates of internal variables.

No perturbations of the constitutive equations are now admitted, so that $\dot{w}^{\mathrm{dist}} \equiv 0$. Rather, arbitrarily small perturbing body forces are considered in connection with the possibility of spontaneous localization of deformation. The corresponding virtual deformation modes within a material element are non-uniform and are described by a velocity field $\tilde{\mathbf{v}}$ considered in the *reference* configuration.

Exclusion condition in integral form. Consider a material element of a reference volume V and with fixed displacements over its boundary ∂V. Under the assumption that a velocity field $\tilde{\mathbf{v}}$ within the element is continuous and piecewise continuously differentiable, along a non-uniform deformation path of departure from equilibrium we have

$$\dot{W} - \dot{W}^{\text{ext}} = \dot{W}(\tilde{\mathbf{v}}) - 0 = \int_V \mathbf{S} \cdot \dot{\mathbf{F}}\, dV \qquad \text{for } \mathbf{v} = \mathbf{0} \text{ over } \partial V\,, \tag{4.17}$$

with $\dot{\mathbf{F}} = (\bar{\nabla}\mathbf{v})\mathbf{F} = \nabla\mathbf{v}$, the velocity gradient in the reference configuration. By the divergence theorem, $\int_V \mathbf{S}^0 \cdot \ddot{\mathbf{F}}\, dV = 0$ for an equilibrium stress $\mathbf{S}^0$ and under the boundary constraint $\dot{\mathbf{v}} = 0$. On using this after differentiating (4.17), the second-order condition (4.8) of directional stability of equilibrium takes the form (cf. Hill, 1958)

$$\int_V \dot{\mathbf{S}} \cdot \dot{\mathbf{F}}\, dV > 0 \qquad \text{for all } \tilde{\mathbf{v}} \neq \tilde{\mathbf{0}} \text{ with } \mathbf{v} = \mathbf{0} \text{ over } \partial V\,. \tag{4.18}$$

Here and in the following, $\dot{\mathbf{S}}$ is assumed to be constitutively related to $\dot{\mathbf{F}}$ through (3.17).

If we start from the stronger stability condition (4.5) with

$$d^2 = \int_V \Delta\mathbf{F} \cdot \Delta\mathbf{F}\, dV\,, \qquad a(d) = \frac{1}{2} b d^2\,, \qquad b > 0 \tag{4.19}$$

then we arrive at the following condition sufficient for *directional* stability of equilibrium with respect to the distance d:

$$\int_V \dot{\mathbf{S}} \cdot \dot{\mathbf{F}}\, dV \geq \int_V b\dot{\mathbf{F}} \cdot \dot{\mathbf{F}}\, dV \qquad \text{for all } \tilde{\mathbf{v}} \text{ with } \mathbf{v} = \mathbf{0} \text{ over } \partial V\,. \tag{4.20}$$

It is pointed out again that this condition without additional assumptions does not necessarily ensure stability for arbitrarily circuitous paths. For instance, (4.4) is always violated on certain deformation cycles within a continuum obeying a non-associative plasticity law (3.18) with different directions of $\bar{\boldsymbol{\lambda}}^{\text{p}}$ and $\bar{\boldsymbol{\lambda}}$ (Petryk, 1985b).

Local condition for stability of equilibrium. Suppose that $\dot{\mathbf{F}}$ is effectively concentrated in a disk-like volume B of predominantly uniform thickness h which is small compared with the dimensions of the element (Fig. 5). This is possible if $\nabla\mathbf{v} = \mathbf{g} \otimes \mathbf{n}$ within the disk, where $\mathbf{n}$ is the normal to the disk mid-plane in the reference configuration and $\mathbf{g}$ is an arbitrary vector. In this case, the velocity gradient outside the disk can be made arbitrarily small as $h \to 0$ at given $\dot{\mathbf{F}}$ within the disk. (Any other form of $\dot{\mathbf{F}}$ in B gives a finite contribution $\dot{\mathbf{F}}\mathbf{s}$ to the velocity gradient *outside* the disk in some direction $\mathbf{s}$ tangent to the disk boundary, by continuity of $\mathbf{v}$). By proceeding as in (Hill, 1962, Petryk, 1992a), from (4.18) in the limit as $h \to 0$ we obtain[10]

$$\dot{\mathbf{S}} \cdot \dot{\mathbf{F}} = (\mathbf{g} \otimes \mathbf{n}) \cdot \mathsf{C}(\mathbf{g} \otimes \mathbf{n}) \cdot (\mathbf{g} \otimes \mathbf{n}) \geq 0 \qquad \text{for all } \dot{\mathbf{F}} = \mathbf{g} \otimes \mathbf{n}\,. \tag{4.21}$$

Similarly, as a consequence of (4.20) we obtain

$$\dot{\mathbf{S}} \cdot \dot{\mathbf{F}} = (\mathbf{g} \otimes \mathbf{n}) \cdot \mathsf{C}(\mathbf{g} \otimes \mathbf{n}) \cdot (\mathbf{g} \otimes \mathbf{n}) > 0 \qquad \text{for all } \dot{\mathbf{F}} = \mathbf{g} \otimes \mathbf{n} \neq \mathbf{0} \tag{4.22}$$

as the exclusion condition for spontaneous dynamic localization of deformation in a *vanishingly thin* band within a finite material element (or in a band of finite thickness in an infinite continuum).

[10] The implication (4.18) $\Longrightarrow$ (4.21) follows also from the theorem due to Graves (1939), as noted in (Petryk, 1985a).

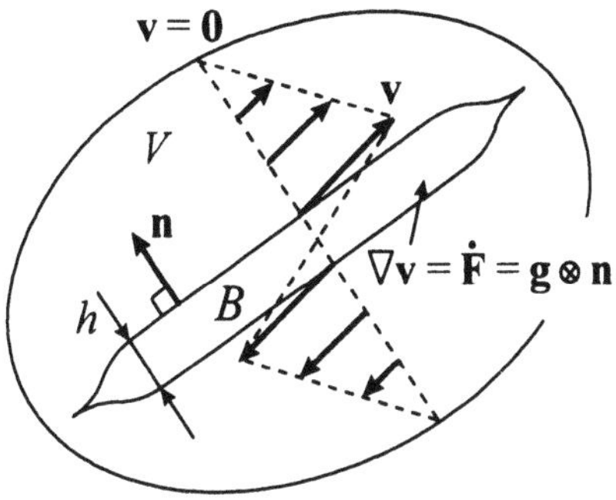

Figure 5. A perturbed velocity field corresponding to incipient localization of deformation.

It is clear that the above conditions refer only to the symmetric part $\frac{1}{2}(\mathbf{C}+\mathbf{C}^{\mathrm{T}})$ of the moduli tensor $\mathbf{C}$. The condition (4.21) is *distinct* from the *semi-strong ellipticity* condition, which has an analogous form but with the difference that a moduli tensor $\mathbf{C}$ is independent of $\mathbf{n}, \mathbf{g}$. However, in the special case of classical elastoplasticity with two linear constitutive branches as in (3.18), the condition (4.21) *implies* (and is obviously implied by) semi-strong ellipticity of the moduli from each branch. For, the sense of a velocity gradient $\mathbf{g}\otimes\mathbf{n}$ can always be adjusted to fall within a chosen constitutive half-space. In conclusion, semi-strong ellipticity of the incremental stiffness moduli from both constitutive branches in the classical elastoplasticity follows as a consequence of the restricted form (4.18) of Hill's condition of stability of equilibrium (Petryk, 1985a, Ryzhak, 1987).

Similar remarks apply to the relation between (4.22) and the condition of *strong ellipticity*.

From (3.14) it follows that the second-order effect of geometric changes should not be neglected *in advance* when evaluating the above left-hand expression. Nevertheless, with the help of (3.7) or (3.14) it is easy to show that (4.22) is equivalent to

$$\dot{\sigma}\cdot\mathbf{D}>0 \qquad \text{for all } \dot{\mathbf{F}}=\mathbf{g}\otimes\mathbf{n}\neq\mathbf{0}\,. \tag{4.23}$$

In contrast to the Drucker condition (3.9), the constraint on the velocity gradient in (4.23) provides a definite coupling between the material spin and strain-rate, removing thus ambiguities raised when the material spin is unspecified.

Condition sufficient for instability. Fulfillment of the condition

$$\dot{\mathbf{S}}\mathbf{n}=\mu\mathbf{g} \qquad \text{for some } \dot{\mathbf{F}}=\mathbf{g}\otimes\mathbf{n}\neq\mathbf{0}\,,\ \mu<0\,. \tag{4.24}$$

may be regarded as sufficient for the local instability of equilibrium, related to spontaneous dynamic localization of deformation into a planar band in a homogeneous, uniformly stressed infinite continuum. This was demonstrated by Mandel (1966) for an incrementally *linear* material in the small-strain framework, with application to classical elastoplasticity without normality. An extension to an incrementally nonlinear material at finite deformations was given in (Petryk,

1992a)[11] in the potential case (3.51) with the left-hand side in (4.24) written as $\partial U(\mathbf{g} \otimes \mathbf{n})/\partial \mathbf{g}$. The present extension to non-symmetric $\mathbf{C}$ is immediate.

To see better both the analogies and differences between the linear and nonlinear cases, (4.24) can be rewritten as[12]

$$\dot{\mathbf{S}}\mathbf{n} \equiv \mathbf{Q}(\mathbf{n}, \dot{\mathbf{F}})\mathbf{g} = \mu \mathbf{g} \qquad \text{for some } \dot{\mathbf{F}} = \mathbf{g} \otimes \mathbf{n} \neq \mathbf{0}\,,\ \mu < 0\,, \tag{4.25}$$

where $\mathbf{Q}$ of components

$$Q_{ik}(\mathbf{n}, \dot{\mathbf{F}}) = C_{ipkq}(\dot{\mathbf{F}}) n_p n_q \tag{4.26}$$

is known as the (elastoplastic) 'acoustic tensor'. In contrast to the linear case, here $\mathbf{Q}$ is generally dependent on $\dot{\mathbf{F}}$, and thus on both $\mathbf{n}$ and $\mathbf{g}$, due to the dependence of $\mathbf{C}$ on the direction of strain-rate. The equation (4.25) represents thus a *nonlinear* eigenvalue problem, while in the classical elastoplasticity (4.25) reduces to a *linear* eigenvalue problem. The distinction is essential; for instance, the critical condition associated with (4.25) for some $\mathbf{n}$ is *not* the same as $\det \mathbf{Q}(\mathbf{n}, \mathbf{g} \otimes \mathbf{n}) = 0$ for a given $\mathbf{g}$, since the vector $\mathbf{g}$ and an eigenvector $\mathbf{g}^*$ corresponding to the zero eigenvalue of $\mathbf{Q}(\mathbf{n}, \mathbf{g} \otimes \mathbf{n})$ need not coincide. The critical condition associated with (4.25) is

$$\dot{\mathbf{S}}\mathbf{n} = \mathbf{Q}(\mathbf{n}, \mathbf{g} \otimes \mathbf{n})\mathbf{g} = \mathbf{0} \qquad \text{for some } \dot{\mathbf{F}} = \mathbf{g} \otimes \mathbf{n} \neq \mathbf{0}\,, \tag{4.27}$$

which corresponds to the quasi-static bifurcation in a band under the restriction to rigid-body motion outside the band (Kolymbas, 1981). A more general case of bifurcation in a band will be examined in Sect. 5.3 and 6.3.

For a class of incrementally nonlinear models proposed for geomaterials, the condition (4.27) has been found to be the most critical in the sense that it can be met earlier along a deformation path than any other bifurcation in a band (Chambon et al., 2000).

Note that (4.27) *differs* from the critical condition $\mathbf{Q}_\mathrm{s}(\mathbf{n}, \mathbf{g} \otimes \mathbf{n})\mathbf{g} = \mathbf{0}$ associated with the failure of (4.22), unless $\mathbf{Q}$ is symmetric, i.e. $\mathbf{Q} = \mathbf{Q}_\mathrm{s}$. Failure of (4.22) along a deformation path cannot appear *later* than fulfillment of (4.27) since (4.27) obviously excludes (4.22). However, failure of (4.22) without satisfying (4.25) or (4.27) has not been shown, to the author's knowledge, to imply a dynamic instability of equilibrium for arbitrarily small disturbances.

Material instability associated with fulfillment of (4.25) can also be seen by comparison with the condition of propagation of an *acceleration wave* into *non-deforming* material (Hill, 1962, Petryk, 1992a):

$$\dot{\mathbf{S}}\mathbf{n} = \mathbf{Q}(\mathbf{n}, \dot{\mathbf{F}})\mathbf{g} = \rho c^2 \mathbf{g} \qquad \text{for } \dot{\mathbf{F}} = \mathbf{g} \otimes \mathbf{n} \neq \mathbf{0}\,, \tag{4.28}$$

where $\dot{\mathbf{F}}$ represents the velocity gradient on the deforming side of the discontinuity surface propagating with normal speed c into the material being at rest, and ρ is the mass density in the reference configuration. It follows that the normal speed of propagation of all acceleration waves into the material being at rest is real (positive or zero) only if (4.25) is excluded.

[11] As mentioned in the reference, the proof is not fully rigorous, not even under the assumptions of the first approximation, since a rigid body motion outside the band is not a free inertial motion. However, the meaning of instability associated with the fulfillment of (4.24) seems to be clear.

[12] The left-hand expression in (4.21) and (4.22) takes the form $\mathbf{g}\mathbf{Q}(\mathbf{n}, \mathbf{g} \otimes \mathbf{n})\mathbf{g}$.

4.4 Stability Under Deformation-Sensitive Loading

Examine now the kind of instability of equilibrium, announced in Sect. 1.3, which may appear when the environment of a stressed material element is not sufficiently 'stiff', also when (4.14) and (4.20) are satisfied. Attention is restricted again to uniform straining within the element, and for simplicity the *rotational* equilibrium is assumed to be not disturbed. The constitutive equations for external loading are assumed in the spin-invariant form (3.78), which at constant λ reduces to $\dot{\mathbf{t}}^{\text{ext}} = -\mathbf{L}^* \cdot \dot{\mathbf{e}}$. This includes, of course, the particular case when *some* measure of external stress $\mathbf{t}^{\text{ext}}$ is fixed.

Stability of an equilibrium state at $\mathbf{t} = \mathbf{t}^{\text{ext}}$ is investigated, assuming that $\dot{w}^{\text{dist}} \equiv 0$. The constitutive rate equations for the material are taken in the form (3.16). Along a path of departure from equilibrium we now have

$$\frac{1}{V}(\dot{W} - \dot{W}^{\text{ext}}) = \dot{w} - \dot{w}^{\text{ext}} = (\mathbf{t} - \mathbf{t}^{\text{ext}}) \cdot \dot{\mathbf{e}}\,. \tag{4.29}$$

The second-order condition (4.8) sufficient for directional stability of equilibrium takes the form

$$(\dot{\mathbf{t}} - \dot{\mathbf{t}}^{\text{ext}}) \cdot \dot{\mathbf{e}} \equiv \dot{\mathbf{e}} \cdot (\mathbf{L}(\dot{\mathbf{e}}) + \mathbf{L}^*) \cdot \dot{\mathbf{e}} > 0 \qquad \text{for all } \dot{\mathbf{e}} \neq \mathbf{0}\,. \tag{4.30}$$

If $\mathbf{t}^{\text{ext}}$ is fixed then (4.30) reduces to (3.11). However, as already stated, the inequality (3.11) is not measure-invariant, and under the transition to another work-conjugate measures we must return to the more general form (4.30) (Hill, 1978). From (3.30) and (3.79) it follows that (4.30) is measure-invariant as required.

If $\mathbf{L}$ and $\mathbf{L}^*$ are diagonally symmetric and, moreover, the second-order work inequality (3.72) holds then (4.30) becomes sufficient for material stability of the type examined here also for arbitrary circuitous paths.

Exercise: Prove the above statement by showing validity of (4.5) with $d = |\Delta \mathbf{e}|$.

5 Instability of a Quasi-Static Process

5.1 The Energy Criterion

Henceforth, we shall examine possible instability of a quasi-static deformation *process* (or *path*, in equivalent terminology). It was already shown by the illustrative example in Sect. 2 that the process or path going through stable equilibrium states may itself be unstable. The *fundamental process* (path) whose stability is to be studied is regarded as known, and all corresponding quantities are distinguished by a superscript '0'. The fundamental path is obtained as a quasi-static solution for a continuous solid body subjected to a loading program parameterized by a scalar loading parameter $\lambda = \lambda(t)$ (which need not be a monotonic function).

Throughout this Section 5, time-independent constitutive equations given in Sect. 3.3 are assumed jointly with *the symmetry and normality conditions* from Sect. 3.5. Since the normality rule may be regarded as a consequence of symmetry of incremental stiffness moduli if a single internal mechanism of plastic deformation is operating, the symmetry conditions are in effect most important. A thermodynamic argument supporting the symmetry assumptions was presented in Sect. 3.5.

Moreover, the incremental loading is assumed to be *conservative* in an overall sense, implying that a potential energy $\Omega(\tilde{\mathbf{u}}, \lambda)$ of the loading device can be defined (at least to second order) as a functional of the displacement field $\tilde{\mathbf{u}}$ and λ. For the external loading from Sect. 3.6, this reduces to another symmetry restriction $\mathbf{C}^* = \mathbf{C}^{*\mathrm{T}}$.

Under these symmetry assumptions, the energy criterion of stability of equilibrium can be extended to deformation processes. [13] For this purpose, *the energy functional* (Petryk, 1982)

$$E = E^{\text{suppl}} + W^{\text{dist}} = W + \Omega \tag{5.1}$$

is defined for a class of kinematically admissible processes of isothermal deformation. An increment ΔE in the value of E represents the amount of energy to be supplied from external sources to the mechanical *system*, consisting of the deformed body *and* the loading device, in order to produce quasi-statically a deformation increment, generally with the help of additional disturbing forces. In the former decomposition of E, E^{suppl} is the energy supplied from external sources *to the loading device*, and W^{dist} is the work done directly by all disturbing forces. In turn, $E^{\text{suppl}} = \Omega + W^{\text{ext}}$, where the part Ω is stored in the loading device and the part W^{ext} is transmitted to the deformed body. The mechanical energy balance involved in (5.1) is completed by the equality $W^{\text{dist}} = W - W^{\text{ext}}$, where W is the work to be supplied to the deforming body according to the constitutive law, as defined by (4.2). It is not essential here what part of E is stored in the material, dissipated, or transformed into heat. A further insight into the meaning of E is gained if the mechanical system is imagined to be placed in a heat reservoir. Then an increment of E can be identified with a change in the internal energy of the resulting compound thermodynamic system {deforming body + loading device + heat reservoir} (cf. Petryk, 1993b). The choice of the system is not arbitrary: at a constant loading parameter λ and in the absence of disturbances, the compound thermodynamic system can be treated as isolated.

Under the symmetry assumptions indicated above, the following energy criterion has been proposed by Petryk (1982, 1985a) for concluding whether a quasi-static process of elastoplastic deformation can be stable or not.

The energy criterion of path instability. *In a stable fundamental process of quasi-static deformation, the increment of E calculated with accuracy to second-order terms is minimized within the class of all kinematically admissible deformation increments*, viz.

$$\Delta E \geq \Delta E^0 \qquad \text{(to second order)}\,. \tag{5.2}$$

This may be regarded as a specification of the intuitive engineering hypothesis that a real deformation mode exhibits a tendency to minimize the energy consumption.

Justification of the criterion by relating it to a necessity condition for stability (and a sufficiency condition for instability) in the first approximation against persistent disturbances was given by Petryk (1991b) for the constitutive law (3.51), and extended to perturbations of thermodynamic forces in multi-mode plasticity in (Petryk, 1999a). Consistency of the criterion with Hill's bifurcation theory was shown earlier in (Petryk, 1985a) (cf. also Petryk, 1993a), and with the initial imperfection approach in (Petryk, 1997a). Thermodynamic aspects of the criterion

[13] This concept is described here only in outline; for more details, see (Petryk, 1997a) and the relevant references therein.

were discussed by Petryk (1993b) and Fedelich and Ehrlacher (1997). Other relevant references can be found in these cited above.

From the virtual work principle it follows that in an equilibrium state where all disturbing forces vanish and are not discontinuous in time, $\dot{E} = \dot{E}^0$ identically. Accordingly, the above criterion can be reformulated as follows: *In a stable fundamental process of quasi-static deformation,*

$$\ddot{E} \geq \ddot{E}^0 \tag{5.3}$$

for every admissible direction of a deformation increment.

Detailed specification of (5.2) and (5.3) for material instability problems will be presented below. Fulfillment of (5.3) has been shown to imply that the fundamental velocity field represents automatically a quasi-static solution. This essential conclusion requires the symmetry assumptions introduced above. In turn, not every quasi-static solution minimizes $\ddot{E}$. It follows that if a solution is not unique then (5.3) can provide a criterion of choice within the solution class, which is lacking in the bifurcation theory alone. The energy criterion (5.3) constitutes a basis for the computational method proposed by Petryk and Thermann (1992) where a minimization procedure is used to exclude unstable solution paths in an automatic way.

The above energy criterion can be used to examine stability of a degenerate deformation process in which an equilibrium state does not undergo any changes. This is so if the loading parameter λ is kept constant in an arbitrarily long time interval, which can always be assumed on account of the freedom in defining a time-like parameter t for quasi-static deformation of a time-independent material. Then $\Delta E^0 = 0$, $\Delta E^{\mathrm{suppl}} = 0$, $\Delta\Omega = -\Delta W^{\mathrm{ext}}$ and (5.2) reduces to

$$\Delta E = \Delta W - \Delta W^{\mathrm{ext}} \geq 0 \qquad \text{at fixed } \lambda \tag{5.4}$$

within that interval, which is nothing else than the energy criterion (4.4) of stability of *equilibrium* in a somewhat weakened form. Under the symmetry assumptions and certain additional reservations, the weakened form (5.4) may be treated as a *necessity* condition for stability (Petryk, 1991b), while (4.4) was rather interpreted as a *sufficiency* condition for directional stability. Nevertheless, the critical state associated with both conditions may be expected to be the same in sufficiently regular processes. In conclusion, the condition (5.2) may be regarded as an extension of the energy criterion of stability of equilibrium to quasi-static deformation processes.

5.2 Instability of a Process of Internal Rearrangements

This section represents in a sense a continuation of Sect. 4.2. Here, as the object of stability investigations we take the fundamental *process* of internal rearrangements within a material element of reference volume V under the assumption of uniform macroscopic strain. Two cases are examined when the process is induced by prescribed variations in time of either strain $\mathbf{e}(\lambda)$ or a certain measure of stress $\mathbf{t} \equiv \mathbf{t}^{\mathrm{ext}}(\lambda)$, which corresponds to $\Omega = 0$ or $\Omega = -V\mathbf{t}^{\mathrm{ext}} \cdot \mathbf{e}$, respectively. The study in Sect. 4.2 corresponded to λ kept fixed; here $\lambda = \lambda(t)$ is a given function. The framework of multi-mode plasticity from Sect. 3.3 is adopted, with the normality (3.49) and symmetry (3.52) assumptions. The rates $(\dot{\mathbf{e}}^0, \dot{\mathbf{t}}^0, \mathring{\gamma}_K^0)$ in fundamental process are obtained as solutions to (P1) or (P1)^{dual} at every instant.

For any other admissible mode of deformation within the element at a micro-level in the presence of disturbances of thermodynamic forces, the rate of the energy functional (5.1) is specified as follows

$$\begin{aligned} \dot{E}/V &= \dot{w} + \dot{w}^{\rm dist} + \dot{\Omega}/V \\ &= \mathbf{t}\cdot\dot{\mathbf{e}} - \sum_K f_K \overset{\circ}{\gamma}_K \qquad \text{for controlled } \mathbf{e}\,, \\ &= -\dot{\mathbf{t}}\cdot\mathbf{e} - \sum_K f_K \overset{\circ}{\gamma}_K \qquad \text{for controlled } \mathbf{t}\,. \end{aligned} \tag{5.5}$$

Stability of the fundamental process can be investigated in a neighborhood of any given equilibrium state $\mathcal{G}_0$ reached at some λ. To establish *necessary* conditions for path stability, consider deviations from the fundamental process along direct paths initiated in all possible directions corresponding to infinitesimal disturbing forces. Such directions are described by vectors $[\overset{\circ}{\gamma}]$ (of non-negative components $\overset{\circ}{\gamma}_K$ with $K \in \mathcal{P}$), while $\overset{\circ}{\gamma}_K$ vanishes if $f_K(\mathcal{G}_0) < 0$, also at subsequent instants. Then $\dot{E} = \dot{E}^0$ in an equilibrium state $\mathcal{G}_0$.

By differentiating (5.5) outside equilibrium, substituting the constitutive rate equations (3.20) or (3.21) along with the normality rule (3.49), and comparing the terms in the equilibrium state, we obtain the identity

$$\frac{1}{2V}(\ddot{E} - \ddot{E}^0) = \mathcal{J}([\overset{\circ}{\gamma}]) - \mathcal{J}([\overset{\circ}{\gamma}{}^0]) \qquad \text{for all } [\overset{\circ}{\gamma}], [\overset{\circ}{\gamma}{}^0] \quad \text{in } \mathcal{G}_0\,, \tag{5.6}$$

where

$$\mathcal{J}([\overset{\circ}{\gamma}]) = \begin{cases} \dfrac{1}{2}\displaystyle\sum_{K,L\in\mathcal{P}} g_{KL}\overset{\circ}{\gamma}_K\overset{\circ}{\gamma}_L - \dot{\mathbf{e}}\cdot\sum_{K\in\mathcal{P}} \boldsymbol{\lambda}_K\overset{\circ}{\gamma}_K & \text{for given } \dot{\mathbf{e}}\,, \\ \dfrac{1}{2}\displaystyle\sum_{K,L\in\mathcal{P}} h_{KL}\overset{\circ}{\gamma}_K\overset{\circ}{\gamma}_L - \dot{\mathbf{t}}\cdot\sum_{K\in\mathcal{P}} \boldsymbol{\mu}_K\overset{\circ}{\gamma}_K & \text{for given } \dot{\mathbf{t}}\,. \end{cases} \tag{5.7}$$

The symmetry condition (3.52) and the fact that $[\overset{\circ}{\gamma}{}^0]$ is a solution to (P1) or (P1)$^{\rm dual}$ are not needed for validity of (5.6).

From (5.3) and (5.6) we arrive at the energy condition for stability of the fundamental process of internal rearrangements described by $\overset{\circ}{\gamma}{}^0_K(t)$, in the form

$$\mathcal{J}([\overset{\circ}{\gamma}]) \geq \mathcal{J}([\overset{\circ}{\gamma}{}^0]) \qquad \text{for all } [\overset{\circ}{\gamma}] \geq [0]\,. \tag{5.8}$$

The interpretation of (5.8) as a condition *necessary* for stability of a process is supported by the following result proven in (Petryk, 1999a).

Theorem on instability of a process of internal rearrangements. *Suppose that along a segment of a solution path* $(\dot{\mathbf{e}}^0, \overset{\circ}{\gamma}{}^0_K)(t)$ *the condition (5.8) is not satisfied while the condition (4.14) or (4.15) of stability of equilibrium holds.*

Then at every point of that segment there exists another solution to (P1) or (P1)$^{\rm dual}$, *respectively, that is energetically preferable to* $[\overset{\circ}{\gamma}]^0$ *in the sense of assigning to* $\ddot{E}$ *a minimum value* •

For any solution $[\overset{\circ}{\gamma}]$ to (P1) or (P1)$^{\rm dual}$, the expression for $\mathcal{J}([\overset{\circ}{\gamma}])$ simplifies on account of $\dot{f}_K\overset{\circ}{\gamma}_K = 0$ to :

$$\mathcal{J}([\overset{\circ}{\gamma}]) = \begin{cases} \frac{1}{2}\dot{\mathbf{t}}^{\mathrm{p}} \cdot \dot{\mathbf{e}} & \text{if } [\overset{\circ}{\gamma}] \text{ is a solution to (P1) for controlled } \mathbf{e}\,, \\ -\frac{1}{2}\dot{\mathbf{t}} \cdot \dot{\mathbf{e}}^{\mathrm{p}} & \text{if } [\overset{\circ}{\gamma}] \text{ is a solution to (P1)}^{\mathrm{dual}} \text{ for controlled } \mathbf{t}\,, \end{cases} \tag{5.9}$$

with $\dot{\mathbf{t}}^{\mathrm{p}}$, $\dot{\mathbf{e}}^{\mathrm{p}}$ defined as in (3.48).

Hence, among all solutions to (P1) or (P1)^{dual}, the particular solution $[\overset{\circ}{\gamma}{}^0]$ which satisfies (5.8) also:

minimizes $\dot{\mathbf{t}} \cdot \dot{\mathbf{e}}$ (and $\dot{\mathbf{t}}^{\mathrm{p}} \cdot \dot{\mathbf{e}}$) for controlled **e**,
maximizes $\dot{\mathbf{t}} \cdot \dot{\mathbf{e}}$ (and $\dot{\mathbf{t}} \cdot \dot{\mathbf{e}}^{\mathrm{p}}$) for controlled **t**.

This corresponds to the criterion of selection of an 'optimal' solution which was proposed by Franciosi and Zaoui (1991) in relation to the energy criterion but without assuming the symmetry property (3.52).

A necessary condition for $[\overset{\circ}{\gamma}{}^0]$ to satisfy (5.8) is that

$$\begin{aligned} &\sum_{K,L\in\mathcal{A}^0} g_{KL} w_K w_L \geq 0 && \text{for controlled } \mathbf{e}, \\ &\qquad\qquad \text{for every } [w],\ \mathcal{A}^0 = \{K : \overset{\circ}{\gamma}{}^0_K > 0\} \\ &\sum_{K,L\in\mathcal{A}^0} h_{KL} w_K w_L \geq 0 && \text{for controlled } \mathbf{t}\,. \end{aligned} \tag{5.10}$$

More details can be found in (Petryk, 1999a). We quote a simple illustrative example of double slipping in a single crystal with the effective hardening matrix for a selected pair of stress and strain measures given by

$$\begin{pmatrix} h_{11} & h_{12} \\ h_{21} & h_{22} \end{pmatrix} = \begin{pmatrix} 1 & 1+q \\ 1+q & 1 \end{pmatrix} h\,, \qquad h > 0, \qquad q > 0. \tag{5.11}$$

It is clear that the conditions (4.15) and (4.14) (for $\mathbf{L}^{\mathrm{e}}$ positive definite) of stability of equilibrium under both strain and stress control are satisfied, although the matrix (h_{KL}) is indefinite. The process of uniform double slipping is unstable under stress control according to the condition $(5.10)_2$, and may be stable under strain control. According to the Theorem quoted above, an incremental solution under stress control exists which does satisfy (5.8), and in the loading case this will be single slipping $\overset{\circ}{\gamma}{}^0_1 > 0$, $\overset{\circ}{\gamma}{}^0_2 = 0$ or $\overset{\circ}{\gamma}{}^0_1 = 0$, $\overset{\circ}{\gamma}{}^0_2 > 0$.

5.3 Instability of Uniform Straining

In this section, the energy criterion of intrinsic stability of equilibrium, examined in Sect. 4.3, is extended to a quasi-static *process* of uniform straining and related to the bifurcation approach. A potential constitutive law (3.51) is assumed. Stability of the fundamental process of uniform deformation within a homogeneous material element is examined under controlled *overall* deformation gradient varying in time, so that $\Delta\Omega \equiv 0$. In particular, the case of a displacement history prescribed over the element boundary is considered. In comparison with Sect. 5.2, the restriction to strain uniformity within the material element is now removed, and path instability associated with spontaneous deviation from uniform straining is studied.

The energy criterion of instability of uniform straining. The integral conditions of path stability that result from the energy criterion formulated in the Sect. 5.1 have been discussed in detail elsewhere (Petryk, 1985a, 1991b, 1993a, 1993b). Following (Petryk, 1992a), as the basic condition necessary for stability of the fundamental, uniform deformation process $\dot{\mathbf{F}}^0(t)$ within a homogeneous material element we take

$$\int_V U(\dot{\mathbf{F}}^0 + \nabla\mathbf{w})\,\mathrm{d}V \geq V\,U(\dot{\mathbf{F}}^0) \qquad \text{for every } \tilde{\mathbf{w}} : \ \mathbf{w} = \mathbf{0} \text{ over } \partial V \tag{5.12}$$

which is a consequence of (5.3). Here, $\tilde{\mathbf{w}}$ belongs to the space of continuous and piecewise smooth vector fields, and the integration domain may be chosen arbitrarily. (5.12) is called the condition of *quasi-convexity of* $U(\dot{\mathbf{F}})$ *at* $\dot{\mathbf{F}}^0$.

From the Green theorem and the boundary condition $\mathbf{w} = \mathbf{0}$ over ∂V it follows that

$$\int_V U(\dot{\mathbf{F}}^0 + \nabla\mathbf{w})\,\mathrm{d}V = V\,U(\dot{\mathbf{F}}^0) + \int_V \mathcal{E}(\dot{\mathbf{F}}^0, \dot{\mathbf{F}}^0 + \nabla\mathbf{w})\,dV\,, \tag{5.13}$$

where $\mathcal{E}$ is the Weierstrass function associated with the constitutive potential $U(\dot{\mathbf{F}})$:

$$\mathcal{E}(\dot{\mathbf{F}}, \dot{\mathbf{F}}^*) = U(\dot{\mathbf{F}}^*) - U(\dot{\mathbf{F}}) - \dot{\mathbf{S}} \cdot (\dot{\mathbf{F}}^* - \dot{\mathbf{F}})\,. \tag{5.14}$$

Let $\nabla\mathbf{w}$ be effectively concentrated in a disk-like volume of predominantly uniform thickness h, taken sufficiently small. As already discussed, such perturbation has a form $\nabla\mathbf{w} = \mathbf{g}\otimes\mathbf{n}$ which previously represented the velocity gradient itself (Fig. 5) and now it represents the *difference* between the perturbed and fundamental velocity gradients (Fig. 6). By using (5.13), in the limit as $h \to 0$ as a consequence of (5.12) we obtain that

$$\mathcal{E}(\dot{\mathbf{F}}^0, \dot{\mathbf{F}}^0 + \mathbf{g}\otimes\mathbf{n}) \geq 0 \qquad \text{for all vectors } \mathbf{g}, \mathbf{n}\,, \tag{5.15}$$

that is, U must be *rank* 1 *convex at* $\dot{\mathbf{F}}^0$.[14] Hence, if (5.15) does not hold then the concentration of deformation within a band (or bands) of orientation $\mathbf{n}$ is shown to be *energetically preferable* to uniform straining with the velocity gradient $\dot{\mathbf{F}}^0$, in the sense of requiring a smaller value of ΔE (to second-order).

In the limit as $|\mathbf{g}||\mathbf{n}| \longrightarrow 0$, (5.15) implies semi-strong ellipticity of the current tangent moduli:

$$(\mathbf{g}\otimes\mathbf{n}) \cdot \mathbf{C}(\dot{\mathbf{F}}^0) \cdot (\mathbf{g}\otimes\mathbf{n}) \geq 0 \qquad \text{for every } \mathbf{g}, \mathbf{n}\,. \tag{5.16}$$

If $|\mathbf{g}||\mathbf{n}| \longrightarrow \infty$ then as a consequence of (5.15) we obtain another condition (Petryk, 1985a)

$$U(\mathbf{g}\otimes\mathbf{n}) \geq 0 \qquad \text{for every } \mathbf{g}, \mathbf{n}\,, \tag{5.17}$$

which already appeared above as the condition (4.21) of stability of equilibrium.

The conditions (5.15), (5.16) and (5.17) are interpreted as *necessary* for stability of the fundamental process since they are derived from the global energy condition (5.3). This also makes explicit the connection between material instability and the instability in a finite solid body with specified boundary conditions.

[14] The implication (5.12) $\Rightarrow$ (5.15) follows also as a corollary from the theorem due to Graves (1939).

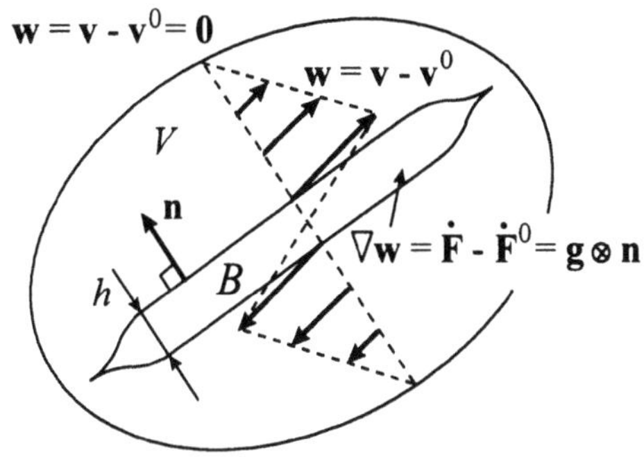

Figure 6. A perturbation superimposed on the fundamental velocity field

The condition (5.15) is necessary for the speed of dynamic propagation of all acceleration waves into the material *deforming with the rate* $\dot{\mathbf{F}}^0$ to be real (Petryk, 1992a). In particular, fulfillment of (5.16) is necessary for this (recall that $\mathbf{C}$ is symmetric here). The distinction between the two conditions is that (5.15) concerns arbitrary acceleration waves, while (5.16) concerns only those acceleration waves which carry a jump in $\dot{\mathbf{F}}$ sufficiently small to remain on *both* sides of the discontinuity surface within the fundamental constitutive cone where $\mathbf{C}^0$ apply.

In turn, if (5.17) fails then the condition (4.24) for dynamic localization of deformation is satisfied. According to the discussion in Sect. 4.3, the condition (5.17) may thus be regarded as necessary for local stability of equilibrium under arbitrary boundary data. Moreover, the normal speed of propagation of all acceleration waves into the material being at rest is real (positive or zero) if and only if (5.17) holds.

Examine now the critical instants on a smooth deformation path beyond which the conditions (5.15), (5.16) or (5.17) fail.

The first critical instant is met when (5.15) still holds but with equality for some nonzero vectors $\mathbf{g}, \mathbf{n}$. Then the partial derivative of $\mathcal{E}(\dot{\mathbf{F}}^0, \dot{\mathbf{F}}^0 + \mathbf{g} \otimes \mathbf{n})$ with respect to $\mathbf{g}$ vanishes, which reads

$$\mathcal{E}_{,\mathbf{g}}(\dot{\mathbf{F}}^0, \dot{\mathbf{F}}^0 + \mathbf{g} \otimes \mathbf{n}) = (\dot{\mathbf{S}} - \dot{\mathbf{S}}^0)\mathbf{n} = \mathbf{0} \qquad \text{for some } \dot{\mathbf{F}} = \dot{\mathbf{F}}^0 + \mathbf{g} \otimes \mathbf{n} \neq \dot{\mathbf{F}}^0 . \tag{5.18}$$

This coincides with the condition for bifurcation in a band, to be discussed below.

Analogously, by appealing to the assumed symmetry of $\mathbf{C}$ and using the definition (4.26) of the acoustic tensor, the critical condition associated with (5.16) reads

$$\mathbf{Q}(\mathbf{n}, \dot{\mathbf{F}}^0)\mathbf{g} = \mathbf{0} \qquad \text{for some } \mathbf{n}, \mathbf{g} \neq \mathbf{0} , \tag{5.19}$$

which is the *linearized* condition for bifurcation in a band.

The critical condition associated with (5.17) coincides, for symmetric $\mathbf{C}$, with (4.27) discussed in Sect. 4.3.

If the constitutive inequality (3.69) holds then (5.12) ⇔ (5.15) ⇔ (5.16) (Petryk, 1992a). However, even then the condition (5.17) represents a different requirement, less restrictive on account of the implication (5.15) $\Rightarrow$ (5.17). This illustrates once again the fundamental distinction between the instability conditions for a deformation process and for an equilibrium state.

In the special case of classical elastoplasticity, the distinction between (5.16) for all $\dot{\mathbf{F}}^0$ and (5.17) disappears, as already discussed in Sect. 4.3, and all the conditions necessary for local (intrinsic) stability of uniform straining reduce to the single requirement of semi-strong ellipticity of the tangent moduli.

Bifurcation in a band. To eliminate the influence of boundary conditions, a material element may also be treated, in another scale but under the current uniform stress, as an infinite homogeneous continuum. The loading conditions at infinity are imagined to be such that uniform and constitutively compatible quasi-static rate fields $\dot{\mathbf{F}}^0, \dot{\mathbf{S}}^0$ represent the fundamental solution. Then, non-trivial solutions that correspond to a material instability in the form of bifurcation within a planar band can be investigated. The gradient $\nabla\mathbf{w}$ of admissible variations in velocities (the meaning of $\mathbf{w}$ is analogous to that in Fig. 6) is assumed to vanish outside the band whose unit normal in the reference configuration is denoted by $\mathbf{n}$. Within the band, $\mathbf{w}$ is assumed constant on planes $\boldsymbol{\xi}\cdot\mathbf{n} = \text{const}$ parallel to the band boundaries. The traction rates applied along the band boundaries result from the uniform fundamental solution $(\dot{\mathbf{F}}^0, \dot{\mathbf{S}}^0)$. From the spatial continuity of $\tilde{\mathbf{w}}$ and from the conditions of continuing equilibrium, we obtain the well-known kinematic and static conditions to be satisfied by a pair $(\dot{\mathbf{F}}, \dot{\mathbf{S}})$ within the bifurcation band (cf. Rice, 1977), viz.

$$(\dot{\mathbf{S}} - \dot{\mathbf{S}}^0)\mathbf{n} = \mathbf{0} \qquad \text{and} \qquad \dot{\mathbf{F}} - \dot{\mathbf{F}}^0 = \mathbf{g}\otimes\mathbf{n}\,, \tag{5.20}$$

respectively, where $\mathbf{g} \in \mathcal{R}^3$ is some function of $(\boldsymbol{\xi}\cdot\mathbf{n})$. Existence of a nontrivial solution to (5.20) is referred to as *bifurcation in a band*. The common term 'shear band bifurcation' is also suitable but it concerns, strictly speaking, the special case when $\mathbf{g}\otimes\mathbf{n}$ represents simple shearing.

The problem of bifurcation in a band in the framework of multi-mode plasticity will be examined in Sect. 6.3 without any symmetry assumptions. The discussion below concerns an incrementally nonlinear constitutive law obeying a potential (3.51).

In the case when the fundamental tangent stiffness moduli tensor $\mathsf{C}^0 = \mathsf{C}(\dot{\mathbf{F}}^0)$ applies also inside the band (which is possible for a piecewise-linear law) then the bifurcation condition takes the commonly assumed form (Hill, 1962, Rice, 1977)

$$\det \mathbf{Q}^0(\mathbf{n}) = 0, \qquad \mathbf{Q}^0(\mathbf{n}) = \mathbf{Q}(\mathbf{n}, \dot{\mathbf{F}}^0)\,, \tag{5.21}$$

equivalent to (5.19). However, it should be noted that (5.19) represents merely a *linearization* of the general condition (5.20). Generally, bifurcation in a band can take place along a deformation path before (5.21) is met, cf. e.g. examples calculated in (Petryk, 1992a, 1992b). An additional constitutive restriction is needed to justify the use of (5.21). For instance, the use of the classical elastoplasticity (3.18) represent such a restriction also if normality does not hold (Rice and Rudnicki, 1980). In incrementally nonlinear plasticity, the justification for using (5.21) is provided by fulfillment of the constitutive inequality (3.69). Then positive definiteness of $\mathbf{Q}^0$ excludes bifurcation in a band (Petryk, 1989); to see this, it suffices to observe that any two pairs $(\dot{\mathbf{F}}, \dot{\mathbf{S}})$ and $(\dot{\mathbf{F}}^0, \dot{\mathbf{S}}^0)$ satisfying (5.20) make the left-hand expression in (3.69) equal to zero.

While the fulfillment of (5.21) may be treated as being *sufficient* for bifurcation in a band,[15] in incrementally nonlinear materials it is *not* necessary. The following statement has been proven in (Petryk, 1992a).[16]

Theorem on continuous spectrum of bifurcation points. *Along a given deformation path* $\mathbf{F}^0(t)$ *there is (energetically preferable) bifurcation in a band of a normal* $\mathbf{n}$ *at every instant at which*

$$\mathcal{E}(\dot{\mathbf{F}}^0, \dot{\mathbf{F}}^0 + \mathbf{g} \otimes \mathbf{n}) < 0 \quad \text{for some } \mathbf{g} \qquad \textit{and} \qquad U(\mathbf{g} \otimes \mathbf{n}) > 0 \qquad \text{for every } \mathbf{g}\,. \tag{5.22}$$

Of course, the conclusion holds true if the former condition in (5.22) is replaced by negation of the condition (5.16). The critical instants (5.21) (e.g., when (3.69) holds) and (4.27) may constitute the limiting points of the bifurcation range (5.22).

The above Theorem jointly with the previous interpretation of (5.17) provide the following justification for treating (5.15) and (5.16) as conditions *necessary* for material stability :

Interpretation of instability of uniform straining. *If (5.15) (or (5.16)) does not hold then formation of localization bands can be initiated either as energetically preferable bifurcation of a quasi-static path if (5.17) holds, or as a dynamic process if (5.17) fails* •

Exercise: Show that within the range (5.22) there exists a continuous range of band orientations for which shear band bifurcation occurs in a given state.

Incipient volume fraction of localization zone. Consider a smooth fundamental path $\mathbf{F}^0(t)$ of uniform deformation of an infinite homogeneous continuum obeying an incrementally nonlinear constitutive law in the potential form (3.51). Suppose that the range of validity of (5.15) and (5.16) along the fundamental path terminates at the critical instant t_0 at which (5.21) is met. Fully localized deformations need not occur immediately beyond that instant, unless the critical instant defined by (5.21) happens to coincide with that defined by (4.27). The coincidence is enforced in the classical elastoplasticity, which can be seen by adjusting appropriately the sense of $\mathbf{g}$. On the contrary, in an incrementally nonlinear material the critical instants may be distinct when $\mathbf{g} \otimes \mathbf{n}$ in (5.19) is directed outside the constitutive domain of $\mathbf{C}^0$.

Suppose that this is so and that (4.22) still holds in a certain range beyond t_0. Then, concentration of post-critical deformation can develop gradually, and can also saturate as in the post-critical solution with a single band analysed by Hutchinson and Tvergaard (1981). On the other hand, a quasi-static solution with a single bifurcation band in an infinite continuum violates the stability requirement (5.16) outside the band if the uniform solution does, and may analogously be regarded as unrealizable in practice.

[15] For a thoroughly nonlinear relation between $\dot{\mathbf{S}}$ and $\dot{\mathbf{F}}$, fulfillment of (5.21) implies bifurcation in a band in *accelerations* rather than in velocities (Petryk, 1989).

[16] The existence of a continuous spectrum of bifurcations points along a deformation path has been established for a general incrementally-nonlinear discretized system by Petryk (1991b); another version of the proof was given by Petryk and Thermann (1992). Examples of a continuous range of shear band bifurcations were calculated in (Petryk, 1992a, 1992b).

Another difficulty, more widely recognized, is associated with an indeterminate band width. It might seem that this difficulty cannot be overcome without enhancing the material model to incorporate an internal length scale, as it is recently being developed in several different ways in the literature in the context of classical elastoplasticity. Nevertheless, a definite (although not complete) conclusion concerning the incipient post-critical deformation pattern can be drawn from the energy criterion (5.3), in particular from its consequence (5.16), still using the constitutive framework for time-independent incrementally nonlinear materials not possessing any internal length scale. Namely, the incipient *volume fraction* of the whole localization zone and the velocity gradient therein can be determined, although the width, number and spatial arrangement of individual localization bands are indeterminate. The analysis below follows closely the author's paper (Petryk, 1997a).

A class of velocity solutions in the infinite continuum at the critical instant t_0 for a pair of unit vectors $(\mathbf{n}, \mathbf{g})$ that satisfy (5.19) can be constructed as follows

$$\dot{\mathbf{F}}(\boldsymbol{\xi}) = \dot{\mathbf{F}}^0 + s(\xi_\mathbf{n})\, \mathbf{g} \otimes \mathbf{n}\,, \qquad \xi_\mathbf{n} = \boldsymbol{\xi} \cdot \mathbf{n} \in (-\infty, +\infty) \tag{5.23}$$

with the restrictions

$$s^- \leq s \leq s^+\,, \qquad \left| \int_{-\infty}^{+\infty} s\, \mathrm{d}\xi_\mathbf{n} \right| < \infty\,. \tag{5.24}$$

The limiting values s^- and s^+ are chosen such that $\mathsf{C}(\dot{\mathbf{F}}) = \mathsf{C}(\dot{\mathbf{F}}^0)$ within the corresponding range of $\dot{\mathbf{F}}$. The restrictions (5.24) ensure that the *mean* velocity gradients in the fundamental and bifurcation solutions are the same, and that the pointwise difference in velocities is infinitesimal relative to fundamental velocities at the re-scaled element boundary if $\dot{\mathbf{F}}^0 \neq \mathbf{0}$. The bifurcation solutions represent thus the branching of the material response under *macroscopically* uniform kinematic boundary data. The sense of $\mathbf{g}$ is chosen below such that $\mathbf{D}^* \cdot \mathbf{D}^0 \geq 0$, where $\mathbf{D}^0$ is the fundamental Eulerian strain-rate and $\mathbf{D}^* = \frac{1}{2}(\mathbf{g} \otimes \mathbf{n} + \mathbf{n} \otimes \mathbf{g})$. The current configuration is taken for simplicity as reference so that $\mathbf{F}^0(t_0) = \mathbf{I}$, and $\mathbf{n}$ may be interpreted as the current normal to a band.

In the particular case when $s \neq 0$ only in a finite interval of $\xi_\mathbf{n}$, the solution is obtained which was analyzed in many papers after (Rice, 1977). That solution violates the path stability conditions (5.15) and (5.16) outside the localization band just beyond the critical instant, assuming that (5.16) then fails along the fundamental path. It is a natural assumption that the *rate* of the left-hand expression in (5.16), starting from $\mathbf{g}, \mathbf{n}$ as above, can be made negative at t_0 not only for $\dot{\mathbf{F}} = \dot{\mathbf{F}}^0$ but also for all $\dot{\mathbf{F}}$ defined by (5.23) with (5.24). By the continuity argument, this is so if the rate of this expression is negative along the fundamental path and if the current constitutive cone is sufficiently sharp. If this is so then (5.16) can be satisfied just beyond the critical instant only due to the transition to a more 'stiff' constitutive branch, which is possible if the strain-rate $\mathbf{D} = \mathbf{D}^0 + s\mathbf{D}^*$, wherever defined, lies at t_0 precisely *on* the boundary of the constitutive cone containing $\mathbf{D}^0$. This means that $s(\xi_\mathbf{n})$ in (5.23) can only take two values, s^- or s^+, which define the intersection points, $\mathbf{D}^-$ and $\mathbf{D}^+$, respectively, of a straight line $\mathbf{D}(s) = \mathbf{D}^0 + s\mathbf{D}^*$ in $\mathbf{D}$-space with the cone boundary (Fig. 7a). The bifurcation solution takes the form of a mixture of layers of a constant velocity gradient, .../$\dot{\mathbf{F}}^-$/$\dot{\mathbf{F}}^+$/$\dot{\mathbf{F}}^-$/$\dot{\mathbf{F}}^+$/... (Fig. 7b), in some analogy to fine phase mixtures discussed by Ball and James (1987). The initial localization zone, which occupies a volume fraction η^+, say, is defined as the collection of all layers where $\mathbf{D} = \mathbf{D}^+$.

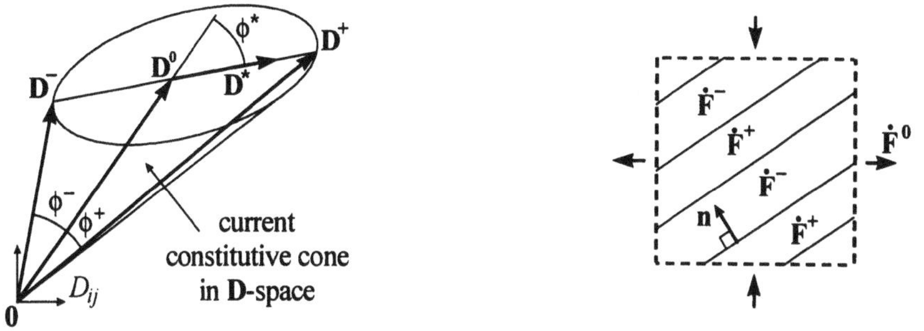

(a) (b)

Figure 7. Schematic sketch (a) of the strain rates and (b) of spatial distribution of the velocity gradient, in a bifurcation solution within a material element.

From the integral condition in (5.24) we obtain the equation

$$\eta^+ s^+ + (1-\eta^+)s^- = 0 \qquad \text{or} \qquad \eta^+ \mathbf{D}^+ + (1-\eta^+)\mathbf{D}^- = \mathbf{D}^0 \,. \tag{5.25}$$

On substituting trigonometric relationships within the hyperplane spanned by $(\mathbf{D}^0, \mathbf{D}^*)$ in **D**-space , the *incipient* volume fraction occupied by the localization zone is determined (Petryk, 1997a) as

$$\eta^+ = \frac{\cot(\phi^+) - \cot(\phi^*)}{\cot(\phi^+) + \cot(\phi^-)} \qquad \text{for } \phi^+ < \phi^* \,. \tag{5.26}$$

Here, ϕ^+, ϕ^- and ϕ^* stand for positive angles, according to a definition based on a scalar product, between $\mathbf{D}^0$ and $\mathbf{D}^+, \mathbf{D}^-$ and $\mathbf{D}^*$, respectively (Fig. 7a). In the particular case when $\phi^+ = \phi^-$, (5.26) reduces to the formula given by Petryk and Thermann (1996) and verified numerically in the context of strain localization in biaxially stretched sheets.

Exercise: Derive the formula (5.26) from (5.25) by using trigonometric relationships which result from Fig. 7a.

Exercise: Show that the value of η^+ is independent of the choice of an angular measure in **D**-space.

Instability in heterogeneous materials. The energy criterion in its general form (5.2) can also be applied to heterogeneous materials (as metal polycrystals or composites) to investigate not only local but also overall instabilities. The remarks below (cf. also Petryk, 1997a) are intended only to illustrate the possibilities.

Let curly brackets $\{\cdot\}$ denote an unweighted volume average over a representative sample of a heterogeneous material in a reference configuration. Let a continuous field $\tilde{\mathbf{w}}$ have the same meaning as above and vanish over the sample boundary. The path stability condition (5.3) applied to the quasi-static velocity field $\tilde{\mathbf{v}}^0$ within the sample gives

$$\{U(\nabla\mathbf{v}^0 + \nabla\mathbf{w})\} \geq \{U(\nabla\mathbf{v}^0)\} \qquad \text{for every } \tilde{\mathbf{w}}\,. \tag{5.27}$$

Accordingly, a minimization procedure: $\{U(\nabla\mathbf{v})\} \to \min$ can be used to determine numerically the material response in the non-uniqueness range, as shown by K. Thermann in another Chapter of this volume. Fulfillment of (5.27) implies (cf. Petryk, 1991b)

$$\{\nabla\mathbf{w}\cdot\mathbf{C}(\nabla\mathbf{v}^0)\cdot\nabla\mathbf{w}\} \geq 0 \qquad \text{and} \qquad \{U(\nabla\mathbf{w})\} \geq 0 \qquad \text{for every } \tilde{\mathbf{w}}\,. \tag{5.28}$$

If $\nabla\mathbf{w}$ is effectively concentrated in a disk-like volume of predominantly uniform thickness, taken sufficiently small as in Sect. 5.3 but now within a heterogeneous material element, (5.28) implies

$$(\mathbf{g}\otimes\mathbf{n})\cdot\{\mathbf{C}(\nabla\mathbf{v}^0)\}_{\mathrm{n}}\cdot(\mathbf{g}\otimes\mathbf{n}) \geq 0 \qquad \text{and} \qquad \{U(\mathbf{g}\otimes\mathbf{n})\}_{\mathrm{n}} \geq 0 \qquad \text{for every } \mathbf{n},\mathbf{g}\,, \tag{5.29}$$

where $\{\cdot\}_{\mathrm{n}}$ denotes averaging over a plane section of reference normal $\mathbf{n}$ through the element.

The conditions (5.29) can be used, for instance, to detect instability in a solution of a layered structure as in Fig. 7b in a scale larger than that set by a width of individual bands.

5.4 Instability Under Deformation-Sensitive Loading

Stability of a process of uniform straining under controlled overall strain does not imply stability when the environment of a stressed material element is not sufficiently 'stiff'. As in Sect. 4.4, attention is restricted in this section to uniform straining within a material element at unperturbed rotational equilibrium and in the absence of disturbances of internal thermodynamic forces. The constitutive equations for the material and external loading are assumed in the spin-insensitive form (3.16) and (3.78), respectively, with the symmetry restrictions $\mathbf{L} = \mathbf{L}^{\mathrm{T}}$ and $\mathbf{L}^* = \mathbf{L}^{*\mathrm{T}}$. Stability of the fundamental process $\mathbf{e}^0(t)$ for infinitesimal perturbations of the external surface tractions associated with $\mathbf{t}^{\mathrm{ext}}$ is investigated by using the energy criterion (5.2).

The potential energy $\Omega(\mathbf{e},\lambda)$ of the (unperturbed) external loading of the element satisfies, by definition, $V\mathbf{t}^{\mathrm{ext}} = -\partial\Omega(\mathbf{e},\lambda)/\partial\mathbf{e}$. Specification of (5.1) yields

$$\dot{E} = V\,\dot{w} + \dot{\Omega} = V\,(\mathbf{t} - \mathbf{t}^{\mathrm{ext}})\cdot\dot{\mathbf{e}} + \Omega_{,\lambda}\dot{\lambda} \tag{5.30}$$

and

$$\ddot{E} = V\,(\dot{\mathbf{t}} - \dot{\mathbf{t}}^{\mathrm{ext}})\cdot\dot{\mathbf{e}} - V\dot{\lambda}\mathbf{t}^{\mathrm{ext}}_{,\lambda}\cdot\dot{\mathbf{e}} + V\,(\mathbf{t} - \mathbf{t}^{\mathrm{ext}})\cdot\ddot{\mathbf{e}} + \Omega_{,\lambda\lambda}(\dot{\lambda})^2 + \Omega_{,\lambda}\ddot{\lambda}\,. \tag{5.31}$$

In any given equilibrium state $\mathcal{G}_0$, the term with $(\mathbf{t} - \mathbf{t}^{\mathrm{ext}})$ vanishes and the last two terms are given. On substituting (3.16) and (3.78) with $\dot{\mathbf{t}}^* = \mathbf{t}^{\mathrm{ext}}_{,\lambda}\dot{\lambda}$, we obtain the identity

$$\frac{1}{2V}(\ddot{E} - \ddot{E}^0) = \mathcal{J}(\dot{\mathbf{e}}) - \mathcal{J}(\dot{\mathbf{e}}^0) \qquad \text{for all } \dot{\mathbf{e}}, \dot{\mathbf{e}}^0 \quad \text{in } \mathcal{G}_0\,, \tag{5.32}$$

where

$$\mathcal{J}(\dot{\mathbf{e}}) = \frac{1}{2}\dot{\mathbf{e}}\cdot(\mathbf{L}(\dot{\mathbf{e}}) + \mathbf{L}^*)\cdot\dot{\mathbf{e}} - \dot{\mathbf{t}}^*\cdot\dot{\mathbf{e}}\,. \tag{5.33}$$

The condition (5.3) necessary for stability of the fundamental path reduces to

$$\mathcal{J}(\dot{\mathbf{e}}) \geq \mathcal{J}(\dot{\mathbf{e}}^0) \qquad \text{for all } \dot{\mathbf{e}}\,. \tag{5.34}$$

The interpretation of the instability associated with failure of (5.34) *under the symmetry assumptions* follows from the following statement, which is a particular case of a more general conclusion obtained in (Petryk, 1991b).

Interpretation of path instability under deformation-sensitive loading. *If (5.34) fails then either the inequality sign in (4.30) is reversed for some $\dot{\mathbf{e}}$, which is interpreted as dynamic instability of equilibrium (in the first approximation),*

or there exists another rate-solution $\dot{\mathbf{e}}$ consistent with $\dot{\mathbf{t}} = \dot{\mathbf{t}}^{\text{ext}}$ which represents constitutive branching energetically preferable to $\dot{\mathbf{e}}^0$ in the sense of a smaller value of $\ddot{E}$ •

Note that without the present symmetry restrictions, failure of (4.30) was not necessarily connected with an instability, and (4.30) was treated only as a condition *sufficient* for directional stability of equilibrium. Under the symmetry assumptions, the critical state associated with (4.30) is an *eigenstate* (cf. Hill, 1983) where a non-trivial solution exists to the rate-problem $\dot{\mathbf{t}} = -\mathbf{L}^* \cdot \dot{\mathbf{e}}$. Beyond the primary eigenstate on the fundamental path, the inequality sign in (4.30) is usually reversed for some $\dot{\mathbf{e}}$. Then it can be shown (cf. Petryk, 1991b) that there exists a direct path of departure from equilibrium that satisfies (in the first approximation and discretized) equations of motion and follows an arbitrarily small initial disturbance.

The range of validity of (5.34) along a smooth fundamental path usually terminates either at an eigenstate or at the point of primary constitutive branching. At the latter critical point, the minimum of $\mathcal{J}(\dot{\mathbf{e}})$ is still attained at $\dot{\mathbf{e}}^0$ but also at some other $\dot{\mathbf{e}}^* \neq \dot{\mathbf{e}}^0$, say. It follows that $\dot{\mathbf{e}}^*$ is a stationarity point of $\mathcal{J}(\dot{\mathbf{e}})$, and hence another solution to the rate-problem $\dot{\mathbf{t}} = \dot{\mathbf{t}}^{\text{ext}}$ under the deformation-sensitive loading (3.78). A typical point of primary constitutive branching can be determined from the following *linearized* condition with $\mathbf{L}^0 = \mathbf{L}(\dot{\mathbf{e}}^0)$:

$$(\mathbf{L}^0 + \mathbf{L}^*) \cdot \delta\dot{\mathbf{e}} = \mathbf{0} \qquad \text{for some } \delta\dot{\mathbf{e}} \neq 0 \tag{5.35}$$

related to the quadratic condition necessary for (5.34):

$$\delta\dot{\mathbf{e}} \cdot (\mathbf{L}^0 + \mathbf{L}^*) \cdot \delta\dot{\mathbf{e}} \geq 0 \qquad \text{for all } \delta\dot{\mathbf{e}}\,. \tag{5.36}$$

Constitutive branching may precede an eigenstate, as expected from Hill's (1959, 1978) general bifurcation theory. In turn, the spectrum of points of constitutive branching in incrementally nonlinear materials in the range where (5.34) fails at stable equilibrium is *continuous* along a strain path, as follows from the theorem on discretized bifurcation problems in incrementally nonlinear solids (Petryk, 1991b, Petryk and Thermann, 1992).

5.5 Energy Conditions for Material Instability: a Summary

The consequences of the energy condition (5.2) and of its reduced form (5.4) presented above may be interpreted as *necessity* conditions for path stability and for stability of equilibrium, respectively. By specifying the incremental energy consumption to second-order terms and determining the circumstances in which it fails to be minimized along the fundamental deformation path, the onset of material instability of a selected type can be estimated.

Six selected conditions, to be interpreted as *sufficient for material instability*, are reformulated below for the three distinguished types of instability. Each of these conditions implies

$$\ddot{E} < \ddot{E}^0 \qquad \text{for some admissible direction of a deformation increment}\,, \tag{5.37}$$

which represents, as the negation of (5.3), a unified form of sufficiency conditions for instability in the energy sense. The left column below refers to instability of equilibrium ($\dot{\mathbf{F}}^0 = \mathbf{0}$), the right column refers to instability in a deforming material ($\dot{\mathbf{F}}^0 \neq \mathbf{0}$). The interpretation of instability in the right column refers to the cases when equilibrium is stable.

Instability for internal rearrangements.

$$\sum_{K,L\in\mathcal{P}} g_{KL}\mathring{\gamma}_K\mathring{\gamma}_L < 0 \quad \text{for some } \mathring{\gamma}_K \geq 0 \qquad \textit{or} \qquad \sum_{K,L\in\mathcal{A}^0} g_{KL}\delta\mathring{\gamma}_K\delta\mathring{\gamma}_L < 0 \quad \text{for some } \delta\mathring{\gamma}_K$$

↓ ↓

snap-through at micro-level — non-uniqueness of active mechanisms

under assumptions $\boldsymbol{\lambda}^{\mathrm{P}}_K = \boldsymbol{\lambda}_K$ and $g_{KL} = g_{LK}$ (cf. Sect. 4.2 and 5.2).

Instability of uniform straining.

$$(\mathbf{g}\otimes\mathbf{n})\cdot\mathsf{C}(\mathbf{g}\otimes\mathbf{n})\cdot(\mathbf{g}\otimes\mathbf{n}) < 0 \quad \text{for some } \mathbf{g},\, \mathbf{n} \qquad \textit{or} \qquad (\mathbf{g}\otimes\mathbf{n})\cdot\mathsf{C}^0\cdot(\mathbf{g}\otimes\mathbf{n}) < 0 \quad \text{for some } \mathbf{g},\, \mathbf{n}$$

↓ ↓

dynamic localization of deformation — quasi-static bifurcation in a band

under assumption $\mathsf{C} = \mathsf{C}^{\mathrm{T}}$ (cf. Sect. 4.3 and 5.3).

Instability under deformation-sensitive loading.

$$\dot{\mathbf{e}}\cdot(\mathsf{L}(\dot{\mathbf{e}}) + \mathsf{L}^*)\cdot\dot{\mathbf{e}} < 0 \quad \text{for some } \dot{\mathbf{e}} \qquad \textit{or} \qquad \delta\dot{\mathbf{e}}\cdot(\mathsf{L}^0 + \mathsf{L}^*)\cdot\delta\dot{\mathbf{e}} < 0 \quad \text{for some } \delta\dot{\mathbf{e}}$$

↓ ↓

snap-through or collapse — constitutive branching

under assumptions $\mathsf{L} = \mathsf{L}^{\mathrm{T}}$ and $\mathsf{L}^* = \mathsf{L}^{*\mathrm{T}}$ (cf. Sect. 4.4 and 5.4).

6 Uniqueness in Multi-Mode Plasticity Without Symmetry

The framework of multi-mode plasticity from Sect. 3.3 is now adopted without any additional assumptions concerning symmetry or normality. Attention is restricted to the problem of incremental uniqueness of the material response under incremental loading of three different types.

In Sect. 6.1, the internal response under fully controlled deformation gradient is represented by plastic multipliers only. In Sect. 6.2, the material response has more freedom under the assumed deformation-sensitive loading, but the macroscopic strain-rate $\dot{\mathbf{e}}$, to be determined, is still restricted to remain uniform. In Sect. 6.3, the problem of bifurcation in a band is addressed in a novel way, without using an incremental stiffness moduli tensor other than purely elastic one.

The presentation below follows closely the author's recent work (Petryk, 2000), where an extension to partial kinematic constraints can also be found.

6.1 Constitutive Uniqueness

In this section, the problem of uniqueness of plastic multipliers is addressed under the assumption of fully controlled, uniform deformation gradient. Accordingly, the rate-problem (P1) is considered.

Denote by M the number of the elements of $\mathcal{P}$, i.e. the number of potentially active mechanisms ($f_K = 0$) in a given material state. Introduce the matrix notation, where vectors from the real M-dimensional space R^M and square $M \times M$ real matrices, both of the indices from $\mathcal{P}$, are denoted by square brackets. In that notation, (3.24) takes the following form

$$[-\dot{f}] = [g^*][\overset{\circ}{\gamma}] + [-b^*], \quad [\overset{\circ}{\gamma}] \geq [0], \quad [-\dot{f}] \geq [0], \quad [-\dot{f}]^{\mathrm{T}}[\overset{\circ}{\gamma}] = 0, \tag{6.1}$$

where inequalities between vectors should be read componentwise. (3.24) is recovered after the identification $[g^*] = (g_{KL})$ and $[b^*] = (\boldsymbol{\lambda}_K \cdot \dot{\mathbf{e}})$ with $K, L \in \mathcal{P}$.

The mathematical structure of the system (6.1) is known as *the linear complementarity problem* (Cottle at al., 1992). The fundamental correspondence between the incremental constitutive law of multi-mode plasticity in the small-strain framework and the linear complementarity problem was emphasized by Maier (1969).

The fundamental theorem concerning the existence and uniqueness of a solution to the linear complementarity problem (cf. Cottle at al., 1992) in the present notation reads

Theorem 6.1. *The problem (6.1) has a unique solution* $[\overset{\circ}{\gamma}]$ *for all vectors* $[b^*] \in R^M$ *if and only if*

$$\textit{all principal minors of the matrix } [g^*] \textit{ are positive} \;\bullet \tag{6.2}$$

Recall that for an index set $\mathcal{A} \subseteq \mathcal{P}$, the determinant of a principal submatrix (a_{KL}), $K, L \in \mathcal{A}$, of a matrix $[a]$ is called a principal minor of $[a]$.

Generally, (6.2) is less restrictive than the requirement (3.67) of positive definiteness of $[g^*]$. It is well known that if the matrix $[g^*]$ is symmetric then these conditions are equivalent to each other.

From the above Theorem we directly obtain the following

Corollary 6.1. *The problem (P1) has a unique solution for all* $\dot{\mathbf{e}}$ *if all principal minors of the matrix* (g_{KL}), $K, L \in \mathcal{P}$, *are positive* $\bullet$

This result is sharper than that given in (Hill and Rice, 1972), in view of the implication (6.2) $\Leftarrow$ (3.67) for $[g^*] = (g_{KL})$ with $K, L \in \mathcal{P}$.

The 'only if' part of Theorem 6.1 requires some care since it concerns *all* vectors $[b^*] \in R^M$. Vectors $[b^*] = (\boldsymbol{\lambda}_K \cdot \dot{\mathbf{e}})$ span the whole space R^M if and only if $\boldsymbol{\lambda}_K$ are linearly independent for $K \in \mathcal{P}$. From Theorem 6.1 we thus obtain:

Corollary 6.2. *Suppose that $\boldsymbol{\lambda}_K$, $K \in \mathcal{P}$, are linearly independent symmetric tensors. Then the problem (P1) has a unique solution for all $\dot{\mathbf{e}}$ if and only if all principal minors of the matrix (g_{KL}), $K, L \in \mathcal{P}$, are positive* •

Both requirements in Corollary 6.2 are measure-invariant conditions, as follows from (3.31).

The above statements may be regarded as extensions of those given by Maier (1969) to finite strain. The extension is formally immediate, however, the question of measure-invariance of the conditions did not appear in the small-strain framework.

6.2 Uniqueness Under Deformation-Sensitive Loading

In this section, the spin-insensitive form (3.78) of the loading condition is taken with $\mathbf{L}^*$ and $\dot{\mathbf{t}}^*$ regarded as known. In comparison with the problem discussed in the preceding section, $\dot{\mathbf{e}}$ represents an additional unknown. The quasi-static rate-problem is defined as follows:

Problem (P2). *In a given material state, find $(\dot{\mathbf{e}}, \overset{\circ}{\gamma}_K)$ such that*

$$\dot{\mathbf{t}} = \dot{\mathbf{t}}^{\text{ext}} \qquad \textit{at unconstrained strain-rate}\,, \tag{6.3}$$

where $\dot{\mathbf{t}}$ and $\dot{\mathbf{t}}^{\text{ext}}$ are defined by $(3.20)_1$ and (3.78), respectively, and the $\overset{\circ}{\gamma}_K$'s are subjected to the constraints (3.19) and (3.23) •

On eliminating $\dot{\mathbf{t}}$ between (6.3) and $(3.20)_1$ and substituting (3.78), we obtain

$$(\mathbf{L}^{\mathrm{e}} + \mathbf{L}^*) \cdot \dot{\mathbf{e}} = \dot{\mathbf{t}}^* + \sum_K \boldsymbol{\lambda}^{\mathrm{p}}_K \overset{\circ}{\gamma}_K\,. \tag{6.4}$$

It is worth noting that an identical equation is obtained when $(3.21)_1$ (multiplied by $\mathbf{L}^{\mathrm{e}}$) is used instead of $(3.20)_1$. The tensor $(\mathbf{L}^{\mathrm{e}} + \mathbf{L}^*)$ is related to the moduli tensors $\bar{\mathbf{L}}^{\mathrm{e}}$ and $\bar{\mathbf{L}}^*$ by the formula

$$\mathbf{L}^{\mathrm{e}} + \mathbf{L}^* = \mathbf{B}_{\mathrm{w}} \cdot (\bar{\mathbf{L}}^{\mathrm{e}} + \bar{\mathbf{L}}^*) \cdot \mathbf{A}^{-1} \tag{6.5}$$

which follows from (3.31) and (3.79).

Although $\mathbf{L}^{\mathrm{e}}$ or $\mathbf{L}^*$ taken separately may be singular, we shall assume that their sum $(\mathbf{L}^{\mathrm{e}} + \mathbf{L}^*)$ possesses a unique inverse $(\mathbf{L}^{\mathrm{e}} + \mathbf{L}^*)^{-1}$ when regarded as a linear operator acting in $\mathcal{G}$, cf. the definition of $\mathbf{A}^{-1}$. It is evident that this condition is necessary and sufficient for the uniqueness of a solution to the purely elastic rate-problem (6.4) with all $\overset{\circ}{\gamma}_K$ given in advance (e.g., set equal to zero), the consistency conditions (3.23) being then disregarded. An essential conclusion drawn from (6.5) is that the invertibility of $(\mathbf{L}^{\mathrm{e}} + \mathbf{L}^*)$ is a measure-invariant property.

The elimination of $\dot{\mathbf{e}}$ between (6.4) and $(3.20)_2$ now yields

$$\dot{f}_K = \boldsymbol{\lambda}_K \cdot (\mathbf{L}^{\mathrm{e}} + \mathbf{L}^*)^{-1} \cdot \dot{\mathbf{t}}^* - \sum_{L \in \mathcal{P}} (g_{KL} - \boldsymbol{\lambda}_K \cdot (\mathbf{L}^{\mathrm{e}} + \mathbf{L}^*)^{-1} \cdot \boldsymbol{\lambda}^{\mathrm{p}}_L) \overset{\circ}{\gamma}_L\,. \tag{6.6}$$

Introduce the notation

$$\mathbf{G}^* = (\mathbf{L}^{\mathrm{e}} + \mathbf{L}^*)^{-1}, \qquad \mathbf{H}^* = \mathbf{L}^* \cdot (\mathbf{L}^{\mathrm{e}} + \mathbf{L}^*)^{-1} \cdot \mathbf{L}^{\mathrm{e}} \tag{6.7}$$

and denote the expressions in (6.6) by

$$\begin{aligned} b^*_K &= \boldsymbol{\lambda}_K \cdot \mathbf{G}^* \cdot \dot{\mathbf{t}}^* & \qquad g^*_{KL} &= g_{KL} - \boldsymbol{\lambda}_K \cdot \mathbf{G}^* \cdot \boldsymbol{\lambda}^{\mathrm{p}}_L \\ &= \boldsymbol{\mu}_K \cdot \mathbf{L}^{\mathrm{e}} \cdot \mathbf{G}^* \cdot \dot{\mathbf{t}}^*, & &= h_{KL} + \boldsymbol{\mu}_K \cdot \mathbf{H}^* \cdot \boldsymbol{\mu}^{\mathrm{p}}_L . \end{aligned} \tag{6.8}$$

The dual expressions in (6.8) involving $\boldsymbol{\mu}_K$ follow from (3.22) and require the assumption of the invertibility of $\mathbf{L}^{\mathrm{e}}$. The equivalence of the two expressions for b^*_K is obvious, and the equivalence of the second pair can be verified as follows:

$$\begin{aligned} g^*_{KL} &= h_{KL} + \boldsymbol{\mu}_K \cdot \mathbf{L}^{\mathrm{e}} \cdot \boldsymbol{\mu}^{\mathrm{p}}_L - \boldsymbol{\mu}_K \cdot \mathbf{L}^{\mathrm{e}} \cdot \mathbf{G}^* \cdot \mathbf{L}^{\mathrm{e}} \cdot \boldsymbol{\mu}^{\mathrm{p}}_L \\ &= h_{KL} + \boldsymbol{\mu}_K \cdot ((\mathbf{L}^{\mathrm{e}} + \mathbf{L}^*) \cdot \mathbf{G}^* - \mathbf{L}^{\mathrm{e}} \cdot \mathbf{G}^*) \cdot \mathbf{L}^{\mathrm{e}} \cdot \boldsymbol{\mu}^{\mathrm{p}}_L \\ &= h_{KL} + \boldsymbol{\mu}_K \cdot \mathbf{H}^* \cdot \boldsymbol{\mu}^{\mathrm{p}}_L . \end{aligned} \tag{6.9}$$

The same expressions appear in the formula for $\dot{f}_K$ obtained by starting from $(3.21)_2$ instead of $(3.20)_2$.

Further transformations yield

$$\begin{aligned} \mathbf{H}^* &= \mathbf{L}^* \cdot (\mathbf{I}_{\mathrm{s}} + \mathbf{M}^{\mathrm{e}} \cdot \mathbf{L}^*)^{-1} & \qquad &\text{if } \mathbf{M}^{\mathrm{e}} = (\mathbf{L}^{\mathrm{e}})^{-1} \text{ exists,} \\ &= (\mathbf{M}^{\mathrm{e}} + \mathbf{M}^*)^{-1} & &\text{if } \mathbf{M}^* = (\mathbf{L}^*)^{-1} \text{ exists.} \end{aligned} \tag{6.10}$$

From (3.23), (6.6) and (6.8) we arrive at the following rate problem to be solved with respect to $\overset{\circ}{\gamma}_K$'s in a given material state :

$$\dot{f}_K = b^*_K - \sum_{L\in\mathcal{P}} g^*_{KL}\overset{\circ}{\gamma}_L , \quad \overset{\circ}{\gamma}_K \geq 0 , \quad \dot{f}_K \leq 0, \quad \dot{f}_K \overset{\circ}{\gamma}_K = 0 \qquad \text{for } K \in \mathcal{P}. \tag{6.11}$$

If this is done then a full solution to (6.3) is completed by using the expression for $\dot{\mathbf{e}}$:

$$\dot{\mathbf{e}} = \mathbf{G}^* \cdot \dot{\mathbf{t}}^* + \mathbf{G}^* \cdot \sum_K \boldsymbol{\lambda}^{\mathrm{p}}_K \overset{\circ}{\gamma}_K \tag{6.12}$$

obtained from (6.4).

In a given state of the material element and for given incremental constitutive laws (3.20) or (3.21) for the material and (3.78) for the loading, the quantities b^*_K and g^*_{KL} are known. The substitution of the transformation formulae (3.31) and (3.79) into (6.7) and (6.8) shows that both $J^{-1}b^*_K$ and $J^{-1}g^*_{KL}$ are measure-invariant, as expected from (3.29).

On applying Theorem 6.1 with the identification of $[g^*]$ and $[b^*]$ as above, we obtain (Petryk, 2000):

Theorem 6.2. *Suppose that the problem (P2) is uniquely solvable for purely elastic rates (so that the inverse tensor* $\mathbf{G}^*$ *defined by (6.7) exists).*
(i) If (6.2) holds then the problem (P2) has a unique solution for every $\dot{\mathbf{t}}^*$.
(ii) Suppose that $\boldsymbol{\lambda}_K$, $K \in \mathcal{P}$, *are linearly independent; then the problem (P2) has a unique solution for every* $\dot{\mathbf{t}}^*$ *if and only if (6.2) holds* •

6.3 Bounds to Bifurcation in a Band

Exclusion conditions. We return to the problem (5.20) of bifurcation in a band of a unit normal $\mathbf{n}$, which is now addressed in a different way. To simplify some formulae, the current configuration is taken as reference, so that $\mathbf{F}$ coincides instantaneously with the unit tensor $\mathbf{I}$ (and $\mathsf{A} = \mathsf{B}_{\mathrm{w}} = \mathsf{I}_{\mathrm{s}}$). The fundamental incremental solution is defined by $(\dot{\mathbf{F}}^*, \mathring{\gamma}^*_K)$ uniform in space; note that a superscript '*' is used in place of previous '0'. The problem of bifurcation in a band is formulated as follows.

Problem (P3). *In a given material state and for given* $(\dot{\mathbf{F}}^*, \mathring{\gamma}^*_K)$ *and* $\mathbf{n}$, *find* $(\mathbf{g}, \mathring{\gamma}_K)$ *such that*

$$(\dot{\mathbf{S}} - \dot{\mathbf{S}}^*)\mathbf{n} = \mathbf{0} \qquad \textit{and} \qquad \dot{\mathbf{F}} - \dot{\mathbf{F}}^* = \mathbf{g} \otimes \mathbf{n}\,, \tag{6.13}$$

where $\dot{\mathbf{S}}$ *and* $\dot{\mathbf{S}}^*$ *are constitutively related through (3.33) to* $(\dot{\mathbf{F}}, \mathring{\gamma}_K)$ *and* $(\dot{\mathbf{F}}^*, \mathring{\gamma}^*_K)$, *respectively, and the* $\mathring{\gamma}_K$*'s and* $\mathring{\gamma}^*_K$*'s are subjected to the constraints (3.19) and (3.23)* **.**

The problem (P3) admits a trivial solution $\mathbf{g} = \mathbf{0}$, $\mathring{\gamma}_K = \mathring{\gamma}^*_K$. Existence of a non-trivial solution to (P3) is identified with bifurcation in a band.

By combining (6.13) and (3.33) and substituting $\mathbf{F} = \mathbf{I}$, we obtain

$$\mathbf{Q}^{\mathrm{e}}\mathbf{g} = \sum_K (\boldsymbol{\lambda}^{\mathrm{p}}_K \mathbf{n})(\mathring{\gamma}_K - \mathring{\gamma}^*_K)\,, \tag{6.14}$$

where the second-order tensor $\mathbf{Q}^{\mathrm{e}}$ is the acoustic tensor (dependent on $\mathbf{n}$) for the purely elastic constitutive branch,

$$\mathbf{Q}^{\mathrm{e}} = \mathbf{n}\bar{\mathbf{L}}^{\mathrm{e}}\mathbf{n} - \frac{1}{2}(\mathbf{n}\boldsymbol{\sigma} \otimes \mathbf{n} + \mathbf{n} \otimes \boldsymbol{\sigma}\mathbf{n} + \boldsymbol{\sigma} - (\mathbf{n}\boldsymbol{\sigma}\mathbf{n})\mathbf{I})\,, \qquad Q^{\mathrm{e}}_{ik} = C^{\mathrm{e}}_{ipkq} n_p n_q\,. \tag{6.15}$$

Suppose that

$$\det \mathbf{Q}^{\mathrm{e}} \neq 0, \tag{6.16}$$

which is the classical ellipticity condition for elastic moduli which excludes a non-trivial solution to the problem (P3) posed for purely elastic rates. Then $\mathbf{g}$ can be expressed in terms of unknown $\mathring{\gamma}_K$ from (6.14). On substituting that $\mathbf{g}$ into (6.13) and next into $(3.33)_2$ with $\mathbf{F} = \mathbf{I}$, we obtain

$$f_K = \boldsymbol{\lambda}_K \cdot \dot{\mathbf{F}}^* - (\boldsymbol{\lambda}_K \mathbf{n})(\mathbf{Q}^{\mathrm{e}})^{-1} \sum_L (\boldsymbol{\lambda}^{\mathrm{p}}_L \mathbf{n})\mathring{\gamma}^*_L - \sum_L g^{\mathrm{band}}_{KL} \mathring{\gamma}_L\,, \tag{6.17}$$

where

$$\begin{aligned} g^{\mathrm{band}}_{KL} &= g_{KL} - \mathbf{n}\boldsymbol{\lambda}_K (\mathbf{Q}^{\mathrm{e}})^{-1} \boldsymbol{\lambda}^{\mathrm{p}}_L \mathbf{n} \\ &= h_{KL} + \boldsymbol{\mu}_K \cdot \mathsf{L}^{\mathrm{e}} \cdot \boldsymbol{\mu}^{\mathrm{p}}_L - (\boldsymbol{\mu}_K \cdot \mathsf{L}^{\mathrm{e}}\mathbf{n})(\mathbf{Q}^{\mathrm{e}})^{-1}(\mathbf{n}\mathsf{L}^{\mathrm{e}} \cdot \boldsymbol{\mu}^{\mathrm{p}}_L)\,. \end{aligned} \tag{6.18}$$

All the quantities that appear in (6.18), and also the set $\mathcal{P}$ of the indices of potentially active mechanisms, are known in the material state in which the problem (P3) is posed. Hence, the matrix $[g^{\mathrm{band}}]$ is known and independent on $(\dot{\mathbf{F}}^*, \mathring{\gamma}^*_K)$.

By applying Theorem 6.1, we arrive at the following conclusion (Petryk, 2000).

Exclusion conditions for bifurcation in a band. *Problem (P3) admits only a trivial solution if simultaneously :*
(a) *the elastic moduli satisfy the ellipticity condition* (6.16),
(b) *all principal minors of the matrix* $[g^{\mathrm{band}}]$ *defined by* (6.18) *with* $K, L \in \mathcal{P}$ *are positive* •

It is pointed out that (a) and (b) jointly exclude the bifurcation in a band for any possible unloading within or outside the band at the bifurcation instant.

For convenience, the condition (b) is rewritten as

$$\det(g_{\mathcal{A}}^{\mathrm{band}}) > 0 \qquad \text{for all } \mathcal{A} \subseteq \mathcal{P}\,, \tag{6.19}$$

where $(g_{\mathcal{A}}^{\mathrm{band}})$ stands for a principal submatrix of $[g^{\mathrm{band}}]$ with indices from $\mathcal{A} \subseteq \mathcal{P}$.

Lower and upper bounds for the primary bifurcation. Consider now a (rather typical) homogeneous deformation path $(\dot{\mathbf{F}}^{*}, \overset{\circ}{\gamma}{}_{K}^{*})(t)$, parameterized by a time-like scalar $t \geq t_0$, along which the following conditions are satisfied:

(i) the condition (a) above holds true at all t,
(ii) the set $\mathcal{P}$ of the indices of potentially active mechanisms is constant,
(iii) $[\overset{\circ}{\gamma}{}^{*}] > [0]$,
(iv) the matrix $[g^{\mathrm{band}}]$ for any $\mathbf{n}$ varies continuously with t,
(v) the condition (b) above is satisfied for all $\mathbf{n}$ at an initial instant $t = t_0$.

The following bounds to the instant of primary bifurcation in a band are obtained (Petryk, 2000).

Bounds for the primary bifurcation. *Along a deformation path that satisfies the conditions (i)-(v) above, the instant* t^c *of the primary bifurcation (in the first-order rates) in a band is bounded by*

$$t_{\mathcal{A}} \leq t^c \leq t_{\mathcal{P}}\,, \tag{6.20}$$

where the lower bound is determined from

$$t_{\mathcal{A}} = \min t \ : \quad \det(g_{\mathcal{A}}^{\mathrm{band}})(t) = 0 \quad \textit{for some } \mathcal{A} \subseteq \mathcal{P} \textit{ and some } \mathbf{n}\,, \tag{6.21}$$

and the upper bound from

$$t_{\mathcal{P}} = \min t \ : \quad \det[g^{\mathrm{band}}](t) = 0 \quad \textit{for some } \mathbf{n}\,. \tag{6.22}$$

The proof of the lower bound follows readily from the exclusion condition (6.19) by the continuity argument applied to $\det(g_{\mathcal{A}}^{\mathrm{band}})$. The upper bound is obtained as a particular bifurcation point at which none of the potentially active mechanisms undergoes unloading; an explicit proof is given in the reference.

There is a variety of possible applications of the above results to specific material models.

If $M = 1$ (only one mechanism of a given index K is considered) then the lower and upper bounds coincide. If $\mathbf{M}^{\mathrm{e}} = (\mathbf{L}^{\mathrm{e}})^{-1}$ exists for the measures chosen then the uniqueness range

$t < t^c$ can be expressed from (6.18) as $h_{KK} > h_K^c$. Here, h_{KK} is the effective hardening modulus for the mechanism, and its critical value h_K^c reached at $t = t^c$ reads

$$
\begin{aligned}
h_K^c &= -\boldsymbol{\lambda}_K \cdot \mathbf{M}^e \cdot \boldsymbol{\lambda}_K^p + \max_{\mathbf{n}} \left(\mathbf{n}\boldsymbol{\lambda}_K (\mathbf{Q}^e)^{-1} \boldsymbol{\lambda}_K^p \mathbf{n} \right) \\
&= -\boldsymbol{\mu}_K \cdot \mathbf{L}^e \cdot \boldsymbol{\mu}_K^p + \max_{\mathbf{n}} \left((\boldsymbol{\mu}_K \cdot \mathbf{L}^e \mathbf{n})(\mathbf{Q}^e)^{-1} (\mathbf{n}\mathbf{L}^e \cdot \boldsymbol{\mu}_K^p) \right).
\end{aligned}
\tag{6.23}
$$

The formula (6.23) is equivalent to that derived by Rice (1977) and Asaro and Rice (1977) for the linearized problem without unloading. The validity of the formula (6.23) for the critical hardening modulus was extended to the case of possible unloading by Rice and Rudnicki (1980); cf. also (Raniecki and Bruhns, 1981, Bigoni and Zaccaria, 1992).

If $M \geq 2$ then the upper bound (6.22) can be reduced to the formula given by Peirce (1983) for finite strain elastoplasticity (of single crystals) with the normality flow rule, and without the restriction to normality in (Rizzi at al., 1996) and (Steinmann, 1996) in the small strain and finite strain framework, respectively. In those investigations, the *linearized* condition (5.21) for the bifurcation in a band has been taken as a starting point, which was not used above. The resulting coincidence of the instants of $\det[g^{\text{band}}] = 0$ and $\det \mathbf{Q} = 0$ is not unexpected since both conditions, although obtained on different routes, correspond to the same bifurcation in a band *without unloading*.

It is pointed out that the exclusion condition (6.19) and the related lower bound (6.21), for the bifurcation in a band in the incrementally piecewise-linear material with multiple ($M \geq 2$) mechanisms of time-independent inelastic deformation, are valid for any possible unloading at the bifurcation instant.

If $M = 2$ and $h_{KK} > h_K^c$ for each mechanism at every instant then (6.19) reduces to $\det[g^{\text{band}}] > 0$, and the bounds (6.21) and (6.22) coincide. It follows that in that case and along a deformation path that satisfies (i)-(v) above, the condition (6.22) defines the exact instant of the primary bifurcation without unloading, any earlier bifurcation with unloading being excluded. This particular result is of interest for certain material models being in use (double slip, classical plasticity coupled with damage or phase transformation, etc.).

Finally, consider the case of symmetry $g_{KL}^{\text{band}} = g_{LK}^{\text{band}}$. Then the condition (6.19) coincides with the requirement (3.67) of positive definiteness of $[g^{\text{band}}]$. Under the assumptions (v) and (iv) above, this requirement is satisfied up to the critical instant at which $[g^{\text{band}}]$ becomes positive *semi*-definite, i.e. singular. But this marks the upper bound. We have thus the corollary that if $g_{KL}^{\text{band}} = g_{LK}^{\text{band}}$ then the above lower and upper bounds coincide for arbitrary $M \geq 1$, and the condition (6.22) defines the exact instant of the primary bifurcation in a band. This corollary follows also from the relative convexity property (3.65) with the comparison moduli C^{L} defined by (3.35) with $\mathcal{A} = \mathcal{P}$.

Appendix

We will use below the formalism with internal state variables $\boldsymbol{\alpha}$ defined in a fixed reference configuration and assume that a small parameter is provided by a norm $||\Delta\boldsymbol{\alpha}||$ in a Hilbert space containing all $\boldsymbol{\alpha}$. We shall consider a class of short piecewise-smooth paths in the $\boldsymbol{\alpha}$-space, leading from $\boldsymbol{\alpha}_0 = \boldsymbol{\alpha}(t_0)$ to a given $\boldsymbol{\alpha}_0 + \Delta\boldsymbol{\alpha} = \boldsymbol{\alpha}(t)$, whose length is of order $||\Delta\boldsymbol{\alpha}||$. The *virtual* dissipation $\Delta\mathcal{D}$ along such a path, defined by time integration of the dissipation function (3.57), can be decomposed as follows:

$$\Delta\mathcal{D} \equiv \int_{t_0}^{t} D(\boldsymbol{\alpha}(\theta), \dot{\boldsymbol{\alpha}}(\theta))\, d\theta = \Delta_1\mathcal{D} + \Delta_2\mathcal{D} + o((\|\Delta\boldsymbol{\alpha}\|)^2)\,, \tag{A.1}$$

where

$$\Delta_1\mathcal{D} = \int_{t_0}^{t} D(\boldsymbol{\alpha}_0, \dot{\boldsymbol{\alpha}}(\theta))\, d\theta \tag{A.2}$$

is the *path-dependent* first-order dissipation, while $\Delta_2\mathcal{D}$ defined by

$$\Delta_2\mathcal{D} = \int_{t_0}^{t} D_1(\hat{\boldsymbol{\alpha}}(\theta), \dot{\boldsymbol{\alpha}}(\theta))\, d\theta\,, \qquad \hat{\boldsymbol{\alpha}} = \boldsymbol{\alpha} - \boldsymbol{\alpha}_0\,, \qquad D_1(\hat{\boldsymbol{\alpha}}, \dot{\boldsymbol{\alpha}}) \equiv \hat{\boldsymbol{\alpha}} \cdot D_{,\boldsymbol{\alpha}}(\boldsymbol{\alpha}_0, \dot{\boldsymbol{\alpha}})\,, \tag{A.3}$$

is the *path-dependent* second-order dissipation, an analog of the path-dependent second-order work, cf. (Petryk, 1991a). Note that $D_1(\hat{\boldsymbol{\alpha}}, r\dot{\boldsymbol{\alpha}}) = rD_1(\hat{\boldsymbol{\alpha}}, \dot{\boldsymbol{\alpha}})$ for every $r > 0$ and that $D_1(\hat{\boldsymbol{\alpha}}, \dot{\boldsymbol{\alpha}})$ is linear with respect to $\hat{\boldsymbol{\alpha}}$. Recall that $D_{,\boldsymbol{\alpha}}$ is a Fréchet derivative of $D(\boldsymbol{\alpha}, \dot{\boldsymbol{\alpha}})$ with respect to $\boldsymbol{\alpha}$.

Evaluation of the first- and second-order dissipation for a class of *direct* paths in $\boldsymbol{\alpha}$-space reduces to the integration in (A.2) and (A.3) along a straight path, which on account of (3.57) yields

$$\Delta_1\mathcal{D}^{\mathrm{d}} = D(\boldsymbol{\alpha}_0, \Delta\boldsymbol{\alpha})\,, \qquad \Delta_2\mathcal{D}^{\mathrm{d}} = \frac{1}{2} D_1(\Delta\boldsymbol{\alpha}, \Delta\boldsymbol{\alpha})\,. \tag{A.4}$$

The following simple lemma holds (cf. Nguyen, 1993).

Lemma A1. *Under the assumptions (3.57), the first-order dissipation (A.2) is always minimized on a straight path, that is,*

$$\Delta_1\mathcal{D} \geq \Delta_1\mathcal{D}^{\mathrm{d}} \tag{A.5}$$

if and only if

$$D(\boldsymbol{\alpha}, \dot{\boldsymbol{\alpha}}) \text{ is convex in } \dot{\boldsymbol{\alpha}}\ \bullet \tag{A.6}$$

Proof. (*Sufficiency*). Suppose that (A.6) holds. Then the convexity and homogeneity of $D(\boldsymbol{\alpha}, \dot{\boldsymbol{\alpha}})$ with respect to $\dot{\boldsymbol{\alpha}}$ imply

$$D(\boldsymbol{\alpha}, \dot{\boldsymbol{\alpha}}) + D(\boldsymbol{\alpha}, \dot{\boldsymbol{\alpha}}^*) \geq 2D(\boldsymbol{\alpha}, \frac{1}{2}(\dot{\boldsymbol{\alpha}} + \dot{\boldsymbol{\alpha}}^*)) = D(\boldsymbol{\alpha}, \dot{\boldsymbol{\alpha}} + \dot{\boldsymbol{\alpha}}^*) \qquad \text{for all } \dot{\boldsymbol{\alpha}}, \dot{\boldsymbol{\alpha}}^*, \boldsymbol{\alpha}\,. \tag{A.7}$$

This can be extended to any finite or infinite sum of any $\dot{\boldsymbol{\alpha}}$'s, which implies

$$\int_{t_0}^{t} D(\boldsymbol{\alpha}_0, \dot{\boldsymbol{\alpha}}(\theta))\, d\theta \geq D(\boldsymbol{\alpha}_0, \int_{t_0}^{t} \dot{\boldsymbol{\alpha}}(\theta)\, d\theta) = D(\boldsymbol{\alpha}_0, \Delta\boldsymbol{\alpha})\,. \tag{A.8}$$

This proves validity of (A.5).

(*Necessity*). Suppose that (A.5) always holds. This means that (A.8) holds for any $(\boldsymbol{\alpha}_0, \Delta\boldsymbol{\alpha})$, which implies (A.7) as a particular case of (A.8) for a path consisting of two arbitrary straight segments. This implies (A.6), which completes the proof of Lemma A1.

For the dissipation evaluated to *second* order to be minimized on a straight path, the convexity property (A.6) alone is not sufficient. Denote by $\mathcal{G}_0$ a material state corresponding to $\boldsymbol{\alpha}_0$. If the path in (A.1) is chosen such that $\dot{\boldsymbol{\alpha}}(\theta) \in \mathcal{L}(\mathcal{G}_0)$ at each θ (cf. (3.59)) then the difference in the first-order dissipation vanishes, while the difference in the second-order dissipation may be of either sign in general. However, we have the following lemma (cf. Petryk, 1993b).

Lemma A2. *Under the assumptions (3.57), (A.6), (3.59) and (3.60), the dissipation (A.1) evaluated to second-order is always minimized on a straight path, that is,*

$$\Delta\mathcal{D} \geq \Delta_1\mathcal{D}^{\rm d} + \Delta_2\mathcal{D}^{\rm d} + o((\|\Delta\boldsymbol{\alpha}\|)^2) \tag{A.9}$$

if and only if

$$D_1(\hat{\boldsymbol{\alpha}},\dot{\boldsymbol{\alpha}}) = D_1(\dot{\boldsymbol{\alpha}},\hat{\boldsymbol{\alpha}}) \qquad \textit{for all } \dot{\boldsymbol{\alpha}},\hat{\boldsymbol{\alpha}} \in \mathcal{L}(\mathcal{G}_0)\,. \tag{A.10}$$

Proof. The integrand in (A.3) is transformed as follows

$$\begin{aligned} 2D_1(\hat{\boldsymbol{\alpha}}(\theta),\dot{\boldsymbol{\alpha}}(\theta)) &= D_1(\hat{\boldsymbol{\alpha}}(\theta),\dot{\boldsymbol{\alpha}}(\theta)) + \textstyle\int_{t_0}^{\theta} D_1(\dot{\boldsymbol{\alpha}}(\zeta),\dot{\boldsymbol{\alpha}}(\theta))\,d\zeta \\ &= D_1(\hat{\boldsymbol{\alpha}}(\theta),\dot{\boldsymbol{\alpha}}(\theta)) + \textstyle\int_{t_0}^{\theta} D_1(\dot{\boldsymbol{\alpha}}(\theta),\dot{\boldsymbol{\alpha}}(\zeta))\,d\zeta + 2D_1^{\rm a}(\theta) \\ &= \tfrac{d}{d\theta}\textstyle\int_{t_0}^{\theta} D_1(\hat{\boldsymbol{\alpha}}(\theta),\dot{\boldsymbol{\alpha}}(\zeta))\,d\zeta + 2D_1^{\rm a}(\theta)\,, \end{aligned} \tag{A.11}$$

where

$$D_1^{\rm a}(\theta) = \int_{t_0}^{\theta} \frac{1}{2}\left(D_1(\dot{\boldsymbol{\alpha}}(\zeta),\dot{\boldsymbol{\alpha}}(\theta)) - D_1(\dot{\boldsymbol{\alpha}}(\theta),\dot{\boldsymbol{\alpha}}(\zeta))\right) d\zeta\,. \tag{A.12}$$

On substituting (A.11) into (A.3) we obtain

$$\Delta_2\mathcal{D} = \int_{t_0}^{t} \left(D_1(\frac{1}{2}\Delta\boldsymbol{\alpha},\dot{\boldsymbol{\alpha}}(\theta)) + D_1^{\rm a}(\theta)\right) d\theta\,. \tag{A.13}$$

By combining this with the expression for $\Delta_1\mathcal{D}$, from (A.1) we arrive at

$$\Delta\mathcal{D} = \int_{t_0}^{t} D(\boldsymbol{\alpha}_0 + \frac{1}{2}\Delta\boldsymbol{\alpha},\dot{\boldsymbol{\alpha}}(\theta))\,d\theta + \int_{t_0}^{t} D_1^{\rm a}(\theta)\,d\theta + o((\|\Delta\boldsymbol{\alpha}\|)^2)\,. \tag{A.14}$$

(*Sufficiency*). Suppose that (A.6) and (A.10) hold. If $\dot{\boldsymbol{\alpha}}(\theta) \notin \mathcal{L}(\mathcal{G}_0)$ for some θ along a piecewise smooth path then (A.5) holds with strict inequality, implying (A.9). It remains thus to consider the case $\dot{\boldsymbol{\alpha}}(\theta) \in \mathcal{L}(\mathcal{G}_0)$ at every θ. On using (A.8) with $\boldsymbol{\alpha}_0 + \frac{1}{2}\Delta\boldsymbol{\alpha}$ in place of $\boldsymbol{\alpha}_0$ and (A.14) with $D_1^{\rm a} = 0$, it follows that

$$\begin{aligned} \Delta\mathcal{D} &= \textstyle\int_{t_0}^{t} D(\boldsymbol{\alpha}_0 + \frac{1}{2}\Delta\boldsymbol{\alpha},\dot{\boldsymbol{\alpha}}(\theta))\,d\theta + o((\|\Delta\boldsymbol{\alpha}\|)^2) \\ &\geq D(\boldsymbol{\alpha}_0 + \tfrac{1}{2}\Delta\boldsymbol{\alpha},\Delta\boldsymbol{\alpha}) + o((\|\Delta\boldsymbol{\alpha}\|)^2) \\ &= D(\boldsymbol{\alpha}_0,\Delta\boldsymbol{\alpha}) + \tfrac{1}{2}D_1(\Delta\boldsymbol{\alpha},\Delta\boldsymbol{\alpha}) + o((\|\Delta\boldsymbol{\alpha}\|)^2)\,, \end{aligned} \tag{A.15}$$

which shows validity of (A.9).

(*Necessity*). Suppose that (A.9) holds universally. Take $\Delta\boldsymbol{\alpha}$ proportional to $\dot{\boldsymbol{\alpha}}^{\rm a}$ in (3.60) corresponding to fully active loading in $\mathcal{G}_0$, and consider the class of paths in (A.1) such that $\dot{\boldsymbol{\alpha}}(\theta) \in \mathcal{L}(\mathcal{G}_0)$ at every θ. Then (A.5) holds with equality sign by (3.59), and D_1 in (A.13) becomes linear in $\dot{\boldsymbol{\alpha}}(\theta) \in \mathcal{L}(\mathcal{G}_0)$ as follows from (3.60). Hence, from (A.1),

$$\begin{aligned} \Delta\mathcal{D} &= D(\boldsymbol{\alpha}_0,\Delta\boldsymbol{\alpha}) + \tfrac{1}{2}D_1(\Delta\boldsymbol{\alpha},\Delta\boldsymbol{\alpha}) + o((\|\Delta\boldsymbol{\alpha}\|)^2) \\ &\quad + \textstyle\int_{t_0}^{t} D_1^{\rm a}(\theta)\,d\theta \qquad \text{if } \dot{\boldsymbol{\alpha}}(\theta) \in \mathcal{L}(\mathcal{G}_0)\,. \end{aligned} \tag{A.16}$$

The integral of $D_1^{\rm a}$ can be of second order and of either sign unless (A.10) holds, e.g., along a path consisting of two arbitrary straight segments directed within $\mathcal{L}(\mathcal{G}_0)$. It follows that (A.10) is necessary for (A.9), which completes the proof of Lemma A2.

Note that the assumption (3.60) is not used in the sufficiency part of the proof.

References

Asaro, R.J. (1983). Micromechanics of crystals and polycrystals. In *Advances in Applied Mechanics, Vol.* 23. New York: Acad. Press. 1-115.

Asaro, R.J. and Rice, J.R. (1977). Strain localization in ductile single crystals. *J. Mech. Phys. Solids* 25:309-338.

Ball, J.M., and James, R.D. (1987). Fine phase mixtures as minimizers of energy. *Arch. Rat. Mech. Anal.* 100:13-52.

Bigoni, D., and Loret, B. (1999). Effects of elastic anisotropy on strain localization and flutter instability in plastic solids. *J. Mech. Phys. Solids* 47:1409-1436.

Bigoni, D., and Zaccaria, D. (1992). Loss of strong ellipticity in non-associative elastoplasticity. *J. Mech. Phys. Solids* 40:1313-1331.

Biot, M.A. (1965). *Mechanics of Incremental Deformations.* New York: John Wiley & Sons.

de Borst, R., and van der Giessen, E., eds. (1998). *Material Instabilities in Solids.* Chichester: John Wiley & Sons.

Chambon, R., Crochepeyre, S., and J. Desrues (2000). Localization criteria for non-linear constitutive equations of geomaterials. *Mech. Cohesive Frictional Mat.* 5:61-82.

Cottle, R.W., Pang, J.-S., and Stone, R.E. (1992). *The Linear Complementarity Problem.* San Diego: Acad. Press.

Drucker, D.C. (1950). Some implications of work hardening and ideal plasticity. *Q. Appl. Math.* 7:411-418.

Drucker, D.C. (1964). On the postulate of stability of material in the mechanics of continua. *J. Mécanique* 3:235-249.

Drugan, W.J. (1998). Thermodynamic equivalence of steady-state shocks and smooth waves in general media; applications to elastic-plastic shocks and dynamic fracture. *J. Mech. Phys. Solids* 46:313-336.

Drugan, W.J., and Shen, Y. (1990). Finite deformation analysis of restrictions on moving strong discontinuity surfaces in elastic-plastic materials: quasi-static and dynamic deformations. *J. Mech. Phys. Solids* 38:553-574.

Eshelby, J.D. (1957). The determination of the elastic field of an ellipsoidal inclusion, and related problems. *Proc. Roy. Soc. London* A 241:376-396.

Fedelich, B., and Ehrlaher, A. (1997). An analysis of stability of equilibrium and of quasi-static transformations on the basis of the dissipation function. *Eur. J. Mech. A/Solids* 16:833-855.

Fleck, N.A., and Hutchinson, J.W. (1997). Strain gradient plasticity. In *Advances in Applied Mechanics, Vol.* 33. New York: Acad. Press. 295-361.

Franciosi, P., and Zaoui, A. (1991). Crystal hardening and the issue of uniqueness, *Int. J. Plasticity* 7:295-311.

Graves, L. (1939). The Weierstrass condition for multiple integral variation problem. *Duke Math. J.* 5:656-660.

Hadamard, J. (1903). *Leçons sur la Propagation des Ondes et les Equations de l'Hydro-dynamique.* Paris: Hermann.

Havner, K.S. (1992). *Finite Plastic Deformation of Crystalline Solids.* Cambridge: University Press.

Hill, R. (1958). A general theory of uniqueness and stability in elastic-plastic solids. *J. Mech. Phys. Solids* 6:236-249.

Hill, R. (1959). Some basic principles in the mechanics of solids without a natural time. *J. Mech. Phys. Solids* 7:209-225.

Hill, R. (1962). Acceleration waves in solids. *J. Mech. Phys. Solids* 10:1-16.

Hill, R. (1967a). The essential structure of constitutive laws for metal composites and polycrystals. *J. Mech. Phys. Solids* 15:79-95.

Hill, R. (1967b). Eigenmodal deformations in elastic/plastic continua. *J. Mech. Phys. Solids* 15:371-386.

Hill, R. (1967c). On the classical constitutive relations for elastic/plastic solids. In Broberg, B. at al., eds., *Recent Progress in Applied Mechanics, The Folke Odkvist Volume*. Stockholm: Almqvist & Wiksell. 241-249.

Hill, R. (1968). On constitutive inequalities for simple materials. *J. Mech. Phys. Solids* 16:229-242, 315-322.

Hill, R. (1978). Aspects of invariance in solids mechanics. In *Advances in Applied Mechanics, Vol.* 18. New York: Acad. Press. 1-75.

Hill, R. (1983). On intrinsic eigenstates in plasticity with generalized variables. *Math. Proc. Camb. Phil. Soc.* 93:177-189.

Hill, R., and Rice, J.R. (1972). Constitutive analysis of elastic- plastic crystals at arbitrary strain. *J. Mech. Phys. Solids* 20:401-413.

Hill, R., and Rice, J.R. (1973). Elastic potentials and the structure of inelastic constitutive laws. *SIAM J. Appl. Math.* 25:448-461.

Hutchinson, J.W. (1970). Elastic-plastic behaviour of polycrystalline metals and composites. *Proc. Roy. Soc. Lond.* A 319, 247-272.

Hutchinson, J.W., and Tvergaard, V. (1981). Shear band formation in plane strain. Int. J. Solids Structures 17:451-470.

de Groot, S.R., and Mazur, P. (1962). *Non-equilibrium Thermodynamics*. Amsterdam: North Holland.

Kolymbas, D. (1981). Bifurcation analysis for sand samples with a non-linear constitutive equation. *Ing.-Archiv* 50:131-140.

Loret, B. (1992). Does deviation from deviatoric associativity lead to the onset of flutter instability ? *J. Mech. Phys. Solids* 40:1363-1375.

Maier, G. (1969). "Linear" flow-laws of elastoplasticity: a unified general approach. *Rend. Acc. Naz. Lincei, Ser. VIII*, XLVII:266-276.

Mandel, J. (1966). Conditions de stabilité et postulat de Drucker. In Kravtchenko, J., ed., *Rheology and Soil Mechanics*. Berlin: Springer. 58-67.

Needleman, A., and Tvergaard, V. (1992). Analyses of plastic flow localization in metals. *Appl. Mech. Rev.* 45: no 3, part 2, S3-S18.

Nguyen, Q.S. (1993). Bifurcation and stability of time-independent standard dissipative systems. In Nguyen, Q.S., ed., *CISM Lecture Notes No.* 327. Wien - New York: Springer. 45-94.

Nguyen, Q.S., and Petryk, H. (1991). A constitutive inequality for time-independent dissipative solids. *C. R. Acad. Sci. Paris, Serie II* 312:7-12.

Nguyen, Q.S., and Radenkovic, D. (1976). Stability of equilibrium in elastic plastic solids. In *Lecture Notes in Mathematics, Vol.* 503. Berlin: Springer. 403-414.

Ogden, R.W. (1984a). On Eulerian and Lagrangean objectivity in continuum mechanics. *Arch. Mech.* 36:207-218.

Ogden, R.W. (1984b). *Non-linear Elastic Deformations*. Chichester: Ellis Horwood.

Peirce, D. (1983). Shear band bifurcations in ductile single crystals. *J. Mech. Phys. Solids* 31:133-153.

Petryk, H. (1982). A consistent energy approach to defining stability of plastic deformation processes. In Schroeder, F.H., ed., *Stability in the Mechanics of Continua, Proc. IUTAM Symp. Nümbrecht* 1981. Berlin - Heidelberg: Springer. 262-272.

Petryk, H. (1985a). On energy criteria of plastic instability. In *Plastic Instability, Proc. Considère Memorial*. Paris: Ecole Nat. Ponts Chauss. 215-226.

Petryk, H. (1985b). On stability and symmetry conditions in time-independent plasticity. *Arch. Mech.* 37:503-520.

Petryk, H. (1989). On constitutive inequalities and bifurcation in elastic-plastic solids with a yield-surface vertex. *J. Mech. Phys. Solids* 37:265-291.

Petryk, H. (1991a). On the second-order work in plasticity, *Arch. Mech.* 43:377-397.

Petryk, H. (1991b). The energy criteria of instability in time-independent inelastic solids. *Arch. Mech.* 43:519-545.

Petryk, H. (1992a). Material instability and strain-rate discontinuities in incrementally nonlinear continua. *J. Mech. Phys. Solids* 40:1227-1250.

Petryk, H. (1992b). On stability of time-independent materials at finite strain. In Besdo, D., and Stein, E., eds., *Finite Inelastic Deformations - Theory and Applications*. Berlin - Heidelberg: Springer. 189-196.

Petryk, H. (1993a). Theory of bifurcation and instability in time-independent plasticity. In Nguyen, Q.S., ed., *CISM Lecture Notes No.* 327. Wien - New York: Springer. 95-152.

Petryk, H. (1993b). Stability and constitutive inequalities in plasticity. In Muschik, W., ed., *CISM Lecture Notes No.* 336. Wien - New York: Springer. 259-329.

Petryk, H. (1995). Thermodynamic stability of equilibrium in plasticity. *J. Non-Equil. Thermodyn.* 20:132-149.

Petryk, H. (1997a). Instability of plastic deformation processes. In Tatsumi, T., et al., eds., *Theoretical and Applied Mechanics 1996, Proc. XIXth IUTAM Congress, Kyoto*. Amsterdam: Elsevier. 497-516.

Petryk, H. (1997b). Plastic instability: criteria and computational approaches. *Arch. Comp. Meth. Engng.* 4:111-151.

Petryk, H. (1999a). On the micro-macro transition and hardening moduli in plasticity. In Bruhns, O.T., and Stein, E., eds., *Proc. IUTAM Symp. Micro- and Macrostructural Aspects of Thermoplasticity, Bochum 1997*. Dordrecht: Kluwer. 219-230.

Petryk, H. (1999b). (Manuscript in preparation).

Petryk, H. (2000). General conditions for uniqueness in materials with multiple mechanisms of inelastic deformation. *J. Mech. Phys. Solids* 48:367-396.

Petryk, H., and Thermann, K. (1992). On discretized plasticity problems with bifurcations. *Int. J. Solids Structures* 29:745-765.

Petryk, H., and Thermann, K. (1996). Post-critical plastic deformation of biaxailly stretched sheets. *Int. J. Solids Structures* 33:689-705.

Raniecki, B. (1979). Uniqueness criteria in solids with non-associated plastic flow laws at finite deformations. *Bull. Acad. Polon. Sci., Sér. sci. techn.* 27:391-399.

Raniecki, B., and Bruhns, O.T. (1981). Bounds to bifurcation stresses in solids with non-associated plastic flow law at finite strain. *J. Mech. Phys. Solids* 29:153-172.

Rice, J.R. (1971). Inelastic constitutive relations for solids: an internal-variable theory and its application to metal plasticity. *J. Mech. Phys. Solids* 19:433-455.

Rice, J.R. (1977). The localization of plastic deformation. In Koiter, W.T., ed., *Theoretical and Applied Mechanics*. Amsterdam: North- Holland. 207-220.

Rice, J.R., and Rudnicki, J.W. (1980). A note on some features of the theory of localization of deformation. *Int. J. Solids Structures* 16:597-605.

Rizzi, E., Maier, G., and Willam, K. (1996). On failure indicators in multi-dissipative materials. *Int. J. Solids Structures* 33:3187-3214.

Ryzhak, E.I. (1987). Necessity of Hadamard conditions for stability of elastic-plastic solids (in Russian). *Mekh. Tv. Tela No.* 4:101-104.

Sewell, M.J. (1972). A survey of plastic buckling. In Leipholz, H.H.E., ed., *Stability*. Waterloo, Ontario: Univ. of Waterloo. 85-197.

Steinmann, P. (1996). On localization analysis in multisurface hyperelasto-plasticity. *J. Mech. Phys. Solids* 44:1691-1713.

Tomita, Y. (1994). Simulation of plastic instabilities in solid mechanics. *Appl. Mech. Rev.* 47: no 6, part 1, 171-205.

POST-CRITICAL PLASTIC DEFORMATION PATTERN IN INCREMENTALLY NONLINEAR MATERIALS AT FINITE STRAIN

K. Thermann
University of Dortmund, Dortmund, Germany

Abstract. Deformation pattern in a simple, time-independent material with a yield-surface vertex effect are investigated in plane stress and plane strain beyond the critical instant of ellipticity loss under isothermal, quasistatic loading. The energy criterion of path instability applied to a family of post-critical solutions eliminates unstable deformation paths and enforces the return to the elliptic regime or its boundary. In that way an automatic branch switching algorithm for the discretized problem is used to determine the overall deformation pattern, although the solutions remain locally indeterminate due to the absence of an internal length scale. The incipient volume fraction of multiple planar bands has a well-defined value insensitive to the mesh size in finite element calculations. The observed formation of coarse, differently aligned secondary shear bands in isochoric plain strain compression at later stages of the numerical simulation is theoretically supported by the energy criterion.

1 Introduction

Isothermal quasi-static uniform straining of time-independent elastic-plastic materials is frequently limited by strain localization phenomena. Originating from the analyses by Hill and Hutchinson (1975), Rudnicki and Rice (1975), and Rice (1977), the onset of localization is commonly understood as a bifurcation problem at the material level, admitting a velocity-gradient discontinuity across a planar band of vanishingly small or indeterminate width when the governing linearized partial differential equations loose ellipticity. Numerical simulations of the material behaviour in the non-elliptic range lead to serious difficulties, e.g. in applications of the finite element method a pathological mesh dependence, making the results useless, has frequently been reported. The deficiencies have been attributed to a missing internal length scale in the simple, time-independent material so that the width of a planar band (or shear band) is governed by the resolution of the mesh and not by the material. Numerous proposals have been made during the last twenty years to overcome these deficiencies, usually by enriching the material model such that ellipticity is retained and a characteristic length scale becomes available (cf. e.g. the surveys Tomita, 1994, Aifantis, 1995, Fleck and Hutchinson, 1997, de Borst and van der Giessen, 1998).

The computational approach to localization as a material instability presented here deviates from the main route in the literature. In contrast to many recent constitutive proposals associated with numerical studies of localization phenomena, no secondary effects are introduced into the material model and the traditional framework of time independent plasticity is adopted as laid down by Hill (1959) already 40 years ago as a basis for his pioneering work on uniqueness and

* The help from Christian Lulla in performing finite element computations and preparing the figures is gratefully acknowledged.

stability of solids without a natural time. Instantaneous moduli relating rates of strain and stress are restricted to be symmetric, or, stated otherwise, the constitutive rate equations are assumed to be derivable from a velocity-gradient potential.

Incremental nonlinearity, predicted by micromechanical studies and in agreement with the essential structure (Hill, 1967) of rate constitutive equations for metal polycrystals is, however, fully exploited. It is taken into account in order to describe the behaviour at a yield-surface vertex, which is formed at the current loading point on a theoretical yield-surface bounding a purely elastic domain. In a vertex model, a transitional regime of strain-rate directions exists, where the moduli gradually increase from their values in the total loading cone to those for elastic unloading. Due to this stiffening effect of partial unloading, which is absent in the classical elastic-plastic solid with a smooth yield-surface and two incrementally linear branches of loading and unloading, a physically meaningful analysis of quasi-static deformations in the post-critical range is frequently enabled under circumstances where the classical model fails.

Such a numerical analysis is still faced by difficulties related with non-unique solutions of the governing rate equations so that the problem arises, which continuation of a deformation path should be followed when alternatives exist. The notion of stability of a deformation process, introduced by Petryk (1982, 1985, 1991) offers a physically motivated criterion of choice. This concept, encompassing stability of an equilibrium state as a particular case, leads to a unifying approach to plastic instability and is applicable to discretized plasticity problems in a rather straightforward manner. Together with the underlying constitutive assumptions, it constitutes the theoretical foundation for the numerical analysis of post-critical deformation pattern studied here: necking in sheets (Section 3) and shear band formation under plane-strain compression (Section 4).

The essence of the computational scheme proposed by Petryk and Thermann (1992), and described in Section 2 together with the constitutive framework and the numerical implementation, is simple. The velocity functional in Hill's stationarity principle is minimized on the space of kinematically admissible fields obtained from a standard finite element discretization (in the actual implementation of the method). In this way, the total energy increment of the system is also minimized, so that the selected velocity field is energetically preferable in cases where mathematically equally well acceptable competitors exist; for a general discussion of the energy criterion, in particular its relation to stability in a dynamic or quasi-static sense, the reader is referred to Petryk's original work and his lecture notes in this Course.

As an outcome of the material instability analysis under plane-stress and plane-strain conditions, the formation of many planar bands of indeterminate individual width, but with a well defined total volume fraction of the created heterogeneous material, is theoretically predicted and supported by finite element results. The time-independent, simple elastic-plastic solid still offers interesting perspectives in this context, which merit further study.

The symbolic notation used throughout these lecture notes generally follows Gurtin (1972). In particular, a dot between tensor symbols always denotes an inner product, e.g. $\mathbf{a} \cdot \mathbf{b} = a_i b_i$, $\mathbf{A} \cdot \mathbf{B} = A_{ij} B_{ij}$ with the usual summation convention for repeated Latin subscripts. Fourth-order tensors are denoted by boldface sans-serif majuscules and the linear mapping that assigns to each second order tensor $\mathbf{B}$ a second order tensor $\mathbf{A} = \mathsf{C}\,\mathbf{B}$ reads $A_{ij} = C_{ijkl} B_{kl}$ in components on an orthonormal basis.

2 Discretized Plasticity Problems with Bifurcations: A Computational Approach Based on the Energy Criterion

2.1 Constitutive Framework and Specification

We consider isothermal quasi-static deformations of a simple time-independent material. Its constitutive rate equations can be expressed as

$$\dot{\mathbf{S}} = \dot{\mathbf{S}}(\dot{\mathbf{F}}) = \mathsf{C}(\dot{\mathbf{F}})\,\dot{\mathbf{F}}\,, \qquad \mathsf{C} \equiv \frac{\partial \dot{\mathbf{S}}}{\partial \dot{\mathbf{F}}}\,, \tag{1}$$

where $\mathbf{S}$ is the first Piola-Kirchhoff stress, related to Cauchy stress $\boldsymbol{\sigma}$ and Kirchhoff stress $\boldsymbol{\tau}$ by $\mathbf{S}\mathbf{F}^T = \det(\mathbf{F})\boldsymbol{\sigma} = \boldsymbol{\tau}$, $\mathbf{F}$ is the deformation gradient and a superimposed dot denotes the forward material derivative with respect to a time-like parameter t. Instantaneous moduli C may depend on the direction (but not on the magnitude) of $\dot{\mathbf{F}}$ in a nonlinear and piecewise continuous manner. They are, however, restricted here by the symmetry requirement

$$\mathsf{C} = \mathsf{C}^T, \qquad \text{or in components} \quad C_{ijkl} = C_{klij}\,. \tag{2}$$

i.e. it is assumed that the relationship (1) admits a potential $U(\dot{\mathbf{F}})$ so that

$$\dot{\mathbf{S}} = \frac{\partial U}{\partial \dot{\mathbf{F}}}\,, \qquad \mathsf{C} = \frac{\partial^2 U}{\partial \dot{\mathbf{F}} \partial \dot{\mathbf{F}}}\,, \qquad U = \tfrac{1}{2}\dot{\mathbf{S}} \cdot \dot{\mathbf{F}}\,. \tag{3}$$

The dependence of potential, stress-rate and moduli on the current stress and the deformation gradient history from some initial state up to the considered current configuration is to be understood. It is taken into account by a set of parameters, say $\mathcal{G} = \{\mathbf{S}, \mathbf{F}, \mathbf{G}^K\}$, which has been omitted for the sake of notational simplicity in the equations above. Whereas variables $\dot{\mathbf{S}}$ and $\dot{\mathbf{F}}$ allow a convenient formulation of the rate boundary value problem, work-conjugate pairs of stress and strain measures are better suited to ensure material objectivity in the original formulation of constitutive laws. The needed transformation rules can be found in Hill (1978). Whereas the constitutive framework (3) is sufficient for general considerations, specifications are given below for two vertex models which will later be used in numerical applications.

***J_2* corner theory** (Christoffersen and Hutchinson, 1979). At a conical vertex, the Eulerian strain rate $\mathbf{D}$ is related to the Jaumann flux of Kirchhoff stress $\overset{\triangledown}{\boldsymbol{\tau}}$ by

$$\mathbf{D} = \frac{\partial \Psi(\overset{\triangledown}{\boldsymbol{\tau}})}{\partial \overset{\triangledown}{\boldsymbol{\tau}}} = \mathsf{M}(\overset{\triangledown}{\boldsymbol{\tau}})\,\overset{\triangledown}{\boldsymbol{\tau}}\,, \qquad \mathsf{M}(\overset{\triangledown}{\boldsymbol{\tau}}) = \frac{\partial^2 \psi(\overset{\triangledown}{\boldsymbol{\tau}})}{\partial \overset{\triangledown}{\boldsymbol{\tau}} \partial \overset{\triangledown}{\boldsymbol{\tau}}}\,, \tag{4}$$

where the stress-rate potential

$$\Psi(\overset{\triangledown}{\boldsymbol{\tau}}) = \tfrac{1}{2}\overset{\triangledown}{\boldsymbol{\tau}} \cdot \mathsf{M}^{\mathrm{e}}\,\overset{\triangledown}{\boldsymbol{\tau}} + \tfrac{1}{2} f(\theta)\,\overset{\triangledown}{\boldsymbol{\tau}} \cdot \mathsf{M}^{\mathrm{p}}\,\overset{\triangledown}{\boldsymbol{\tau}} \tag{5}$$

depends on an angular measure θ between the cone axis (taken parallel to the deviatoric stress $\boldsymbol{\tau}'$) and the stress rate $\overset{\triangledown}{\boldsymbol{\tau}}'$, defined by

$$\cos\theta = \frac{\boldsymbol{\tau}' \cdot \mathsf{M}^{\mathrm{p}}\,\overset{\triangledown}{\boldsymbol{\tau}}'}{(\boldsymbol{\tau}' \cdot \mathsf{M}^{\mathrm{p}}\,\boldsymbol{\tau}')^{1/2}(\overset{\triangledown}{\boldsymbol{\tau}}' \cdot \mathsf{M}^{\mathrm{p}}\,\overset{\triangledown}{\boldsymbol{\tau}}')^{1/2}}\,. \tag{6}$$

M^e and M^p are linear elastic and plastic total loading compliances, respectively, such that $\mathsf{M}^e + \mathsf{M}^p$ is the compliance tensor of the hyperelastic version of J_2 deformation theory. $f(\theta)$ is a smooth transition function, equal to unity in the total (fully active) loading range $0 \leq \theta < \theta_0$, vanishing in the total unloading range $\theta_c < \theta \leq \pi$ and decreasing monotonically in the transition regime of partial unloading $\theta_0 \leq \theta \leq \theta_c$ such that the potential $\Psi(\overset{\triangledown}{\boldsymbol{\tau}})$ is continuously differentiable and strictly convex. The reader is referred to Christoffersen and Hutchinson (1979) for more details concerning the theory and to Tvergaard et al. (1981) for specifications of the transition function and the hyperelastic material law, which have also been used in the present computations.

Two-surface model (Petryk and Thermann, 1997). The recently proposed yield-vertex modification of two-surface models of metal plasticity combines features of micromechanical models with the classical flow theory. The incrementally nonlinear constitutive law (1) is specified as

$$\mathbf{D} = \frac{\partial \Psi}{\partial \overset{\triangledown}{\boldsymbol{\tau}}}\,, \qquad \Psi = \tfrac{1}{2}\overset{\triangledown}{\boldsymbol{\tau}} \cdot \mathsf{M}^e \overset{\triangledown}{\boldsymbol{\tau}} + \tfrac{1}{2}F(\beta)\,|\overset{\triangledown}{\boldsymbol{\tau}}'|^2\,, \qquad \cos\beta = \frac{\boldsymbol{\tau}' \cdot \overset{\triangledown}{\boldsymbol{\tau}}'}{|\boldsymbol{\tau}'|\,|\overset{\triangledown}{\boldsymbol{\tau}}'|}\,, \tag{7}$$

$$F(\beta) = \bar{M} \cdot \begin{cases} \pi - 2\beta_0 + \sin 2\beta_0 \cos 2\beta & \text{for } 0 \leq \beta \leq \beta_0 \quad \text{(total loading)} \\ \pi - (\beta + \beta_0) + \frac{1}{2}\sin 2(\beta + \beta_0) & \text{for } \beta_0 \leq \beta \leq \pi - \beta_0 \quad \text{(partial unloading)} \\ 0 & \text{for } \pi - \beta_0 \leq \beta \leq \pi \quad \text{(total unloading)}. \end{cases} \tag{8}$$

β_0 and $\bar{M}$ are state-dependent parameters. If the range of partial unloading is entered, then (7) is no longer valid for an arbitrary stress-rate, but may still be used in the vicinity of the *actual* $\overset{\triangledown}{\boldsymbol{\tau}}$ to determine the current moduli tensor C, under certain restrictions on the straining path (cf. *op. cit.*). The above incremental model differs from J_2 corner theory in several aspects. The most essential one, besides differences in the definition of the function $F(\beta)$ which generates the potential, is that the parameters β_0 and $\bar{M}$ in (8) are assumed to depend on the relative distance of the current deviatoric Kirchhoff stress $\boldsymbol{\tau}'$ from an *extremal surface*, understood in the sense of Hill (1967). The extremal surface is assumed to be described by the standard yield condition of the Huber-Mises type:

$$\bar{\tau} = k\,, \qquad \bar{\tau} \equiv \sqrt{\tfrac{1}{2}(\boldsymbol{\tau}' - \boldsymbol{\alpha}) \cdot (\boldsymbol{\tau}' - \boldsymbol{\alpha})}\,, \tag{9}$$

where $\boldsymbol{\alpha}$ denotes the deviatoric backstress and k is the yield shear stress. Contrary to classical elastoplasticity, the surface (9) is not allowed to be reached, and plastic deformation can take place when the current stress point $\boldsymbol{\tau}$ constitutes a vertex on an *inner* yield surface. $\boldsymbol{\alpha}$ and k can vary with the plastic deformation according to the usual rules. In the following applications, an isotropic power hardening law will be used, i.e. $\boldsymbol{\alpha} \equiv 0$ and $k = k_0(1 + \bar{\varepsilon}^p E/\tau_0)^N$, where $\bar{\varepsilon}^p$ is the equivalent plastic strain, k_0 and $\tau_0 = k_0\sqrt{3}$ denote the initial yield stresses for shear and uniaxial tension, respectively, and N is the hardening exponent. β_0 and $\bar{M}$ are expressed in terms of the ratio $\bar{\tau}/k$ as

$$\beta_0 = \arcsin\frac{k \sin\beta_c^{\max}}{\bar{\tau}}\,, \qquad \frac{\bar{\tau}}{k} \in (\sin\beta_c^{\max}, 1)\,, \qquad \beta_c^{\max} = \text{const} \in (\frac{\pi}{2}, \pi)\,, \tag{10}$$

$$\bar{M}(\beta_0) = \frac{1}{E}\frac{2}{1 - \chi(\beta_0)/\chi(\pi - \beta_c^{\max})}\,, \qquad \chi(\beta_0) = \frac{\pi - 2\beta_0 - \sin 2\beta_0}{\sin\beta_0}\,, \tag{11}$$

where $\beta_c^{\max}$ is a material constant and E is the elastic Young modulus. In the range $\bar{\tau}/k \leq \sin\beta_c^{\max}$, we substitute $\bar{M} = 0$ with β_0 undefined. (10) means that generators of the elastic unloading cone are tangent to a sphere with centre $\boldsymbol{\alpha}$ and radius $k\sqrt{2}\sin\beta_c^{\max}$ in $\boldsymbol{\tau}'$-space. In comparison with the standard elastoplastic solid, the present computational model requires only one additional material constant $\beta_c^{\max}$.

2.2 Discretized Rate Boundary Value Problem in Velocities

Formulation of the rate boundary value problem. Stresses, displacements and material parameters are assumed to be known at the current equilibrium state of the material body. Let V and S be the body volume and surface, respectively, in the fixed reference configuration, where a position vector is denoted by $\boldsymbol{\xi}$. Displacements $\mathbf{u}$ and nominal surface tractions $\mathbf{T}$ are prescribed on complementary parts S_u and S_T of the body surface, respectively, while in V nominal body forces $\mathbf{b}$ are given. The conditions of continuing equilibrium are conveniently expressed by the rate form of the virtual power principle,

$$\int_V \dot{\mathbf{S}} \cdot \nabla \mathbf{w}\, dV = \int_V \dot{\mathbf{b}} \cdot \mathbf{w}\, dV + \int_S \dot{\mathbf{T}} \cdot \mathbf{w}\, dS\,, \tag{12}$$

where $\nabla() = \partial()/\partial\boldsymbol{\xi}$ denotes a gradient in the reference configuration, and $\mathbf{w} \equiv \delta\mathbf{v}$ is any kinematically admissible variation of velocities $\mathbf{v}$. The spatially discretized problem is obtained by reducing the class of kinematically admissible velocity fields to a finite dimensional set $\mathcal{V}$ whose elements, assumed to be continuous and piecewise differentiable, are of the form

$$\mathbf{v}(\boldsymbol{\xi}) = \bar{\mathbf{v}}(\boldsymbol{\xi}) + v_\alpha \boldsymbol{\Phi}_\alpha(\boldsymbol{\xi})\,, \tag{13}$$

where $\bar{\mathbf{v}}(\boldsymbol{\xi})$ is an arbitrarily chosen but fixed function which assumes prescribed values on S_u, and $\boldsymbol{\Phi}_\alpha(\boldsymbol{\xi}), \alpha = 1, \ldots, M$ are given, linear independent shape functions which vanish on S_u. They form a basis of the finite M-dimensional linear space $\mathcal{W}$ of admissible variations, so that any field $\tilde{\mathbf{w}} \in \mathcal{W}$ has the representation $\tilde{\mathbf{w}} = w_\alpha \tilde{\boldsymbol{\Phi}}_\alpha$ with scalar coefficients w_a. The summation convention over the range from 1 to M is adopted for repeated Greek subscripts and a tilde over a symbol is used to distinguish a spatial field over V from its value at a material point. A field $\tilde{\mathbf{v}} \in \mathcal{V}$, defined by the vector $[\mathrm{v}]$ of components v_α, is called a (discretized) velocity solution if it satisfies the nonlinear algebraic equations

$$\dot{Q}_\alpha(\tilde{\mathbf{v}}) = \dot{P}_\alpha\,, \qquad \alpha = 1, \ldots, M\,, \qquad \tilde{\mathbf{v}} \in \mathcal{V}\,, \tag{14}$$

where

$$\dot{Q}_\alpha(\tilde{\mathbf{v}}) = \int_V \dot{\mathbf{S}}(\nabla\mathbf{v}) \cdot \nabla\boldsymbol{\Phi}_\alpha\, dV \tag{15}$$

and

$$\dot{P}_\alpha = \int_V \dot{\mathbf{b}} \cdot \boldsymbol{\Phi}_\alpha\, dV + \int_{S_T} \dot{\mathbf{T}} \cdot \boldsymbol{\Phi}_\alpha\, dS \tag{16}$$

may be interpreted as rates of internal and external forces, respectively. (14) follows in the standard manner from (12) on substituting (1) and (13). At a given state, the tangent stiffness matrix $K_{\alpha\beta}$, defined by

$$K_{\alpha\beta}(\tilde{\mathbf{v}}) \equiv \frac{\partial \dot{Q}_\alpha}{\partial v_\beta} = \int_V \nabla\boldsymbol{\Phi}_\alpha \cdot \mathbf{C}(\nabla\mathbf{v})\nabla\boldsymbol{\Phi}_\beta\, dV\,, \qquad \alpha, \beta = 1, \ldots, M\,, \tag{17}$$

depends on $\tilde{\mathbf{v}}$ and is symmetric. Both properties are direct consequences of the potential form of the constitutive law (3), which, moreover, allows the variational formulation of (14) in the form

$$\frac{\partial J(\tilde{\mathbf{v}})}{\partial v_\alpha} = \dot{Q}_\alpha(\tilde{\mathbf{v}}) - \dot{P}_\alpha = 0\,, \qquad \alpha = 1, \ldots, M\,, \tag{18}$$

where

$$J(\tilde{\mathbf{v}}) = \int_V U(\nabla \mathbf{v})\, dV - \dot{P}_\alpha v_\alpha\,. \tag{19}$$

Solution properties in the post-critical range. Important solution properties, which are the basis of the computational approach, are briefly listed below; for details, the reader is referred to the cited original papers, in particular Petryk (1991), Petryk and Thermann (1992).

(a) Existence of solutions and stability of equilibrium. *The system of equations* (14) *has a solution which assigns to J an absolute minimum value in $\mathcal{V}$ provided that*

$$I(\tilde{\mathbf{w}}) \equiv \tfrac{1}{2}\,\dot{Q}_\alpha(\tilde{\mathbf{w}}) w_\alpha = \tfrac{1}{2}\,K_{\alpha\beta}(\tilde{\mathbf{w}})\; w_\alpha w_\beta > 0 \qquad \text{for every non-zero } \tilde{\mathbf{w}} \in \mathcal{W}\,. \tag{20}$$

The condition (20) is the discrete version of Hill's well known sufficient condition for (directional) stability. When (20) holds, an energy barrier is imposed in all directions, so that a departure from equilibrium on a straight path in a dynamic motion induced by vanishingly small disturbances is excluded (Hill, 1958). Conversely, if

$$I(\tilde{\mathbf{w}}) < 0 \qquad \text{for some } \tilde{\mathbf{w}} \in \mathcal{W} \tag{21}$$

then such a departure is energetically possible; moreover, the equilibrium state can also be shown to be unstable in a dynamic sense (Petryk, 1991), so that a quasi-static continuation of the deformation path would not be meaningful. The existence of a velocity solution is thus guaranteed under an assumption which has a clear physical meaning.

(b) Uniqueness. Suppose that a fundamental velocity solution $\tilde{\mathbf{v}}^0$ is known. *Uniqueness of the solution in velocities (as well as fulfilment of* (20)) *is ensured so long as $[K^0] \equiv [K](\tilde{\mathbf{v}}^0)$ is positive definite, provided that an additional constitutive inequality is satisfied, either in the form of the relative convexity property* (Hill, 1959), *or in the similar, but less restrictive form* (Petryk, 1989)

$$(\dot{\mathbf{S}} - \dot{\mathbf{S}}^0) \cdot (\dot{\mathbf{F}} - \dot{\mathbf{F}}^0) \geq (\dot{\mathbf{F}} - \dot{\mathbf{F}}^0) \cdot \mathsf{C}(\dot{\mathbf{F}}^0)\, (\dot{\mathbf{F}} - \dot{\mathbf{F}}^0) \qquad \text{for every } \dot{\mathbf{F}}\,, \tag{22}$$

where $\dot{\mathbf{S}}^0$ and $\dot{\mathbf{F}}^0$ correspond to $\tilde{\mathbf{v}}^0$, and $\dot{\mathbf{S}}$ is constitutively related to $\dot{\mathbf{F}}$.

(c) Continuous range of bifurcation points. *There is a bifurcation in velocities at every point on a segment of a solution path along which*

$$K_{\alpha\beta}(\tilde{\mathbf{v}})\; w_\alpha w_\beta < 0 \qquad \text{for some } \tilde{\mathbf{w}} \in \mathcal{W} \tag{23}$$

holds simultaneously with (20). Such a path segment is considered not to be realizable in a physical system, since the path can be left quasi-statically at *every* instant, even without disturbing forces. Contrary to incrementally linear (e.g. elastic) solids, where the two inequalities contradict each other ($K_{\alpha\beta}$ is constant), a range of continuous bifurcation points is typically met in elastic-plastic solids on a segment of the fundamental path $\tilde{\mathbf{v}}^0$ just beyond the primary bifurcation point. At that point, the smallest eigenvalue of $[K](\tilde{\mathbf{v}}^0)$ ceases to be positive for the first time, but the equilibrium states may still be stable. A famous example is the Shanley column (Shanley, 1947) in the load range between the tangent and the reduced modulus load.

(d) Energy criterion of path instability. Whereas the energy criterion in studies of stability of *equilibrium* has a long tradition, the broader concept of stability of a *deformation process* or path (Petryk, 1982, 1985), encompassing stability of equilibrium, is less well known. In the latter approach, a deformation path is considered as an entity, not merely as a sequence of equilibrium states. For any kinematically admissible deformation path, define the energy functional as

$$E = W + \Omega\,, \tag{24}$$

where W is the sum of the deformation work in a deformation process up to the current instant and Ω is the current potential energy of the loading device. It has been shown that $\dot{E}$ at an equilibrium state is independent of $\tilde{\mathbf{v}} \in \mathcal{V}$,

$$\dot{E}(\tilde{\mathbf{v}}) = \text{const in } \mathcal{V}\,, \tag{25}$$

and that the second-order rate $\ddot{E}$ is related to the velocity functional by the simple identity

$$\tfrac{1}{2}\,\ddot{E}(\tilde{\mathbf{v}}^*) - \tfrac{1}{2}\,\ddot{E}(\tilde{\mathbf{v}}) = J(\tilde{\mathbf{v}}^*) - J(\tilde{\mathbf{v}}) \qquad \text{for every } \tilde{\mathbf{v}}, \tilde{\mathbf{v}}^* \in \mathcal{V}\,. \tag{26}$$

Hence, a minimizer $\tilde{\mathbf{v}}^*$ of J also minimizes $\ddot{E}$, and, on account of (25), also the increment of energy $\Delta E = \dot{E}\ \Delta t + \frac{1}{2}\,\ddot{E}\ (\Delta t)^2$, calculated with accuracy to second-order. A unifying approach to plastic instability is obtained with Petryk's energy criterion:

A deformation process is said to be unstable in the energy sense at some stage of the deformation, if the respective velocity solution $\tilde{\mathbf{v}}^0$ does not minimize the value of $\ddot{E}$ in $\mathcal{V}$, i.e. if

$$\ddot{E}(\tilde{\mathbf{v}}) < \ddot{E}(\tilde{\mathbf{v}}^0) \qquad \text{for some } \tilde{\mathbf{v}} \in \mathcal{V}\,. \tag{27}$$

Deformation paths which are unstable in the energy sense may be rejected in advance as apparently unrealizable in a physical system. The instability is to be understood either as instability of the traversed equilibrium states in a dynamic sense at fixed external loading (if (20) fails), or in the quasi-static sense of a persisting possibility of energetically preferable deviations from the path at varying loading (if (20) holds). In the first case, when (21) holds, J is unbounded from below, the minimum does not exist, and a spontaneous departure from the equilibrium position is to be expected (Petryk, 1991). Both kinds of instability are avoided only if $J(\tilde{\mathbf{v}})$ attains its absolute minimum in $\mathcal{V}$ at the actual velocity solution $\tilde{\mathbf{v}}_0$, i.e.

$$J(\tilde{\mathbf{v}}_0) \le J(\tilde{\mathbf{v}}) \qquad \text{for every } \tilde{\mathbf{v}} \in \mathcal{V}\,. \tag{28}$$

The existence of such a velocity solution is ensured in the range of directionally stable equilibrium states. On this theoretical basis, a computational algorithm which selects a velocity solution by minimizing $J(\tilde{\mathbf{v}})$ in $\mathcal{V}$ has been proposed by Petryk and Thermann (1992).

2.3 Computational Approach

Numerical method to minimize the velocity functional. The main step of the proposed algorithm is to find a solution of the nonlinear system of equations (18) by minimizing the velocity functional (19) within the set of kinematically admissible velocity fields of the form (13). Unconstrained minimization of a functional on a finite-dimensional space is a basic and well explored problem in optimization. By selecting one of the known numerical methods, it has to be taken into account that (i) the problem has to be solved repeatedly during time integration along a deformation path, (ii) that the number of variables in a finite element discretization is generally large, and (iii) that the tangent stiffness matrix is large and sparse. At a trial point, not only the functional, but also its first and second derivatives are available from (18) and (17), respectively.

Newton's method would be ideally suited under these circumstances; it must, however, be combined with some other numerical technique, in order to guarantee convergence to the required minimum of the generally non-convex functional. One possibility is to continue with a negative curvature line search (cf. Fiacco and McCormick, 1968), when Newton's method stops at a point where the tangent stiffness matrix is not positive definite. In this way, the example from Petryk and Thermann (1992) in Section 2.4 has been calculated. In later applications, where a number of bifurcation points had to be crossed along the deformation path (Petryk and Thermann, 1995), a trust region method (cf. Fletcher, 1987) was tried and found to be more efficient. On the basis of available experience, this method can be recommended as reliable and efficient. It retains the quadratic convergence of Newton's method in the neighbourhood of a minimum and is known to be globally convergent.

The underlying idea is simple. A quadratic model, obtained from a truncated Taylor series,

$$J(\mathbf{v}^{(k)} + \boldsymbol{\delta}) \approx q^{(k)}(\boldsymbol{\delta}) = J^{(k)} + \mathbf{g}^{(k)} \cdot \boldsymbol{\delta} + \tfrac{1}{2}\, \boldsymbol{\delta} \cdot \mathbf{K}^{(k)} \boldsymbol{\delta}\,, \qquad |\boldsymbol{\delta}| \leq r^{(k)}\,, \tag{29}$$

is accepted as an approximation to J in the neighbourhood of the current point $\mathbf{v}^{(k)}$ in the kth iteration (and not on the whole space as in Newton's method)[1]. $\mathbf{g}^{(k)}$ and $\mathbf{K}^{(k)}$ denote gradient and tangent stiffness matrix evaluated at $\mathbf{v}^{(k)}$, respectively, and $|\boldsymbol{\delta}|$ is the Euclidean norm of $\boldsymbol{\delta} = \mathbf{v} - \mathbf{v}^{(k)}$. The radius $r^{(k)}$ of the trust region is adapted in the course of the iteration, so that the trust region may grow or shrink according to the agreement between the actual reduction in f and the predicted reduction in $q^{(k)}$, obtained from the solution to

$$\text{minimize } q^{(k)}(\boldsymbol{\delta}) \quad \text{subject to} \quad |\boldsymbol{\delta}| \leq r^{(k)}\,. \tag{30}$$

The following adaptive scheme has been implemented (cf. Fletcher, 1987):

1. given $\mathbf{v}^{(k)}$ and $r^{(k)}$, calculate $\mathbf{g}^{(k)}$ and $\mathbf{K}^{(k)}$,
2. solve (30) for $\boldsymbol{\delta}^{(k)}$,
3. evaluate $J(\mathbf{v}^{(k)} + \boldsymbol{\delta}^{(k)})$ and $\rho^{(k)} = \dfrac{J^{(k)} - J(\mathbf{v}^{(k)} + \boldsymbol{\delta}^{(k)})}{J^{(k)} - q^{(k)}(\boldsymbol{\delta}^{(k)})}$,

[1] Since confusions with spatial fields can not arise here, we write $\mathbf{v}$ for the M-vector denoted above as $[v]$ and $\mathbf{K}$ instead of $[K]$

4. set

$$r^{(k+1)} = \begin{cases} |\delta^{(k)}|/4 & \text{if } \rho^{(k)} < 0.25 & \text{(model is bad)} \\ 2r^{(k)} & \text{if } \rho^{(k)} > 0.75 \text{ and } |\delta| = r^{(k)} & \text{(model is good)} \\ r^{(k)} & \text{otherwise} \end{cases}$$

$$\mathbf{v}^{(k+1)} = \begin{cases} \mathbf{v}^{(k)} & \text{if } \rho^{(k)} \leq 0 & \text{(step failed)} \\ \mathbf{v}^{(k)} + \boldsymbol{\delta}^{(k)} & \text{otherwise}\,. \end{cases}$$

The remaining subproblem (30) has been shown (Gay, 1981, Moré and Sorensen, 1981) to be equivalent to finding $\lambda \geq 0$ and $\boldsymbol{\delta}$ such that $(\mathbf{K}^{(k)} + \lambda\mathbf{I})$ is positive semidefinite,

$$(\mathbf{K}^{(k)} + \lambda\mathbf{I})\boldsymbol{\delta} = -\mathbf{g}^{(k)} \qquad \text{and} \qquad |\boldsymbol{\delta}| = r^{(k)} \text{ if } \lambda > 0\,. \tag{31}$$

If $\mathbf{K}^{(k)}$ is positive definite then the Newton step $\boldsymbol{\delta}^{(k)} = -(\mathbf{K}^{(k)})^{-1}\mathbf{g}^{(k)}$ solves (30) (and (31) with $\lambda = 0$), provided that the radius of the trust region is sufficiently large, i.e. $|\boldsymbol{\delta}^{(k)}| < r^{(k)}$. Otherwise, the minimum is located on the boundary, and $\lambda > 0$ is obtained by zero-finding from the nonlinear equation (31) in one variable $|\boldsymbol{\delta}(\lambda)| = r^{(k)}$. Available iterative methods exploit the particular structure of this equation and converge on average in less than two steps. When, however, $\mathbf{g}^{(k)}$ is orthogonal to the eigenspace corresponding to the smallest eigenvalue $\lambda_1 < 0$ of $\mathbf{K}^{(k)}$, modifications of the solution strategy are required and have been suggested (cf. Gay, 1981, Moré and Sorensen, 1981), in order not to degrade the performance of the algorithm. The present implementation incorporates these ideas and is partially based on the code described by Gay (1983).

General aspects. By minimizing J as described above, a deformation path is *automatically* left as soon as the tangent stiffness matrix ceases to be positive definite and becomes indefinite along the path. This is justified, since the smooth continuation of the path would be unstable in a physically acceptable sense. A secondary branch is then automatically followed since the existence of another velocity solution minimizing (19) has been ensured by (20). A *nonlinear* bifurcation problem in velocities, not the linearized problem for a comparison solid, is solved in that way, so that the bifurcation mode cannot be freely scaled. It is pointed out that the branch switching does not require any imperfections to be introduced, nor another technique when a bifurcation is met. Computations can not be continued, when directional stability of equilibrium is violated. In this case, the velocity functional unboundedly decreases and its minimum does not exist. The algorithm is thus *not* suited for crossing bifurcation points in purely elastic solids.

Once the velocity solution has been found, the *second-order* rate boundary value problem can optionally be solved as a linear problem with relatively little additional cost (Petryk and Thermann, 1992). Quasi-static accelerations can then be used for updating purposes similarly as in a dynamic response analysis to increase efficiency of the whole algorithm. Generally, the presently implemented explicit time integration scheme has the well known deficiencies and should be improved. Numerical instabilities are avoided here by taking sufficiently small time steps. This is not entirely satisfactory, but not a serious drawback for the applications to be discussed with a number of bifurcations to be localized along a deformation path, restricting the step size in any case. A more fundamental problem concerns the question whether the required absolute minimum of J has been found. Unfortunately, for non-convex mathematical programming problems there is no general affirmative answer. Some additional remarks are made in Section 4.2.

2.4 Illustrative Example

The computational approach is illustrated by an example of plane-strain tension (cf. Petryk and Thermann, 1992). The lateral sides of a homogeneous rectangular specimen are traction free, while on the ends normal displacements and zero shear tractions are prescribed. Contrary to the study of the imperfect specimen (Tvergaard et al., 1981), the perfect system is considered here so that the fundamental path, corresponding to uniform straining, is known. This knowledge can be advantageously used by starting the numerical calculations from the homogeneous state at some later instant.

The smallest eigenvalue is plotted in Fig. 1, showing that the tangent stiffness matrix ceases to be positive definite at a relative end displacement $u/l_0 = 0.1405$. As soon as the matrix became indefinite in the course of the computation, a secondary solution in velocities, corresponding to diffuse necking, was found. The unstable fundamental post-bifurcation path was left and the secondary branch was followed, in full agreement with the theory. The again positive definite tangent stiffness matrix on the secondary branch is generally due to growth of an unloading zone where the total loading moduli become replaced by stiffer moduli from the partial or total unloading regime of the incrementally nonlinear material. Hence, the secondary post-bifurcation branch can exhibit analogous qualitative properties as a typical pre-bifurcation branch.

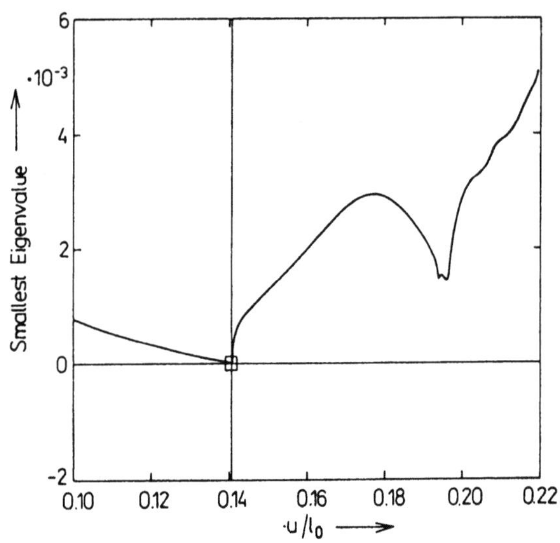

Figure 1. The smallest eigenvalue of the tangent stiffness matrix vs relative end displacement in an example of plain-strain tension of an initially perfect specimen. J_2 corner theory: $\beta_c^{\max} = 135°$, $N = 0.1$, $E = 500\,\tau_0$, $\nu = 0.3$, $l_0/h_0 = 3$. (From Petryk and Thermann, 1992)

Further deformations in the considered range up to $u/l_0 = 0.220$ take place with positive definite tangent stiffness, although a definite decrease of its smallest eigenvalue is observed at $u/l_0 = 0.196$ in correlation with a concentration of deformation rate in the central elements at the lateral surfaces and in shear bands propagating from these elements. Somewhat later, cf. Fig. 2b, these shear band pattern are also visible in the distribution of principal strain with largest magnitude. In comparison with Fig. 2a, the typical initial distribution at diffuse necking has been completely rearranged in the central part of the specimen.

The observed shear band deformation pattern in Fig. 2b suggests that a *material instability* at the level of a material element had occurred. The non-uniformities induced by diffuse necking aid strain localization to develop here without further bifurcations. This subject will be pursued further in Section 4 as a central theme of this Course.

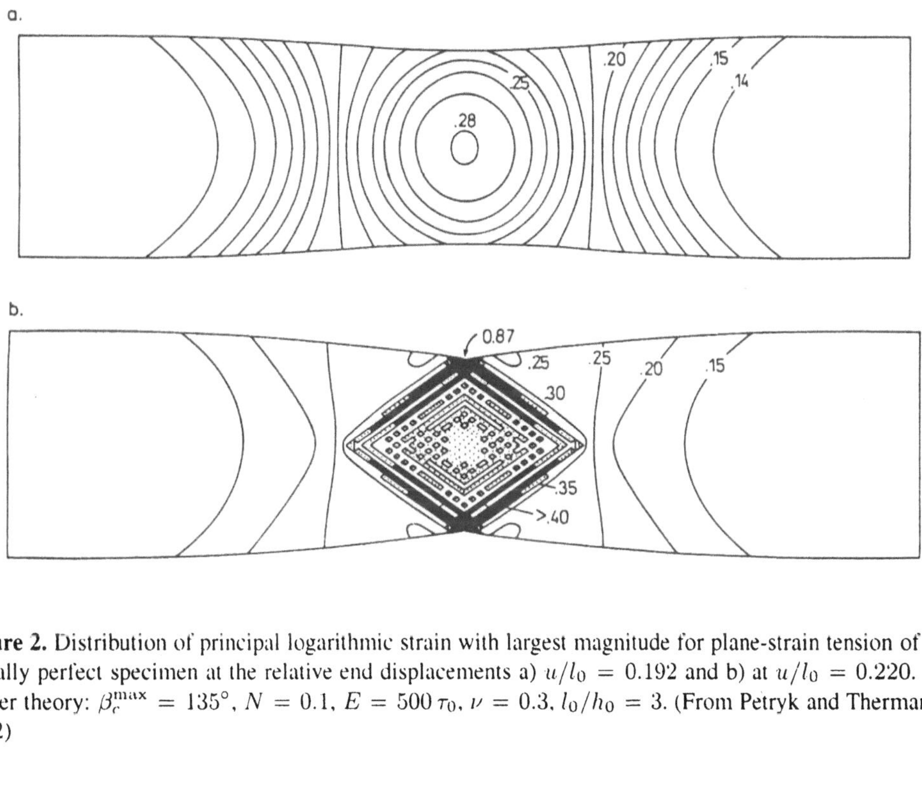

Figure 2. Distribution of principal logarithmic strain with largest magnitude for plane-strain tension of an initially perfect specimen at the relative end displacements a) $u/l_0 = 0.192$ and b) at $u/l_0 = 0.220$. J_2 corner theory: $\beta_c^{max} = 135°$, $N = 0.1$, $E = 500\,\tau_0$, $\nu = 0.3$, $l_0/h_0 = 3$. (From Petryk and Thermann, 1992)

3 Necking in Biaxially Stretched Sheets

3.1 Problem Formulation

Consider a rectangular plane sheet, initially homogeneous isotropic and of uniform thickness, subject to quasi-static biaxial tension in the directions x_1 and x_2 (Fig. 3). Zero shear tractions and normal velocities over the sheet end planes $x_1 = \pm l$ and $x_2 = 0, x_2 = l_2$ are prescribed, so that averaged values $\bar{D}_1$, $\bar{D}_2$ of Eulerian strain-rates D_{11}, D_{22} are given. The sheet surface is traction free.

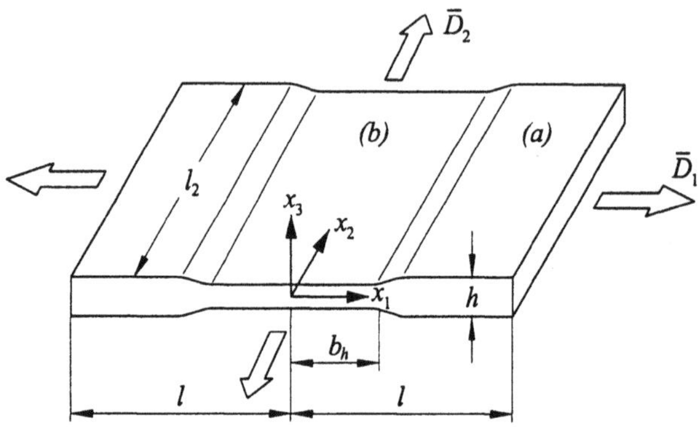

Figure 3. A stretched sheet in the current configuration

According to experimental observations and theory (cf. Stören and Rice, 1975, Petryk and Thermann, 1996), localized necking occurs along the direction of minor principal stretch, taken as the x_2 axis. The analysis is, therefore, restricted to a generalized plane problem in the (x_1, x_3) plane: All quantities of interest are independent of x_2, the shear components F_{21}, F_{23} and σ_{21}, σ_{23} of deformation gradient $\mathbf{F}$ and Cauchy stress $\boldsymbol{\sigma}$, respectively, vanish identically, and D_{22} always coincides with the given value $\bar{D}_2 = \rho \bar{D}_1$, where ρ, $0 < \rho \leq 1$, is the given stretching ratio.

The problem is analogous to the plane-strain tension example, however, generalized by a prescribed non-zero strain-rate component $\bar{D}_2$ perpendicular to the (x_1, x_3) plane. It will be further pursued in Section 3.3.

3.2 Plane-Stress Idealization

It is now additionally assumed that the sheet is sufficiently thin so that the plane-stress idealization is justified. Then, the dependence of stress and strain on the thickness coordinate x_3 can be neglected, and the problem is reduced to one dimension x_1. Furthermore, all shear components of Eulerian strain rate $\mathbf{D}$ and first Piola-Kirchhoff stress $\mathbf{S}$ vanish, and it is convenient to work with principal values D_k and S_k of $\mathbf{D}$ and $\mathbf{S}$, respectively. Taking the current sheet configuration as reference, the constitutive rate equations (1) and (3) can be rewritten as

$$\dot{S}_i = \frac{\partial \hat{U}(D_1, D_2)}{\partial D_i} = \hat{C}_{ik}\, D_k\,, \qquad \hat{U} = \tfrac{1}{2}\, \dot{S}_i D_i\,, \qquad i, k = 1, 2\,, \tag{32}$$

with plane-stress moduli $\hat{C}_{ik} = \hat{C}_{ki}$, related to $\mathbf{C}$ in the standard manner. A real-valued continuous and piecewise continuously differentiable velocity function $v_1(x_1)$ on $[0, l]$ satisfying the boundary conditions $v_1(0) = 0$ and $v_1(l) = l\bar{D}_1$ is called kinematically admissible, and the set of these functions is denoted as $\mathcal{V}_1$. $\tilde{v}_1 \in \mathcal{V}_1$ is a velocity solution if the condition of continuing equilibrium,

$$h\dot{S}_1 = \text{const} \qquad \text{for } x_1 \in (0, l)\,, \tag{33}$$

holds on substituting $D_1 = v_1' \equiv dv_1/dx_1$ into (32). Since $\dot{S}_1$ is derivable from the plane-stress potential $\hat{U}$, a velocity solution makes

$$J(\tilde{v}_1) = \int_0^l h\hat{U}(v_1', \bar{D}_2)\, dx_1 \tag{34}$$

stationary in $\mathcal{V}_1$. This is a specialized version of Hill's (1959) variational theorem and analogous to its finite dimensional variant (18)–(19) (with $\dot{P}_\alpha = 0$ in the absence of body forces and nominal surface tractions).

Energy condition for path stability. Moreover, according to the energy criterion (Petryk, 1982, 1985, 1991) and again analogously to its discretized counterpart (28), a solution $\tilde{v}_1^0$ can correspond to a stable deformation path only if it minimizes (34), that is

$$\int_0^l h\hat{U}(v_1', \bar{D}_2)\, dx_1 \geq \int_0^l h\hat{U}(v_1^{0\prime}, \bar{D}_2)\, dx_1 \qquad \text{for every } \tilde{v}_1 \in \mathcal{V}_1\,. \tag{35}$$

It has been shown (Petryk and Thermann, 1996) that (35) is *equivalent* to the pointwise condition

$$\hat{\mathcal{E}}(D_1^0, D_1; \bar{D}_2) \geq 0 \qquad \text{for every } D_1 \text{ and at every } x_1 \in (0, l)\,, \tag{36}$$

where the Weierstrass function

$$\hat{\mathcal{E}}(D_1^0, D_1; \bar{D}_2) = \hat{U}(D_1, \bar{D}_2) - \hat{U}(D_1^0, \bar{D}_2) - \dot{S}_1^0 \cdot (D_1 - D_1^0) \tag{37}$$

represents the excess of $\hat{U}$ at D_1 over its linear approximation constructed at D_1^0. Quantities with a superscript zero are associated with v_1^0. The inequality (36) has to be satisfied in particular for $D_1 \to \infty$ and for $D_1 \to D_1^0$. On substituting $\hat{U} = \frac{1}{2}\hat{C}_{ik} D_i D_k$ and using the continuity and homogeneity of the constitutive law, this implies

$$\hat{C}_{11}(\pm 1, 0) \geq 0 \qquad \text{at any } x_1 \in (0, l)\,, \tag{38}$$

$$\hat{C}_{11}(D_1^0, \bar{D}_2) \geq 0 \qquad \text{at any } x_1 \in (0, l)\,. \tag{39}$$

The inequalities (36), (38) and (39) can be regarded as one-dimensional plane-stress specifications of Petryk's (1992) general conditions for material stability. Accordingly, (38) is necessary for stability of equilibrium (in the dynamic sense), and (39), the familiar Legendre condition for the minimization problem (35), can be interpreted as necessary for stability of the deformation path (in a quasi-static sense).

Conversely, it has been proved (Petryk and Thermann, 1996) that no bifurcations in velocities are possible along a path of uniform stretching so long as (39) holds as strict inequality, provided that the material law satisfies the constitutive inequality (22). Under this material restriction, uniqueness of uniform stretching is lost exactly at the point of ellipticity loss determined for the tangent moduli, that is at the critical instant $t^\star$ where the linearized bifurcation condition

$$\hat{C}_{11}(\bar{D}_1, \bar{D}_2) = 0 \tag{40}$$

is met; this is the critical point examined by Stören and Rice (1975).

Incipient width of the necking band. Suppose that $(\bar{D}_1, \bar{D}_2)$ corresponds to proportional stretching and represents the symmetry axis of a constitutive cone of total loading within which the modulus $\hat{C}_{11}$ is constant, and at $t^\star$ equal to zero by (40). In the (D_1, D_2) plane, introduce the angular measure ϕ for the deviation of a strain rate vector (D_1, D_2) from the cone axis, and denote the value of ϕ for the direction $(1,0)$ by $\phi^\star$, and the cone angle by ϕ_0 (cf. Fig. 4a).

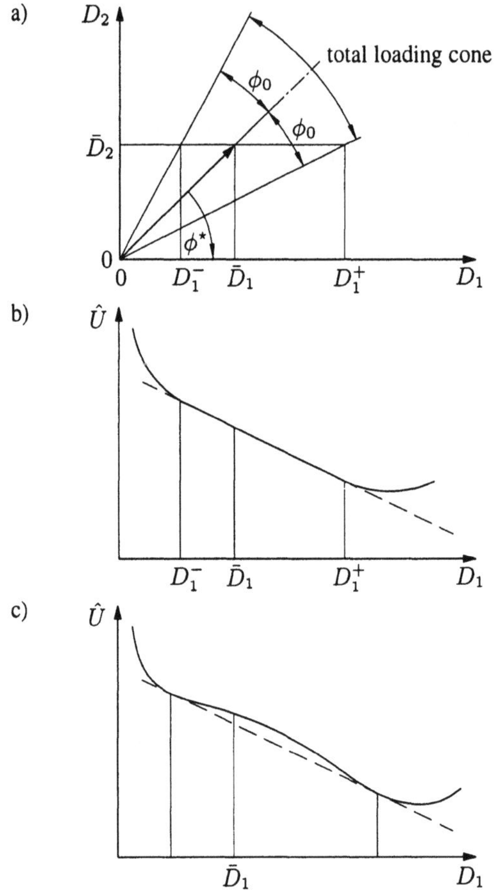

Figure 4. a) Strain rates and angles associated with the constitutive cone of total loading. b) Schematic graph of the constitutive potential $\hat{U}$ vs D_1 for fixed D_2 at the critical stage, and c) beyond the critical stage. (From Petryk and Thermann, 1996)

Obviously, if $\phi^\star < \phi_0$ then stability of equilibrium (38) and ellipticity are simultaneously lost at the critical instant $t^\star$. (This is always the case in classical plasticity with an incrementally linear loading branch, so that $\phi_0 = \pi/2$.)

Suppose thus that at the critical state $\phi^\star > \phi_0$ holds, and, moreover $\hat{C}_{11} > 0$ for $\phi > \phi_0$. The graph of $\hat{U}$ as a function of D_1 at given $D_2 = \bar{D}_2$ in this state has the form visualized in Fig. 4b, with the linear range corresponding to total loading. By taking an arbitrary pair of strain rates $D_1^{(b)}$ and $D_1^{(a)}$, say, from the linear range of $\hat{U}$ including the end point D_1^+ and D_1^- such that $D_1^{(b)} > \bar{D}_1 > D_1^{(a)}$, infinitely many velocity solutions can be constructed, e.g. by $v_1'(x_1) = D_1^{(b)}$ for $0 < x_1 < b$ and $v_1'(x_1) = D_1^{(a)}$ for $b < x_1 < l$ with b/l determined from the kinematic boundary condition $v_1(l) = l\bar{D}_1 = bD_1^{(b)} + (l-b)D_1^{(a)}$ as

$$\frac{b}{l} = \frac{\bar{D}_1 - D_1^{(a)}}{D_1^{(b)} - D_1^{(a)}} \,. \tag{41}$$

All these solutions fulfil the condition (36) *at* the critical instant (as equality). However, in typical circumstances indicated by Petryk and Thermann (1996), the velocity solution v_1^0 which satisfies (39) just *beyond* $t^\star$ must satisfy $\phi = \phi_0$ at $t^\star$, not only in the localization zone, (b) say, but also in the remaining portion (a) of the sheet, so that $D_1^{(b)} = D_1^+$, $D_1^{(a)} = D_1^-$. On substituting these values in (41) and by using trigonometric relationships with the help of Fig. 4a, the incipient relative width of the necking band, b/l, is uniquely obtained as

$$\frac{b}{l} = \frac{1}{2}\left(1 - \frac{\tan\phi_0}{\tan\phi^\star}\right) . \tag{42}$$

Rearrangements of the single band $0 < x_1 < b$ into separated individual bands of arbitrary number and width, but so that their total width equals b, are obviously possible. Apart of this trivial kind of non-uniqueness, all indeterminacies in velocity solutions are removed with the help of the necessary condition (39) for stability of a deformation path according to the energy criterion. Formula (42) is the plane stress specialization of a general expression for the incipient volume fraction of the localization zone (Petryk, 1997). It remains valid when the angle ϕ between $\mathbf{D}$ and $\bar{\mathbf{D}}$ is defined in terms of a scalar product generated by an arbitrary positive definite fourth order tensor, as in J_2 corner theory.

In passing we note that the remaining kind of non-uniqueness just mentioned is inevitably met in finite element calculations. Figure 5 shows two velocity solutions for a biaxially stretched circular sheet slightly beyond the critical instant of ellipticity loss (Petryk and Thermann, 1995). The incipient localization zone in both cases consists of a number of necking bands (black) along the direction of minor stretch, distributed more or less randomly over the sheet. The only difference between the two calculations was that the minimization procedure for the discretized problem started from different trial fields. Although the solutions are locally different, they exhibit similar overall characteristics. For example, the volume fraction of the localization zone, determined from the thickness strain-rate distribution in the central part of the sheet is 0.346 and 0.349 for the solutions from Fig. 5a and 5b, whereas the theoretical value found from (42) is 0.338.

Post-critical behaviour. We return to the original one-dimensional problem. For the examples from J_2 corner theory considered by Petryk and Thermann (1996), the analysis showed that a post-critical solution with a fixed nominal width of the necking band does not satisfy (36) in the zone (b) and must be rejected as being unstable.[2] However, a decreasing nominal width of the

[2] Similarly, if a single, vanishingly narrow band ($b = 0$) is assumed, as in Christoffersen and Hutchinson (1979) but without imperfection, the stability requirement is violated outside the band.

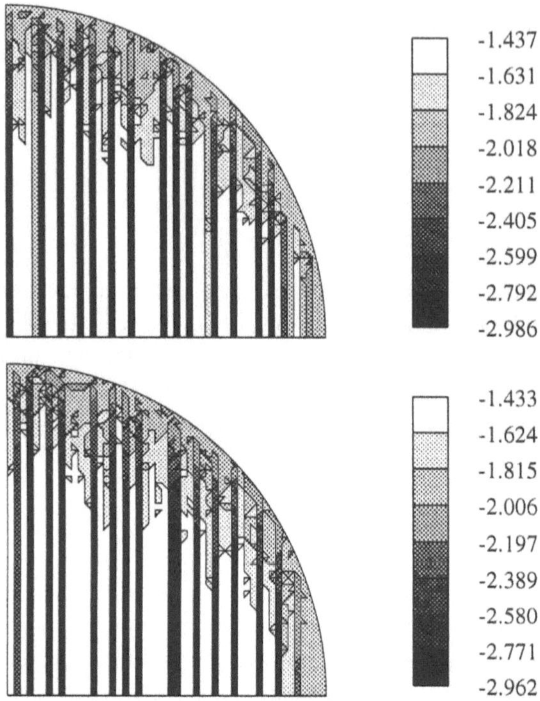

Figure 5. Thickness strain-rate distribution in two post-critical solutions for a circular sheet with prescribed boundary velocities corresponding to $\rho = 0.99$. Plane stress idealization, 2D FE solution with 50 elements along the radius. J_2 corner theory: $\theta_c = 135°$, $\theta_0 = 22.5°$, $N = 0.2$, $E = 500\,\tau_0$, $\nu = 0.3$. (From Petryk and Thermann, 1995)

necking band was found to be in agreement with the stability requirements. The boundary of zone (b) is then moving with respect to the material and the strain-rate suffers a jump from $D_1^{(b)}$ to $D_1^{(c)}$, say, at the moving end point. The two unknowns $D_1^{(b)}$ and $D_1^{(c)}$, and thus the deformation history in zone (b), can be calculated from the conditions

$$\dot{S}_1(D_1^{(c)}, \bar{D}_2) = \dot{S}_1(D_1^{(b)}, \bar{D}_2) \qquad \text{and} \qquad \hat{\mathcal{E}}(D_1^{(b)}, D_1^{(c)}, \bar{D}_2) = 0\,. \tag{43}$$

The $\hat{U}$-graph in zone (b) is non-convex, analogous to that for the fundamental post-critical path sketched in Fig. 4c. $D_1^{(b)}$ and $D_1^{(c)}$ satisfying (43) correspond to the tangent points of this graph to a common straight line. Note that along any post-critical path involving a homogeneous zone where the $\hat{U}$-graph has the form in Fig. 4b or 4c, inifinitely many velocity solutions exist at every instant. This indeterminacy is removed by (43) with the help of the stability condition (36).

Once $D_1^{(b)}$ is known, $D_1^{(a)}$ is determined by the statical condition

$$h^{(a)}\dot{S}_1(D_1^{(a)}, \bar{D}_2) = h^{(b)}\dot{S}_1(D_1^{(b)}, \bar{D}_2)\,, \tag{44}$$

b follows from (41), and $\dot{h}/h = D_3$ is obtained from the continuing plane-stress condition $\dot{S}_3 = 0$. The current width b_h of the neck, corresponding to the sheet thickness distribution, is defined by a formula analogous to (41), but with D_1 replaced by h. The averaged sheet thickness $\bar{h}$ can be approximately calculated from the incompressibility condition. Hence, important quantities of interest can be computed by analysing the material state in the two uniformly deforming zones (a) and (b). A knowledge of the deformation history in other parts of the sheet is not needed. The evolution of the transitory zone (c), separating the zones (a) and (b), remains, of course, undetermined.

The governing equations have been solved in a step-by-step manner (Petryk and Thermann, 1996), using the finite strain version of J_2 corner theory to describe the incrementally nonlinear plastic response at the current yield vertex. As a partial result from that work, the post-critical evolution of b and b_h with overall logarithmic strain $\bar{e}_1$ is shown in Fig. 6 by broken lines. The graphs correspond to balanced biaxial tension ($\rho = 1$). Material parameters are given in the figure, however, sufficiently large elastic moduli have been taken to model the incompressible rigid-plastic material (considered in Christoffersen and Hutchinson (1979)).

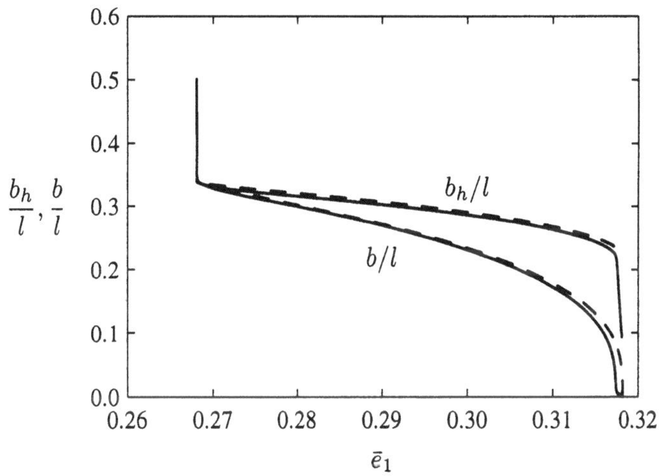

Figure 6. Relative width of the current necking band, b/l, and of the neck, b_h/l, vs overall logarithmic strain $\bar{e}_1$. J_2 corner theory: $\theta_c = 135°$, $\theta_0 = 22.5°$, $N = 0.2$, $E = 500\,\tau_0$, $\nu = 0.3$. Solid lines correspond to the 2D FE solution with distributions shown in Fig. 7, broken lines have been determined for negligible elastic compliances from the analytical model under plane-stress conditions. (From Petryk and Thermann, 1996)

3.3 Finite Element Results for a Sheet of Finite Thickness

The problem formulated in Section 3.1 is resumed here. Without the plane-stress idealization, a sheet of finite, but small initial thickness $h_0 = 2l_0/100$ is numerically analysed by the finite element method, using the computational procedure proposed in Section 2. Balanced biaxial stretching ($\rho = 1$) is considered for the same material as above, however, elasticity is not neglected.

During the calculations (Petryk and Thermann, 1996), only one bifurcation point was found, and the instant and mode of the bifurcation for the discretized problem (600 × 3 quadrilateral elements on one quadrant, each consisting of four crossed triangles) practically coincided with the continuum results. Although the strain-rate field just beyond the bifurcation point has the typical sinusoidal form, during subsequent deformation it undergoes rapid redistribution. Two zones of practically constant thickness strain-rate $\bar{D}_3 \equiv \dot{h}/h$, separated by a transition region, are already visible in Fig. 7a when the overall logarithmic strain $\bar{e}_1$ at the bifurcation point is incremented by 0.0001. $|\bar{D}_3|$ in the necking zone is rapidly growing, whereas the zone itself is shrinking with continuing deformation. This gives rise to a similar, but more moderate growth of the magnitude of the thickness strain $\bar{e}_3 \equiv \ln(h/h_0)$ in the respective zone, shown in Fig. 7b.

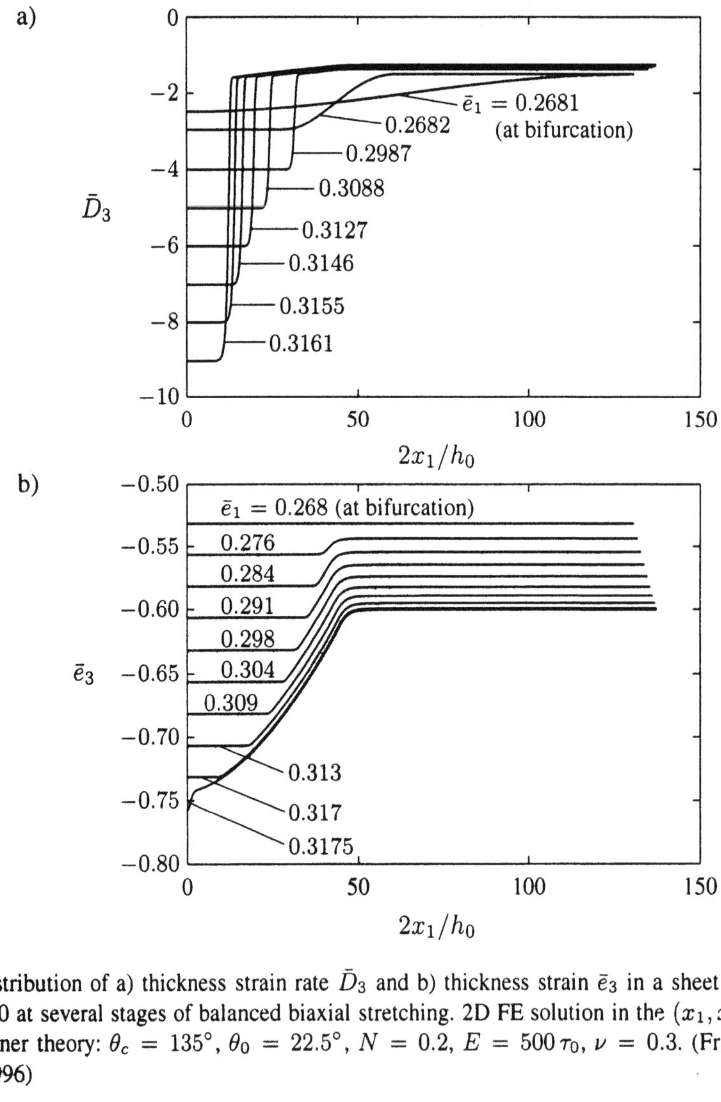

Figure 7. Distribution of a) thickness strain rate $\bar{D}_3$ and b) thickness strain $\bar{e}_3$ in a sheet of initial ratio $2l_0/h_0 = 100$ at several stages of balanced biaxial stretching. 2D FE solution in the (x_1, x_3) plane from Fig. 3. J_2 corner theory: $\theta_c = 135°$, $\theta_0 = 22.5°$, $N = 0.2$, $E = 500\,\tau_0$, $\nu = 0.3$. (From Petryk and Thermann, 1996)

The computations have been stopped at $\bar{e}_1 = 0.3175$, where the limit of stability of equilibrium was approached.

The strain-rate distribution qualitatively agrees with all predictions from the analysis in Section 3.2 for the one-dimensional plane-stress model, and gives additional information in the transitory zone. Moreover, the quantities

$$\frac{b}{l} = \frac{\bar{D}_3^{\text{mean}} - \bar{D}_3^{\text{max}}}{\bar{D}_3^{\text{min}} - \bar{D}_3^{\text{max}}}, \qquad \frac{b_h}{l} = \frac{h^{\text{max}} - \bar{h}}{h^{\text{max}} - h^{\text{min}}}, \tag{45}$$

based on the distributions in Fig. 7, evolve in time in the same manner as the corresponding quantities from the plane-stress theory (Fig. 6).

4 Evolution of Shear Bands Under Plane-Strain Compression

4.1 Introduction

The formation and evolution of shear bands in the post-critical range of an elastic-plastic solid under plane-strain conditions will be discussed by a number of numerical studies. The underlying elasto-plastic material is assumed to be time-independent and incrementally nonlinear as described in section 2.1. In most of the computations J_2 corner theory is used, complemented by sample results for the recently proposed yield vertex modification of two-surface models.

In contrast to buckling and bulging phenomena, where boundary conditions play an essential role, the notion of material instability is used to describe deviations from a fundamental deformation path inside a material body. To study such instabilities, the influence of boundary conditions should be eliminated or at least reduced. For computations with the finite element method, assuming that the overall deformation is specified by its gradient $\bar{\mathbf{F}}(t)$ as a given function of time t, two natural approaches suggest themselves.

A first possibility is to consider a finite domain with

- kinematically controlled motion of the boundary,

so that the current position of a material point on the boundary initially at $\boldsymbol{\xi}$ is given by

$$\mathbf{x}(\boldsymbol{\xi}, t) = \bar{\mathbf{F}}(t)\boldsymbol{\xi}. \tag{46}$$

Smooth bifurcations of the fundamental mode of uniform straining $\mathbf{F}^0 = \bar{\mathbf{F}}$ have not been met in the computations. Provided that a sufficiently fine discretization by finite elements is employed, the influence of the boundary on the post-critical deformation is expected to be confined to a small region in its vicinity, leaving the central part of the domain unaffected. In this inner region the material behaviour should consequently be largely independent of the shape of the boundary.

The second approach tries to simulate the behaviour of an infinite continuum under a given mean deformation $\bar{\mathbf{F}}$. The corresponding discretized problem can be solved on a finite domain, when

- periodic boundary conditions

are prescribed. In the undeformed body the unit cell is given as a rectangular domain. Its corner points experience the motion prescribed by (46) to enforce the given overall deformation, and displacements on opposite edges are coupled so that in the deformed state the whole plane can still be filled with a periodical arrangement of unit cells. As a disadvantage in comparison with purely kinematically constrained boundary conditions, the much stronger coupling of unknowns due to periodicity and the resulting memory requirements for data storage in the numerical analysis should be mentioned. On the other hand, even with a coarse mesh reasonable results can be obtained (at early stages in the post-critical range), when the mesh is properly designed and the aspect ratio of the unit cell is appropriately selected.

Figure 8 shows such a mesh at various stages of deformation under isochoric compression along the horizontal coordinate axis x_1. As long as the compression is less than $\approx 27\%$, the deformation remains uniform. Then, the solution bifurcates and two zones with markedly different spatial velocity gradients $\mathbf{V}$ appear. In these zones the stress and deformation state develops uniformly but differently with a spatial arrangement that remains unchanged up to $\approx 32\%$ compression. At this stage, a shear band is already clearly visible. After a second bifurcation, three uniformly deformed zones are observed up to the depicted final state with 40% compression. The mesh is too coarse to properly describe these later stages of deformation, where a propagation of discontinuity surfaces and more complicated spatial deformation patterns are to be expected.

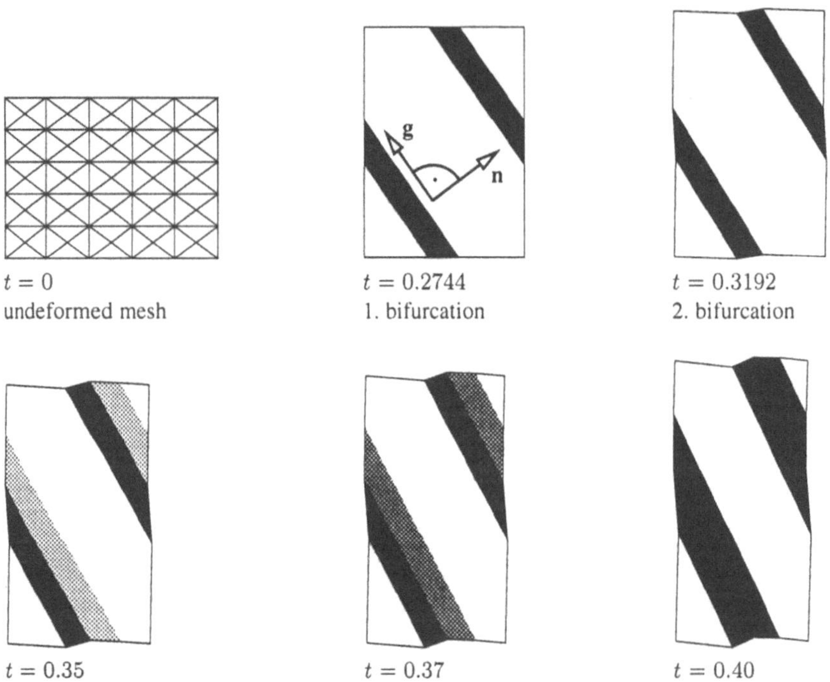

Figure 8. Evolution of shear rate fields $V_{gn} = \mathbf{g} \cdot \mathbf{V}\mathbf{n}$ on a coarse 5×5 mesh with periodic boundary conditions. Overall plane-strain compression along the horizontal axis x_1: $\bar{F}_{11} = 1 - t$, $\bar{F}_{22} = 1/\bar{F}_{11}$, $\bar{F}_{33} = 1$, J_2 corner theory: $E = 500\,\tau_0$, $\nu = 0.3$, $N = 0.1$, $\beta_c = 126.2°$.

A typical feature, the existence of a post-critical range with a one-dimensional stress and deformation state, is, however, correctly predicted. In this range, the computational effort can be drastically reduced by

- one-dimensional finite element modelling,

requiring only n elements to describe one row of the mesh in Fig. 8 instead of $4n^2$ for a full two-dimensional analysis.

Still further simplifications are possible. At the early stages in the post-critical range, the material is simply in one of two states, and it is sufficient to analyse these uniformly deforming zones (the collection of black and white areas, respectively, in Fig. 8) using the kinematical and statical conditions at their interfaces together with the material law. This

- semi-analytical approach

also allows an approximative study of propagating discontinuities, where the two zones are separated by a transitory region (grey areas in Fig. 8). It has been proposed by Petryk (1997) on purely theoretical grounds. His results will be presented here as a necessary prerequisite for mesh design and as a basis for comparisons.

The main purpose of this study is to illustrate the possibilities and power of the proposed computational finite element approach and to relate the results to findings from a general theory of stability of a deformation process. By using different models, a rather complete picture of shear band development over a broad range after primary bifurcation is obtained (cf. Petryk and Thermann, 1999a, 1999b), Later stages of post-critical deformation are at present less well understood and require further numerical and theoretical work.

4.2 Uniform Deformation and the Onset of Path Instability

Since our computational procedure is able to follow the stable, energetically preferable deformation path, there is no need to introduce geometrical or material imperfections to trigger localization. An analysis of the perfect system has the advantage that finite element computations may be started at any state of uniform deformation, e.g. near the critical instant where shear bands become possible for the first time. Only a time integration of the constitutive equations is needed to obtain the stress and hardening state of the uniformly deforming material. Any standard ordinary differential equation solver can be used for this purpose, and in the present study a predictor-corrector method from the NAG library has been employed. Besides an error controlled variable step-size selection, it incorporates a root-finding algorithm for user specified functions of the variables so that the endpoint of the time integration can be specified implicitly as a root of such an event-function.

In this way, local conditions for material stability, available from Petryk's (1992) analysis for incrementally nonlinear solids, can be conveniently checked. In general, rank 1 convexity of the constitutive potential U at $\dot{\mathbf{F}}^0$,

$$\mathcal{E}(\dot{\mathbf{F}}^0, \dot{\mathbf{F}}^0 + \mathbf{g} \otimes \mathbf{n}) \geq 0 \qquad \text{for all vectors } \mathbf{g}, \mathbf{n}\,, \tag{47}$$

where $\mathcal{E}$ is the Weierstrass function associated with $U(\dot{\mathbf{F}})$,

$$\mathcal{E}(\dot{\mathbf{F}}, \dot{\mathbf{F}}^*) = U(\dot{\mathbf{F}}^*) - U(\dot{\mathbf{F}}) - \dot{\mathbf{S}} \cdot (\dot{\mathbf{F}}^* - \dot{\mathbf{F}})\,,$$

has to be examined to assure stability of the fundamental path of uniform straining with the spatial velocity gradient $\dot{\mathbf{F}}^0$. The Legendre-Hadamard or semi-strong ellipticity condition

$$(\mathbf{g} \otimes \mathbf{n}) \cdot \mathsf{C}(\dot{\mathbf{F}}^0)\,(\mathbf{g} \otimes \mathbf{n}) \geq 0 \qquad \text{for every } \mathbf{g}, \mathbf{n} \tag{48}$$

and the local condition necessary for stability of equilibrium

$$U(\mathbf{g} \otimes \mathbf{n}) \geq 0 \qquad \text{for every } \mathbf{g}, \mathbf{n} \tag{49}$$

are implied by (47). Minima of the left hand expressions in (47)–(49) with respect to unit vectors $\mathbf{g}$ and $\mathbf{n}$ are calculated at each time step along the fundamental path. Critical instants, where these functions pass through zero, may be determined automatically by the time integration subroutine.

If the constitutive inequality (22) is satisfied then (48) is equivalent to (47) and bifurcations into shear bands are energetically preferable as soon as (48) fails to hold, i.e. when ellipticity is lost (cf. Petryk, 1992 for a detailed discussion). Hence, the familiar, linearized bifurcation condition (Hill, 1962, Rice, 1977)

$$\mathbf{A}(\mathbf{n}^\star)\mathbf{g}^\star = \mathbf{0} \qquad \text{with} \qquad A_{ik} = C_{ijkl}(\dot{\mathbf{F}}^0) n_j^\star n_l^\star \tag{50}$$

can be used to determine the critical instant and the corresponding unit vectors $\mathbf{g}^\star$ and $\mathbf{n}^\star$ for which the left-hand side in (48) assumes a minimum with value zero, being positive for other non-zero vectors.

Critical instants $t^\star$ and related quantities are recorded in Table 1 for both material models subjected to isochoric plane-strain compression:

$$F_{11}^0 = 1 - t\,, \qquad F_{22}^0 = 1/F_{11}^0\,, \qquad F_{33}^0 = 1\,.$$

Table 1. Data at primary bifurcation

Material	β_c^{max}	$t^\star$	φ_n	φ_g	φ_{n_0}	η
J_2 corner theory	110°					0.2918
(Christoffersen and Hutchinson, 1979)	115°					0.2579
$E = 500\ \tau_0$	120°	0.2744	36.00°	125.93°	54.07°	0.2294
$\nu = 0.3$	125°					0.2051
$N = 0.1$	126.158°					0.2
	130°					0.1841
as above	110°	0.2750	35.94°	125.94°	54.06°	0.2921
but with $\nu = 0.499$	130°					0.1846
Two-surface model	110°	0.3801	39.03°	128.96°	64.64°	0.2301
(Petryk and Thermann, 1997)	115°	0.3143	37.37°	127.30°	58.38°	0.2430
$E = 500\ \tau_0$	117.365°	0.2896	36.54°	126.47°	55.75°	0.25
$\nu = 0.3$	120°	0.2656	35.58°	125.51°	52.99°	0.2566
$N = 0.1$	125°	0.2279	33.63°	123.56°	48.13°	0.2770
	130°	0.1978	31.43°	121.36°	43.52°	0.2985
	130.315°	0.1961	31.29°	121.21°	43.23°	0.3
as above	110°	0.3809	38.97°	128.97°	64.65°	0.2308
but with $\nu = 0.499$	130°	0.1982	31.36°	121.36°	43.47°	0.2990

The cone-angle $\beta_c^{\max}$ is varied and all other material parameters are kept fixed as indicated. φ_n and φ_{n_0} are the angles between the compression axis and the normals to the shear band in the current and initial configuration, respectively, and the direction of shear is $\mathbf{g}^\star = (\cos\varphi_g, \sin\varphi_g)$. $\mathbf{g}^\star$ and $\mathbf{n}^\star$ are almost orthogonal with a slight deviation from 90°, which is less than 0.1°. Elastic compressibility has generally little influence on the results, as indicated in Table 1 by sample data for Poisson's ratio equal to 0.499. It is obvious from the symmetry of plane-strain compression that $\mathbf{n}^\star$ and $\mathbf{g}^\star$, corresponding to φ_n and φ_g taken with negative sign, constitute a second pair of minimizing directions. The last entry η records the incipient volume fraction of the highly strained material as predicted by Petryk's (1997) analysis. Simple formula are available (not for η) from Hutchinson and Tvergaard (1981) in the case of of J_2 corner theory when compressibility is neglected.

Finite element results. It is emphasized that the bifurcation results, collected here for later comparisons, can be conveniently used to reduce the computational burden, but are in principle not needed, when a finite element analysis with the method described in Section 2.2 is envisaged. The mesh has, however, to be designed in such a manner that the formation of bands in the theoretically predicted directions is possible. In this sense the angles φ_{n0} or φ_n are the most useful quantities for the FE analysis.

In the following two-dimensional FE-computations quadrilateral elements, divided into four constant strain 'crossed triangles', have been used. When the quadrilaterals are oriented parallel to the coordinate axes, bands can most easily be formed parallel to the element diagonals so that the angles, say φ_0 (in the initial, unstressed configuration), between vertical lines and diagonals determine preferred band orientations. Another possibility is to construct the mesh with all rectangular elements rotated by φ_0. Both techniques are used here, the first in the case of periodic boundary conditions, the second for an initially circular domain with kinematically controlled motion of the boundary. The influence of φ_0 on the critical compressive strain, obtained from FE computations with n quadrilateral elements along the sides of the unit cell or on the radius of the circular domain, is shown in Fig. 9. The discretized system is generally stiffer than the continuum and has, therefore, higher bifurcation strains. When the number of elements is increased, the smallest bifurcation strain as well as the corresponding φ_0 converge to the continuum values discussed in the preceeding paragraph. Note that for the rectangular domain with periodic boundary conditions these theoretical values, i.e. $t^\star = 0.2744$, $\varphi_{n0} = 54.1°$ for the material data from Fig. 9 and Table 1, are already exactly obtained for a 2×2 mesh.

Besides $t^\star$ and φ_0 the volume fraction η of the localization zone, also simply denoted as bandwidth in the following, is a third quantity which may be used directly to compare the theoretical and numerical results. In the above computations the cone angle of the material has been specified such that the theoretical bandwidth, predicted by Petryk's (1997) theory, is $0.2 = 1/5$. This means that a periodic 5×5 mesh should be sufficient to reproduce this volume fraction exactly, what is indeed the case, as illustrated in Fig. 10. The final velocity field, which is returned by the minimization procedure, depends on the nodal velocities used to start the trust region method. Figure 10 shows 20 different distributions of the principal strain rate with largest magnitude, obtained from 35 randomly selected initial velocity vectors at $t = t^\star + 10^{-7}$. Grey scales from white (lowest intensity) to black (highest intensity) are used to visualize strain-rate levels in the elements. All exclusively 'black/white' distributions are equivalent in the sense that the minimized velocity functional attains the same minimum value $J = 0.416983$, being smaller

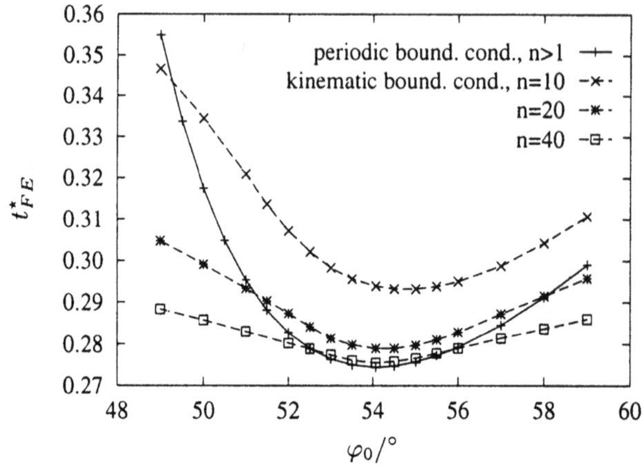

Figure 9. Dependence of the critical compressive strain on mesh design. J_2 corner theory $\beta_c^{\max} = 126.2°$. Overall plane-strain compression: $\bar{F}_{11} = 1 - t$, $\bar{F}_{22} = 1/\bar{F}_{11}$

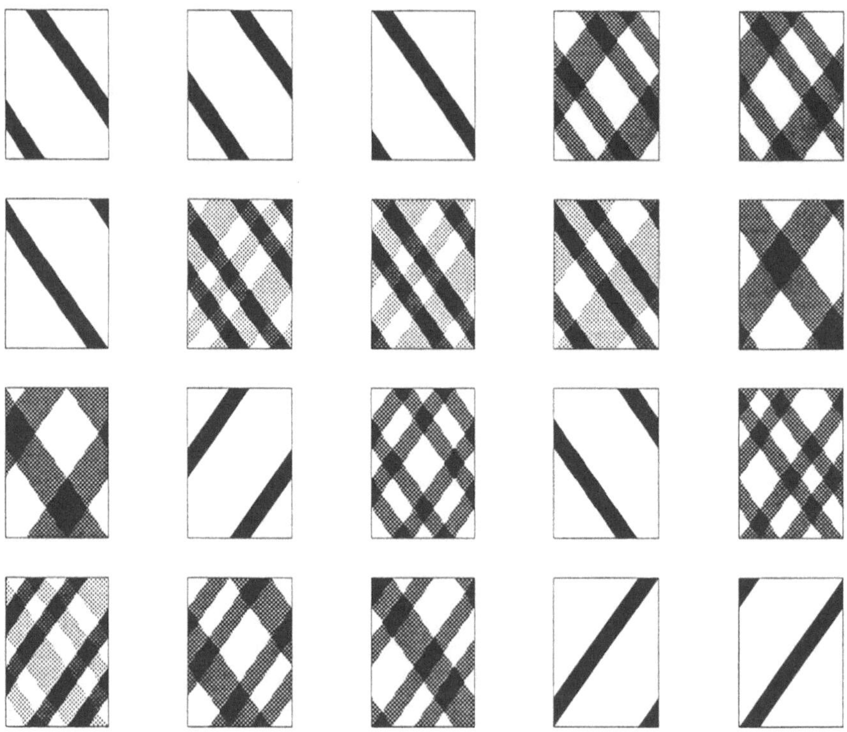

Figure 10. Distributions of principal strain-rate with largest magnitude at primary bifurcation on a 5 × 5 mesh with periodic boundary conditions. Randomly selected starting velocity fields. J_2 corner theory: $\beta_c^{\max} = 126.2°$. Overall plane-strain compression: $\bar{F}_{11} = 1 - t$, $\bar{F}_{22} = 1/\bar{F}_{11}$

than for any other distribution. Obviously, the bandwidth for these solutions is 1/5, in agreement with the theoretical prediction. When the bandwidth is a rational number, say k/n, there exist altogether $2\binom{n}{k} = 2\frac{n!}{k!(n-k)!}$ equivalent distributions of this type and 8 out of 10 have been found after 35 random trials for the sample computation from Fig. 10. Patterns with bands in two directions represent only local minima, with values slightly above the absolute minimum ($\Delta J = 0.000124$ or 0.000119), and must be rejected.

This demonstrates a general difficulty. The employed trust region method works efficiently and returns at least a local minimum, but there is no guarantee that the absolute minimum, which we are looking for, is obtained. However, the situation is not as bad as it might look at first glance. From (limited) numerical experience it is found that at some later stage of overall deformation ($\Delta t \approx 0.002$) the local minimum of a non-optimal bifurcation pattern has turned into a stationary point or local maximum and is then automatically rejected by the minimization procedure. After another bifurcation the solution always returned to the supposed global optimum, yielding bands in only one direction with the correct volume fraction.

The 5×5 mesh in Fig. 10 is ideally adapted to the theoretical solution with a bandwidth of 1/5. In general, for a given subdivision of the unit cell in $n \times n$ quadrilateral elements the best approximation of the discretized problem to the continuum solution with a volume fraction η of the localization zone is found to consist of k bands (with a high velocity gradient), where k is the largest integer so that $k/n \leq \eta$. If equality in this relation is not possible, the inability to exactly achieve the bandwidth is compensated by the formation of one additional band with an adjusted velocity gradient. This is best illustrated by plotting the component $V_{gn} = \mathbf{g}^\star \cdot \mathbf{V}\mathbf{n}^\star$ of the velocity gradient over the accumulated element volume. Figure 11 shows such a plot for different values of n, when the shear rates V_{gn} are sorted in descending numerical order and the total volume is normalized to unity. The total area under each distribution is the same, namely $\bar{V}_{gn} = \mathbf{g}^\star \cdot \bar{\mathbf{V}}\mathbf{n}^\star$, where $\bar{\mathbf{V}}$ is the velocity gradient of the prescribed overall deformation.

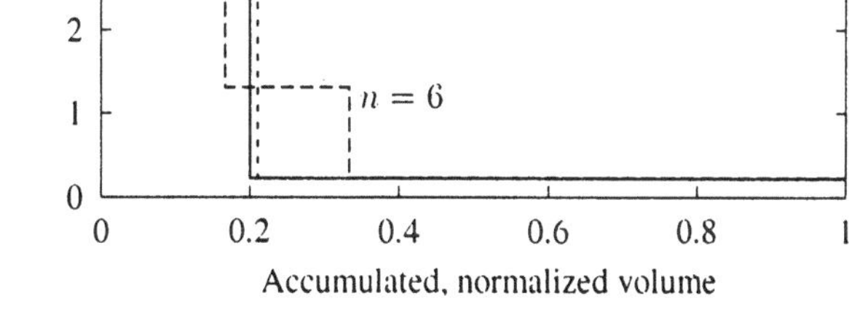

Figure 11. Approximations of the shear rate distribution at primary bifurcation on an $n \times n$ periodic mesh. J_2 corner theory: $\beta_c^{\max} = 126.2°$. Overall plane-strain compression: $\bar{F}_{11} = 1 - t, \bar{F}_{22} = 1/\bar{F}_{11}$. (From Petryk and Thermann, 1999a)

Moreover, the high and low shear rates are not influenced by n (when $n \geq 1/\eta$) so that an improved approximation η_{FE} to the bandwidth, with an accuracy better than the resolution $1/n$ of the mesh, can be computed from

$$V_{gn}^{\max}\eta_{FE} + V_{gn}^{\min}(1-\eta_{FE}) = \bar{V}_{gn}\,. \tag{51}$$

It coincides with the theoretical value.

4.3 Semi-Analytical Approach

The analysis presented here follows Petryk's (1997,1998) work with minor modifications (cf. also Petryk and Thermann, 1999b). The slight degree of elastic compressibility of the material is neglected from the very beginning and for the integration of the governing equations the predictor-corrector method mentioned above has been used.

The post-critical solution for an infinite continuum is assumed to be one-dimensional with the deformation gradient

$$\mathbf{F}(\boldsymbol{\xi},t) = \bar{\mathbf{F}}(t)\,(\mathbf{I} + \gamma(\xi,t)\,\boldsymbol{\gamma}\otimes\boldsymbol{\nu}^\star)\,, \qquad \xi = \boldsymbol{\xi}\cdot\boldsymbol{\nu}^\star\,. \tag{52}$$

$\boldsymbol{\gamma}$ and $\boldsymbol{\nu}^\star$ are orthogonal unit vectors, known from the bifurcation analysis, representing the direction of shear and its normal in the initial configuration with coordinates $\boldsymbol{\xi}$. Using the inverse $\mathbf{F}^{-1} = (\mathbf{I} - \gamma\boldsymbol{\gamma}\otimes\boldsymbol{\nu}^\star)\bar{\mathbf{F}}^{-1}$ the spatial velocity gradient $\mathbf{V} = \dot{\mathbf{F}}\mathbf{F}^{-1}$ is obtained as

$$\mathbf{V} = \bar{\mathbf{V}} + g\,\mathbf{g}\otimes\mathbf{n}^\star\,, \tag{53}$$

where $\bar{\mathbf{V}} = \dot{\bar{\mathbf{F}}}\bar{\mathbf{F}}^{-1}$ is the velocity gradient of the given overall deformation,

$$\mathbf{g} = \frac{\bar{\mathbf{F}}\boldsymbol{\gamma}}{|\bar{\mathbf{F}}\boldsymbol{\gamma}|}\,, \qquad \mathbf{n}^\star = \frac{\bar{\mathbf{F}}^{-T}\boldsymbol{\nu}^\star}{|\bar{\mathbf{F}}^{-T}\boldsymbol{\nu}^\star|} \tag{54}$$

are the shear direction and its normal in the current configuration with coordinates $\mathbf{x}$, and

$$g = \dot{\gamma}|\bar{\mathbf{F}}\boldsymbol{\gamma}||\bar{\mathbf{F}}^{-T}\boldsymbol{\nu}^\star| \tag{55}$$

is the shear rate. g may be considered as a function of the coordinate $x = \mathbf{x}\cdot\mathbf{n}^\star$ in the direction of the current normal and is taken as the basic unknown quantity, subjected to the restriction

$$\left|\int_{-\infty}^{+\infty} g(x)dx\right| < \infty\,. \tag{56}$$

The analysis shows that two zones, (a) and (b), of uniform velocity gradient can still exist in the post-critical range, in general accompanied by a developing transitory zone where γ is not uniform (as illustrated by Fig. 8).

The volume fraction $\eta^{(b)}$ of the uniform localization zone (b) is allowed to be fixed or decreasing in time. For simplicity we assume that the constitutive inequality (22) is satisfied. A condition necessary for stability of the post-critical deformation path then is

$$C_{gn}^{(b)} \equiv (\mathbf{g}\otimes\mathbf{n}^\star)\cdot\mathsf{C}(\mathbf{V}^{(b)})(\mathbf{g}\otimes\mathbf{n}^\star) \geq 0\,, \tag{57}$$

as a consequence of (48) with the velocity gradient $\mathbf{V}^{(b)} = \bar{\mathbf{V}} + g^{(b)}\mathbf{g} \otimes \mathbf{n}^\star$ substituted in place of $\dot{\mathbf{F}}^0$. As long as $C_{gn}^{(b)} > 0$, $\eta^{(b)}$ may remain constant. If $C_{gn}^{(b)}$ has decreased to zero and further deformation with fixed $\eta^{(b)}$ would make $C_{gn}^{(b)}$ negative, then $\eta^{(b)}$ can decrease in accordance with the condition for quasi-static propagation of a velocity gradient discontinuity, cf. Hill (1962). Thus, the evolution criterion for $\eta^{(b)}$ takes the form

$$C_{gn}^{(b)}\dot{\eta}^{(b)} = 0\,, \qquad C_{gn}^{(b)} \geq 0\,, \qquad \dot{\eta}^{(b)} \leq 0 \tag{58}$$

at each instant.

Analogous conclusions for the current shear modulus and the volume fraction in zone (a) are obtained when (b) is replaced by (a) in the above paragraph.

(58) represents the additional condition derived from the requirement of path stability. Together with the usual kinematical and statical conditions, it can be used to determine *overall* characteristics of post-critical deformation, although the spatial arrangement of different zones at a micro-level is indeterminate.

Continuing equilibrium requires the nominal shear stress rate $\dot{S}_{gn} = \mathbf{g} \cdot \dot{\mathbf{S}}\mathbf{n}^\star$ to be constant along x and imposes the statical condition

$$\dot{S}_{gn}^{(a)} = \dot{S}_{gn}^{(b)} \tag{59}$$

on the stress rates in zones (a) and (b). With the constitutive law and the velocity gradient substituted, this is a first equation for the unknown shear rates $g^{(b)}$ and $g^{(a)}$ of opposite sign. At the onset of path instability the incipient volume fraction is known. When $\eta^{(b)}$ remains fixed subsequently, a second equation for the shear rates is obtained from the kinematical condition

$$\eta^{(b)}g^{(b)} + (1 - \eta^{(b)})g^{(a)} = 0\,, \tag{60}$$

implied by (56).

When $\eta^{(b)}$ decreases, the propagation condition (57), $C_{gn}^{(b)} = 0$, is available as a second equation to determine $\eta^{(a)}$ and $\eta^{(b)}$. Now, two decoupled nonlinear equations have to be solved, $g^{(b)}$ is obtained from (57), and then $g^{(a)}$ from (59). The actual variation of $\eta^{(b)}$ is not computed, rather its approximation η is found as a conventional volume fraction of the localization zone from the formula

$$\eta = \frac{-g^{(a)}}{g^{(b)} - g^{(a)}} = \frac{\mathbf{g} \cdot (\bar{\mathbf{V}} - \mathbf{V}^{(a)})\mathbf{n}^\star}{\mathbf{g} \cdot (\mathbf{V}^{(b)} - \mathbf{V}^{(a)})\mathbf{n}^\star}\,, \tag{61}$$

motivated by (60).

With this approximation, it is sufficient to determine the stress and deformation state only in the two uniformly strained zones. As time integration procedure the same predictor-corrector method already used to find the stress and hardening state in the case of uniform straining can be conveniently applied with little changes. The difference is that the number of differential equations is now twice as large and that the shear rates $g^{(a)}, g^{(b)}$ have to be found iteratively in an inner loop by Newton's method. The amount of additional shear is simultaneously obtained by integrating $\dot{\gamma}^{(a)}$ and $\dot{\gamma}^{(b)}$ from (55), specified for zone (a) and (b). The volume fraction (61) depends on the spatial distribution of the velocity-gradient; a similar quantity η_F is also calculated,

$$\eta_F = \frac{-\gamma^{(a)}}{\gamma^{(b)} - \gamma^{(a)}} = \frac{\mathbf{g} \cdot (\bar{\mathbf{F}} - \mathbf{F}^{(a)})\boldsymbol{\nu}^\star}{\mathbf{g} \cdot (\mathbf{F}^{(b)} - \mathbf{F}^{(a)})\boldsymbol{\nu}^\star}\,, \tag{62}$$

reflecting the non-uniformity of the deformation gradient.

The interaction between $C_{gn}^{(b)}$ and η is illustrated in Fig. 12 for the two-surface model with a

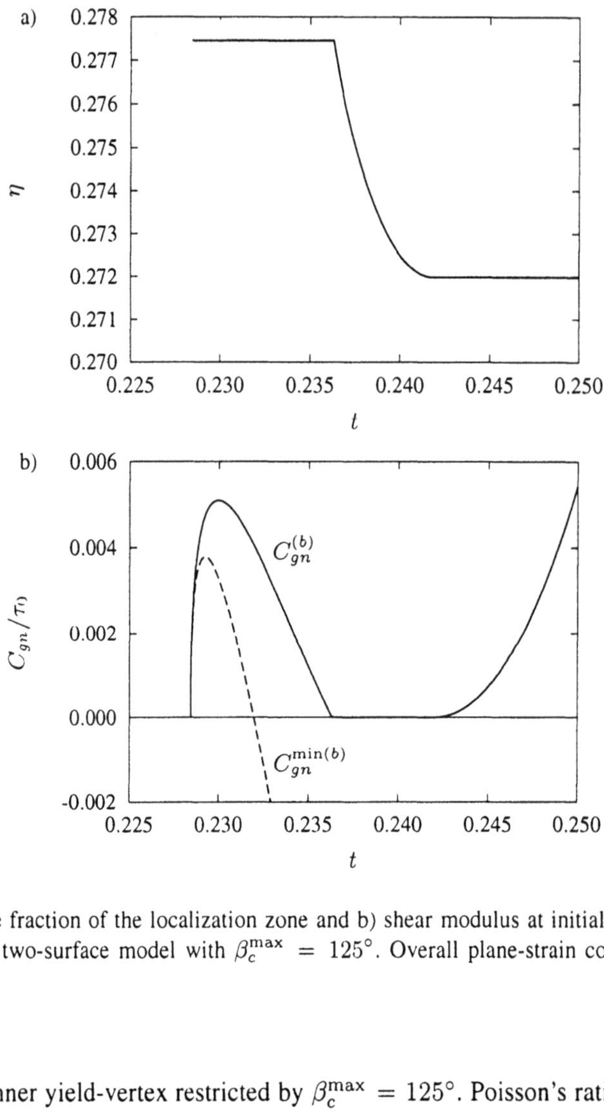

Figure 12. a) Volume fraction of the localization zone and b) shear modulus at initial stages in the post-critical range for the two-surface model with $\beta_c^{\max} = 125°$. Overall plane-strain compression: $\bar{F}_{11} = 1 - t, \bar{F}_{22} = 1/\bar{F}_{11}$

cone-angle of the inner yield-vertex restricted by $\beta_c^{\max} = 125°$. Poisson's ratio has been taken as 0.499 and other material parameters are as given in the previous section. Immediately beyond the critical instant, the solution returns to the strongly elliptic regime and η remains equal to the incipient volume fraction. Shortly later, $C_{gn}^{(b)}$ again falls to zero and then η starts to decrease in accordance with (58), that is, an undetermined number of boundary planes of shear bands propagate such that the volume fraction of the localization zone is decreasing in time. This only happens during a short interval of overall compression and $C_{gn}^{(b)}$ soon becomes positive again, limiting the decrease in η to a small amount. Up to large further deformations, η can remain constant since the condition (57) is satisfied.

However, by (57) the content of the general condition (48) for stability of a deformation path has only been partly exhausted. In fact, (48) should be satisfied for *any* band orientation $\mathbf{n}$, not only for the particular direction $\mathbf{n}^*$. An absolute minimum value, $C_{gn}^{\min(b)}$, of the left-hand expression in (48) for orthonormal vectors $\mathbf{g}, \mathbf{n}$ and with $\dot{\mathbf{F}}^0$ replaced by $\mathbf{V}^{(b)}$ has therefore been computed. It is plotted as the broken line in Fig. 12. Being initially positive, it quickly falls below zero when the overall compressive strain is further increased. This means that the post-critical solution of the assumed form (53) ceases to satisfy the path stability condition (48), and a solution of a more complex spatial structure should be considered instead. However, one may also argue that in order to satisfy with sufficient accuracy the condition of *macroscopic* uniformity of the deformation, the width of any individual shear band must be sufficiently small with respect to the dimensions of the material element under consideration. Then, the formation of differently aligned secondary localization bands inside primary bands is constrained by the boundary conditions along the interfaces of primary bands. This means that violation of (48) in zone (b) for directions other than $\mathbf{n}^*$ need not automatically invalidate a solution of the form (53), at least not in the first instance. A similar conclusion may be drawn for the zone (a).

Possible negative values $C_{gn}^{\min(b)}$ or $C_{gn}^{\min(a)}$ should not, however, be numerically too small. This follows from the condition (48) applied at the macroscopic level to the effectively heterogeneous, layered material that has appeared due to the post-critical non-uniform deformation of an initially homogeneous material. An upper bound to the actual macroscopic stiffness moduli can simply be obtained as a volume average of the moduli. Consequently, if a non-uniform transitory zone is absent or negligible as in the case illustrated in Fig. 12 then the solution (53) may be regarded as stable only if

$$(\mathbf{g}\otimes\mathbf{n})\cdot\left(\eta\,\mathbf{C}(\mathbf{V}^{(b)})+(1-\eta)\mathbf{C}(\mathbf{V}^{(a)})\right)(\mathbf{g}\otimes\mathbf{n})\geq 0 \qquad \text{for every } \mathbf{g},\mathbf{n}\,. \tag{63}$$

Together with $C_{gn}^{\min(b)}$ and $C_{gn}^{\min(a)}$ the absolute minimum value of the left-hand expression in (63), $C_{gn}^{\min(a/b)}$, is plotted in Fig. 13a and the angles between the corresponding minimizing normal vectors $\mathbf{n}^{**}$ and the compression axis are shown in Fig. 13b. The condition (63) for the mixture of layers is satisfied over a remarkably large range of post-critical deformations up to an overall compressive strain of more than 40% (for $\beta_c^{\max}=125°$). Since the minimum values $C_{gn}^{\min(b)}$ and $C_{gn}^{\min(a)}$ are reached for distinct orientations, with angles $\varphi^{(b)}$ and $\varphi^{(a)}$ in Fig. 13b, they may be simultaneously negative without violating (63). Of particular interest is the angle $\varphi^{(a/b)}$ at the second critical instant when the inequality (63) ceases to hold. As explained above, this new critical angle may be associated with the normal to an incipient *macroscopic* shear band in the effectively heterogeneous, layered material created in the post-critical deformation process. It is found to be 35° for the material considered in Fig. 13, whereas the normal to primary bands forms an angle of $\varphi_n = 20°$ with the compression axis at that stage.

The main conclusions stated above are taken from Petryk's (1998) analysis without change and are confirmed by the finite element results presented here in later sections. It is interesting to note that they were originally formulated at a time, when no support by finite element studies was available.

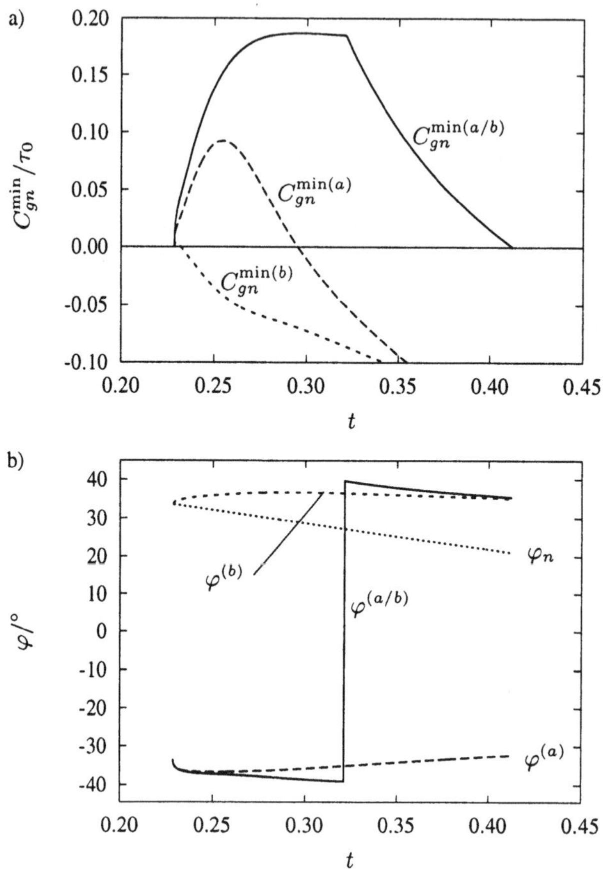

Figure 13. Post-critical evolution of a) the shear modulus $C_{gn}^{\min(a)}$, $C_{gn}^{\min(b)}$ in zones (a), (b) and $C_{gn}^{\min(a/b)}$ for the mixture of layers and b) the corresponding minimizing directions for the two-surface model with $\beta_c^{\max} = 125°$. Overall plane-strain compression: $\bar{F}_{11} = 1 - t, \bar{F}_{22} = 1/\bar{F}_{11}$

The range of validity of the one-dimensional approach is, of course, dependent on the material law, in particular, on the maximum cone angle allowed to develop in the incrementally nonlinear solid. The influence of $\beta_c^{\max}$ is shown in Fig. 14 for the two surface model, and in Fig. 15 for the J_2 corner theory. These results indicate that the shear modulus of the heterogeneous material remains positive over a significant range of post-critical deformations, provided the yield vertex corner is not too blunt. Both cases considered in the analysis, a decreasing volume fraction in either zone (b) or zone (a), are observed, depending on the material model and $\beta_c^{\max}$.

A systematic study of material parameters of the models, including kinematic hardening, is not intended here. One result should, however, be mentioned. It seems that only in extreme circumstances the bandwidth decreases rapidly to zero. This is, for example, the case when J_2 corner theory is applied and a very sharp yield vertex cone, $\beta_c^{\max} = 145°$, is admitted. The qualitative difference to the former cases with more moderate cone angles can be seen by comparing

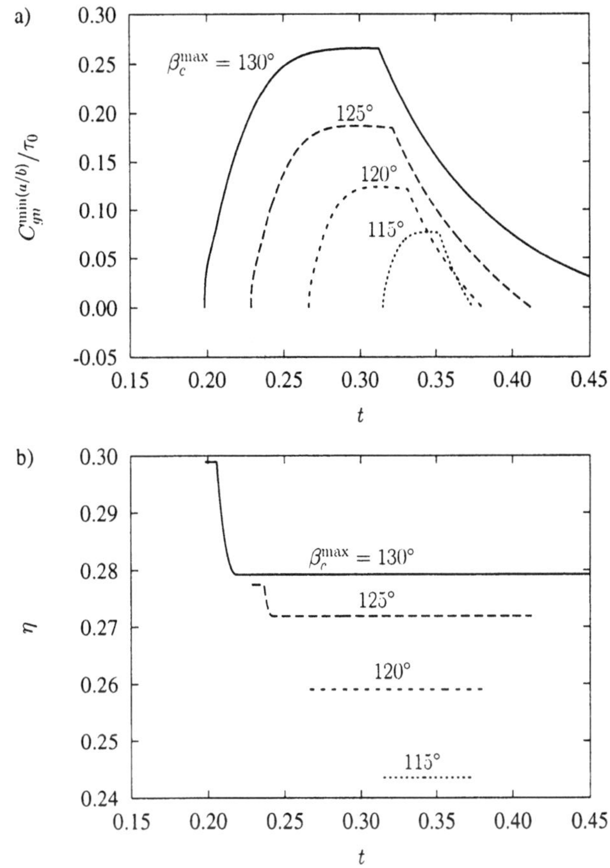

Figure 14. Post-critical evolution of a) the shear modulus for the mixture of layers and b) the volume fraction of the localization zone. Two-surface model. Overall plane-strain compression: $\bar{F}_{11} = 1-t$, $\bar{F}_{22} = 1/\bar{F}_{11}$

Fig. 16b with Fig. 15b. For the sharper corner, i.e. for $\beta_c^{max} = 145°$, the post-critical solution terminates at another critical point at which the condition (49) of stability of *equilibrium* is violated for $\mathbf{n} = \mathbf{n}^\star$. The value of η is rapidly decreasing to zero, terminating with a vertical asymptote. This corresponds to an unlimited growth of the shear rate $g^{(b)}$ in zone (b), as illustrated in Fig. 16a. The whole process from primary bifurcation up to the onset of instability of equilibrium is limited by a rather small amount of additional overall compressive strain, in contrast to the cases considered in Fig. 15b. Saturation or growth of shear deformations, depending on the parameters of J_2 corner theory, has already been reported by Hutchinson and Tvergaard (1981) in their imperfection study, where a vanishingly thin band was assumed.

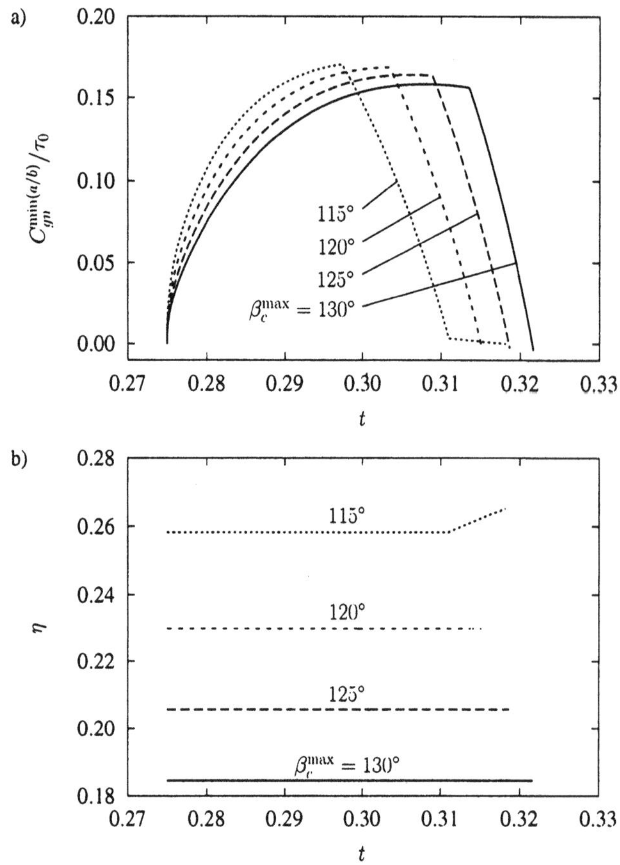

Figure 15. Post-critical evolution of a) the shear modulus for the mixture of layers and b) the volume fraction of the localization zone. J_2 corner theory. Overall plane-strain compression: $\bar{F}_{11} = 1 - t$, $\bar{F}_{22} = 1/\bar{F}_{11}$

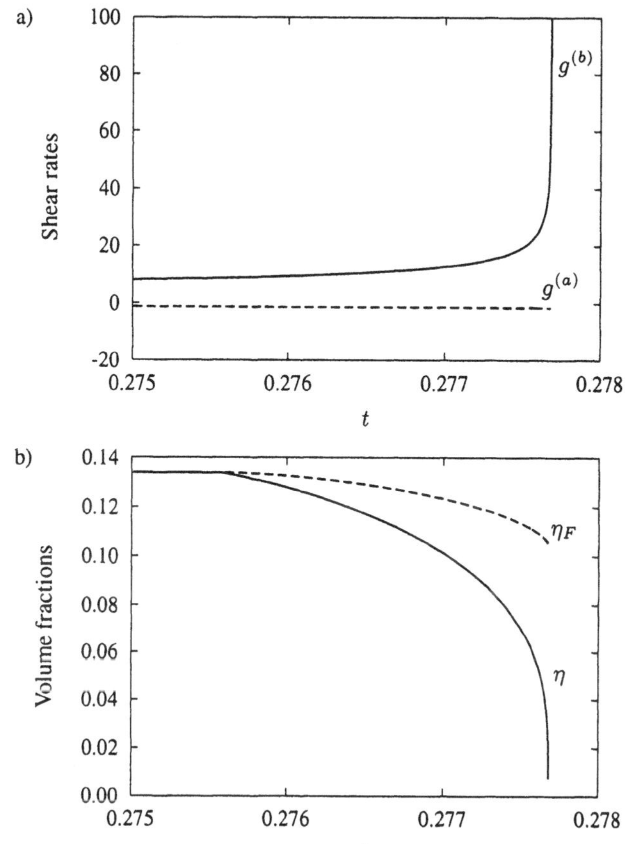

Figure 16. Post-critical evolution of a) the shear rates in zones (a) and (b) and b) the volume fractions of the localization zone, based on velocity gradient, η, and on deformation gradient, η_F. J_2 corner theory: $\beta_c^{\max} = 145°$. Overall plane-strain compression: $\bar{F}_{11} = 1 - t$, $\bar{F}_{22} = 1/\bar{F}_{11}$

4.4 One-Dimensional Finite Element Model

Before discussing two-dimensional finite element computations, we consider a FE modelling with one-dimensional elements as a natural complement to the semi-analytical approach. Since the computational procedure is entirely different, it offers an independent and mutual check of results. Moreover, additional information about the stress and strain history in the transitory zone can be obtained, not available in detail from the analytical approach.

Element description The one-dimensional in-plane motion is modelled by finite elements as illustrated in Fig. 17. In the initial configuration an interval on the coordinate axis $x_2 = 0$ is

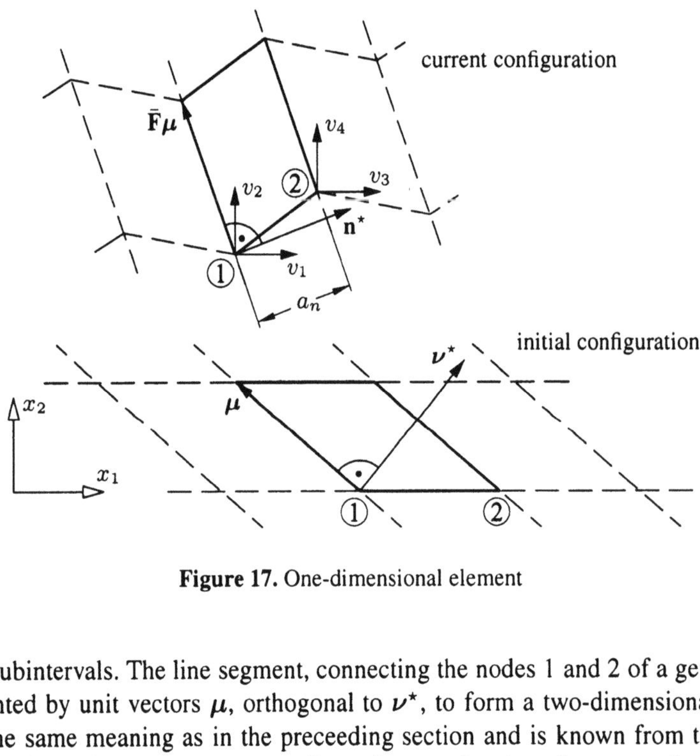

Figure 17. One-dimensional element

divided into subintervals. The line segment, connecting the nodes 1 and 2 of a generic element, is complemented by unit vectors $\boldsymbol{\mu}$, orthogonal to $\boldsymbol{\nu}^*$, to form a two-dimensional domain. $\boldsymbol{\nu}^*$ has exactly the same meaning as in the preceeding section and is known from the bifurcation analysis. The deformation to the current configuration may be viewed as a mapping with gradient $(\mathbf{I} + \boldsymbol{\gamma} \otimes \boldsymbol{\nu}^*)$ to which the given mean deformation with gradient $\bar{\mathbf{F}}$ is applied. The image of $\boldsymbol{\mu}$ is therefore $\bar{\mathbf{F}}\boldsymbol{\mu}$, with the normal $\mathbf{n}^*$, related to $\boldsymbol{\nu}^*$ by the second equation in (54). The distance of lines normal to $\mathbf{n}^*$ passing through the current element nodes, a_n, and the current element volume (for unit element thickness), V^{el},

$$a_n = (\mathbf{x}_2 - \mathbf{x}_1) \cdot \mathbf{n}^* , \qquad V^{\mathrm{el}} = |\bar{\mathbf{F}}\boldsymbol{\mu}|\, a_n , \tag{64}$$

are introduced as geometrical quantities.

In the current configuration, positions $\mathbf{x}_1$ and $\mathbf{x}_2$ of element nodal points 1 and 2, respectively, are known, together with the stress and hardening state.

Basic unknowns of the updated Lagrangian description are the element nodal velocities,

$$\mathbf{v}^{\mathrm{el}} = (\mathbf{v}(\mathbf{x}_1), \mathbf{v}(\mathbf{x}_2))^T = (v_1, v_2, v_3, v_4)^T \,, \tag{65}$$

used to interpolate the element velocity field in the form

$$\mathbf{v}(\mathbf{x}) = (\bar{\mathbf{V}} + \mathbf{g} \otimes \mathbf{n}^\star)(\mathbf{x} - \mathbf{x}_1) + \mathbf{v}(\mathbf{x}_1) \,, \tag{66}$$

where

$$\mathbf{g} = \left(\mathbf{v}(\mathbf{x}_2) - \mathbf{v}(\mathbf{x}_1) - \bar{\mathbf{V}}\,(\mathbf{x}_2 - \mathbf{x}_1)\right)/a_n \,. \tag{67}$$

Obviously, the spatial gradient of (66) yields the mean velocity gradient $\bar{\mathbf{V}}$ to which the dyad $\mathbf{g} \otimes \mathbf{n}^\star$ is added. The additional part would represent a simple shear, if $\mathbf{g}$ were orthogonal to $\mathbf{n}$, what is, however, not assumed here, contrary to the analysis in Section 4.3 (note also that $\mathbf{g}$ is no longer a unit vector). Elastic compressibility is now taken into account, and the vector $\mathbf{g}$ is obtained by (67) directly from the element nodal velocities.

The element contribution to the velocity functional, J^{el}, is obtained by integrating the constitutive potential U over the element volume. Since the velocity gradient is constant within the element, this simply gives

$$J^{\mathrm{el}} = \int_{V^{\mathrm{el}}} U \,\mathrm{d}V = U\, V^{\mathrm{el}} \,. \tag{68}$$

By using the chain rule and the constitutive law, the rate of internal forces,

$$\dot{Q}^{\mathrm{el}}_\alpha \equiv \frac{\partial J^{\mathrm{el}}}{\partial v_\alpha} = \frac{\partial g_i}{\partial v_\alpha}\, \dot{S}_{ik}\, n^\star_k\, V^{\mathrm{el}} \,, \qquad \alpha = 1, \ldots, 4 \,, \tag{69}$$

and the tangent stiffness matrix,

$$K^{\mathrm{el}}_{\alpha\beta} \equiv \frac{\partial \dot{Q}_\alpha}{\partial v_\beta} = \frac{\partial g_i}{\partial v_\alpha}\, A_{ik}\, \frac{\partial g_k}{\partial v_\beta}\, V^{\mathrm{el}} \,, \qquad \alpha, \beta = 1, \ldots, 4 \,, \tag{70}$$

are found on the element level. $\mathbf{A}$ in (70) has components

$$A_{ik} = C_{ijkl}\, n^\star_j n^\star_l \,, \qquad i, k = 1, 2 \,, \tag{71}$$

i.e. the matrix $\mathbf{A}$ in the linearized bifurcation condition (50) is recovered. When scaled by $\pm V^{\mathrm{el}}/a_n^2$ it constitutes symmetric 2×2 blocks of the tangent stiffness matrix, whose explicit form is found from (70) and (67) as

$$\mathbf{K}^{\mathrm{el}} = \begin{bmatrix} \mathbf{A} & -\mathbf{A} \\ -\mathbf{A} & \mathbf{A} \end{bmatrix} V^{\mathrm{el}}/a_n^2 \,. \tag{72}$$

Likewise, (69) leads to the rate of internal forces,

$$\begin{aligned} \dot{Q}^{\mathrm{el}}_3 &= -\dot{Q}^{\mathrm{el}}_1 = (\dot{S}_{11} n^\star_1 + \dot{S}_{12} n^\star_2)\, V^{\mathrm{el}}/a_n \,, \\ \dot{Q}^{\mathrm{el}}_4 &= -\dot{Q}^{\mathrm{el}}_2 = (\dot{S}_{21} n^\star_1 + \dot{S}_{22} n^\star_2)\, V^{\mathrm{el}}/a_n \,, \end{aligned} \tag{73}$$

with an obvious interpretation of the right-hand sides in (73).

Assembling the element contributions in the standard way yields the discretized rate problem, whose nonlinearity enters via the dependence of $\dot{S}_{ik}$ and C_{ijkl} on velocity gradient. This dependence has been suppressed, for the sake of readability, in the formulae above.

Applications. It is sufficient here to state that the one-dimensional model shows the expected behaviour and is in agreement with the semi-analytical approach. Some results will be presented later, together with those from 2D computations.

We concentrate on a study of imperfections. Although the advantages of an analysis of the perfect system have been pointed out, we deviate from this line in order (i) to show that our computational procedure is likewise suited for imperfection studies and (ii) to relate results from both approaches to each other. Concerning the more important second point, it is expected that overall mechanical properties of an imperfect system tend to those for the perfect system when the imperfection size tends to zero. A precise meaning of this statement has recently been given in Petryk (1997). We consider the one-dimensional model with a material imperfection given by initial Kirchhoff yield stresses, τ_0, varying over the elements.[3] The initial yield stress, $\tau_{0,j}$, of element j is prescribed as

$$\tau_{0,j} = \tau_0 \left(1 - \zeta \cos\left(\frac{j-1}{n_{\mathrm{el}}-1}\pi\right)\right) \qquad \text{(half-cosine imperfection)} \tag{74}$$

or as

$$\tau_{0,j} = \tau_0(1 + \zeta \, \mathrm{rndm}(j)) \qquad \text{(random imperfection)}, \tag{75}$$

where n_{el} is the number of elements, ζ the imperfection amplitude and rndm() a uniformly distributed random number in the range $[-1, 1]$.

Computations must now be started from the unstressed initial configuration at $t = 0$. Sample results for a solid obeying the J_2 corner theory with material parameters as specified in Table 1, and for $\beta_c^{\max} = 126.2°$, are shown in Figures 18–21.

In Fig. 18 typical distributions of the (additional) shear rate g are plotted over the accumulated element volume, normalized by the total volume. A random distribution of the yield stress with an imperfection amplitude of $\zeta = 10^{-3}$ and $n_{\mathrm{el}} = 50$ elements have been used. g has the same meaning as the corresponding quantities $g^{(a)}, g^{(b)}$ for the zones introduced in Section 4.3, and is obtained here from the magnitude of $\mathbf{g}$, defined by (67), taken with negative sign when $\mathbf{g} \cdot \bar{\mathbf{V}}\mathbf{n}^\star < 0$. The existence of two zones of uniform deformation, accompanied by an evolving transitory zone, as predicted by the theory for the perfect system, is supported by the strain rate distributions in Fig. 18. Elements belonging to zones (a) and (b) or to the transitory zone are randomly distributed along the horizontal axis. In the depicted range of overall strain, the discontinuity in velocity gradient is propagating into zone (a). This can be simulated in the discretized FE model only in a stepwise manner. During the considered increase of overall compression from 32.9% to 40.2%, four elements from zone (a) have been successively converted into transitory elements and are missing in zone (a) from Fig. 18b. The different zones and the decreasing volume fraction in zone (a) are better visible in Fig. 19, obtained from Fig. 18 by sorting the strain rates in descending numerical order. From Fig. 19 it is readily read off that the bandwidth, i.e. the volume fraction of zone (b), is increasing from 0.22 to 0.30. Generally, at smaller overall strains, the zones are not so pronounced, and initially the sorted strain rate distribution resembles

[3] This means that the material properties are changing in the direction of $\boldsymbol{\nu}^\star$, but remain constant along $\boldsymbol{\mu}$. The question whether this is a meaningful assumption is immaterial here. Since the intention is only to relate the behaviour of perfect and imperfect systems, the model problem does not matter. For the same reason we deliberately extend the computations beyond the expected range of validity of the one-dimensional analysis.

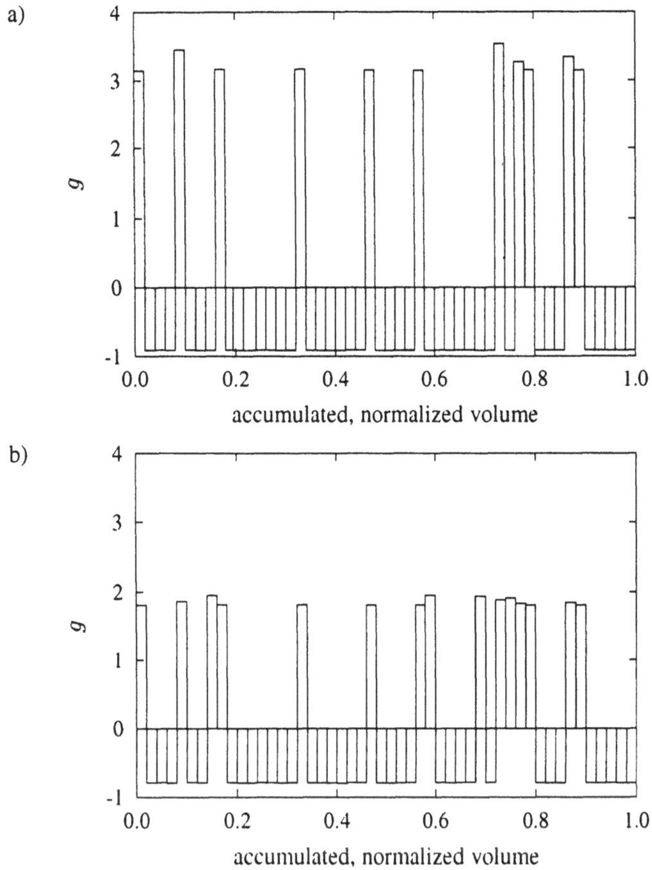

Figure 18. Shear rate distribution in natural element order a) at $t = 0.3291$ and b) at $t = 0.4015$. J_2 corner theory: $\beta_c^{max} = 126.2°$. Overall plane-strain compression: $\bar{F}_{11} = 1 - t$, $\bar{F}_{22} = 1/\bar{F}_{11}$

more a continuously decreasing function. A least-squares fit to the sorted distribution by a piecewise constant function $g(x) = g^{(b)}$ if $x < \eta$, otherwise $g(x) = g^{(a)}$, defines values $g^{(b)}, g^{(a)}$ and η, which may be used as substitutes for the corresponding values of the perfect system. At later stages of overall strain, as in Fig. 19, the least-squares fit closely approximates the actual distribution.

How $g^{(b)}$, $g^{(a)}$ and η evolve in time is shown in Fig. 20. The type of yield stress imperfection is visible, but has no significant influence. For both distributions of initial yield stress, the corresponding graphs for the perfect system, calculated with the method from Section 4.3, are approached as expected, when the imperfection size tends to zero. The progression of the graphs just shown is explained by a closer look at the strain rate distributions, plotted in Fig. 21 at different t for a fine discretization with 500 elements. A cosine imperfection has been prescribed with $\zeta = 10^{-3}$ so that element 1 has the lowest initial yield stress (note the negative sign in (74)). Accumulated element volumes in Fig. 21 are plotted in the original element order from left

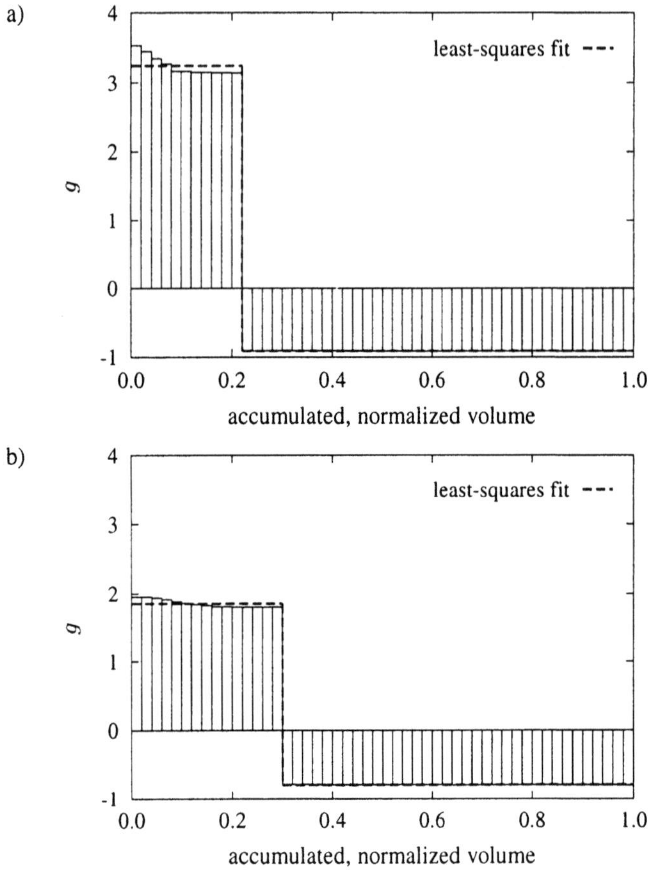

Figure 19. Sorted shear rate distribution a) at $t = 0.3291$ and b) at $t = 0.4015$. J_2 corner theory: $\beta_c^{\max} = 126.2°$. Overall plane-strain compression: $\bar{F}_{11} = 1 - t$, $\bar{F}_{22} = 1/\bar{F}_{11}$

to right. Up to a compression of $\approx 24\%$, the strain rate is generally small and growing slowly. Its distribution follows the cosine shape of the initial yield stress, associated with a substitute value for η of 0.5. Somewhat below the critical instant of the perfect system ($t^\star = 0.274$), this pattern is rapidly changed and the localization zone develops in the weakest elements (on the left in Fig. 21a) with the highest strain rate in element 1. At later stages, above $t^\star$, distinctive zones with uniform strain rates, separated by a transitory region, are formed (Fig. 21b). Qualitatively, all features of the perfect system are recovered.

Note that there is a remarkable similarity between the plots in Fig. 21a and Fig. 7, obtained in an entirely different context. In both cases, an initially sinusoidal distribution is rapidly converted into two zones of practically constant strain rate. In retrospect, the thickness effect in a three-dimensional analysis of a sheet can be interpreted as some kind of small imperfection that changes the mathematical character of the problem and is absent in a two-dimensional analysis as a thin sheet.

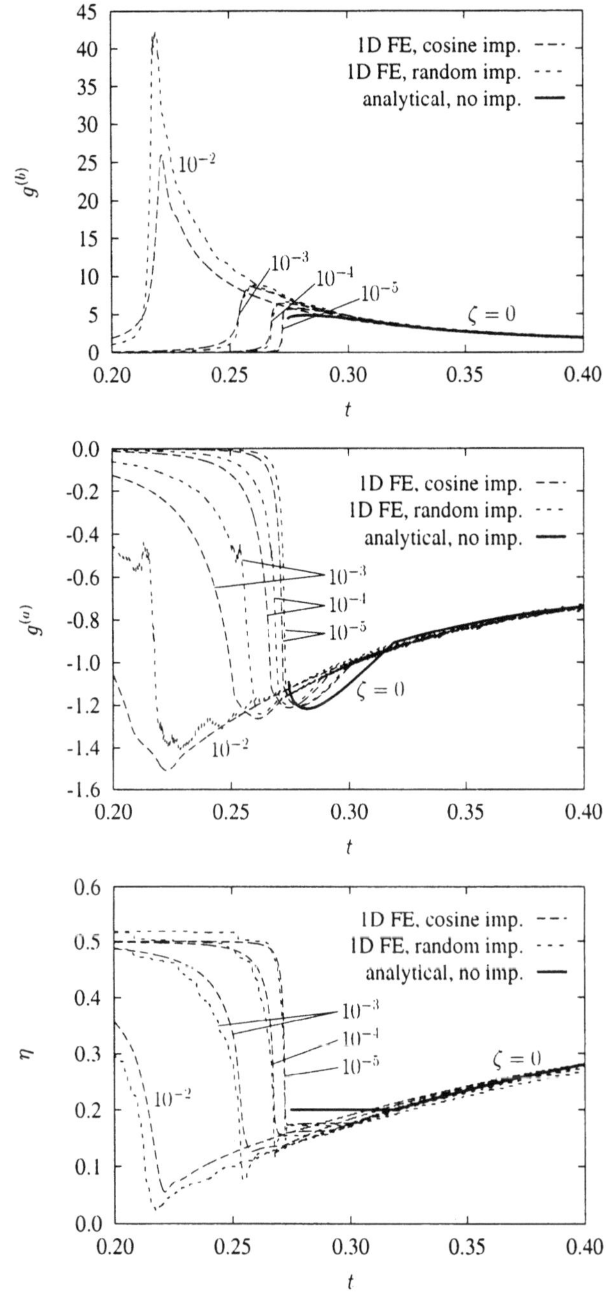

Figure 20. Influence of yield stress imperfections on the shear rates $g^{(b)}$ and $g^{(a)}$ and on the volume fraction η of the localization zone (from a least-squares fit to the 1D FE solution with 500 elements). J_2 corner theory: $\beta_c^{\max} = 126.2°$. Overall plane-strain compression: $\bar{F}_{11} = 1 - t$, $\bar{F}_{22} = 1/\bar{F}_{11}$

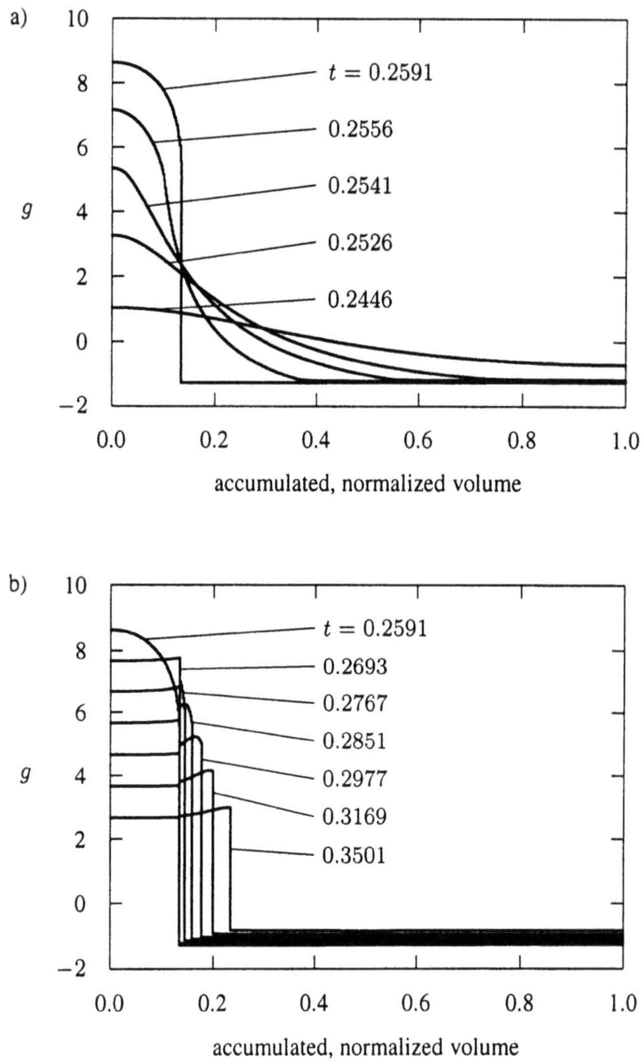

Figure 21. Shear rate evolution for a cosine yield stress imperfection with amplitude $\zeta = 10^{-3}$ in the range of a) increasing, b) decreasing shear rates in the localization zone. 1D FE solution: $n_{\mathrm{el}} = 500$. J_2 corner theory: $\beta_c^{\max} = 126.2°$. Overall plane-strain compression: $\bar{F}_{11} = 1 - t, \bar{F}_{22} = 1/\bar{F}_{11}$

4.5 Two-Dimensional Finite Element Model

The general computational procedure, described in detail in Section 2.2 can be directly applied without further theoretical considerations. Questions of mesh design have already been discussed, and the same two-dimensional discretizations and elements used to study the onset of path instability are employed. We will consider finite element results for boundaries constrained either entirely kinematically or by periodicity conditions. In all numerical studies presented here, J_2 corner theory is used with $\beta_c^{\max} = 126.2°, E = 500\,\tau_0, \nu = 0.3$ and a power-hardening exponent $N = 0.1$.

Periodic boundary conditions. According to the theoretical analysis in Section 4.3 and for the underlying material data, the volume fraction of the localization zone is constant at early stages in the post-critical range (cf. Fig. 15). Reliability of the two-dimensional finite element model, in particular the correct implementation of periodic boundary conditions, can be conveniently checked in this 'fixed band' range since an analytical reference solution is available.

Four different meshes with $n = 5, 10, 20$ and 40 quadrilateral elements along a side of the unit cell have been analysed in the fixed band range. For each mesh any bifurcation solution from a generally large number[4] of equivalent solutions at the critical instant $t^* = 0.2744$ may be used to continue the computations into the post-critical range. Figure 22 shows an arbitrary selection; it is recalled that the initial bifurcation pattern can be controlled by the trial velocity field, used to minimize the velocity functional. This choice being made, the pattern remains unchanged up to an overall compressive strain of 32%. The component V_{gn} of velocity gradient on an orthonormal basis along the current band normal is changing in time. However, like any other statical or kinematical quantity it assumes only two values on each mesh at each stage $t < 0.32$.

All differences between meshes for different n disappear completely, when sorted values are plotted over the accumulated element volume, as in Fig. 23 for V_{gn}. This illustrates that overall quantities are still well defined and meshsize independent, e.g. the volume fraction of the localization zone is found to be the same for all meshes considered, either directly from Fig. 23 or from Fig. 22 by counting the number of black lines along a side of the mesh and dividing by n. This ratio is always 1/5.

Increasing n yields a finer microstructure, but leaves the results otherwise unchanged, identical to those from a one-dimensional FE computation or from the analytical model. This is shown in Figs. 24–26 for the evolution of the superimposed shear rates $g^{(b)}, g^{(a)}$ and the volume fraction η of the localization zone (b). $g^{(b)}$ and $g^{(a)}$ are obtained from the 2D results by subtracting the given mean rate $\bar{V}_{gn}$ from the value V_{gn} in the respective zones. For this comparison, the slight influence of elastic compressibility has been taken into account in the analytical model, by using an extended analysis and program version (Lulla, 1999).

Well known numerical difficulties are met, when finite element computations for the classical elasto-plastic solid with only two constitutive branches of loading and unloading are extended beyond the instant of ellipticity loss of uniform deformation. They are *not* encountered here, because the vertex stiffening effect of the incrementally nonlinear material allows the deformation path to return to the elliptic regime or to move along its boundary. Moreover, the computational

[4] There are e.g. 1 316 016 such solutions, when $n = 40$, cf. the discussion in Section 4.2.

procedure assures not only continuing equilibrium, but returns a minimizer of the velocity functional J, not just nodal velocities rendering J stationary. Possible indeterminacies in velocity solutions are thereby significantly reduced. This is, in particular, important when the computations are continued farther in the post-critical range.

The deformation patterns may then be truly two-dimensional so that coarse meshes are no longer useful. Another kinematical quantity, the principal Eulerian strain-rate of largest magnitude, has been used to visualize the results in Fig. 27 for a 40×40 mesh. Depending only on intrinsic quantities, it is better suited than the shear rate V_{gn}, which is only meaningful when bands with normal $\mathbf{n}^*$ are actually formed. This is still the case at stages of overall deformation directly following the fixed band range. At $t = 0.3193$, we enter the moving band range. A bifurcation occurs and an additional band appears so that the volume fraction of the zone with low strain-rate is decreasing. This process is repeated and the motion remains one-dimensional up to 38% overall compression with a stepwise increase in the bandwidth from 0.2 to 0.275.

At $t = 0.3856$ a new pattern becomes visible for the first time. Secondary bands are gradually formed in a series of further bifurcations starting at this instant and following one another in close sequence. They are fully developed at $t = 0.4099$ after the last of totally 9 bifurcations in this range.

The secondary band pattern persists during subsequent deformations, but is rearranged once more at $t = 0.4856$. There are no other bifurcations met up to $\approx$ 50% overall compression, where the computations were stopped.

The results are in qualitative agreement with all theoretical predictions from Section 4.3. In the early stages of post-critical deformation coincidence between three different approaches was even obtained. The approximation in the moving band range is only stepwise, but the final value for the bandwidth at $t = 0.38$ is in agreement with the theoretical result. Secondary bands should, however, appear already just before the fixed band range ends, according to the predictions from the analytical model for the heterogeneous material with a sufficiently fine micro-structure. For the considered mesh, i.e. on a finite dimensional subspace with 6398 degrees of freedom, secondary bands are observed somewhat later ($\Delta t = 0.06$), deviating by $\approx 2°$ from the theoretically estimated direction, cf. Fig. 28. This is a surprisingly good result in view of the simple mesh with 'crossed triangles', in particular, since no effort was made to capture secondary bands by special mesh design. Quantitative improvements are expected when finer and better adapted meshes or more refined elements are used.

The observed ranges of different behaviour are also found in Fig. 29 where the mean constitutive potential, i.e. essentially the velocity function minimized at each time step, is plotted in the post-critical range.

Kinematically constrained boundary. Results of a sample computation in the case of kinematical boundary conditions are shown in Fig. 30. The boundary motion of an initially circular domain was prescribed so that it corresponds to the given overall isochoric plane-strain compression. Qualitative features of solutions in the post-critical range, described in detail for periodic boundary conditions, are found again, however with some differences. At primary bifurcation ($t = 0.2755$), in the central region a localization zone with a bandwidth of 0.5 is formed, which rapidly spreads out to the boundary. A fourth bifurcation shortly later ($t = 0.2770$) changes this pattern so that subsequently the volume fraction of the localization zone is obtained as 0.25. No further bifurcation points are crossed until $t = 0.4103$. At this stage secondary bands are observed which develop in the same manner as in the case of periodic boundary conditions, passing through a number of bifurcation points.

The theoretical bandwidth of 0.2 is not obtained. Comparison computations for a mesh with only 20 quadrilateral elements along the radius show the same behaviour. It is likely that the mesh structure in connection with the boundary conditions favours the value 0.25 at early stages in the post-critical range. This question requires further study.

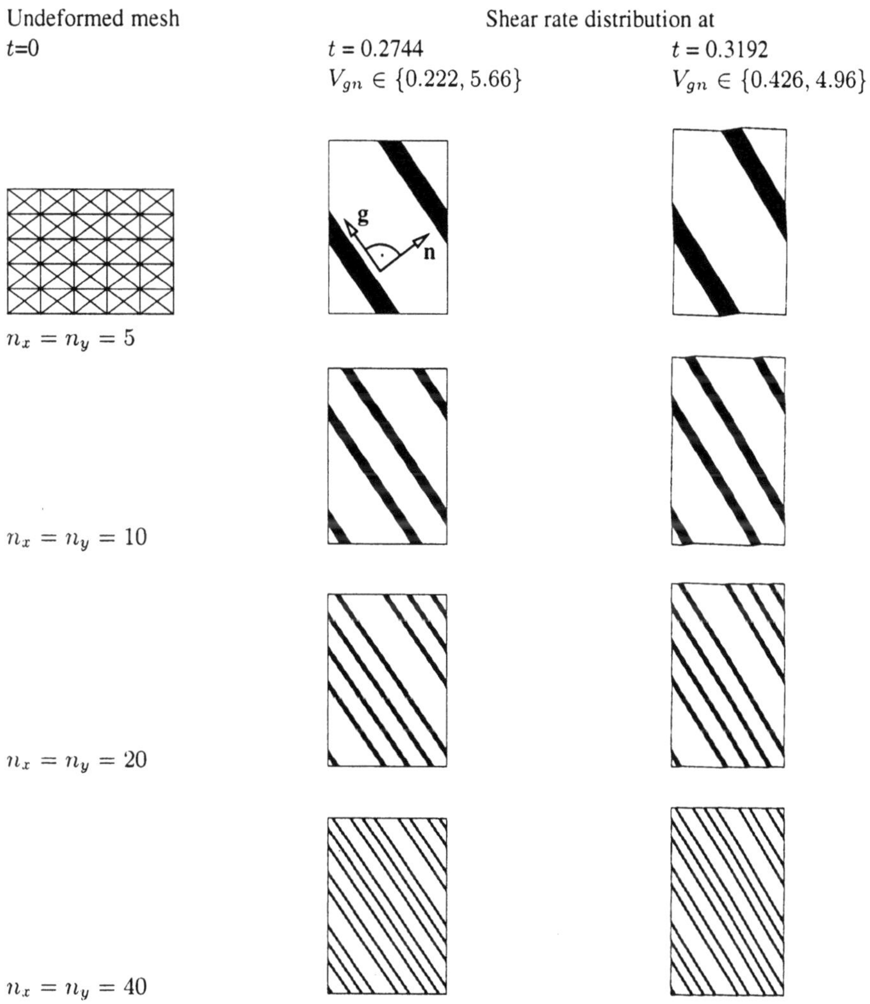

Figure 22. Influence of meshsize on the shear rate distribution $V_{gn} = \mathbf{g} \cdot \mathbf{V}\mathbf{n}$ at early stages in the post-critical range. J_2 corner theory: $\beta_c^{\max} = 126.2°$. Overall plane-strain compression along the horizontal axis x_1: $\bar{F}_{11} = 1 - t$, $\bar{F}_{22} = 1/\bar{F}_{11}$. Periodic boundary conditions.

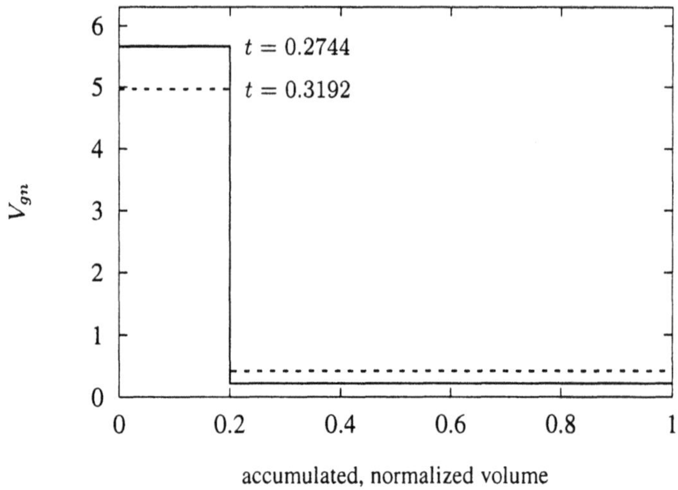

Figure 23. Sorted shear strain-rate vs normalized volume at primary bifurcation ($t = 0.2744$) and at the end of the fixed band range ($t = 0.3192$) for all meshes in Figure 22.

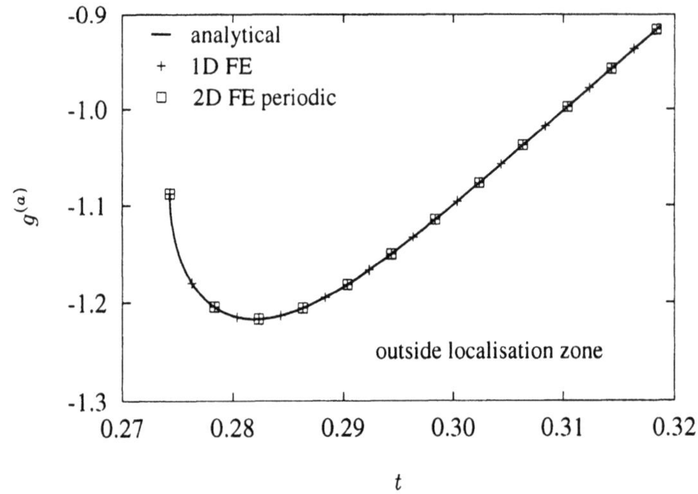

Figure 24. Post-critical evolution of the superimposed shear rate in zone (a). J_2 corner theory: $\beta_c^{\max} = 126.2°$. Overall plane-strain compression: $\bar{F}_{11} = 1 - t$, $\bar{F}_{22} = 1/\bar{F}_{11}$.

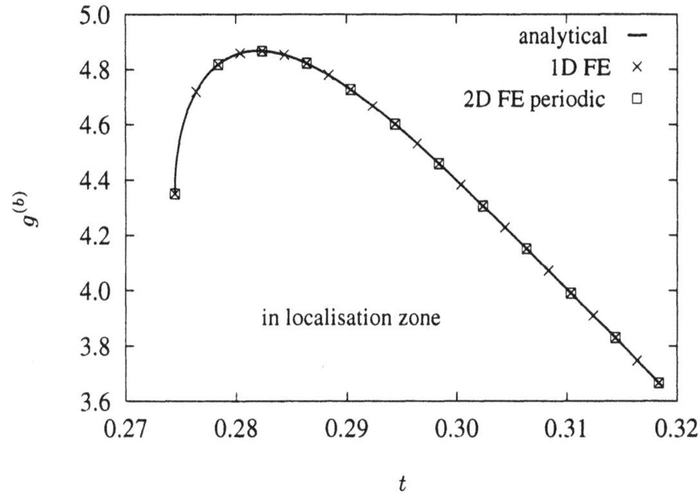

Figure 25. Post-critical evolution of the superimposed shear rate in zone (b). J_2 corner theory: $\beta_c^{\max} = 126.2°$. Overall plane-strain compression: $\bar{F}_{11} = 1 - t, \bar{F}_{22} = 1/\bar{F}_{11}$.

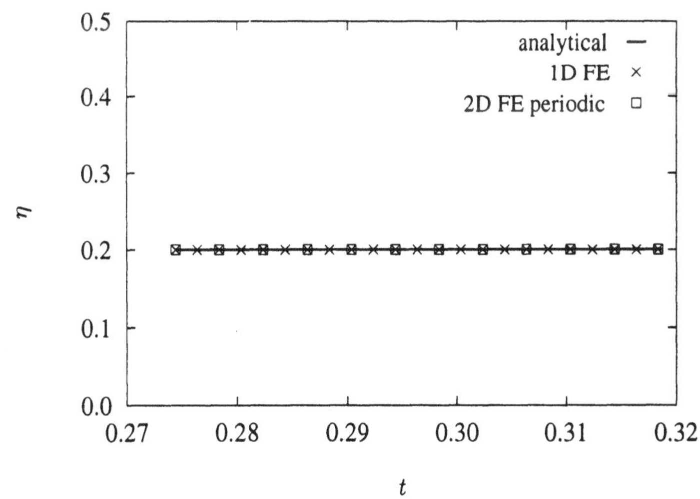

Figure 26. Post-critical evolution of the volume fraction of the localization zone. J_2 corner theory: $\beta_c^{\max} = 126.2°$. Overall plane-strain compression: $\bar{F}_{11} = 1 - t, \bar{F}_{22} = 1/\bar{F}_{11}$.

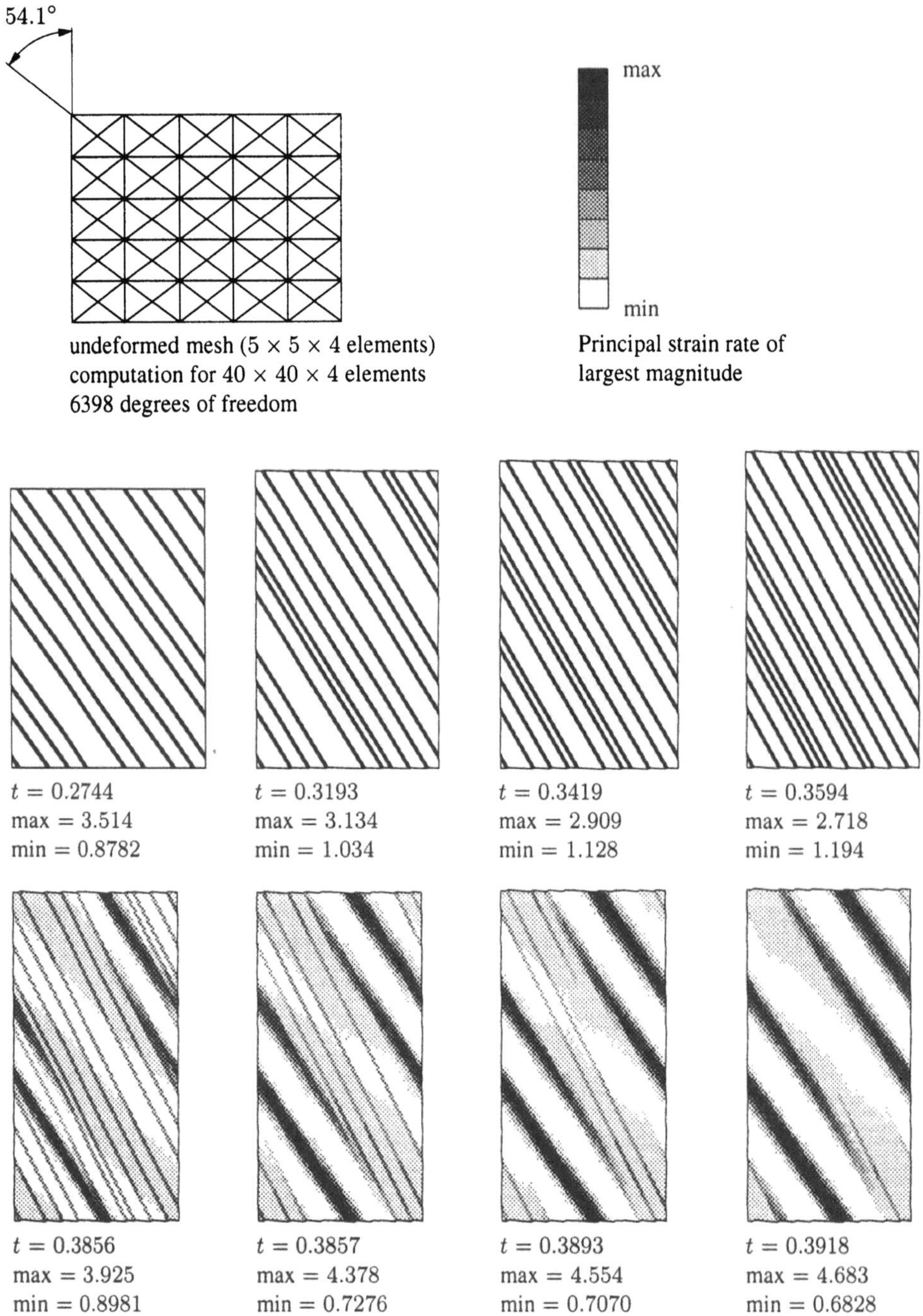

Figure 27. Post-critical evolution of the principal strain-rate of largest magnitude at various stages of overall compression $\bar{F}_{11} = 1 - t$, $\bar{F}_{22} = 1/\bar{F}_{11}$. J_2 corner theory: $\beta_c^{\max} = 126.2°$. Periodic boundary conditions.

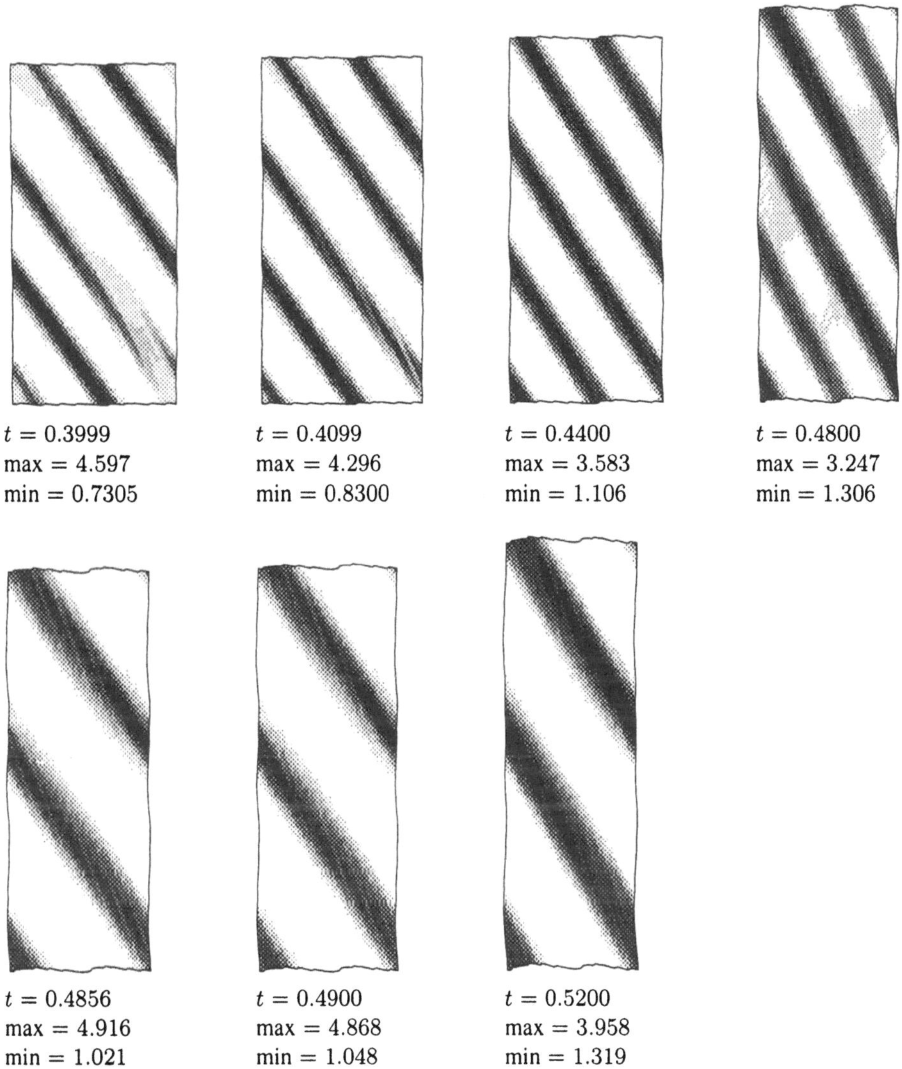

Figure 27. (continued) Distribution of principal strain-rate of largest magnitude at various stages of overall compression $\bar{F}_{11} = 1 - t$, $\bar{F}_{22} = 1/\bar{F}_{11}$. J_2 corner theory: $\beta_c^{\max} = 126.2°$. Periodic boundary conditions.

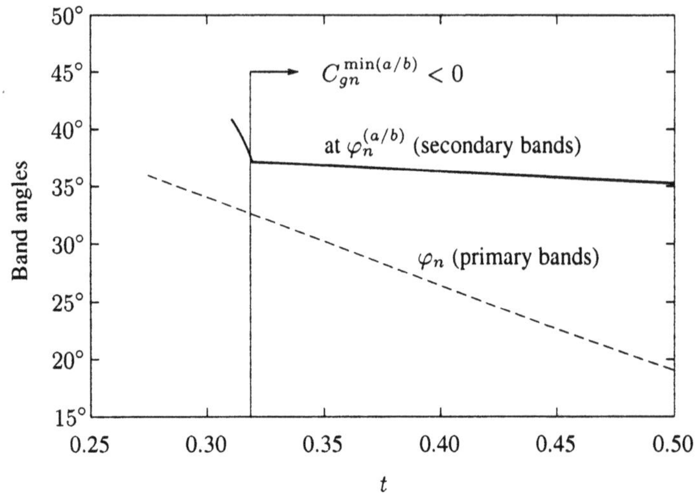

Figure 28. Estimated direction of secondary bands (from semi-analytical model). J_2 corner theory: $\beta_c^{\max} = 126.2°$. Overall plane-strain compression: $\bar{F}_{11} = 1 - t$, $\bar{F}_{22} = 1/\bar{F}_{11}$.

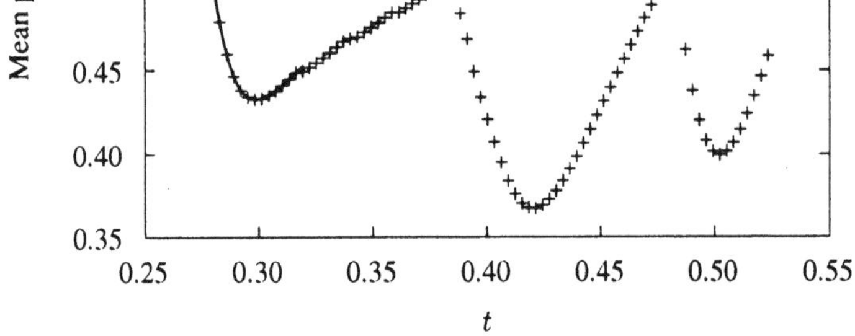

Figure 29. Post-critical evolution of the mean potential per initial volume. J_2 corner theory: $\beta_c^{\max} = 126.2°$. Overall plane-strain compression: $\bar{F}_{11} = 1 - t$, $\bar{F}_{22} = 1/\bar{F}_{11}$. 2D FE computations with periodic boundary conditions: $n_x = n_y = n = 40$.

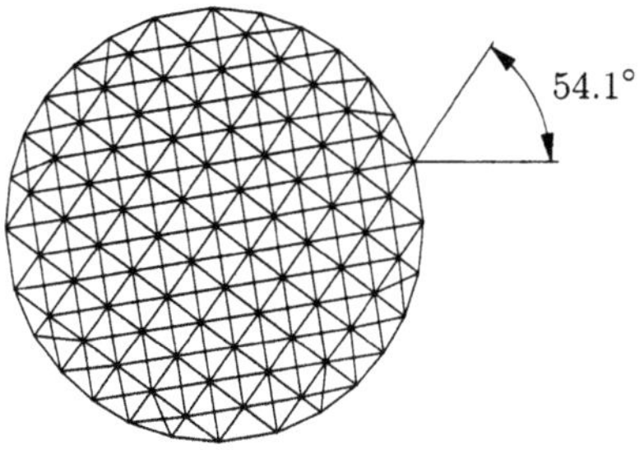

undeformed mesh ($n_x = n_y = 5$)
computations for $n_x = n_y = 40$
20028 elements
19802 degrees of freedom

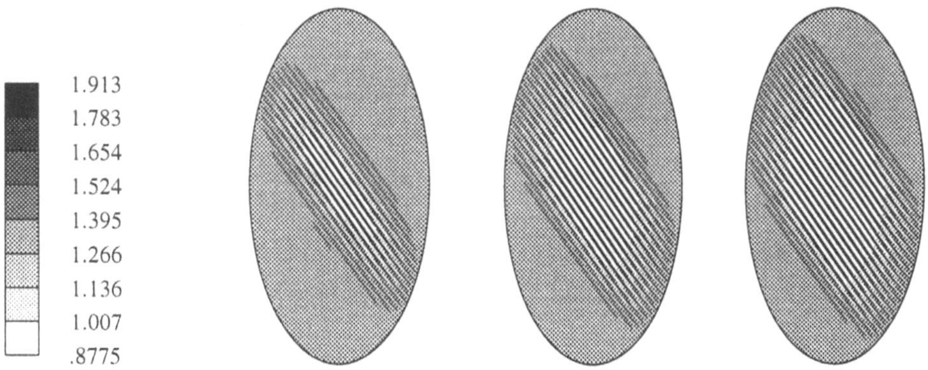

$t = 0.2755$	$t = 0.2755$	$t = 0.2756$	$t = 0.2757$
	max = 1.913	max = 1.923	max = 1.923
	min = 0.8775	min = 0.8706	min = 0.8679

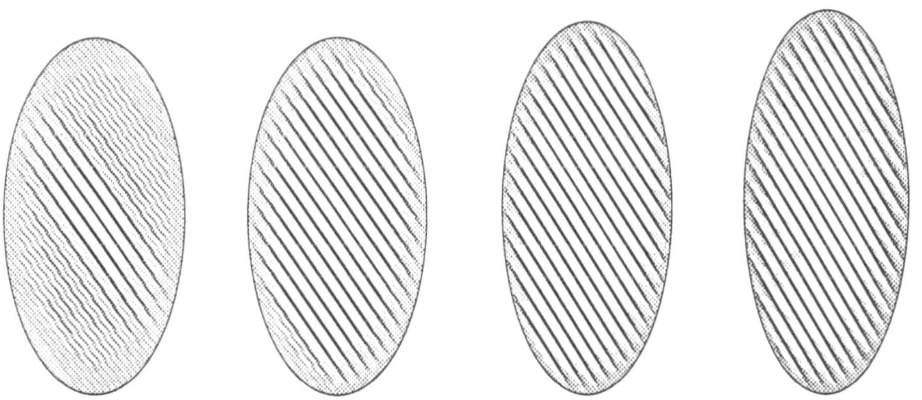

$t = 0.2770$	$t = 0.2800$	$t = 0.3100$	$t = 0.3300$
max = 3.088	max = 3.270	max = 3.175	max = 3.002
min = 0.8134	min = 0.7475	min = 0.7813	min = 0.8271

Figure 30. Post-critical evolution of the principal strain-rate of largest magnitude at various stages of overall compression $\bar{F}_{11} = 1 - t$, $\bar{F}_{22} = 1/\bar{F}_{11}$. J_2 corner theory: $\beta_c^{\max} = 126.2°$. Kinematically constrained boundary: 20028 elements

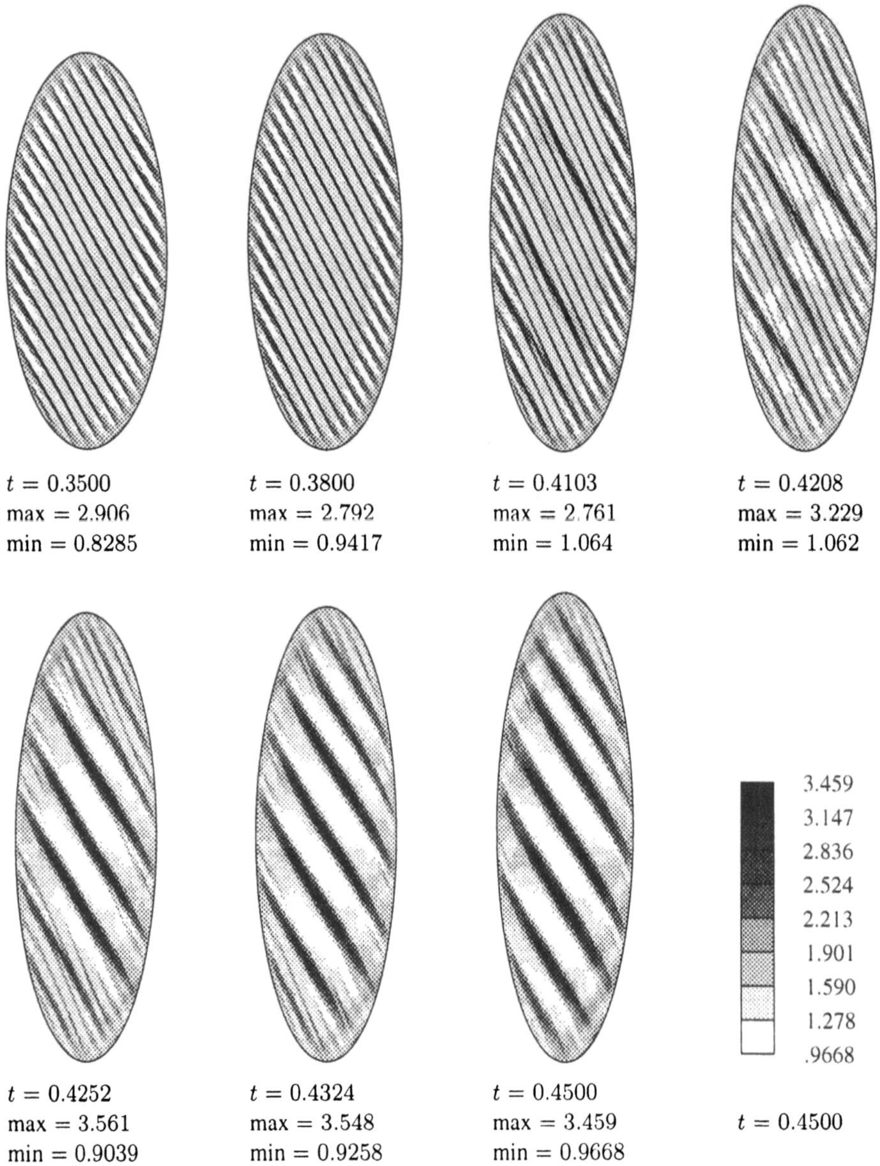

Figure 30. (continued) Post-critical evolution of the principal strain-rate of largest magnitude at various stages of overall compression $\bar{F}_{11} = 1 - t$, $\bar{F}_{22} = 1/\bar{F}_{11}$. J_2 corner theory: $\beta_c^{\max} = 126.2°$. Kinematically constrained boundary: 20028 elements

References

Aifantis, E. C. (1995). Pattern formation in plasticity. *Int. J. Engng. Sci.* 33:2161–2178.

Christoffersen, J., and Hutchinson, J. W. (1979). A class of phenomenological corner theories of plasticity. *J. Mech. Phys. Solids* 27:465–487.

de Borst, R., and van der Giessen, E., eds. (1998). *Material Instabilities in Solids.* Chichester: John Wiley & Sons.

Fiacco, A. V., and McCormick, G. P. (1968). *Nonlinear programming. Sequential Unconstrained Minimization Techniques.* New York: John Wiley.

Fleck, N. A., and Hutchinson, J. W. (1997). Strain gradient plasticity. *Advances in Applied Mechanics* 33:295–361.

Fletcher, R. (1987). *Practical Methods for Optimization.* Chichester: John Wiley.

Gay, D. M. (1981). Computing optimally constrained steps. *SIAM J. Sci. Stat. Comput.* 2:186–197.

Gay, D. M. (1983). Algorithm 611, subroutines for unconstrained minimization using a model/trust-region approach. *ACM Transactions on Mathematical Software* 9:503–524.

Gurtin, M. E. (1972). The linear theory of elasticity. In Flügge, S., ed., *Encyclopedia of Physics, Volume VIa/2*, 1–295. Berlin, Heidelberg, New York: Springer.

Hill, R., and Hutchinson, J. W. (1975). Bifurcation phenomena in the plane tension test. *J. Mech. Phys. Solids* 23:239–264.

Hill, R. (1958). A general theory of uniqueness and stability in elastic-plastic solids. *J. Mech. Phys. Solids* 6:236–249.

Hill, R. (1959). Some basic principles in the mechanics of solids without a natural time. *J. Mech. Phys. Solids* 7:209–225.

Hill, R. (1962). Acceleration waves in solids. *J. Mech. Phys. Solids* 10:1–16.

Hill, R. (1967). The essential structure of constitutive laws for metal composites and polycrystals. *J. Mech. Phys. Solids* 15:79–95.

Hill, R. (1978). Aspects of invariance in solids mechanics. *Advances in Applied Mechanics* 18:1–75. Academic Press, New York.

Hutchinson, J., and Tvergaard, V. (1981). Shear band formation in plane strain. *Int. J. Solids Structures* 17:451–470.

Lulla, C. (1999). Phd thesis, Dortmund University, to appear. Private Communication.

Moré, J. J., and Sorensen, D. C. (1981). Computing a trust region step. *SIAM J. Sci. Stat. Comput.* 4:553–572.

Petryk, H., and Thermann, K. (1992). On discretized plasticity problems with bifurcations. *Int. J. Solids Structures* 29:745–765.

Petryk, H., and Thermann, K. (1995). On plastic strain localisation in the non-elliptic range under plane stress. In Owen, D., and Onate, E., eds., *Computational Plasticity: Fundamentals and Applications*, 647–658. Swansea: Pineridge Press.

Petryk, H., and Thermann, K. (1996). Post-critical plastic deformation of biaxially stretched sheets. *Int. J. Solids Structures* 33:689–705.

Petryk, H., and Thermann, K. (1997). A yield-vertex modification of two-surface models of metal plasticity. *Arch. Mech.* 49:847–863.

Petryk, H., and Thermann, K. (1999a). Post-critical deformation pattern in plane strain plastic flow with yield-surface vertex effect. *Int. J. Mech. Sciences* , in press.

Petryk, H., and Thermann, K. (1999b). Post-critical plastic deformation in incrementally nonlinear materials. to appear.

Petryk, H. (1982). A consistent energy approach to defining stability of plastic deformation processes. In Schroeder, F., ed., *Stability in the Mechanics of Continua, Proc. IUTAM Symposium, Nümbrecht 1981*, 262–272. Berlin: Springer.

Petryk, H. (1989). On constitutive inequalities and bifurcation in elastic-plastic solids with a yield-surface vertex. *J. Mech. Phys. Solids* 37:265–291.

Petryk, H. (1991). The energy criteria of instability in time-independent inelastic solids. *Arch. Mech.* 43:519–545.

Petryk, H. (1992). Material instability and strain-rate discontinuities in incrementally non-linear continua. *J. Mech. Phys. Solids* 40:1227–1250.

Petryk, H. (1997). Instability of plastic deformation processes. In et al., T. T., ed., *Proceedings XIXth Int. Congress of Theoretical and Applied Mechanics, Kyoto 1996*, 497–516. Amsterdam: Elsevier.

Petryk, H. (1998). Evolution of the localization zone in an icrementally non-linear plastic material. Private Communication.

Rice, J. R. (1977). The localization of plastic deformation. In Koiter, W. T., ed., *Theoretical and Applied Mechanics*, 207–220. Amsterdam: North-Holland.

Rudnicki, J. W., and Rice, J. R. (1975). Conditions for the localization of deformation in pressure-sensitive materials. *J. Mech. Phys. Solids* 23:371–394.

Shanley, F. R. (1947). Inelastic column theory. *J. Aero. Sci.* 14:261–267.

Stören, S., and Rice, J. R. (1975). Localized necking in thin sheets. *J. Mech. Phys. Solids* 23:421–441.

Tomita, Y. (1994). Simulation of plastic instabilities in solid mechanics. *Appl. Mech, Rev.* 47:171–205.

Tvergaard, V., Needleman, A., and Lo, K. K. (1981). Flow localization in the plane strain tensile test. *J. Mech. Phys. Solids* 29:115–147.

MIX
Papier aus verantwortungsvollen Quellen
Paper from responsible sources
FSC® C105338

FSC
www.fsc.org

If you have any concerns about our products,
you can contact us on
ProductSafety@springernature.com

In case Publisher is established outside the EU,
the EU authorized representative is:
Springer Nature Customer Service Center GmbH
Europaplatz 3, 69115 Heidelberg, Germany

Printed by Libri Plureos GmbH
in Hamburg, Germany